# FAST TRACK TO A 5

## Preparing for the
## AP® Physics 1 and AP® Physics 2
## Examinations

CENGAGE
Learning

Australia · Brazil · Mexico · Singapore · United Kingdom · United States

National Geographic Learning/Cengage Learning is pleased to offer our college-level materials to high schools for Advanced Placement®, honors, and electives courses. To contact your National Geographic Learning representative, please call us toll-free at **1-888-915-3276** or visit us at **http://ngl.cengage.com**.

For permission to use material from this text or product, submit all requests online at **www.cengage.com/permissions** Further permissions questions can be emailed to **permissionrequest@cengage.com**.

ISBN: 978-1-337-62929-4

**Cengage Learning**
20 Channel Center Street
Boston, MA 02210
USA

Cengage Learning is a leading provider of customized learning solutions with office locations around the globe, including Singapore, the United Kingdom, Australia, Mexico, Brazil, and Japan. Locate your local office at: **www.cengage.com/global**.

Cengage Learning products are represented in Canada by Nelson Education, Ltd.

To learn more about Cengage Learning Solutions, visit **www.cengage.com**.

To find online supplements and other instructional support, please visit **www.cengagebrain.com**.

AP® is a trademark registered by the College Board, which is not affiliated with, and does not endorse, this product.

Printed in the United States of America
Print Number: 01     Print Year: 2017

# CONTENTS

# Part I

## Strategies for the AP® Test

# PREPARING FOR THE AP® PHYSICS 1 AND AP® PHYSICS 2 EXAMINATIONS*

Congratulations! By taking Advanced Placement® Physics and getting this guide, you've set yourself up for an exhilarating course and for success on the test in May. The stage is set for a great intellectual experience.

You've taken a very important step toward mastering an AP® Physics exam; this book will help you prepare for and master the test. If you stick to working through this book alongside your classwork and lab work, there will be no surprises in May. Your hard work will earn you the confidence and tools you need to be a great physics student.

Many AP® students are used to earning an A on every assignment, knowing almost every answer on a test, and feeling great after a classroom assessment. AP® exams are not set up for you to get every question right or earn all of the points. Do not be discouraged if you get stuck on a particular question or even a particular topic. You still need to know your physics, but the most important thing is to earn as many points as you can, not to focus on the two questions you did not understand.

## WHAT ARE THE AP® PHYSICS 1 AND AP® PHYSICS 2 COURSES?

AP® Physics 1 and 2 are algebra-based courses equivalent to the first and second semesters, respectively, of a typical introductory algebra-based university physics course. Your physics course will help you develop a deeper conceptual understanding through lab work, analysis, and writing.

AP® Physics 1 focuses on kinematics, forces, gravitation, impulse-momentum, energy, rotational motion, electricity (simple circuits and charge), and waves (mechanical waves, sound, simple harmonic motion). AP® Physics 2 focuses on fluids, thermodynamics, electrostatics, circuits, magnetism, optics, and modern physics. In both courses, you must have a good understanding of lab work.

AP® Physics is NOT a math course, and the test is NOT a math test. Sure, you are expected to use mathematics as a tool, but these courses are about understanding and explaining. In 2016, none of the five free

---

*AP® is a trademark registered by the College Board, which is not affiliated with, and does not endorse, this product.

response questions on the AP® Physics 1 test involved a calculation. On the 2016 AP® Physics 2 exam, only one of the four questions involved calculating anything.

## WHAT'S IN THIS BOOK

This book follows the College Board's outline for the AP® Physics courses and is compatible with all AP® Physics textbooks. It is divided into three sections. Part I offers suggestions for getting yourself ready—from signing up to take the test and sharpening your pencils to organizing a free-response answer. This is followed by the list of AP® Physics learning objectives from the College Board curriculum—the definitive list of everything AP® Physics students need to know and be able to do.

Part II is made up of 19 chapters matching the College Board's design. These chapters are not a substitute for your textbook and class work; they simply review the topics in AP® Physics 1 and 2. At the end of each chapter, you will find 15 multiple-choice questions and 4 free-response problems based on the material of that chapter. You will find references at the end of each answer directing you to the AP® Physics learning objective number covering that point.

Part III has two complete AP® Physics 1 and AP® Physics 2 level examinations. At the end of each test, you will find the answers, explanations, and references to the AP® Physics learning objective for the multiple-choice questions and the free-response problems.

## SETTING UP A REVIEW SCHEDULE

If you have been doing your homework regularly and keeping up with the coursework, you are in good shape. Organize your notes, home-work, and handouts from class by topic. Reference these materials as well as your textbook and this study guide when you have difficulty in a specific section. Even if you've done all that—or if it's too late to do all that—there are more ways to get it all together.

To begin, read Part I of this book. You will be much more comfort-able going into the test if you understand how the test questions and problems are designed and how best to approach them.

Take out a calendar and set up a schedule for yourself. If you begin studying early, you can chip away at the review chapters in Part II. You'll be surprised—and pleased—by how much material you can cover studying a half hour a day for a month or so before the test. The practice tests in Part III will give you more experience with different kinds of multiple-choice questions and the wide range of free-response problems. Look carefully at the sections of the practice tests; if you missed a number of questions in a particular area, review the chapters that cover that area of the course.

If time is short, skip reading the review chapters (although you might read through the chapter subheadings) and work on the multiple-choice questions and free-response problems at the end of each review.

This will give you a good idea of your understanding of that particular topic. Then take the practice tests in Part III.

If time is *really* short, go straight from Part I to Part III. Taking practice tests over and over again is the fastest, most practical way to prepare. You cannot study physics by reading about it or watching your teacher do it. You must actively do problems to get better at the mental gymnastics necessary to relate the diverse topics in physics. Athletes don't perform well just by reading books about their sport or by watching others; they get up and practice. So you too must practice, practice, practice if you want to be the very best!

## BEFORE THE EXAMINATION

By February, long before the exam, make sure you're registered to take the right exam. Many schools manage the paperwork and fees for their AP® students, but check with your teacher or the AP® coordinator to make sure you are on the list. This is especially important if you have a documented disability and need test accommodations. If you are studying AP® independently, call AP® Services at the College Board for the name of the local AP® coordinator who will help you through the registration process.

The evening before the exam is not a great time for partying, nor is it a great time for cramming. If you like, look over class notes or look through your textbook, concentrating on the broad outlines of the course, not the small details. You also might want to skim through this book and read the AP® tips.

The evening before the exam *is* a great time to get your things together for the next day. Sharpen a fistful of no. 2 pencils with good erasers. You will be writing your free-response answers on newsprint type paper, so be sure you can write dark enough with your pencils for the answers to be read. You can use a pen on the free response, but a pen cannot be erased. Changing your mind with a pen means making your answer messy and harder to grade. The more legible your writing, which includes heavy pencil markings, the better. The readers will make every effort to read what you've written, but small or messy handwriting makes it difficult to decipher your answer. Readers cannot give points for writing they can't read. You may choose to bring a ruler to help drawing straight lines on the test; however, one is not required or even really needed. Put some fresh batteries in a scientific calculator you're comfortable using. You might not need it on the test, but it could come in handy. Certain types of calculators are not allowed, so be sure to verify with your teacher or the College Board that your model is acceptable. For example, you cannot use a calculator with a typewriter-style keyboard or a cell phone. Because cell phones are not allowed in the testing room, you will need a watch (smartwatches are not allowed); make sure you turn off the alarm if your watch has one. Make sure you have your Social Security number and whatever photo identification and admission ticket are required. Then relax and get a good night's sleep.

On the day of the examination, do not skip breakfast; studies show that students who eat a hot breakfast before testing get higher grades than students who do not. Be careful not to drink a lot of liquids so that you don't waste precious minutes in the bathroom. With a break between Section I and Section II, the AP® Physics exams last more than three hours, so be prepared for a long day. You do not want to be distracted by a growling stomach.

Wear comfortable clothes along with an extra layer in case the heating or air conditioning is erratic. Be sure to wear clothes you like—students perform better when they think they look better—and by all means wear your lucky socks.

# TAKING THE AP® PHYSICS 1 AND AP® PHYSICS 2 EXAMINATIONS*

Both exams have two sections: multiple choice and free response. The test comes with an equation sheet you can use along with your calculator for the whole test. Each section is 90 minutes long and is equally weighted to comprise half of your score.

| AP® Physics 1, Section I | 45 multiple choice questions with one answer |
| | 5 multiple choice questions with two answers |
| AP® Physics 1, Section II | A 12-point lab-based question |
| | A 12-point qualitative-quantitative translation question |
| | Three 7-point short answer questions, one of which requires a paragraph length response |
| AP® Physics 2, Section I | 45 multiple-choice questions with one answer |
| | 5 multiple-choice questions with two answers |
| AP® Physics 2, Section II | A 12-point lab-based question |
| | A 12-point qualitative-quantitative translation question |
| | Two 10-point short answer questions, one of which requires a paragraph length response |

## MULTIPLE-CHOICE STRATEGIES

■ **The equations and constants** come with the exam and can be used throughout the whole test. Become familiar with these documents by practicing with them throughout the year. They are freely available for download from the College Board.

■ **Read the whole question carefully.** If you practice with this book and your teacher uses AP® style questions, many of the questions on your test will seem familiar. Although that's great, the question might not be asking what you assume it's asking. Read the question looking for word clues such as *not* and *except* that change the meaning.

---

AP® is a trademark registered by the College Board, which is not affiliated with, and does not endorse, this product.

**7**

- **Pace yourself** to about two minutes (1.8 minutes to be exact) per question. If you've spent a minute and a half on a question and it still hasn't "clicked," move on. Come back later after you've answered the questions you were sure of.

- **Mark and skip** tough questions. If you are hung up on a question, mark it in the margin of the question book. You can come back to it later if you have time. Mark something—anything—in the answer book in case you don't have time to come back to that question.

- **Guess!** There is no guessing penalty. There are four possible answers for each question. Correct answers earn 1 point; incorrect answers are not penalized. Even if you're unsure of the exact answer, narrow the options, guess, and move on.

- **Eliminate answers you know are incorrect.** You can write on the multiple-choice questions in the test book. As you read through the responses, draw a line through any answer you know is wrong. The three incorrect answer options were written to look attractive for some reason; by reading carefully, you can find the flaw that makes one incorrect.

- **Read all of the possible answers;** then choose the one you think is correct. Many answer options look similar, so find the differences between them and consider what those differences mean. AP® examinations are written to test precise understanding, so the differences between answers could be subtle.

- **Calculation questions** may appear in the multiple-choice section, although they are typically designed to be solved with mental math rather than a calculator. If a question seems to require calculation, first make a plan. Grab your calculator only after you have a plan for what to calculate.

- **The acceleration due to gravity** is 10 on test day. You can easily multiply by 10 in your head, and your answer will be less than 2% off a more precise answer. Using 10 m/s² for that magnitude of $g$ is acceptable on all parts of the test, and using 10 m/s² makes calculations considerably faster.

## TYPES OF MULTIPLE-CHOICE QUESTIONS

In both exams, you will encounter a variety of types of multiple-choice questions. Here are some suggestions for approaching each type.

### CLASSIC/BEST RESPONSE TYPE QUESTIONS

**Physics 1 Example:** As shown below, a body is projected upward with some initial velocity. The altitude of the projectile is shown as a function of time.

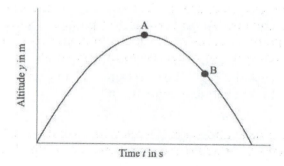

In the absence of air resistance

(A) at point A, both the velocity and the acceleration of the projectile are zero.

(B) at point A, the projectile has a smaller velocity than at point B. The acceleration at point A is less than the acceleration at point B.

(C) the acceleration at point A is zero, but it has a negative value at point B since the projectile is moving downward.

(D) the acceleration is the same at points A and B. The velocity at point A is zero, and the velocity at point B has some negative value.

**Answer: B** In the absence of air resistance, the projectile's acceleration is $g$ downward. This eliminates choices (A) and (C), which call gravity zero. At the apex of the flight, the projectile has no upward velocity, but still has horizontal velocity. For a fraction of a second, the vertical velocity of the projectile is zero as it stops and changes its direction, increasing its velocity in a downward direction. At point B, the velocity is a combination of the same horizontal velocity at point A and a negative vertical velocity, making the total velocity greater than at point A.

## RANKING

**Physics 1 or Physics 2 Example:** A series/parallel electrical circuit is shown below.

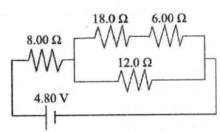

Rank the resistors from the one conducting the highest current to the one conducting the lowest current.

(A) $I_{18.0\,\Omega} > I_{12.0\,\Omega} > I_{8.00\,\Omega} > I_{6.00\,\Omega}$

(B) $I_{6.00\,\Omega} > I_{8.00\,\Omega} > I_{12.0\,\Omega} > I_{18.0\,\Omega}$

(C) $I_{8.00\,\Omega} > I_{12.0\,\Omega} > (I_{18.0\,\Omega} = I_{6.00\,\Omega})$

(D) $I_{8.00\,\Omega} > (I_{12.0\,\Omega} = I_{18.0\,\Omega} = I_{6.00\,\Omega})$

**Answer: C** All of the current in the circuit flows through the 8 Ω resistor that is in series with the battery. Because the 8 Ω resistor has the most current, we can eliminate choices (A) and (B). In the parallel section, more current flows through the 12 Ω resistor, which eliminates (D). The 18 Ω resistor and the 6 Ω resistor are in series, so they have the same current.

## BAR CHARTS

**Physics 1 Example:** A 50 kg skier approaches a hill with a velocity of 10 m/s. The hill has some friction, reducing the initial energy of the skier by 50 J. Which of the following bar charts correctly describes the initial and final energy in the system?

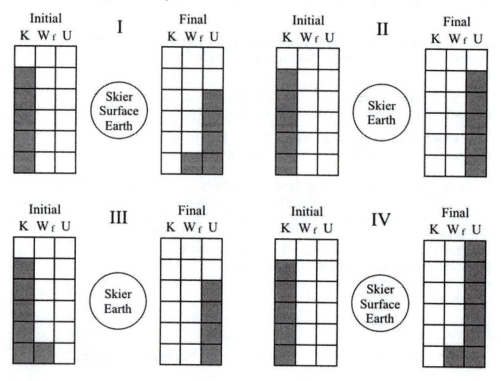

(A) I

(B) II

(C) III

(D) IV

**Answer: A** The correct set of graphs will show the skier's kinetic energy becoming potential energy with some being lost to work done by friction. The 50 kg skier approaches the hill with a velocity of 10 m/s. The initial kinetic energy of the skier is

$$K = \frac{1}{2}mv^2 = \frac{1}{2}(50 \text{ kg})\left(10 \text{ }^m\!/_s\right)^2 = 250 \text{ J}.$$

Each block on the bar chart represents 50 J. Answer A shows friction reducing the mechanical energy by 50 J and the remaining 200 J being gravitational potential energy. Answer A accounts for the conservation of all of the energy in the skier-surface-Earth system and the losses to friction.

## GRAPHING

**Physics 1 Example:** A variable force moves a 2.00 kg body through a displacement of 12.0 m.

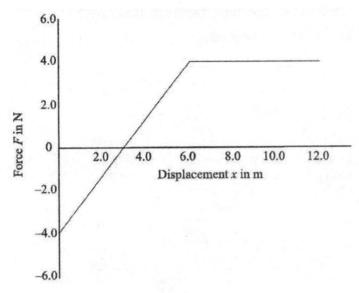

The work done by the force is closest to

(A) 12 J

(B) 18 J

(C) 24 J

(D) 36 J

**Answer: C** The work done is the area under the curve. From the graph, $A_1 = \frac{1}{2}bh$ is negative and $A_2 = \frac{1}{2}bh$ is positive. The two areas have the same magnitude and give a sum of zero. The work done is therefore the area $A_3 = bh$. This is (4.0 N – 0)(12.0 m – 6.0 m) = 24.0 J.

**Physics 1 Example:** An object undergoes simple harmonic motion with an amplitude of 0.03 m. The graph of its motion as a function of time is shown below. What is the frequency of this oscillation?

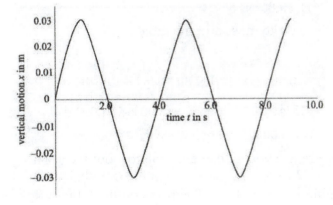

(A) 0.17 Hz

(B) 0.25 Hz

(C) 0.33 Hz

(D) 0.5 Hz

**Answer: B** The wave makes one complete oscillation every 4.0 s. Since frequency and period are related by $f = \dfrac{1}{T}$, the frequency is

$f = \dfrac{1}{4.0}$ Hz $= 0.25$ Hz.

## CALCULATION TYPES

**Physics 1 Example:** A 2.00 kg body, $M_1$, is pulled to the right across a horizontal tabletop by a cord that passes over an ideal pulley to a 1.00 kg body, $M_2$. The coefficient friction of friction $\mu_k$ between the 2.00 kg body and the table is 0.10.

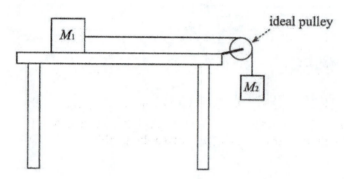

The acceleration of the system is

(A) 2.61 m/s²

(B) 3.27 m/s²

(C) 3.92 m/s²

(D) 4.90 m/s²

**Answer: A** We use Newton's second law to analyze an accelerating system. $\Sigma \left| \vec{F} \right| = m\vec{a}$

The weight of the 1.00 kg hanging body is 9.8 N

$\vec{W}_1 = (1.00 \text{ kg})\left(9.80 \text{ m}\!\!\diagup\!\!_{s^2}\right) = 9.80 \text{ N}$

The friction between the 2.00 kg mass and the table is

$\left| \vec{f} \right| = \mu_k N = \mu_k M_2 \vec{g}$

The frictional force is $\left| \vec{f} \right| = 0.10(2.00 \text{ kg})\left(9.80 \text{ m}\!\!\diagup\!\!_{s^2}\right) = 1.96 \text{ N}$.

Substitution into $\Sigma \left| \vec{F} \right| = m\vec{a}$ is $9.80 \text{ N} - 1.96 \text{ N} = (2.00 \text{ kg} + 1.00 \text{ kg})\vec{a}$..

The acceleration is $7.84 \text{ N} = (3.00 \text{ kg})\vec{a}$. Then $\vec{a} = 2.61 \text{ m}\!\!\diagup\!\!_{s^2}$.

This question might seem like you'll need a calculator, but try it with $g = 10 \text{ m/s}^2$.

The weight of the hanging mass is 10 N. The mass on the table weighs 20 N.

Friction is 0.1 × the weight of the mass on the table.

$a = \Sigma F/m$

$a = (10 \text{ N} - 2 \text{ N})/3 \text{ m}$

$a = 8/3 \text{ m/s}^2$, which is about 2.7 m/s²

## SET QUESTIONS

With this type of multiple-choice question, you are given a leading statement and/or picture that applies to the next two or more questions.

**Physics 1 Example:** Questions 1 and 2 refer to the information below.

A soccer ball is kicked toward a goal and follows a parabolic path in its flight. The mass of the soccer ball is 0.43 kg, and it was kicked at an initial velocity of 30 m/s at 15°.

1.  Which of the following would illustrate a position vs. time graph for the entire flight of the ball in the vertical direction?

a)    b)    c)    d)

2.  If the ball was kicked toward the right, which of the following vectors would represent the direction of the acceleration due to gravity as the ball reached the top of the parabolic path?

    (A) →

    (B) ←

    (C) ↑

    (D) ↓

## CORRECT ANSWERS

1.  **A**  The projectile follows a parabolic path, so it will have direction in both the x and y directions The position in the x direction would move forward as shown in choice D. Choice B shows non-accelerated motion, but a projectile is experiencing acceleration, so the graph must be curved as in choice A.

2.  **D**  The acceleration experienced by an object in free fall is due only to the acceleration due to gravity. It will experience this gravity at every point in its flight in the same direction, which is down as shown by choice D.

## TWO CORRECT ANSWER CHOICES

This is a special type of multiple-choice question that appears at the end of the multiple-choice section of the exam. The exam will give you instructions as to when you have reached this section and how to grid your answers on the answer sheet. You will be given a multiple-choice question with four choices. Of the four answer choices, TWO of them will be correct. You will only get credit if you get BOTH answers correct.

**Physics 1 Example:** A car is speeding up while moving on the highway. Which of the following choices could explain the car's velocity and acceleration? Select TWO answers.

(A) The velocity is positive, and the acceleration is positive.

(B) The velocity is negative, and the acceleration is negative.

(C) The velocity is positive, and the acceleration is negative.

(D) The velocity is negative, and the acceleration is positive.

## CORRECT ANSWERS

**A, B** Both velocity and acceleration determine how an object will change its motion. Remember that both are vector quantities. Positive and negative designations only tell us the direction of the vector, not how the motion changes. Both velocity and acceleration must be in the same direction for the object to speed up.

## STRATEGIES FOR THE FREE-RESPONSE PROBLEM SECTION

- Look through all of the problems first. Start where you feel most confident.
- Do all of your work and show your answers in the answer booklet. Work for each section of a problem should stay within the space given for that section. If you absolutely must use a separate paper, make sure it has your AP® number written on it clearly and label it with enough information to identify the problem the work is for.
- Show all of your work. Points are earned for specific correct concepts or applications of principles. A correct answer in isolation might not be worth any points. In problems involving calculations, box or circle your final answer.
- Points are awarded in the scoring of a problem according to the problem rubric. Points are never taken away.
- Cross out incorrect answers and work with an $X$ rather than spending time erasing or scribbling. A simple $X$ or line through it is fine.
- Be clear and organized in your work. If an AP® reader cannot clearly understand your work, you cannot earn full credit. If your handwriting is too difficult to read, readers cannot assign points.
- Free-response problems will have sets of parts: (a), (b), (c), etc. Attempt to solve each part. Even if your answer to (a) is incorrect, you can be awarded points for the remaining parts.
- Questions that ask for an explanation or justification must be technically correct and answered with complete sentences. Explanations and justifications rely on physics principles, not examples, stories, movies, or personal experience.
- Make sure your work includes the proper units.
- You should be able to recognize that equations of the form $y = mx + b$ are linear with a slope $m$ and a vertical intercept $b$. An equation of the form $x = at^2 + bt + c$ is parabolic. An equation of the form $PV = k$ is hyperbolic. You should be able to distinguish between these curves.
- In graphing $P$ vs. $V$, $P$ is plotted along the vertical axis; $V$, along the horizontal axis. When graphing $s = f(t)$, $s$ is plotted along the vertical axis; $t$, along the horizontal axis.
- An answer by itself is never enough. The reader must see how the solution came to be.

■ Some questions may ask you to put an X on a line to choose between several options and then justify your choice. You will earn points only if you select the correct option and use the correct justification. No points are earned for only checking the correct box.

# TYPES OF FREE-RESPONSE PROBLEMS

## THE EXPERIMENTAL DESIGN QUESTION

You will be given a scenario related to a lab-type experience. Through this scenario, you will be required to do at least one of the following:

■ Using appropriate or given equipment, design a procedure to answer a given question.

■ Justify your developed procedure.

■ State what measurements should be taken and how they will be used. Constants that could be looked up are not considered measurements.

■ Analyze data from an already completed lab. This may mean graphing data, interpreting a graph, or performing error analysis.

## SAMPLE PROBLEM 1

**Physics 2 Example:** During a classroom lab, students are tasked with determining the refractive index of an unknown transparent material. The students are given a block of the unknown material. They have access to laser pointers, rulers, protractors, stopwatches, and other standard lab equipment, all of which they may use as needed.

(a) Write a clear, precise procedure that could be used to find the refractive index of the unknown material.

(b) Describe how measurements made in the procedure listed above could be graphed so that the slope of a best-fit line is equal to the index of refraction.

(c) Should the slope of the graph be greater than 1, less than 1, or equal to 1? Justify your answer.

## SAMPLE PROBLEM 1 SOLUTION

(a) Although there may be different ways of doing this experiment, the most common and easiest would be to lay the block flat on a table. Students could shine the laser through the block at an angle. The students could then measure the angles of incident and refraction of the laser. A diagram of the experimental setup would ensure that the graders understand what you mean. There should be no mention of using the stopwatch, as light is way too fast to be timed.

(b) Only two measurements are taken, the angle of incidence and the angle of refraction. Snell's law should be used to show that the sine of the angle of incidence should be graphed vs. the sine of the angle of refraction. Students should note that the index of refraction of the incident material is $n = 1$ because the laser is incident in air.

(c) Light will slow down because the material will not allow the light to move as fast as it does through air. Therefore, the slope of the graph and the index of refraction it represents should be greater than 1.

## Experimental Design Question Hints

■ Keep it simple and realistic. There is no need to get fancy. Try to solve the problem in a few succinct and well-written steps.

■ One of your procedure steps should NOT be "and now go gather equipment" or "and now go do the calculations." These are not procedural steps. Just write what you do in the actual lab.

■ Use common equipment. Most labs do not have cannons, radar guns, or particle accelerators. Simple metersticks and stopwatches can be used for many lab set-ups.

■ If numerous pieces of equipment are listed, choose the ones you are very familiar with. If you are uncertain how a piece of equipment works, you should not write about it in your procedure. There are usually several ways to answer the question.

■ Label graph axes with units. Give the graph a descriptive title. If the graph is linear, draw a best-fit line and do NOT connect the dots.

## Qualitative-Quantitative Translation

The "qualitative" portion of this question usually requires you to make predictions based on physics principles and equations without using numbers. These questions may read "If quantity X is changed, how is quantity Y affected? Justify your answer." You must be able to explain your reasoning clearly as you will not receive points for merely stating the change.

The "quantitative" portion is what many physics students think will be on the exam. This part of the free-response question will require you to calculate your answer using equations and given numbers OR derive an expression using only symbols. These are your typical calculator questions. It is entirely possible that there will be only one or none of these.

## Sample Problem 2

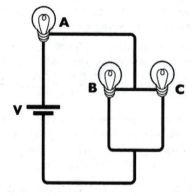

**Physics 1 or Physics 2 Example:** In the circuit above, bulbs A, B, and C all have the same resistance R.

(a) Redraw the circuit to include a fourth bulb D with resistance 2R in a way that would make bulb A brighter.

(b) Explain why adding the resistance of bulb D in such a way would make bulb A brighter.

(c) The initial power output of bulb A is 20 watts; what would its power output be after bulb D is added?

## ANSWER TO SAMPLE PROBLEM 2

(a)

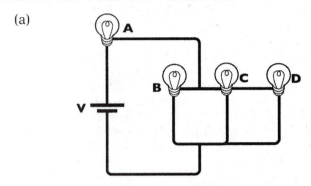

(b) Adding bulb D in parallel with B and C will provide another path through which current can flow and will reduce the resistance of the parallel section. The reduced resistance will allot more current to flow in the circuit. More current flowing through bulb A will make it brighter.

(c) The initial resistance of the circuit is $R + (1/R + 1/R)^{-1}$, which is 1.5 R. The new resistance of the circuit is $R + (1/R + 1/R + 1/2R)^{-1}$, which is 1.4 R. Because the circuit's resistance changed by a factor of 1.4/1.5, the current will change by a factor of 1.5/1.4 since current and resistance are inversely related. Because $P = I^2R$, power will change by a factor of $(1.5/1.4)^2$. 20 W $(1.5/1.4)^2$ = 23 W.

## HINTS FOR QUALITATIVE-QUANTITATIVE TRANSLATION QUESTIONS

When doing calculation problems, show ALL your work. Write the formula. Then show the numbers in the formula. Finally, write the answer with units. Many times it is not the answer that gets you the most points but the work leading up to the answer.

▪ This is not the paragraph length response. Your justification should be clear and to the point. Do not write lengthy statements.

▪ "Explain" means that you need to use words to explain the nature or cause of a phenomenon.

▪ You may be asked to "justify" a previous answer by providing an argument supported by evidence. The evidence could include statements of physical principles, equations, data, graphs, and diagrams.

▪ Read the whole question. Many times parts a and b have very little to do with parts c and d. Do what you can do first.

■ Write only one answer to the question. If you contradict yourself, you will get no points. In the case of multiple answers to the same question, AP® readers will score the most incorrect answer.

■ If you are taking more than the recommended time on a question, you are probably doing it wrong or taking the wrong approach. Come back to that question later if time permits.

## SHORT ANSWER QUESTIONS

The short answer questions are exactly what they sound like. They can be structured in many ways and can address any of the concepts in the course.

One of the short answer questions will have a part that requires you to write a paragraph-length response (PLR). You will be asked to write a short paragraph about a given physics concept, perhaps comparing two quantities in an experiment. Unless directed otherwise, the PLR is the only part of the exam where you are required to write in complete sentences. The PLR will usually make up 5 of the total points on the question. This should clue you in that you are not going to be required to write a multiparagraph essay. You should try to make four or five distinct points about the posed question. Be careful not to contradict yourself. Many times you will earn the "fifth point" on the rubric for providing no incorrect or contradicting statements.

## SAMPLE SHORT ANSWER PROBLEM

**Physics 2 Example:**

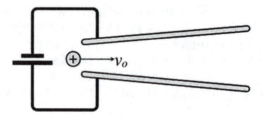

A positive charge traveling with velocity $v_o$ is projected into the region between two non-parallel plates attached to a battery. The separation distance between the plates is greater on the right side than on the left side.

(a) Sketch the electrical field between the two plates with solid lines.

(b) As the charge moves between the plates, it does not make contact with the plates. Sketch the charge's trajectory as it moves through the region between the plates and into the region beyond the plates with a dashed line.

(c) In a coherent paragraph-length response, provide a physical explanation of any changes to the velocity of the positive charge as it travels through the region between the plates and continues outside the plates.

SOLUTION TO SAMPLE SHORT ANSWER PROBLEM

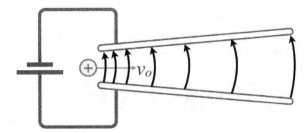

a.  The field is strongest where the plates are closest, so the electric field lines are densest where the plates are closest. The electric field points up because the bottom plate is positive.

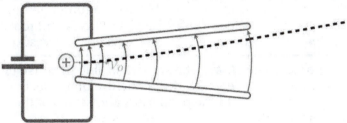

b.  The charge curves upward while it is in the field and travels straight when it is beyond the plates.

c.  The electric field applies an upward force on the particle. The magnitude of the upward force decreases as the charge moves to the right because the plates are farther apart. The particle's path curves as the particle moves to the right, and the curvature decreases along with the force. The particle travels with constant speed and direction when it leaves the region between the plates because there is no field to accelerate the particle.

# AP® PHYSICS
# LEARNING OBJECTIVES*

Following is the list of learning objectives from the College Board curriculum.

| | |
|---|---|
| **PHYSICS 2** | **Learning objective 1.A.2.1** The student is able to construct representations of the differences between a fundamental particle and a system composed of fundamental particles and to relate this to the properties and scales of the systems being investigated. |
| **PHYSICS 2** | **Learning objective 1.A.4.1** The student is able to construct representations of the energy level structure of an electron in an atom and to relate this to the properties and scales of the systems being investigated. |
| **PHYSICS 1** | **Learning objective 1.A.5.1** The student is able to model verbally or visually the properties of a system based on its substructure and to relate this to changes in the system properties over time as external variables are changed. |
| **PHYSICS 2** | **Learning objective 1.A.5.2** The student is able to construct representations of how the properties of a system are determined by the interactions of its constituent substructures. |
| **PHYSICS 1**<br>**PHYSICS 2** | **Learning objective 1.B.1.1** The student is able to make claims about natural phenomena based on conservation of electric charge. |
| **PHYSICS 1**<br>**PHYSICS 2** | **Learning objective 1.B.1.2** The student is able to make predictions, using the conservation of electric charge, about the sign and relative quantity of net charge of objects or systems after various charging processes, including conservation of charge in simple circuits. |
| **PHYSICS 1**<br>**PHYSICS 2** | **Learning objective 1.B.2.1** The student is able to construct an explanation of the two-charge model of electric charge based on evidence produced through scientific practices. |
| **PHYSICS 2** | **Learning objective 1.B.2.2** The student is able to make a qualitative prediction about the distribution of positive and negative electric charges within neutral systems as they undergo various processes. |
| **PHYSICS 2** | **Learning objective 1.B.2.3** The student is able to challenge claims that polarization of electric charge or separation of charge must result in a net charge on the object. |

---

AP® is a trademark registered by the College Board, which is not affiliated with, and does not endorse, this product.

| | |
|---|---|
| **PHYSICS 1**<br>**PHYSICS 2** | **Learning objective 1.B.3.1** The student is able to challenge the claim that an electric charge smaller than the elementary charge has been isolated. |
| **PHYSICS 1** | **Learning objective 1.C.1.1** The student is able to design an experiment for collecting data to determine the relationship between the net force exerted on an object, its inertial mass, and its acceleration. |
| **PHYSICS 1** | **Learning objective 1.C.3.1** The student is able to design a plan for collecting data to measure gravitational mass and to measure inertial mass, and to distinguish between the two experiments. |
| **PHYSICS 2** | **Learning objective 1.C.4.1** The student is able to articulate the reasons that the theory of conservation of mass was replaced by the theory of conservation of mass-energy. |
| **PHYSICS 2** | **Learning objective 1.D.1.1** The student is able to explain why classical mechanics cannot describe all properties of objects by articulating the reasons that classical mechanics must be refined and an alternative explanation developed when classical particles display wave properties. |
| **PHYSICS 2** | **Learning objective 1.D.3.1** The student is able to articulate the reasons that classical mechanics must be replaced by special relativity to describe the experimental results and theoretical predictions that show that the properties of space and time are not absolute. [Students will be expected to recognize situations in which nonrelativistic classical physics breaks down and to explain how relativity addresses that breakdown, but students will not be expected to know in which of two reference frames a given series of events corresponds to a greater or lesser time interval, or a greater or lesser spatial distance; they will just need to know that observers in the two reference frames can "disagree" about some time and distance intervals.] |
| **PHYSICS 2** | **Learning objective 1.E.1.1** The student is able to predict the densities, differences in densities, or changes in densities under different conditions for natural phenomena and design an investigation to verify the prediction. |
| **PHYSICS 2** | **Learning objective 1.E.1.2** The student is able to select from experimental data the information necessary to determine the density of an object and/or compare densities of several objects. |
| **PHYSICS 1**<br>**PHYSICS 2** | **Learning objective 1.E.2.1** The student is able to choose and justify the selection of data needed to determine resistivity for a given material. |
| **PHYSICS 2** | **Learning objective 1.E.3.1** The student is able to design an experiment and analyze data from it to examine thermal conductivity. |
| **PHYSICS 1** | **Learning objective 2.B.1.1** The student is able to apply $\vec{F} = m\vec{g}$ to calculate the gravitational force on an object with mass $m$ in a gravitational field of strength $g$ in the context of the effects of a net force on objects and systems. |

| **PHYSICS 1** | Learning objective 2.B.2.1 The student is able to apply $g = G\dfrac{M}{r^2}$ to calculate the gravitational field due to an object with mass $M$, where the field is a vector directed toward the center of the object of mass $M$. |
|---|---|
| **PHYSICS 1** | Learning objective 2.B.2.2 The student is able to approximate a numerical value of the gravitational field ($g$) near the surface of an object from its radius and mass relative to those of the Earth or other reference objects. |
| **PHYSICS 2** | Learning objective 2.C.1.1 The student is able to predict the direction and the magnitude of the force exerted on an object with an electric charge $q$ placed in an electric field $E$ using the mathematical model of the relation between an electric force and an electric field: $\vec{F} = q\vec{E}$; a vector relation. |
| **PHYSICS 2** | Learning objective 2.C.1.2 The student is able to calculate any one of the variables—electric force, electric charge, and electric field—at a point given the values and sign or direction of the other two quantities. |
| **PHYSICS 2** | Learning objective 2.C.2.1 The student is able to qualitatively and semi-quantitatively apply the vector relationship between the electric field and the net electric charge creating that field. |
| **PHYSICS 2** | Learning objective 2.C.3.1 The student is able to explain the inverse square dependence of the electric field surrounding a spherically symmetric electrically charged object. |
| **PHYSICS 2** | Learning objective 2.C.4.1 The student is able to distinguish the characteristics that differ between monopole fields (gravitational field of spherical mass and electrical field due to single point charge) and dipole fields (electric dipole field and magnetic field) and make claims about the spatial behavior of the fields using qualitative or semiquantitative arguments based on vector addition of fields due to each point source, including identifying the locations and signs of sources from a vector diagram of the field. |
| **PHYSICS 2** | Learning objective 2.C.4.2 The student is able to apply mathematical routines to determine the magnitude and direction of the electric field at specified points in the vicinity of a small set (2–4) of point charges, and express the results in terms of magnitude and direction of the field in a visual representation by drawing field vectors of appropriate length and direction at the specified points. |
| **PHYSICS 2** | Learning objective 2.C.5.1 The student is able to create representations of the magnitude and direction of the electric field at various distances (small compared to plate size) from two electrically charged plates of equal magnitude and opposite signs, and is able to recognize that the assumption of uniform field is not appropriate near edges of plates. |

| **PHYSICS 2** | **Learning objective 2.C.5.2** The student is able to calculate the magnitude and determine the direction of the electric field between two electrically charged parallel plates, given the charge of each plate, or the electric potential difference and plate separation. |
| --- | --- |
| **PHYSICS 2** | **Learning objective 2.C.5.3** The student is able to represent the motion of an electrically charged particle in the uniform field between two oppositely charged plates and express the connection of this motion to projectile motion of an object with mass in the Earth's gravitational field. |
| **PHYSICS 2** | **Learning objective 2.D.1.1** The student is able to apply mathematical routines to express the force exerted on a moving charged object by a magnetic field. |
| **PHYSICS 2** | **Learning objective 2.D.2.1** The student is able to create a verbal or visual representation of a magnetic field around a long straight wire or a pair of parallel wires. |
| **PHYSICS 2** | **Learning objective 2.D.3.1** The student is able to describe the orientation of a magnetic dipole placed in a magnetic field in general and the particular cases of a compass in the magnetic field of the Earth and iron filings surrounding a bar magnet. |
| **PHYSICS 2** | **Learning objective 2.D.4.1** The student is able to use the representation of magnetic domains to qualitatively analyze the magnetic behavior of a bar magnet composed of ferromagnetic material. |
| **PHYSICS 2** | **Learning objective 2.E.1.1** The student is able to construct or interpret visual representations of the isolines of equal gravitational potential energy per unit mass and refer to each line as a gravitational equipotential. |
| **PHYSICS 2** | **Learning objective 2.E.2.1** The student is able to determine the structure of isolines of electric potential by constructing them in a given electric field. |
| **PHYSICS 2** | **Learning objective 2.E.2.2** The student is able to predict the structure of isolines of electric potential by constructing them in a given electric field and make connections between these isolines and those found in a gravitational field. |
| **PHYSICS 2** | **Learning objective 2.E.2.3** The student is able to qualitatively use the concept of isolines to construct isolines of electric potential in an electric field and determine the effect of that field on electrically charged objects. |
| **PHYSICS 2** | **Learning objective 2.E.3.1** The student is able to apply mathematical routines to calculate the average value of the magnitude of the electric field in a region from a description of the electric potential in that region using the displacement along the line on which the difference in potential is evaluated. |

| PHYSICS 2 | Learning objective 2.E.3.2 The student is able to apply the concept of the isoline representation of electric potential for a given electric charge distribution to predict the average value of the electric field in the region. |
| --- | --- |
| PHYSICS 1 | Learning objective 3.A.1.1 The student is able to express the motion of an object using narrative, mathematical, and graphical representations. |
| PHYSICS 1 | Learning objective 3.A.1.2 The student is able to design an experimental investigation of the motion of an object. |
| PHYSICS 1 | Learning objective 3.A.1.3 The student is able to analyze experimental data describing the motion of an object and is able to express the results of the analysis using narrative, mathematical, and graphical representations. |
| PHYSICS 1 PHYSICS 2 | Learning objective 3.A.2.1 The student is able to represent forces in diagrams or mathematically using appropriately labeled vectors with magnitude, direction, and units during the analysis of a situation. |
| PHYSICS 1 | Learning objective 3.A.3.1 The student is able to analyze a scenario and make claims (develop arguments, justify assertions) about the forces exerted on an object by other objects for different types of forces or components of forces. |
| PHYSICS 1 PHYSICS 2 | Learning objective 3.A.3.2 The student is able to challenge a claim that an object can exert a force on itself. |
| PHYSICS 1 PHYSICS 2 | Learning objective 3.A.3.3 The student is able to describe a force as an interaction between two objects and identify both objects for any force. |
| PHYSICS 2 | Learning objective 3.A.3.4 The student is able to make claims about the force on an object due to the presence of other objects with the same property: mass, electric charge. |
| PHYSICS 1 PHYSICS 2 | Learning objective 3.A.4.1 The student is able to construct explanations of physical situations involving the interaction of bodies using Newton's third law and the representation of action-reaction pairs of forces. |
| PHYSICS 1 PHYSICS 2 | Learning objective 3.A.4.2 The student is able to use Newton's third law to make claims and predictions about the action-reaction pairs of forces when two objects interact. |
| PHYSICS 1 PHYSICS 2 | Learning objective 3.A.4.3 The student is able to analyze situations involving interactions among several objects by using free-body diagrams that include the application of Newton's third law to identify forces. |
| PHYSICS 1 | Learning objective 3.B.1.1 The student is able to predict the motion of an object subject to forces exerted by several objects using an application of Newton's second law in a variety of physical situations with acceleration in one dimension. |

| **PHYSICS 1** | **Learning objective 3.B.1.2** The student is able to design a plan to collect and analyze data for motion (static, constant, or accelerating) from force measurements and carry out an analysis to determine the relationship between the net force and the vector sum of the individual forces. |
|---|---|
| **PHYSICS 1** **PHYSICS 2** | **Learning objective 3.B.1.3** The student is able to reexpress a free-body diagram representation into a mathematical representation and solve the mathematical representation for the acceleration of the object. |
| **PHYSICS 2** | **Learning objective 3.B.1.4** The student is able to predict the motion of an object subject to forces exerted by several objects using an application of Newton's second law in a variety of physical situations. |
| **PHYSICS 1** **PHYSICS 2** | **Learning objective 3.B.2.1** The student is able to create and use free-body diagrams to analyze physical situations to solve problems with motion qualitatively and quantitatively. |
| **PHYSICS 1** | **Learning objective 3.B.3.1** The student is able to predict which properties determine the motion of a simple harmonic oscillator and what the dependence of the motion is on those properties. |
| **PHYSICS 1** | **Learning objective 3.B.3.2** The student is able to design a plan and collect data in order to ascertain the characteristics of the motion of a system undergoing oscillatory motion caused by a restoring force. |
| **PHYSICS 1** | **Learning objective 3.B.3.3** The student can analyze data to identify qualitative or quantitative relationships between given values and variables (i.e., force, displacement, acceleration, velocity, period of motion, frequency, spring constant, string length, mass) associated with objects in oscillatory motion to use that data to determine the value of an unknown. |
| **PHYSICS 1** | **Learning objective 3.B.3.4** The student is able to construct a qualitative and/or a quantitative explanation of oscillatory behavior given evidence of a restoring force. |
| **PHYSICS 1** | **Learning objective 3.C.1.1** The student is able to use Newton's law of gravitation to calculate the gravitational force the two objects exert on each other and use that force in contexts other than orbital motion. |
| **PHYSICS 1** | **Learning objective 3.C.1.2** The student is able to use Newton's law of gravitation to calculate the gravitational force between two objects and use that force in contexts involving orbital motion (for circular orbital motion only in Physics 1). |
| **PHYSICS 1** **PHYSICS 2** | **Learning objective 3.C.2.1** The student is able to use Coulomb's law qualitatively and quantitatively to make predictions about the interaction between two electric point charges (interactions between collections of electric point charges are not covered in Physics 1 and instead are restricted to Physics 2). |

| | |
|---|---|
| **PHYSICS 1**<br>**PHYSICS 2** | **Learning objective 3.C.2.2** The student is able to connect the concepts of gravitational force and electric force to compare similarities and differences between the forces. |
| **PHYSICS 2** | **Learning objective 3.C.2.3** The student is able to use mathematics to describe the electric force that results from the interaction of several separated point charges (generally 2 to 4 point charges, though more are permitted in situations of high symmetry). |
| **PHYSICS 2** | **Learning objective 3.C.3.1** The student is able to use right-hand rules to analyze a situation involving a current-carrying conductor and a moving electrically charged object to determine the direction of the magnetic force exerted on the charged object due to the magnetic field created by the current-carrying conductor. |
| **PHYSICS 2** | **Learning objective 3.C.3.2** The student is able to plan a data collection strategy appropriate to an investigation of the direction of the force on a moving electrically charged object caused by a current in a wire in the context of a specific set of equipment and instruments and analyze the resulting data to arrive at a conclusion. |
| **PHYSICS 1**<br>**PHYSICS 2** | **Learning objective 3.C.4.1** The student is able to make claims about various contact forces between objects based on the microscopic cause of those forces. |
| **PHYSICS 1**<br>**PHYSICS 2** | **Learning objective 3.C.4.2** The student is able to explain contact forces (tension, friction, normal, buoyant, spring) as arising from interatomic electric forces and that they therefore have certain directions. |
| **PHYSICS 1** | **Learning objective 3.D.1.1** The student is able to justify the selection of data needed to determine the relationship between the direction of the force acting on an object and the change in momentum caused by that force. |
| **PHYSICS 1** | **Learning objective 3.D.2.1** The student is able to justify the selection of routines for the calculation of the relationships between changes in momentum of an object, average force, impulse, and time of interaction. |
| **PHYSICS 1** | **Learning objective 3.D.2.2** The student is able to predict the change in momentum of an object from the average force exerted on the object and the interval of time during which the force is exerted. |
| **PHYSICS 1** | **Learning objective 3.D.2.3** The student is able to analyze data to characterize the change in momentum of an object from the average force exerted on the object and the interval of time during which the force is exerted. |
| **PHYSICS 1** | **Learning objective 3.D.2.4** The student is able to design a plan for collecting data to investigate the relationship between changes in momentum and the average force exerted on an object over time. |
| **PHYSICS 1** | **Learning objective 3.E.1.1** The student is able to make predictions about the changes in kinetic energy of an object based on considerations of the direction of the net force on the object as the object moves. |

| | |
|---|---|
| **PHYSICS 1** | **Learning objective 3.E.1.2** The student is able to use net force and velocity vectors to determine qualitatively whether kinetic energy of an object would increase, decrease, or remain unchanged. |
| **PHYSICS 1** | **Learning objective 3.E.1.3** The student is able to use force and velocity vectors to determine qualitatively or quantitatively the net force exerted on an object and qualitatively whether kinetic energy of that object would increase, decrease, or remain unchanged. |
| **PHYSICS 1** | **Learning objective 3.E.1.4** The student is able to apply mathematical routines to determine the change in kinetic energy of an object given the forces on the object and the displacement of the object. |
| **PHYSICS 1** | **Learning objective 3.F.1.1** The student is able to use representations of the relationship between force and torque. |
| **PHYSICS 1** | **Learning objective 3.F.1.2** The student is able to compare the torques on an object caused by various forces. |
| **PHYSICS 1** | **Learning objective 3.F.1.3** The student is able to estimate the torque on an object caused by various forces in comparison to other situations. |
| **PHYSICS 1** | **Learning objective 3.F.1.4** The student is able to design an experiment and analyze data testing a question about torques in a balanced rigid system. |
| **PHYSICS 1** | **Learning objective 3.F.1.5** The student is able to calculate torques on a two-dimensional system in static equilibrium, by examining a representation or model (such as a diagram or physical construction). |
| **PHYSICS 1** | **Learning objective 3.F.2.1** The student is able to make predictions about the change in the angular velocity about an axis for an object when forces exerted on the object cause a torque about that axis. |
| **PHYSICS 1** | **Learning objective 3.F.2.2** The student is able to plan data collection and analysis strategies designed to test the relationship between a torque exerted on an object and the change in angular velocity of that object about an axis. |
| **PHYSICS 1** | **Learning objective 3.F.3.1** The student is able to predict the behavior of rotational collision situations by the same processes that are used to analyze linear collision situations using an analogy between impulse and change of linear momentum and angular impulse and change of angular momentum. |
| **PHYSICS 1** | **Learning objective 3.F.3.2** In an unfamiliar context or using representations beyond equations, the student is able to justify the selection of a mathematical routine to solve for the change in angular momentum of an object caused by torques exerted on the object. |
| **PHYSICS 1** | **Learning objective 3.F.3.3** The student is able to plan data collection and analysis strategies designed to test the relationship between torques exerted on an object and the change in angular momentum of that object. |

| | |
|---|---|
| **PHYSICS 1** | **Learning objective 3.G.1.1** The student is able to articulate situations when the gravitational force is the dominant force and when the electromagnetic, weak, and strong forces can be ignored. |
| **PHYSICS 2** | **Learning objective 3.G.1.2** The student is able to connect the strength of the gravitational force between two objects to the spatial scale of the situation and the masses of the objects involved and compare that strength to other types of forces. |
| **PHYSICS 2** | **Learning objective 3.G.2.1** The student is able to connect the strength of electromagnetic forces with the spatial scale of the situation, the magnitude of the electric charges, and the motion of the electrically charged objects involved. |
| **PHYSICS 2** | **Learning objective 3.G.3.1** The student is able to identify the strong force as the force that is responsible for holding the nucleus together. |
| **PHYSICS 1** | **Learning objective 4.A.1.1** The student is able to use representations of the center of mass of an isolated two-object system to analyze the motion of the system qualitatively and semiquantitatively. |
| **PHYSICS 1** | **Learning objective 4.A.2.1** The student is able to make predictions about the motion of a system based on the fact that acceleration is equal to the change in velocity per unit time, and velocity is equal to the change in position per unit time. |
| **PHYSICS 1** | **Learning objective 4.A.2.2** The student is able to evaluate using given data whether all the forces on a system or whether all the parts of a system have been identified. |
| **PHYSICS 1** | **Learning objective 4.A.2.3** The student is able to create mathematical models and analyze graphical relationships for acceleration, velocity, and position of the center of mass of a system and use them to calculate properties of the motion of the center of mass of a system. |
| **PHYSICS 1** | **Learning objective 4.A.3.1** The student is able to apply Newton's second law to systems to calculate the change in the center-of-mass velocity when an external force is exerted on the system. |
| **PHYSICS 1** | **Learning objective 4.A.3.2** The student is able to use visual or mathematical representations of the forces between objects in a system to predict whether or not there will be a change in the center-of-mass velocity of that system. |
| **PHYSICS 1** | **Learning objective 4.B.1.1** The student is able to calculate the change in linear momentum of a two-object system with constant mass in linear motion from a representation of the system (data, graphs, etc.). |
| **PHYSICS 1** | **Learning objective 4.B.1.2** The student is able to analyze data to find the change in linear momentum for a constant-mass system using the product of the mass and the change in velocity of the center of mass. |
| **PHYSICS 1** | **Learning objective 4.B.2.1** The student is able to apply mathematical routines to calculate the change in momentum of a system by analyzing the average force exerted over a certain time on the system. |

| PHYSICS 1 | Learning objective 4.B.2.2 The student is able to perform analysis on data presented as a force-time graph and predict the change in momentum of a system. |
|---|---|
| PHYSICS 1 | Learning objective 4.C.1.1 The student is able to calculate the total energy of a system and justify the mathematical routines used in the calculation of component types of energy within the system whose sum is the total energy. |
| PHYSICS 1 | Learning objective 4.C.1.2 The student is able to predict changes in the total energy of a system due to changes in position and speed of objects or frictional interactions within the system. |
| PHYSICS 1 | Learning objective 4.C.2.1 The student is able to make predictions about the changes in the mechanical energy of a system when a component of an external force acts parallel or antiparallel to the direction of the displacement of the center of mass. |
| PHYSICS 1 | Learning objective 4.C.2.2 The student is able to apply the concepts of conservation of energy and the work-energy theorem to determine qualitatively and/or quantitatively that work done on a two-object system in linear motion will change the kinetic energy of the center of mass of the system, the potential energy of the systems, and/or the internal energy of the system. |
| PHYSICS 2 | Learning objective 4.C.3.1 The student is able to make predictions about the direction of energy transfer due to temperature differences based on interactions at the microscopic level. |
| PHYSICS 2 | Learning objective 4.C.4.1 The student is able to apply mathematical routines to describe the relationship between mass and energy and apply this concept across domains of scale. |
| PHYSICS 1 | Learning objective 4.D.1.1 The student is able to describe a representation and use it to analyze a situation in which several forces exerted on a rotating system of rigidly connected objects change the angular velocity and angular momentum of the system. |
| PHYSICS 1 | Learning objective 4.D.1.2 The student is able to plan data collection strategies designed to establish that torque, angular velocity, angular acceleration, and angular momentum can be predicted accurately when the variables are treated as being clockwise or counterclockwise with respect to a well-defined axis of rotation, and refine the research question based on the examination of data. |
| PHYSICS 1 | Learning objective 4.D.2.1 The student is able to describe a model of a rotational system and use that model to analyze a situation in which angular momentum changes due to interaction with other objects or systems. |
| PHYSICS 1 | Learning objective 4.D.2.2 The student is able to plan a data collection and analysis strategy to determine the change in angular momentum of a system and relate it to interactions with other objects and systems. |

| | |
|---|---|
| **PHYSICS 1** | **Learning objective 4.D.3.1** The student is able to use appropriate mathematical routines to calculate values for initial or final angular momentum, or change in angular momentum of a system, or average torque or time during which the torque is exerted in analyzing a situation involving torque and angular momentum. |
| **PHYSICS 1** | **Learning objective 4.D.3.2** The student is able to plan a data collection strategy designed to test the relationship between the change in angular momentum of a system and the product of the average torque applied to the system and the time interval during which the torque is exerted. |
| **PHYSICS 2** | **Learning objective 4.E.1.1** The student is able to use representations and models to qualitatively describe the magnetic properties of some materials that can be affected by magnetic properties of other objects in the system. |
| **PHYSICS 2** | **Learning objective 4.E.2.1** The student is able to construct an explanation of the function of a simple electromagnetic device in which an induced emf is produced by a changing magnetic flux through an area defined by a current loop (i.e., a simple microphone or generator) or of the effect on behavior of a device in which an induced emf is produced by a constant magnetic field through a changing area. |
| **PHYSICS 2** | **Learning objective 4.E.3.1** The student is able to make predictions about the redistribution of charge during charging by friction, conduction, and induction. |
| **PHYSICS 2** | **Learning objective 4.E.3.2** The student is able to make predictions about the redistribution of charge caused by the electric field due to other systems, resulting in charged or polarized objects. |
| **PHYSICS 2** | **Learning objective 4.E.3.3** The student is able to construct a representation of the distribution of fixed and mobile charge in insulators and conductors. |
| **PHYSICS 2** | **Learning objective 4.E.3.4** The student is able to construct a representation of the distribution of fixed and mobile charge in insulators and conductors that predicts charge distribution in processes involving induction or conduction. |
| **PHYSICS 2** | **Learning objective 4.E.3.5** The student is able to plan and/or analyze the results of experiments in which electric charge rearrangement occurs by electrostatic induction, or is able to refine a scientific question relating to such an experiment by identifying anomalies in a data set or procedure. |
| **PHYSICS 2** | **Learning objective 4.E.4.1** The student is able to make predictions about the properties of resistors and/or capacitors when placed in a simple circuit, based on the geometry of the circuit element and supported by scientific theories and mathematical relationships. |

| | |
|---|---|
| **PHYSICS 2** | **Learning objective 4.E.4.2** The student is able to design a plan for the collection of data to determine the effect of changing the geometry and/or materials on the resistance or capacitance of a circuit element and relate results to the basic properties of resistors and capacitors. |
| **PHYSICS 2** | **Learning objective 4.E.4.3** The student is able to analyze data to determine the effect of changing the geometry and/or materials on the resistance or capacitance of a circuit element and relate results to the basic properties of resistors and capacitors. |
| **PHYSICS 2** | **Learning objective 4.E.5.1** The student is able to make and justify a quantitative prediction of the effect of a change in values or arrangements of one or two circuit elements on the currents and potential differences in a circuit containing a small number of sources of emf, resistors, capacitors, and switches in series and/or parallel. |
| **PHYSICS 2** | **Learning objective 4.E.5.2** The student is able to make and justify a qualitative prediction of the effect of a change in values or arrangements of one or two circuit elements on currents and potential differences in a circuit containing a small number of sources of emf, resistors, capacitors, and switches in series and/or parallel. |
| **PHYSICS 2** | **Learning objective 4.E.5.3** The student is able to plan data collection strategies and perform data analysis to examine the values of currents and potential differences in an electric circuit that is modified by changing or rearranging circuit elements, including sources of emf, resistors, and capacitors. |
| **PHYSICS 1** | **Learning objective 5.A.2.1** The student is able to define open and closed systems for everyday situations and apply conservation concepts for energy, charge, and linear momentum to those situations. |
| **PHYSICS 1** | **Learning objective 5.B.1.1** The student is able to set up a representation or model showing that a single object can only have kinetic energy and use information about that object to calculate its kinetic energy. |
| **PHYSICS 1** | **Learning objective 5.B.1.2** The student is able to translate between a representation of a single object, which can only have kinetic energy, and a system that includes the object, which may have both kinetic and potential energies. |
| **PHYSICS 1** **PHYSICS 2** | **Learning objective 5.B.2.1** The student is able to calculate the expected behavior of a system using the object model (i.e., by ignoring changes in internal structure) to analyze a situation. Then, when the model fails, the student can justify the use of conservation of energy principles to calculate the change in internal energy due to changes in internal structure because the object is actually a system. |
| **PHYSICS 1** | **Learning objective 5.B.3.1** The student is able to describe and make qualitative and/or quantitative predictions about everyday examples of systems with internal potential energy. |

| PHYSICS 1 | Learning objective 5.B.3.2 The student is able to make quantitative calculations of the internal potential energy of a system from a description or diagram of that system. |
|---|---|
| PHYSICS 1 | Learning objective 5.B.3.3 The student is able to apply mathematical reasoning to create a description of the internal potential energy of a system from a description or diagram of the objects and interactions in that system. |
| PHYSICS 1 PHYSICS 2 | Learning objective 5.B.4.1 The student is able to describe and make predictions about the internal energy of systems. |
| PHYSICS 1 PHYSICS 2 | Learning objective 5.B.4.2 The student is able to calculate changes in kinetic energy and potential energy of a system, using information from representations of that system. |
| PHYSICS 1 | Learning objective 5.B.5.1 The student is able to design an experiment and analyze data to examine how a force exerted on an object or system does work on the object or system as it moves through a distance. |
| PHYSICS 1 | Learning objective 5.B.5.2 The student is able to design an experiment and analyze graphical data in which interpretations of the area under a force-distance curve are needed to determine the work done on or by the object or system. |
| PHYSICS 1 | Learning objective 5.B.5.3 The student is able to predict and calculate from graphical data the energy transfer to or work done on an object or system from information about a force exerted on the object or system through a distance. |
| PHYSICS 1 PHYSICS 2 | Learning objective 5.B.5.4 The student is able to make claims about the interaction between a system and its environment in which the environment exerts a force on the system, thus doing work on the system and changing the energy of the system (kinetic energy plus potential energy). |
| PHYSICS 1 PHYSICS 2 | Learning objective 5.B.5.5 The student is able to predict and calculate the energy transfer to (i.e., the work done on) an object or system from information about a force exerted on the object or system through a distance. |
| PHYSICS 2 | Learning objective 5.B.5.6 The student is able to design an experiment and analyze graphical data in which interpretations of the area under a pressure-volume curve are needed to determine the work done on or by the object or system. |
| PHYSICS 2 | Learning objective 5.B.6.1 The student is able to describe the models that represent processes by which energy can be transferred between a system and its environment because of differences in temperature: conduction, convection, and radiation. |
| PHYSICS 2 | Learning objective 5.B.7.1 The student is able to predict qualitative changes in the internal energy of a thermodynamic system involving transfer of energy due to heat or work done and justify those predictions in terms of conservation of energy principles. |

| PHYSICS 2 | **Learning objective 5.B.7.2** The student is able to create a plot of pressure versus volume for a thermodynamic process from given data. |
|---|---|
| PHYSICS 2 | **Learning objective 5.B.7.3** The student is able to use a plot of pressure versus volume for a thermodynamic process to make calculations of internal energy changes, heat, or work, based upon conservation of energy principles (i.e., the first law of thermodynamics). |
| PHYSICS 2 | **Learning objective 5.B.8.1** The student is able to describe emission or absorption spectra associated with electronic or nuclear transitions as transitions between allowed energy states of the atom in terms of the principle of energy conservation, including characterization of the frequency of radiation emitted or absorbed. |
| PHYSICS 1 | **Learning objective 5.B.9.1** The student is able to construct or interpret a graph of the energy changes within an electrical circuit with only a single battery and resistors in series and/or in, at most, one parallel branch as an application of the conservation of energy (Kirchhoff's loop rule). |
| PHYSICS 1 | **Learning objective 5.B.9.2** The student is able to apply conservation of energy concepts to the design of an experiment that will demonstrate the validity of Kirchhoff's loop rule ($\Sigma\Delta V = 0$) in a circuit with only a battery and resistors either in series or in, at most, one pair of parallel branches. |
| PHYSICS 1 | **Learning objective 5.B.9.3** The student is able to apply conservation of energy (Kirchhoff's loop rule) in calculations involving the total electric potential difference for complete circuit loops with only a single battery and resistors in series and/or in, at most, one parallel branch. |
| PHYSICS 2 | **Learning objective 5.B.9.4** The student is able to analyze experimental data including an analysis of experimental uncertainty that will demonstrate the validity of Kirchhoff's loop rule ($\Sigma\Delta V = 0$). |
| PHYSICS 2 | **Learning objective 5.B.9.5** The student is able to use conservation of energy principles (Kirchhoff's loop rule) to describe and make predictions regarding electrical potential difference, charge, and current in steady-state circuits composed of various combinations of resistors and capacitors. |
| PHYSICS 2 | **Learning objective 5.B.9.6** The student is able to mathematically express the changes in electric potential energy of a loop in a multiloop electrical circuit and justify this expression using the principle of the conservation of energy. |
| PHYSICS 2 | **Learning objective 5.B.9.7** The student is able to refine and analyze a scientific question for an experiment using Kirchhoff's Loop rule for circuits that includes determination of internal resistance of the battery and analysis of a non-ohmic resistor. |
| PHYSICS 2 | **Learning objective 5.B.9.8** The student is able to translate between graphical and symbolic representations of experimental data describing relationships among power, current, and potential difference across a resistor. |

| PHYSICS 2 | Learning objective 5.B.10.1 The student is able to use Bernoulli's equation to make calculations related to a moving fluid. |
|---|---|
| PHYSICS 2 | Learning objective 5.B.10.2 The student is able to use Bernoulli's equation and/or the relationship between force and pressure to make calculations related to a moving fluid. |
| PHYSICS 2 | Learning objective 5.B.10.3 The student is able to use Bernoulli's equation and the continuity equation to make calculations related to a moving fluid. |
| PHYSICS 2 | Learning objective 5.B.10.4 The student is able to construct an explanation of Bernoulli's equation in terms of the conservation of energy. |
| PHYSICS 2 | Learning objective 5.B.11.1 The student is able to apply conservation of mass and conservation of energy concepts to a natural phenomenon and use the equation $E = mc^2$ to make a related calculation. |
| PHYSICS 2 | Learning objective 5.C.1.1 The student is able to analyze electric charge conservation for nuclear and elementary particle reactions and make predictions related to such reactions based upon conservation of charge. |
| PHYSICS 2 | Learning objective 5.C.2.1 The student is able to predict electric charges on objects within a system by application of the principle of charge conservation within a system. |
| PHYSICS 2 | Learning objective 5.C.2.2 The student is able to design a plan to collect data on the electrical charging of objects and electric charge induction on neutral objects and qualitatively analyze that data. |
| PHYSICS 2 | Learning objective 5.C.2.3 The student is able to justify the selection of data relevant to an investigation of the electrical charging of objects and electric charge induction on neutral objects. |
| PHYSICS 1 | Learning objective 5.C.3.1 The student is able to apply conservation of electric charge (Kirchhoff's junction rule) to the comparison of electric current in various segments of an electrical circuit with a single battery and resistors in series and in, at most, one parallel branch and predict how those values would change if configurations of the circuit are changed. |
| PHYSICS 1 | Learning objective 5.C.3.2 The student is able to design an investigation of an electrical circuit with one or more resistors in which evidence of conservation of electric charge can be collected and analyzed. |
| PHYSICS 1 | Learning objective 5.C.3.3 The student is able to use a description or schematic diagram of an electrical circuit to calculate unknown values of current in various segments or branches of the circuit. |
| PHYSICS 2 | Learning objective 5.C.3.4 The student is able to predict or explain current values in series and parallel arrangements of resistors and other branching circuits using Kirchhoff's junction rule and relate the rule to the law of charge conservation. |

| | |
|---|---|
| **PHYSICS 2** | **Learning objective 5.C.3.5** The student is able to determine missing values and direction of electric current in branches of a circuit with resistors and NO capacitors from values and directions of current in other branches of the circuit through appropriate selection of nodes and application of the junction rule. |
| **PHYSICS 2** | **Learning objective 5.C.3.6** The student is able to determine missing values and direction of electric current in branches of a circuit with both resistors and capacitors from values and directions of current in other branches of the circuit through appropriate selection of nodes and application of the junction rule. |
| **PHYSICS 2** | **Learning objective 5.C.3.7** The student is able to determine missing values, direction of electric current, charge of capacitors at steady state, and potential differences within a circuit with resistors and capacitors from values and directions of current in other branches of the circuit. |
| **PHYSICS 1** | **Learning objective 5.D.1.1** The student is able to make qualitative predictions about natural phenomena based on conservation of linear momentum and restoration of kinetic energy in elastic collisions. |
| **PHYSICS 1** | **Learning objective 5.D.1.2** The student is able to apply the principles of conservation of momentum and restoration of kinetic energy to reconcile a situation that appears to be isolated and elastic, but in which data indicate that linear momentum and kinetic energy are not the same after the interaction, by refining a scientific question to identify interactions that have not been considered. Students will be expected to solve qualitatively and/or quantitatively for one-dimensional situations and only qualitatively in two-dimensional situations. |
| **PHYSICS 1** | **Learning objective 5.D.1.3** The student is able to apply mathematical routines appropriately to problems involving elastic collisions in one dimension and justify the selection of those mathematical routines based on conservation of momentum and restoration of kinetic energy. |
| **PHYSICS 1** | **Learning objective 5.D.1.4** The student is able to design an experimental test of an application of the principle of the conservation of linear momentum, predict an outcome of the experiment using the principle, analyze data generated by that experiment whose uncertainties are expressed numerically, and evaluate the match between the prediction and the outcome. |
| **PHYSICS 1** | **Learning objective 5.D.1.5** The student is able to classify a given collision situation as elastic or inelastic, justify the selection of conservation of linear momentum and restoration of kinetic energy as the appropriate principles for analyzing an elastic collision, solve for missing variables, and calculate their values. |
| **PHYSICS 2** | **Learning objective 5.D.1.6** The student is able to make predictions of the dynamical properties of a system undergoing a collision by application of the principle of linear momentum conservation and the principle of the conservation of energy in situations in which an elastic collision may also be assumed. |

| | |
|---|---|
| **PHYSICS 2** | **Learning objective 5.D.1.7** The student is able to classify a given collision situation as elastic or inelastic, justify the selection of conservation of linear momentum and restoration of kinetic energy as the appropriate principles for analyzing an elastic collision, solve for missing variables, and calculate their values. |
| **PHYSICS 1** | **Learning objective 5.D.2.1** The student is able to qualitatively predict, in terms of linear momentum and kinetic energy, how the outcome of a collision between two objects changes depending on whether the collision is elastic or inelastic. |
| **PHYSICS 1** | **Learning objective 5.D.2.2** The student is able to plan data collection strategies to test the law of conservation of momentum in a two-object collision that is elastic or inelastic and analyze the resulting data graphically. |
| **PHYSICS 1** | **Learning objective 5.D.2.3** The student is able to apply the conservation of linear momentum to a closed system of objects involved in an inelastic collision to predict the change in kinetic energy. |
| **PHYSICS 1** | **Learning objective 5.D.2.4** The student is able to analyze data that verify conservation of momentum in collisions with and without an external friction force. |
| **PHYSICS 1** **PHYSICS 2** | **Learning objective 5.D.2.5** The student is able to classify a given collision situation as elastic or inelastic, justify the selection of conservation of linear momentum as the appropriate solution method for an inelastic collision, recognize that there is a common final velocity for the colliding objects in the totally inelastic case, solve for missing variables, and calculate their values. |
| **PHYSICS 2** | **Learning objective 5.D.2.6** The student is able to apply the conservation of linear momentum to a closed system of objects involved in an inelastic collision to predict the change in kinetic energy. |
| **PHYSICS 1** | **Learning objective 5.D.3.1** The student is able to predict the velocity of the center of mass of a system when there is no interaction outside of the system but there is an interaction within the system (i.e., the student simply recognizes that interactions within a system do not affect the center of mass motion of the system and is able to determine that there is no external force). |
| **PHYSICS 2** | **Learning objective 5.D.3.2** The student is able to make predictions about the velocity of the center of mass for interactions within a defined one-dimensional system. |
| **PHYSICS 2** | **Learning objective 5.D.3.3** The student is able to make predictions about the velocity of the center of mass for interactions within a defined two-dimensional system. |
| **PHYSICS 1** | **Learning objective 5.E.1.1** The student is able to make qualitative predictions about the angular momentum of a system for a situation in which there is no net external torque. |

| | |
|---|---|
| **PHYSICS 1** | **Learning objective 5.E.1.2** The student is able to make calculations of quantities related to the angular momentum of a system when the net external torque on the system is zero. |
| **PHYSICS 1** | **Learning objective 5.E.2.1** The student is able to describe or calculate the angular momentum and rotational inertia of a system in terms of the locations and velocities of objects that make up the system. Students are expected to do qualitative reasoning with compound objects. Students are expected to do calculations with a fixed set of extended objects and point masses. |
| **PHYSICS 2** | **Learning objective 5.F.1.1** The student is able to make calculations of quantities related to flow of a fluid, using mass conservation principles (the continuity equation). |
| **PHYSICS 2** | **Learning objective 5.G.1.1** The student is able to apply conservation of nucleon number and conservation of electric charge to make predictions about nuclear reactions and decays such as fission, fusion, alpha decay, beta decay, or gamma decay. |
| **PHYSICS 1** | **Learning objective 6.A.1.1** The student is able to use a visual representation to construct an explanation of the distinction between transverse and longitudinal waves by focusing on the vibration that generates the wave. |
| **PHYSICS 1**<br>**PHYSICS 2** | **Learning objective 6.A.1.2** The student is able to describe representations of transverse and longitudinal waves. |
| **PHYSICS 2** | **Learning objective 6.A.1.3** The student is able to analyze data (or a visual representation) to identify patterns that indicate that a particular mechanical wave is polarized and construct an explanation of the fact that the wave must have a vibration perpendicular to the direction of energy propagation. |
| **PHYSICS 1** | **Learning objective 6.A.2.1** The student is able to describe sound in terms of transfer of energy and momentum in a medium and relate the concepts to everyday examples. |
| **PHYSICS 2** | **Learning objective 6.A.2.2** The student is able to contrast mechanical and electromagnetic waves in terms of the need for a medium in wave propagation. |
| **PHYSICS 1** | **Learning objective 6.A.3.1** The student is able to use graphical representation of a periodic mechanical wave to determine the amplitude of the wave. |
| **PHYSICS 1** | **Learning objective 6.A.4.1** The student is able to explain and/or predict qualitatively how the energy carried by a sound wave relates to the amplitude of the wave, and/or apply this concept to a real-world example. |
| **PHYSICS 1** | **Learning objective 6.B.1.1** The student is able to use a graphical representation of a periodic mechanical wave (position versus time) to determine the period and frequency of the wave and describe how a change in the frequency would modify features of the representation. |

| | |
|---|---|
| **PHYSICS 1** | Learning objective 6.B.2.1 The student is able to use a visual representation of a periodic mechanical wave to determine wavelength of the wave. |
| **PHYSICS 2** | Learning objective 6.B.3.1 The student is able to construct an equation relating the wavelength and amplitude of a wave from a graphical representation of the electric or magnetic field value as a function of position at a given time instant and vice versa, or construct an equation relating the frequency or period and amplitude of a wave from a graphical representation of the electric or magnetic field value at a given position as a function of time and vice versa. |
| **PHYSICS 1** | Learning objective 6.B.4.1 The student is able to design an experiment to determine the relationship between periodic wave speed, wavelength, and frequency and relate these concepts to everyday examples. |
| **PHYSICS 1**<br>**PHYSICS 2** | Learning objective 6.B.5.1 The student is able to create or use a wave front diagram to demonstrate or interpret qualitatively the observed frequency of a wave, dependent upon relative motions of source and observer. |
| **PHYSICS 2** | Learning objective 6.C.1.1 The student is able to make claims and predictions about the net disturbance that occurs when two waves overlap. Examples should include standing waves. |
| **PHYSICS 2** | Learning objective 6.C.1.2 The student is able to construct representations to graphically analyze situations in which two waves overlap over time using the principle of superposition. |
| **PHYSICS 2** | Learning objective 6.C.2.1 The student is able to make claims about the diffraction pattern produced when a wave passes through a small opening, and to qualitatively apply the wave model to quantities that describe the generation of a diffraction pattern when a wave passes through an opening whose dimensions are comparable to the wavelength of the wave. |
| **PHYSICS 2** | Learning objective 6.C.3.1 The student is able to qualitatively apply the wave model to quantities that describe the generation of interference patterns to make predictions about interference patterns that form when waves pass through a set of openings whose spacing and widths are small compared to the wavelength of the waves. |
| **PHYSICS 2** | Learning objective 6.C.4.1 The student is able to predict and explain, using representations and models, the ability or inability of waves to transfer energy around corners and behind obstacles in terms of the diffraction property of waves in situations involving various kinds of wave phenomena, including sound and light. |
| **PHYSICS 1** | Learning objective 6.D.1.1 The student is able to use representations of individual pulses and construct representations to model the interaction of two wave pulses to analyze the superposition of two pulses. |

| **PHYSICS 1** | **Learning objective 6.D.1.2** The student is able to design a suitable experiment and analyze data illustrating the superposition of mechanical waves (only for wave pulses or standing waves). |
| --- | --- |
| **PHYSICS 1** | **Learning objective 6.D.1.3** The student is able to design a plan for collecting data to quantify the amplitude variations when two or more traveling waves or wave pulses interact in a given medium. |
| **PHYSICS 1** | **Learning objective 6.D.2.1** The student is able to analyze data or observations or evaluate evidence of the interaction of two or more traveling waves in one or two dimensions (i.e., circular wave fronts) to evaluate the variations in resultant amplitudes. |
| **PHYSICS 1** | **Learning objective 6.D.3.1** The student is able to refine a scientific question related to standing waves and design a detailed plan for the experiment that can be conducted to examine the phenomenon qualitatively or quantitatively. |
| **PHYSICS 1** | **Learning objective 6.D.3.2** The student is able to predict properties of standing waves that result from the addition of incident and reflected waves that are confined to a region and have nodes and antinodes. |
| **PHYSICS 1** | **Learning objective 6.D.3.3** The student is able to plan data collection strategies, predict the outcome based on the relationship under test, perform data analysis, evaluate evidence compared to the prediction, explain any discrepancy and, if necessary, revise the relationship among variables responsible for establishing standing waves on a string or in a column of air. |
| **PHYSICS 1** | **Learning objective 6.D.3.4** The student is able to describe representations and models of situations in which standing waves result from the addition of incident and reflected waves confined to a region. |
| **PHYSICS 1** | **Learning objective 6.D.4.1** The student is able to challenge with evidence the claim that the wavelengths of standing waves are determined by the frequency of the source regardless of the size of the region. |
| **PHYSICS 1** | **Learning objective 6.D.4.2** The student is able to calculate wavelengths and frequencies (if given wave speed) of standing waves based on boundary conditions and length of region within which the wave is confined, and calculate numerical values of wavelengths and frequencies. Examples should include musical instruments. |
| **PHYSICS 1** | **Learning objective 6.D.5.1** The student is able to use a visual representation to explain how waves of slightly different frequency give rise to the phenomenon of beats. |
| **PHYSICS 2** | **Learning objective 6.E.1.1** The student is able to make claims using connections across concepts about the behavior of light as the wave travels from one medium into another, as some is transmitted, some is reflected, and some is absorbed. |

| PHYSICS 2 | Learning objective 6.E.2.1 The student is able to make predictions about the locations of object and image relative to the location of a reflecting surface. The prediction should be based on the model of specular reflection with all angles measured relative to the normal to the surface. |
|---|---|
| PHYSICS 2 | Learning objective 6.E.3.1 The student is able to describe models of light traveling across a boundary from one transparent material to another when the speed of propagation changes, causing a change in the path of the light ray at the boundary of the two media. |
| PHYSICS 2 | Learning objective 6.E.3.2 The student is able to plan data collection strategies as well as perform data analysis and evaluation of the evidence for finding the relationship between the angle of incidence and the angle of refraction for light crossing boundaries from one transparent material to another (Snell's law). |
| PHYSICS 2 | Learning objective 6.E.3.3 The student is able to make claims and predictions about path changes for light traveling across a boundary from one transparent material to another at non-normal angles resulting from changes in the speed of propagation. |
| PHYSICS 2 | Learning objective 6.E.4.1 The student is able to plan data collection strategies, and perform data analysis and evaluation of evidence about the formation of images due to reflection of light from curved spherical mirrors. |
| PHYSICS 2 | Learning objective 6.E.4.2 The student is able to use quantitative and qualitative representations and models to analyze situations and solve problems about image formation occurring due to the reflection of light from surfaces. |
| PHYSICS 2 | Learning objective 6.E.5.1 The student is able to use quantitative and qualitative representations and models to analyze situations and solve problems about image formation occurring due to the refraction of light through thin lenses. |
| PHYSICS 2 | Learning objective 6.E.5.2 The student is able to plan data collection strategies, perform data analysis and evaluation of evidence, and refine scientific questions about the formation of images due to refraction for thin lenses. |
| PHYSICS 2 | Learning objective 6.F.1.1 The student is able to make qualitative comparisons of the wavelengths of types of electromagnetic radiation. |
| PHYSICS 2 | Learning objective 6.F.2.1 The student is able to describe representations and models of electromagnetic waves that explain the transmission of energy when no medium is present. |
| PHYSICS 2 | Learning objective 6.F.3.1 The student is able to support the photon model of radiant energy with evidence provided by the photoelectric effect. |

| | |
|---|---|
| **PHYSICS 2** | **Learning objective 6.F.4.1** The student is able to select a model of radiant energy that is appropriate to the spatial or temporal scale of an interaction with matter. |
| **PHYSICS 2** | **Learning objective 6.G.1.1** The student is able to make predictions about using the scale of the problem to determine at what regimes a particle or wave model is more appropriate. |
| **PHYSICS 2** | **Learning objective 6.G.2.1** The student is able to articulate the evidence supporting the claim that a wave model of matter is appropriate to explain the diffraction of matter interacting with a crystal, given conditions where a particle of matter has momentum corresponding to a de Broglie wavelength smaller than the separation between adjacent atoms in the crystal. |
| **PHYSICS 2** | **Learning objective 6.G.2.2** The student is able to predict the dependence of major features of a diffraction pattern (e.g., spacing between interference maxima), based upon the particle speed and de Broglie wavelength of electrons in an electron beam interacting with a crystal. (de Broglie wavelength need not be given, so students may need to obtain it.) |
| **PHYSICS 2** | **Learning objective 7.A.1.1** The student is able to make claims about how the pressure of an ideal gas is connected to the force exerted by molecules on the walls of the container, and how changes in pressure affect the thermal equilibrium of the system. |
| **PHYSICS 2** | **Learning objective 7.A.1.2** Treating a gas molecule as an object (i.e., ignoring its internal structure), the student is able to analyze qualitatively the collisions with a container wall and determine the cause of pressure, and at thermal equilibrium, to quantitatively calculate the pressure, force, or area for a thermodynamic problem given two of the variables. |
| **PHYSICS 2** | **Learning objective 7.A.2.1** The student is able to qualitatively connect the average of all kinetic energies of molecules in a system to the temperature of the system. |
| **PHYSICS 2** | **Learning objective 7.A.2.2** The student is able to connect the statistical distribution of microscopic kinetic energies of molecules to the macroscopic temperature of the system and to relate this to thermodynamic processes. |
| **PHYSICS 2** | **Learning objective 7.A.3.1** The student is able to extrapolate from pressure and temperature or volume and temperature data to make the prediction that there is a temperature at which the pressure or volume extrapolates to zero. |
| **PHYSICS 2** | **Learning objective 7.A.3.2** The student is able to design a plan for collecting data to determine the relationships between pressure, volume, and temperature, and amount of an ideal gas, and to refine a scientific question concerning a proposed incorrect relationship between the variables. |

| | |
|---|---|
| **Physics 2** | **Learning objective 7.A.3.3** The student is able to analyze graphical representations of macroscopic variables for an ideal gas to determine the relationships between these variables and to ultimately determine the ideal gas law $PV = nRT$. |
| **Physics 2** | **Learning objective 7.B.1.1** The student is able to construct an explanation, based on atomic scale interactions and probability, of how a system approaches thermal equilibrium when energy is transferred to it or from it in a thermal process. |
| **Physics 2** | **Learning objective 7.B.2.1** The student is able to connect qualitatively the second law of thermodynamics in terms of the state function called entropy and how it (entropy) behaves in reversible and irreversible processes. |
| **Physics 2** | **Learning objective 7.C.1.1** The student is able to use a graphical wave function representation of a particle to predict qualitatively the probability of finding a particle in a specific spatial region. |
| **Physics 2** | **Learning objective 7.C.2.1** The student is able to use a standing wave model in which an electron orbit circumference is an integer multiple of the de Broglie wavelength to give a qualitative explanation that accounts for the existence of specific allowed energy states of an electron in an atom. |
| **Physics 2** | **Learning objective 7.C.3.1** The student is able to predict the number of radioactive nuclei remaining in a sample after a certain period of time, and also predict the missing species (alpha, beta, gamma) in a radioactive decay. |
| **Physics 2** | **Learning objective 7.C.4.1** The student is able to construct or interpret representations of transitions between atomic energy states involving the emission and absorption of photons. [For questions addressing stimulated emission, students will not be expected to recall the details of the process, such as the fact that the emitted photons have the same frequency and phase as the incident photon; but given a representation of the process, students are expected to make inferences such as figuring out from energy conservation that since the atom loses energy in the process, the emitted photons taken together must carry more energy than the incident photon.] |

# Part II

## A Review of AP® Physics 1 and AP® Physics 2

# 1

# VECTORS

Vectors are not given a specific learning objective on the AP® exam. However, you must be familiar with them to apply their use to many concepts throughout Physics 1 and Physics 2.

## FRAMES OF REFERENCE

Our study of physics begins with kinematics, the study of motion. To investigate an object's motion, we need to be able to measure its position at different times and compare multiple measurements. When we measure an object's position, we consider where the object is relative to something else, assign the symbol X to our measurement, and check the time. Later, we can take a new position measurement relative to the same point and check a clock again. The difference between the last and first position measurements is called *displacement*, and the difference between the last and first time measurements is called *duration* or *elapsed time*. We use the symbol $\Delta X$ to show displacement and the symbol $\Delta t$ to show how much time passed between measurements. To measure position with a ruler, measuring tape, or meterstick, look at where the object is when the zero mark of the ruler is on the point of reference. The most common way to measure duration is with a stopwatch since its first measurement is always at zero seconds. Stopwatches don't measure what time it is; they measure the time interval $\Delta t$. If a group of physics students observes an object's motion, the group should largely agree on the displacement and the duration so long as everyone measured displacement from the same point and began their stopwatches at the same time.

This set of agreed-upon reference points in space and time is called a *frame of reference*. For the sake of an experiment or a conversation, we assume that the objects used as reference points, such as walls, the ground, or a painted mark on the highway, are not moving. A frame of reference also defines which directions to call positive and negative. In some situations, it might be convenient to call up positive, while calling down positive works better in other situations. "Counterclockwise," "down a ramp," or "toward that stop sign" are all positive directions as long as every observer agrees on the frame of reference.

We define convenient frames of reference and make measurements so that we can describe events, find patterns, and ultimately make predictions and changes. A good physics student can use simple tools and careful measurements to determine an object's motion—its displacement, velocity, and acceleration. A good physics student can explain events in many ways—as a narrative, a set of equations, and a graph.

Graphs are particularly good tools for investigating patterns of cause and effect. They show how one measured quantity on the x-axis affects another measured quantity on the y-axis; they also can show how a single measurement on the y-axis changes with time on the x-axis. Graphs also show a person's frame of reference, with positive and negative values stretching away from the agreed-upon zero points at the origin.

## SCALAR QUANTITIES AND VECTOR QUANTITIES

Any quantity that can be completely specified by a number and a unit is called a *scalar quantity*. Scalars can be added by ordinary arithmetic: 1 m + 2 m + 3 m = 6 m. Mass, volume, temperature, energy, undirected distance, and speed are all classified as scalar quantities, as are the number of gumballs in a jar and the calories in a brownie.

Other quantities such as displacement and velocity specify direction as well as a number and a unit: 30 m/s east and 12 m vertically upward. Quantities with both magnitude and geometrical direction are called *vector quantities*. Further examples of vector quantities are force, acceleration, momentum, and torque.

A vector quantity is conveniently represented by a *vector*, an arrow-tipped line segment. The length of a vector indicates the *magnitude* of the vector quantity; the direction of the arrow specifies the *direction* of the vector. The arrowhead is called the tip, and the beginning of the vector is called the tail.

## VECTOR COMPONENTS

While scalars are treated with ordinary arithmetic, vectors are not; they have their own mathematics, called vector algebra, which accounts for direction. We can do math with straight lines along the x- or y-axis, but we can't do math with vectors pointing at other angles. Vector algebra gives us tools we can use to treat the vertical and horizontal parts of a vector separately, do math with them, and then recombine those parts into a new vector.

Every point in Cartesian or two-dimensional space has a set of two coordinates, and every vector in two-dimensional space has a set of two components: a *horizontal* or *x-component* and a *vertical* or *y-component*.

We will represent a vector as an *italicized* letter with an arrow over it, as with vector $\vec{V}$. Velocity is a vector quantity that tells us how fast something is moving and in what direction. We denote velocity with a $\vec{V}$; its units are meters per second, or m/s. The velocity of an object at some particular moment of time is called its *instantaneous velocity*. The speedometer of a car tells the driver the instantaneous speed of the car.

The magnitude of a vector is a scalar quantity since it does not specify direction. We can show the magnitude of a vector $\vec{A}$ as A. This is the length of a drawn vector. We identify a vector $\vec{V}$ with both magnitude $V$ and direction $\theta$. Vector $\vec{V}$ has two components in two-dimensional space, $V_x$ and $V_y$.

To calculate the $x$-component of a vector $\vec{V}$, we write $V_x = V\cos\theta$. We multiply the magnitude of the vector by the cosine of the angle in degrees the vector makes with respect to the $+x$-axis. To calculate the $y$-component of a vector $\vec{V}$, we write $V_y = V\sin\theta$. In this case, we multiply the sine of the angle expressed in standard position.

Procedures to calculate the components of a vector:

1.  Make a sketch of the situation, attaching a frame of reference. Pick a convenient axis to use as the 0° reference point.
2.  Using $V_x = V\cos\theta$, calculate the $x$-component. This is the component adjacent to the reference angle.
3.  Using $V_y = V\sin\theta$, calculate the $y$-component.

## Sample Problem 1

A water balloon has a velocity of 30.0 m/s at an angle of 50° with respect to the $+x$-axis. We can use a shorthand way to express this vector: $\vec{V}$ = 30.0 m/s @ 50°. Calculate the $x$- and $y$-components of its velocity vector.

## Solution to Problem 1

Sketch the vector showing its magnitude and direction. Sketch and label the components too.

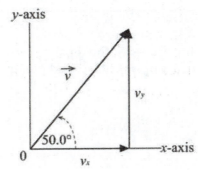

To calculate the $x$-component of a vector $\vec{V}$: $V_x = V\cos\theta$

To calculate the $y$-component of a vector $\vec{V}$: $V_y = V\sin\theta$

$v_x = v\cos\theta = (30.0 \text{ m/s})(\cos 50°) = \textbf{19.3 m/s}$

$v_y = v\sin\theta = (30.0 \text{ m/s})(\sin 50°) = \textbf{23.0 m/s}$

### SAMPLE PROBLEM 2

An object travels along a curved path as shown below. At the moment the particle is at point P, its instantaneous velocity is $\vec{V}$ = 120.0 m/s @ 70.0°. Calculate the x- and y-components of $\vec{V}$.

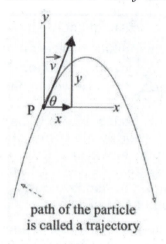

path of the particle
is called a trajectory

### SOLUTION TO PROBLEM 2

To find the components at point P, we write the component equations and substitute

$$v_x = v\cos\theta = (120.0 \text{ m/s})(\cos 70.0°) = \textbf{41.0 m/s}$$

$$v_y = v\sin\theta = (120.0 \text{ m/s})(\sin 70.0°) = \textbf{112.8 m/s}$$

In its simplest sense, a *force, $\vec{F}$,* may be defined as a *push* or a *pull* on an object. Forces are vectors. Forces cause objects to speed up, slow down, or change direction. In the SI, the unit of force is the newton (N).

### SAMPLE PROBLEM 3

A wooden box is being pulled across a horizontal floor. A rope is attached to the box as shown below. In physics, we call the pulling force from a ropy by a special name: *tension, $\vec{T}$*. Tension is a vector. If the rope has a tension of 25.0 N and is acting at 135° with respect to the +x-axis, what are the x- and y-components of the tension vector? Use a frame of reference in which "to the right" is considered positive.

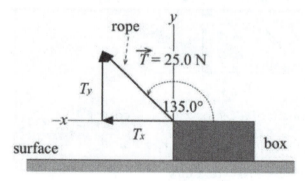

## SOLUTION TO PROBLEM 3

Write the component equations and substitute in to them.

$$T_x = T\cos\theta = (25.0\ \text{N})(\cos 135°) = \textbf{−17.7 N}$$

$$T_y = T\sin\theta = (25.0\ \text{N})(\sin 135°) = \textbf{17.7 N}$$

The *x*-component of this force is drawn to the left, so it is negative. We can use either the 135° angle it makes with respect to the +*x*-axis or the 45° angle it makes with the −*x*-axis. Either way, the *x*-component has a magnitude of 17.1N and points in the −*x*-direction.

---

### AP® Tip

The force in strings, ropes, wires, cables, and chains is called tension. Strings, ropes, cables, wires, and chains can only pull. They cannot push.

---

## DETERMINING A VECTOR FROM ITS COMPONENTS

Vector algebra has two purposes: it can express a vector as two components so that we can do math with components, or it can resolve components into vectors once the math has been done. When a vector's *x*- and *y*-components are known, they can be resolved into an angled vector using the *Pythagorean theorem* and the *inverse tangent function* (tan⁻¹).

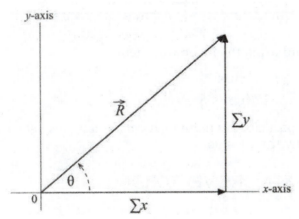

The magnitude of vector $\vec{R}$ is found from its $\Sigma x$ and $\Sigma y$ components using the Pythagorean theorem:

$$R = |R| = \left| \sqrt{(\Sigma x)^2 + (\Sigma y)^2} \right|$$

The absolute value symbols are a reminder that the magnitude of the resultant is a scalar quantity.

From the inverse tangent function, the angle $\theta$ of $R$ is:

$$\theta = \left| \tan^{-1} \frac{\Sigma y}{\Sigma x} \right|$$

## SAMPLE PROBLEM 4

A particle following a trajectory is illustrated below. At point Q, the particle has an instantaneous velocity $v$. The horizontal and vertical components of the velocity vector are $v_x = 12.0$ m/s and $v_y = -7.0$ m/s. Determine the instantaneous velocity of the particle at Q.

## SOLUTION TO PROBLEM 4

Notice that the vertical component is negative and that the horizontal component is positive.

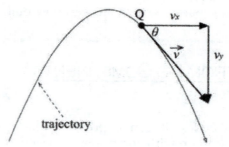

The magnitude of the instantaneous velocity vector is found using the Pythagorean theorem. We write the theorem and make the substitution,

$$v = \left| \sqrt{\left(v_x\right)^2 + \left(v_y\right)^2} \right| = \left| \sqrt{\left(12.0 \text{ m/s}\right)^2 + \left(-7.0 \text{ m/s}\right)^2} \right| = \mathbf{13.9 \text{ m/s}}$$

The direction of $v$ is found using the inverse tangent

$$\theta = \left| \tan^{-1} \frac{v_y}{v_x} \right| = \left| \tan^{-1}\left( \frac{-7 \text{ m/s}}{12.0 \text{ m/s}} \right) \right| = 30.3°$$

At the very moment the particle is at point Q, its instantaneous velocity is **13.9 m/s @ 30.3° below the +x-axis**.

## RESULTANTS OF TWO OR MORE VECTORS

Being able to express a vector in terms of its $x$- and $y$-components is a powerful tool. We can't add or subtract vectors at different angles, but we can perform math on components along the same direction. To add vectors, we add their $x$-components separately from their $y$-components. The sum of the $x$-components of two or more vectors is the $x$-component of the *resultant*. In the same sense, the sum of the $y$-components is the $y$-component of the *resultant*. The resultant is what we call the *result* of adding vectors using vector algebra. This many components and vectors can be difficult to organize; it can be helpful to construct a table like the following:

| Force | x-component | y-component |
|---|---|---|
| $F_1 @ \theta_1$ | $F_1 \cos\theta_1$ | $F_1 \sin\theta_1$ |
| $F_2 @ \theta_2$ | $F_2 \cos\theta_2$ | $F_2 \sin\theta_2$ |
| Sum of the components | $\Sigma x = x_1 + x_2$ | $\Sigma y = y_1 + y_2$ |

## SAMPLE PROBLEM 5

The *displacement* of a body is the straight-line distance from the starting point to the ending point. It is the comparison of a final position and an initial position. Displacement, $\vec{S}$, is a vector since it has both magnitude and direction. During a soccer game, you run 24 m in a straight line at an angle of 30° with respect to the sideline. Then you run 10 m more at 60° with respect to the same sideline. What is your displacement?

## SOLUTION TO PROBLEM 5

To find the displacement, we start by making a diagram. First, we attach a frame of reference, a Cartesian coordinate system, to the starting point, the origin. This diagram is not to scale.

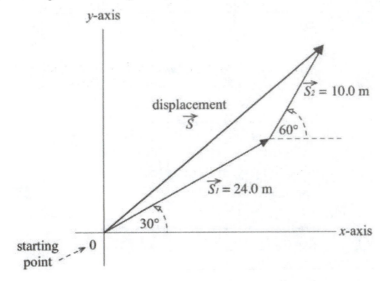

In the diagram, the displacements $\vec{S}_1$, $\vec{S}_2$, and the resultant displacement, $\vec{S}$, are vectors with magnitude and direction. To find the displacement, $\vec{S}$, we first calculate the x- and y-components.

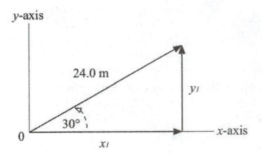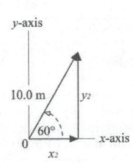

$$X_1 = (24 \text{ m}) \cos 30° = 20.8 \text{ m} \qquad X_2 = (10 \text{ m}) \cos 60° = 5.0 \text{ m}$$

$$y_1 = (24 \text{ m}) \sin 30° = 12.0 \text{ m} \qquad y_2 = (10 \text{ m}) \sin 60° = 8.7 \text{ m}$$

Now we find the sum of the $x$ displacements and the sum of the $y$ displacements.

| $\vec{S_1} = 24.0 \text{ m} @ 30°$ | $S_{1x}$<br>$X_1 = S_1 \cos 30°$<br>$X_1 = 20.8 \text{ m}$ | $S_{1y}$<br>$y_1 = S_1 \sin 30°$<br>$y_1 = 12.0 \text{ m}$ |
|---|---|---|
| $\vec{S_2} = 10.0 \text{ N} @ 60°$ | $S_{2x}$<br>$X_2 = S_2 \cos 60°$<br>$X_2 = 5.0 \text{ m}$ | $S_{2y}$<br>$y_2 = S_2 \sin 135°$<br>$y_2 = 8.7 \text{ m}$ |
| Sum of the components | $\Sigma x = X_1 + X_2$<br>$\Sigma x = 20.8 \text{ m} + 5.0 \text{ m}$<br>$\mathbf{\Sigma x = 25.8 \text{ m}}$ | $\Sigma y = y_1 + y_2$<br>$\Sigma y = 12.0 \text{ m} + 8.7 \text{ m}$<br>$\mathbf{\Sigma y = 20.7 \text{ m}}$ |

The sum of all of the $x$-components is the $x$-component of the resultant. The same is true of the $y$-components. Now we can calculate the resultant displacement, $\vec{R}$.

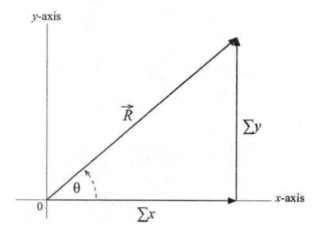

$$R = \sqrt{(\Sigma x)^2 + (\Sigma y)^2} = \sqrt{(25.8 \text{ m})^2 + (20.7 \text{ m})^2} = 33.1 \text{ m}$$

$$\theta = \tan^{-1}\left(\frac{\Sigma y}{\Sigma x}\right) = \tan^{-1}\left(\frac{20.7 \text{ m}}{25.8 \text{ m}}\right) = \tan^{-1}(0.8023) = 38.7°$$

You displaced yourself 33.1 m at an angle of 38.7°. We can write the displacement vector as $\vec{S}$ = 33.1 m @ 38.7°.

## SAMPLE PROBLEM 6

Two forces, $\vec{A}$ = 100 N @ 30° and $\vec{B}$ = 60 N @ 135°, act on an object. Calculate the following:
(a)  $x$- and $y$-components of each force
(b)  magnitude of the resultant
(c)  angle of the resultant

## SOLUTION TO PROBLEM 6

First, make a vector diagram sketching all of the forces and their components.

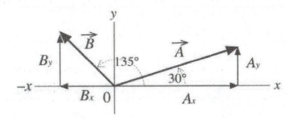

(a) Construct a table. List all of the forces and calculate the components.

| $\vec{A} = 100$ N @ 30° | $A_x$ | $A_y$ |
|---|---|---|
| | $A_x = A\cos 30°$ | $A_y = A\sin 30°$ |
| | $A_x = (100$ N$) \cos 30°$ | $A_y = (100$ N$) \sin 30°$ |
| | $A_x = 86.6$ N | $A_y = 50.0$ N |
| $\vec{B} = 60.0$ N @ 135° | $B_x$ | $B_y$ |
| | $B_x = B\cos 135°$ | $B_y = B\sin 135°$ |
| | $B_x = (60.0$ N$) \cos 135°$ | $B_y = (60.0$ N$) \sin 135°$ |
| | $B_x = -42.4$ N | $B_y = 42.4$ N |
| Sum of the components | $\Sigma x = A_x + B_x$ | $\Sigma y = A_y + B_y$ |
| | $\Sigma x = 86.6$ N $+ (-42.4$ N$)$ | $\Sigma y = 50.0$ N $+ 42.4$ N |
| | $\Sigma x = 44.2$ N | $\Sigma y = 92.4$ N |

(b) The magnitude of the resultant, |R|, is

$$|R| = \sqrt{(\Sigma x)^2 + (\Sigma y)^2} = \sqrt{(44.2 \text{ N})^2 + (92.4 \text{ N})^2} = \textbf{102.4 N}$$

(c) The angle of the resultant is found by

$$\theta = \tan^{-1}\left|\frac{\Sigma y}{\Sigma x}\right| = \tan^{-1}\left|\frac{92.4 \text{ N}}{44.2 \text{ N}}\right| = \textbf{64.4°}$$

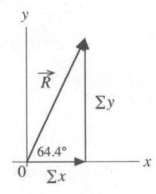

So $\vec{R} = $ **102.4 N @ 64.4°**. The body behaves as if a single force of 102.4 N at 64.4° is acting on it.

## VECTORS: STUDENT OBJECTIVES FOR THE AP® EXAM

- You should be able to define a frame of reference.
- You should be able to define both a scalar and a vector and give examples of both.
- You should know what vector components are and how to calculate them.
- You should be able to express angles relative to an axis.
- You should be able to find a vector from its components.
- You should be able to use the Pythagorean theorem and the inverse tangent in problem solving.
- You should be able to define and calculate the resultant.
- You should be able to add two or more vectors.

## MULTIPLE-CHOICE QUESTIONS

1. Which of the following is a vector quantity?
   (A) Temperature
   (B) Mass
   (C) Distance
   (D) Force

2. Four objects are moving to the right with velocity $v$. Each object experiences two forces of equal magnitude in different directions. In which situation is the vector sum of the force vectors, $\Sigma F$, the greatest?

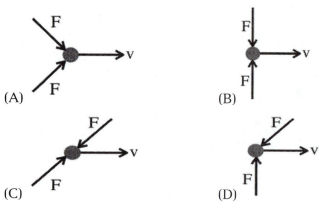

3. Students are shown two vectors and asked to determine the magnitude of the resultant vector. The two vectors have magnitudes of 4 m/s and 3 m/s. The students make claims regarding vector addition. Which student is correct?

(A) Alexandra: "We cannot determine the resultant vectors because we have not been given the directions of the vectors."

(B) Blaze: "We are only asked about magnitude of the resultant, so we have enough information. All we need to do is add the magnitudes to find the resultant, so the answer is 7 m/s."

(C) Camden: "These are vectors, so we cannot just add them together. We should use the Pythagorean theorem to find the resultant, so the answer is 5 m/s."

(D) Deasha: "You have to add the first vector to the opposite of the second vector to find the resultant, so the answer is 1 m/s."

## FREE-RESPONSE PROBLEMS

1.  Using a table of vectors, calculate the resultant force acting on the object shown below. Vectors $\vec{A}$ and $\vec{B}$ are simultaneous and concurrent.

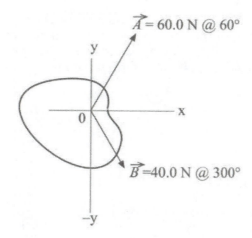

2.  A proton travels through space along its trajectory, its path. At point Q, it has an instantaneous velocity $\vec{V}$ with components $v_x = -500.0$ m/s and $v_y = -275.0$ m/s. Calculate the instantaneous velocity of the proton at point Q.

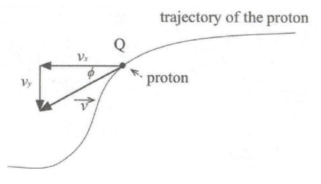

# Answers

## Multiple-Choice Questions

1. **D** Force is the only vector listed. All other quantities are scalar because they only have magnitude. Force should have magnitude and direction, making it a vector quantity.

2. **A** The $x$-components of these force vectors point in the same direction, so their sum will have a greater magnitude than either individual component. Because the $y$-components of these force vectors are equal in magnitude and in opposite directions, they sum to zero.

3. **A** Alexandra is correct by saying that we must know the magnitude AND direction of the vectors. We do not know whether the vectors are in the same direction, opposite directions, or at angles. Therefore, we have no way of finding the components of the vectors to add them.

## Free-Response Problems

1. List the vectors and calculate their components.

| $\vec{A} = 60.0$ N @ 60° | $A_x$<br>$A_x = A \cos 60°$<br>$A_x = (60.0$ N$) \cos 60°$<br>$A_x = 30.0$ N | $A_y$<br>$A_y = A \sin 300°$<br>$A_y = (60.0$ N$) \sin 60°$<br>$A_y = 52.0$ N |
|---|---|---|
| $\vec{B} = 40.0$ N @ 300° | $B_x$<br>$B_x = B \cos 300°$<br>$B_x = (40.0$ N$) \cos 300°$<br>$B_x = 20.0$ N | $B_y$<br>$B_y = B \sin 300°$<br>$B_y = (40.0$ N$) \sin 300°$<br>$B_y = -34.6$ N |
| Sum of the components | $\Sigma x = A_x + B_x$<br>$\Sigma x = 30.0$ N $+ 20.0$ N<br>$\Sigma x = 50.0$ N | $\Sigma y = A_y + B_y$<br>$\Sigma y = 52.0$ N $+ (-34.6$ N$)$<br>$\Sigma y = 17.4$ N |

Next, sketch the components of the resultant, $\vec{R}$.

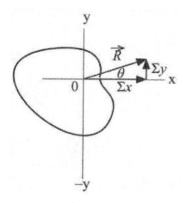

Use the Pythagorean theorem and the inverse tangent to calculate the resultant. Note that both components are positive, placing the resultant in the first quadrant.

$$R = \sqrt{(\Sigma x)^2 + (\Sigma y)^2} = \sqrt{(50.0 \text{ N})^2 + (17.4 \text{ N})^2} = 52.9 \text{ N}$$

$$\theta = \tan^{-1}\left|\frac{\Sigma y}{\Sigma x}\right| = \tan^{-1}\left|\frac{17.4 \text{ N}}{50.0 \text{ N}}\right| = 19.2°$$

The resultant force is $\vec{R} = $ **52.9 N @ 19.2°** .

2. Use the Pythagorean theorem to calculate the magnitude of the instantaneous velocity vector.

$$v = \sqrt{[(500 \text{ m/s})^2 + (275 \text{ m/s})^2]} = 570.6 \text{ m/s}$$

$$v = \sqrt{v_x^2 + v_y^2} = \sqrt{(-500.0\text{m/s})^2 + (-275.0 \text{ m/s})^2} = 570.6 \text{ m/s}$$

$$\theta = \tan^{-1}\left|\frac{v_y}{v_x}\right| = \tan^{-1}\left|\frac{275.0 \text{ m/s}}{500.0 \text{ m/s}}\right| = 28.8°$$

To calculate the angle, we find the inverse tangent.

At the very moment the proton is at point Q, it is traveling 570.6 m/s 28.8° below the −*x*-axis.

# 2

# EQUILIBRIUM

## MASS AND INERTIA

*Mass* is an *intrinsic property* shared by all objects. It is a measure of the quantity of material present in an object, the sum of all of the protons, neutrons, and electrons that make the object. In our everyday world, the mass of a body is considered to be constant. Mass $m$ is a scalar quantity, and its SI unit is the kilogram, kg. Mass is not dependent on gravity, so if you travel to the moon one day, your mass will remain the same.

Mass is also a measure of the *inertia* of an object. Inertia can be defined as the resistance to any change in the motion, stopping or starting the movement, of an object. You can think of inertia and mass almost interchangeably. An ant has a small mass, so it has a small inertia. A boulder has a large mass as well as a large inertia. It is easy to stop the motion of an ant, but you would have a harder time stopping a moving boulder from rolling. When we refer to how difficult it is to move an object, we're referring to its *inertial mass*.

Picture a large truck, a tractor-trailer, having a mass of 18,000 kilograms at rest in neutral with the parking brake released. It sits in a large, empty, flat parking lot. If weighed at a truck scale, the truck would weigh about 40,000 pounds. This truck has quite a large inertia. Twenty very strong football players get behind the truck and push in unison. Slowly, the truck begins to move. To make it move, the force all of the football players simultaneously exert on it overcomes its inertia, and it moves on its own across the parking lot at five miles an hour. It will now be hard to stop the large moving truck. Its inertia must be overcome to bring it to rest. Because the truck had a large inertia, it was hard to start it moving, and once it began to move, it was hard to stop it. The truck did not "want" to change its current state of motion.

# WEIGHT

We sometimes say we "weigh" something on a balance when we are asked to find the mass of an object, but weight and mass are *NOT* the same thing. Weight, $\vec{w}$, is the force of gravitational attraction on an object and depends on the strength of a gravitational field. Your weight is a vector, and its direction is always vertically downward toward the center of Earth. Your weight can change based on your location. When you take that trip to the moon, your mass will not change since you're still made of the same components, but your weight certainly will because the gravitational field on the moon is not the same as it is on Earth.

The weight of a body on the surface of Earth is the gravitational attraction Earth and the body exert on each other. If the object weighs 100 N, it means that the gravitational attraction between the body and Earth is 100 N. We stated previously that in our everyday world, mass is a constant. Weight, however, is not. Mass has another important property. Masses generate *gravitational fields*, $\vec{g}$-fields. Small masses have a small $\vec{g}$-field associated with them, and very large masses generate very large gravitational fields. The reason is not fully understood. Compared to you or me, the $\vec{g}$-field of Earth is huge and extends out a great distance into space. It is the interaction between the $\vec{g}$-fields of Earth and the moon that brings about the gravitational attraction between them.

Assuming that Earth is a homogeneous sphere, the $\vec{g}$-field at its surface has a constant value. For the most part, it truly is constant. The *strength* or *intensity* of the $\vec{g}$-field of Earth at or within a few kilometers of the surface is

$$\vec{g} = -9.8 \ \frac{\text{m}}{\text{s}^2}$$

The intensity, $\vec{g}$, is a vector and is directed vertically downward toward the center of Earth. This intensity weakens as you move further from the center of Earth. In dealing with falling bodies and projectile motion, the intensity is also called the acceleration due to gravity, and within a few kilometers of the surface, it has the value

$$\vec{g} = -9.8 \ \frac{\text{m}}{\text{s}^2}$$

Weight, $\vec{W}$ or $\vec{w}$, or the force of gravity, $\vec{F_g}$, may be defined as the product of the mass of a body and the acceleration due to gravity. That is, weight is an interaction between two masses, not a property of a single mass. Weight is a vector, mass is a scalar, and $\vec{g}$ is a vector. When we discuss mass this way and consider how it interacts with gravity, we are discussing *gravitational mass*:

$$\vec{W} = \vec{w} = \vec{F}_g = m\vec{g}$$

Knowing that weight is a vector and is always directed vertically downward toward the center of Earth, we can write $w = mg$ to calculate the scalar magnitude of the weight.

## Sample Problem 1

Determine the weight of a 75.0 kg body at rest on the surface of Earth.

## Solution to Problem 1

Weight is defined as $w = mg = (75.0 \text{ kg})\left(9.8 \ ^{m}/_{s^2}\right) = \textbf{735.0 N}$

> ## AP® Tip
>
> Weight is a vector and is expressed in N. Weight is measured on scales and is dependent on location.

## Sample Problem 2

A small pickup truck has a weight of $14.7 \times 10^3$ N. What is the mass of the truck?

## Solution to Problem 2

On Earth, mass and weight are related by

$$m = \frac{w}{g} = \frac{\left(14.7 \times 10^3 \text{ N}\right)}{9.8 \ ^{m}/_{s^2}} = \frac{14.7 \times 10^3 \text{ kg} \cdot ^{m}/_{s^2}}{9.8 \ ^{m}/_{s^2}} = \textbf{1.5} \times \textbf{10}^3 \textbf{ kg}$$

> ## AP® Tip
>
> Mass is a scalar and is expressed in kg. Mass is measured on a balance. Mass in our everyday experience is constant and is independent of location.

## Sample Problem 3

Assuming that the pickup truck of Problem 2 were placed on the surface of Mars and knowing that the gravitational intensity for the surface of Mars is $\vec{g}_{Mars} = 3.7 \ ^{m}/_{s^2}$, how would the mass and weight of the truck change?

## Solution to Problem 3

The mass of the truck would remain unchanged because mass is a property of the truck itself. The weight of the truck would change because the gravitational strength compared to that of Earth is different. The weight of the pickup truck on Mars is

$$w_{Mars} = mg_{Mars} = \left(1.5 \times 10^3 \text{ kg}\right)\left(3.7 \ ^{m}/_{s^2}\right) = \textbf{5.6} \times \textbf{10}^3 \textbf{ N}$$

Mars is a smaller planet with a smaller $\vec{g}$-field. Weight varies from place to place, whereas mass remains constant.

## CENTER OF MASS

All objects have a unique point called the *center of mass,* also known as c.m. In terms of their gravitational pull, objects behave as if all of their mass is concentrated at the center of mass. For shapes like cubes and spheres, the c.m. is inside the object; for boomerangs and hula-hoops, the c.m. is outside the object. Your center of mass is located roughly near your belly button.

Objects on Earth experience a common force, weight. Objects are made of many particles with weight vectors that each point to the center of Earth. Since Earth acts as if all of its mass is concentrated at its c.m., the $\vec{g}$-field of Earth appears to "sink" into its center.

The terms *center of mass* and *center of gravity* are often used interchangeably, yet "center of mass" is preferable in the present context because gravity is not involved.

### AP® Tip

If a single force is applied to the c.m. of a body, purely straight-line motion will occur.

### AP® Tip

A body is any object, and any object is a body. A particle is a tiny body—so small in fact that we can treat it as a point. Under certain conditions, we can treat a body as large as the sun as a particle and treat a baseball as one too.

## FORCE

The previous chapter defined a force as a push or a pull on a body. The chapter also discussed how to add up the forces acting on a body by adding them as vectors. In our everyday lives, we encounter two classes of forces, *contact forces* and *field forces.* In the case of contact forces, the object producing the force actually touches the object being acted upon. With field forces, the force is experienced over a distance, such as Earth's gravity acting on a satellite in orbit. Something does not have to be touching something else to experience a field force with it.

Gravity is one of the four fundamental forces in nature. Even though you experience gravity all the time, it is the weakest of the four. Gravity is the primary force influencing the motion of astronomical bodies. It is always an attractive force. The force of gravity between two ordinary objects on Earth is negligibly small. You and your physics book are attracted to each other by the force of gravity between you, but you would never know it because the force is so small.

On the atomic level, physics recognizes that there are three fundamental forces: (1) gravitational force, (2) electro-weak force, and (3) the strong nuclear force. The contact forces we experience in daily life only appear to be contact forces. Actually, most contact forces are the results of electromagnetic forces acting at very small distances.

## AP® Tip

It is a common practice to treat bodies as a point located at the c.m. of the body and to attach the origin of a frame of reference there.

## NEWTON'S FIRST LAW OF MOTION

The part of physics we call mechanics is based on Newton's three laws of motion clearly stated for the first time by Sir Isaac Newton (1642–1727). These laws were published in 1686 in his *Philosophiae Naturalis Principia Mathematica* (The Mathematical Principles of Natural Philosophy).

## AP® Tip

A major effect of a force is to alter the state of motion of a body.

Newton's first law can be stated as follows: *an object at rest remains at rest; a moving object will maintain its velocity unless an external unbalanced force acts on it. Constant velocity means constant speed in a straight line in one direction.* Newton's first law is also called the law of inertia. We have already seen that the property of a body that allows it to remain at rest or maintain a constant state of motion is called inertia.

## NEWTON'S THIRD LAW OF MOTION

*For every force that acts on an object, there is always a reaction force that is equal in magnitude and is opposite in direction.* This is Newton's third law of motion. Newton's third law fundamentally redefined how scientists looked at forces. Newton defined forces as interactions between two objects; forces do not exist in isolation. Therefore, an object cannot exert an external force on itself. When you push on a wall, the wall pushes back on you. Yet, you can't push yourself and cause yourself to move.

Action-reaction forces act on two different bodies, and that means that they do not cancel each other out. The forces two objects mutually exert on each other are called action-reaction pairs.

Consider the drawing of a table with a box resting on its surface. The box pushes downward on the tabletop. The tabletop pushes vertically upward on the box. The two forces are an action-reaction pair.

The table pushes vertically downward on the floor, and the floor in return pushes vertically upward on the table. The forces are action-reaction pairs.

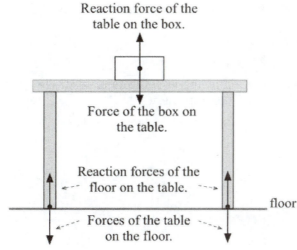

## FORCES IN EQUILIBRIUM

If forces are acting on an object but the object is not moving, the object is in a state of *static equilibrium*. In the example above, the box sitting on the table is in static equilibrium. If an object travels in a straight line with constant velocity when acted on by forces, it is in a state of *translational equilibrium*. A hockey puck sliding across the ice in a straight line at constant velocity is in translational equilibrium. There is no acceleration when a body or a system is in a state of equilibrium. In such a system, the resultant force acting is always zero, $\Sigma \vec{F} = \vec{R} = 0$.

### SAMPLE PROBLEM 4

Two forces, $\vec{F}_1 = 20.0$ N at $10.0°$ and $\vec{F}_2 = 15.0$ N at $110.0°$, act on an object as shown on the next page.

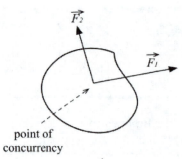

point of
concurrency

(a)  Determine the resultant force, $\vec{R}$, the body experiences.
(b)  Explain why this object should not be considered to be in
     equilibrium.
         What single force would we add to place the system in a
     state of equilibrium?

## SOLUTION TO PROBLEM 4

(a)  The first thing we do is to attach a frame of reference. Then we
     sketch in and identify the x- and y-components.

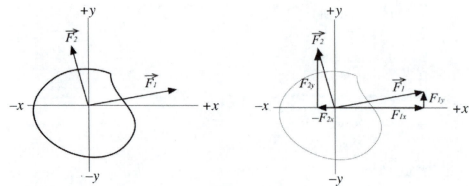

Next, we add the x- and y-components.

$$\Sigma x = F_1 \cos\theta_1 + F_2 \cos\theta_2 = (20.0 \text{ N})(\cos 10.0°) + (15.0 \text{ N})(\cos 110.0°) = 14.57 \text{ N}$$

$$\Sigma y = F_1 \sin\theta_1 + F_2 \sin\theta_2 = (20.0 \text{ N})(\sin 10°) + (15.0 \text{ N})(\sin 110.0°) = 17.57 \text{ N}$$

Recall that the sums of the components are the x- and
y-component of the resultant.

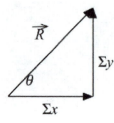

Calculating the magnitude of the resultant:

$$R = \sqrt{(\Sigma x)^2 + (\Sigma y)^2} = \sqrt{(14.57 \text{ N})^2 + (17.57 \text{ N})^2} = \mathbf{22.8 \text{ N}}$$

Finding the angle,

$$\theta = \tan^{-1}\left(\frac{\Sigma y}{\Sigma x}\right) = \tan^{-1}\left(\frac{17.57\ N}{14.57\ N}\right) = \mathbf{50.3°}$$

The object would move as if a 22.8 N force were pulling it at a 50.3° angle.

(b)    The object is not in equilibrium because the sum of the forces, the resultant, is not zero.

(c)    We can add a force that is equal in magnitude to the resultant but is directed 180° away from the resultant.

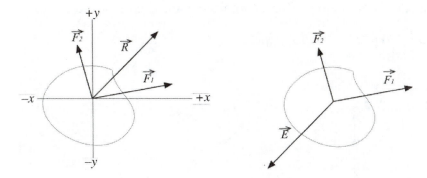

Forces that are not in equilibrium may be put in equilibrium by the addition of a force equal and opposite to the resultant. This means that the force must have a magnitude of E = 22.8 N at an angle of $\theta = 50.3° + 180° = 230.3°$.

When there is no resultant force, all of the $x$- and $y$-components in the system must sum to zero. This is a statement of the First Condition for Equilibrium, $\Sigma x = 0$ and $\Sigma y = 0$. A system is in a state of static equilibrium or in a state of translational equilibrium if and only if the sum of forces acting on the system is zero.

## FREE-BODY DIAGRAMS

Sometimes drawing a picture or sketch of a word problem helps you understand what is happening in the problem, allowing you to focus on the most important parts. The *free-body diagram* is a special type of picture used in physics to represent the forces acting on one object. You can draw one regardless of your art skills because instead of drawing the actual object, you draw a dot to represent its center of mass. A free-body diagram only shows the forces acting on the object. The forces are shown as vectors with their tails on the dot and their arrowheads in the direction the forces act. Each object in a problem gets its own free-body diagram even if the objects are touching.

Your physics book sitting on the desk has two forces acting on it: the book's weight and the force of the table pushing up on it. So you would draw one arrow up and one arrow down and label the arrows with the symbol of the force they represent.

A man is pushing a lawn mower on a flat, frictionless surface. The mower's handle makes a 45° angle with the horizontal. We need to identify the forces acting on both the lawn mower and the man. The lawn mower has three forces acting on it: its weight, the force from the surface that holds it up, and the push force from the man. The man also has three forces: his weight, the force from the surface, and the pushback from the lawn mower. Both the man and the lawn mower need a free-body diagram.

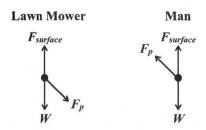

Following are a few special notes to remember when drawing free-body diagrams on the AP® exam:

- Do not draw a force's x- or y-component unless the question specifically directs you to do so. Free-body diagrams should show just the forces, not the components.
- Do not get too caught up in how to label the arrows. Just make sure you label them so that the exam readers understand what you mean.
- Weight is the force of gravity on an object, and the arrow should always be drawn pointing straight down. The preferred label for weight is $W$, but you can label that arrow $F_g$ too. Do not label the arrow $G$.
- All force vectors need to begin on the dot and point away from the dot.

## ROPES SUPPORTING A LOAD

Ropes supply tension forces, which can have horizontal and vertical components. For a system at rest, the sum of the horizontal force components is zero and the sum of the vertical force components is zero.

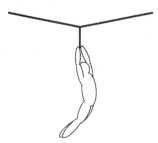

## SAMPLE PROBLEM 5

A trapeze artist hangs at rest from a vertical rope that hangs from another nearly horizontal rope. The vertical rope is tied to the horizontal rope with a knot, which is a convenient point to base a frame of reference.

(a) Draw one free-body diagram for the knot where the vertical rope meets the nearly horizontal rope.

(b) Draw another free-body diagram for the trapeze artist.

## SOLUTION TO PROBLEM 5

(a) The vertical rope pulls vertically downward on the knot. Each side of the nearly horizontal rope pulls in a direction along its length with force components in both directions. Because the system is in equilibrium, we know that the sum of the upward tension from each side of the nearly horizontal rope is equal in magnitude to the tension supplied by the vertical rope. We also know that the horizontal tensions must add to zero. Because of the shallow angle, the horizontal tension must be substantial to supply enough vertical tension. This situation is frequently addressed on the AP® exam.

(b) The trapeze artist is pulled downward by weight and pulled upward by tension. Because the system is in equilibrium, we know that these forces must be equal in magnitude.

## THE NORMAL FORCE AND THE FRICTION FORCE

When two objects apply contact forces to each other, there are two components of that force. The component of a contact force that is perpendicular to the objects' surfaces is called normal force and is always applied at a right angle to the surface. The word *normal* actually means perpendicular. The component of a contact force that is applied parallel to the surfaces is called friction. These contact forces can be described macroscopically as the push of one object against another and the roughness of a surface. They can be described microscopically as the electrical repulsion between the atoms that make up the objects.

## SAMPLE PROBLEM 6

A 200.0 N wooden crate rests on the floor of a storeroom. Determine the normal force exerted on the crate.

## SOLUTION TO PROBLEM 6

First, we make a free-body or vector diagram. When drawing a free-body diagram on the AP® test, unless otherwise specified, you should draw all force vectors acting on an object as radiating from the object's center of mass. Even though the normal force acts between the floor and the bottom of the crate, we show the vector radiating from the crate's center of mass.

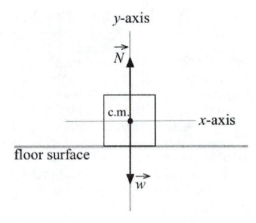

Since the crate is in equilibrium vertically, $\Sigma y = N - w = 0$. Transposing, we have $N = w = $ **200.0 N**.

Friction force is directly proportional to the normal force between two surfaces and the amount of drag that exists between those surfaces. There is a property called the coefficient of friction, and it is unique for different pairs of surfaces. The coefficient of friction is represented by $\mu$, the Greek letter mu. Friction does not depend on the area of contact between the surfaces. If a small and large cube have the same coefficient of friction and normal force and are pushed with the same force on the same surface, then they experience the same frictional force even though they have different areas of contact with the surface.

When two objects are in contact and their surfaces are sliding over each other, the friction force is called kinetic friction and the coefficient is called the coefficient of kinetic friction. Kinetic friction is drawn as a force vector that points opposite the direction of an object's motion. The force of kinetic friction is then $f_k = \mu kN$. When the surfaces are not in motion, the force is called static friction. Static friction keeps an object from moving. Depending on how hard an object is being pushed and the coefficient of static friction, the force of static friction can vary from very small up to some maximum value. Leaning gently against a stationary refrigerator or pushing against a refrigerator might not be enough to cause the refrigerator to move. The refrigerator won't move until enough force is applied. The force of static friction is then $f_s \leq \mu_s N$. Static friction is drawn as a force vector that points opposite the direction of an applied force. If no force is applied to a stationary object on a level surface, there will be no friction force between the object and the surface.

We define the maximum value of friction as follows:

$$\vec{f} = \mu\vec{N}$$

In a state of translational or static equilibrium, $\Sigma\vec{R} = 0$, there is no resultant force. An object in equilibrium either is at rest or is moving with constant velocity.

$$\vec{f}_s \le \mu_s\vec{N}$$

## SAMPLE PROBLEM 7

The diagram below shows a 100 N suitcase on a rough level floor with $\mu_s = 0.4$ and a variable tension force pulling the suitcase to the right. The situation changes depending on the value of the friction force.

## AP® Tip

The word *smooth* on the AP® exam means that a pair of surfaces has no friction. The word *rough* means that the surfaces do have friction.

(a) What maximum possible static friction force could the floor apply to the suitcase?
(b) What friction force does the floor apply if $T = 0$ N?
(c) What friction force does the floor apply if $T$ is greater than zero but less than your answer to (a)?
(d) What friction force does the floor apply if $T$ equals your answer to (a)?
(e) What friction force does the floor apply if $T$ is greater than your answer to (a)?

## SOLUTION TO PROBLEM 7

(a) The maximum possible friction force is $f_s = \mu_s N = (0.4)(100 \text{ N}) = 40$ N. This is the greatest force the floor is capable of providing to prevent the suitcase from moving.
(b) If there is no tension, the floor applies zero friction force to the suitcase. The only forces acting on the suitcase are its weight pulling downward and the normal force from the floor pushing upward.
(c) If the tension is greater than 0 and less than 40, $f_s < \mu_s N$. The friction force will vary from zero up to nearly 40 N and balance the tension exactly. The suitcase will remain at rest, subject to no net force.
(d) If the tension is exactly 40 N, the suitcase will be about to move. The slightest jostle will set it in motion. We call this the state of *impending motion*. The floor will be providing all of the friction force it can.

(e) If the tension force exceeds 40 N, the suitcase will not be in equilibrium and will begin moving to the right.

Once the suitcase above is set in motion, the frictional force between the contact surfaces decreases. This lower frictional force is proportional to the normal force and a proportionality factor called the coefficient of kinetic or sliding friction, $\mu_k$. The kinetic friction force $f_k$ is found by $\vec{f}_k = \mu_k \vec{N}$.

Kinetic frictional forces are parallel to the contact surfaces and directly oppose motion of the surfaces across each other.

## AP® Tip

The force of static friction is greater than the force of sliding friction for the same materials. In symbols, $\vec{f}_s > \vec{f}_k$ and $\mu_s > \mu_k$.

The coefficient of sliding friction is independent of the velocity. The coefficient of static friction between two surfaces will always be higher than the coefficient of kinetic friction because surfaces that are mutually at rest develop electrical attraction.

## Two-Body System

### Sample Problem 8

A block of mass $M = 4.00$ kg is at rest on a laboratory tabletop in a state of impending motion. A light cord of negligible mass is attached to block $M$ and runs over an ideal pulley to a hanging mass $m$. The coefficient of friction for the contact surfaces is $\mu_s = 0.30$. What value must the hanging mass $m$ have to keep the system in its state of impending motion?

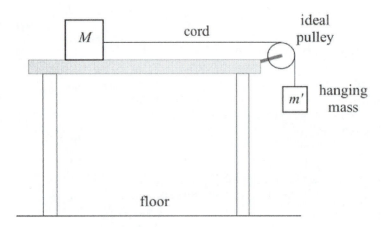

### SOLUTION TO PROBLEM 8

First, we pick a convenient frame of reference in which the clockwise rotation of the pulley is "positive." The system is in equilibrium and in a state of impending motion. Just touching the cord or the masses will start the system in motion, and it will no longer be at rest or in equilibrium. Make separate free-body diagrams for the masses $M$ and $m'$. Now we can identify the unbalanced external forces that will affect the motion of the system.

The normal force acting on $M$ is balanced by the weight of $M$, and tension is an internal force. The only positive external force on the system is the weight of the hanging mass. The only negative force is friction. These must be equal because the system is in equilibrium.

$$\Sigma F = 0$$

$$mg - \mu Mg = 0$$

$$m = \mu M = (0.3)(4 \text{ kg}) = 1.2 \text{ kg}$$

Only if $m = 1.2$ kg will the system be in both a system of equilibrium and a state of impending motion.

## THE INCLINED PLANE

The inclined plane is a simple machine. Inclined planes are tilted at some angle $\theta$.

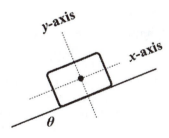

We attach our frame of reference to the c.m. of the body and make the $x$-axis parallel to the surface of the inclined plane and the $y$-axis perpendicular to that plane. It is convenient to call the direction of motion or impending motion the positive direction.

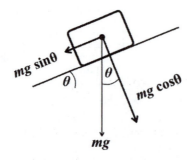

The weight of an object is a force *w* directed vertically downward. On an inclined plane, weight both holds the block against the plane and pulls it down the plane. Because the weight serves two functions, one normal to the plane and one parallel to it, we look at the weight's components.

The component parallel to the plane and directed down the plane is $mg \sin\theta$. This is the force that *"pulls"* the object down the inclined plane, sometimes called the *effective weight*.

The component that is normal to the plane is $mg \cos\theta$. This component is balanced by the plane's normal force $N$. The *normal force* that the inclined plane exerts on the object is $N = mg \cos\theta$.

Earlier, we defined the force of friction as $f = \mu N$. On inclined planes, the normal is $N = mg \cos\theta$, so the frictional force an object experiences on an inclined plane is $f = \mu mg \cos\theta$.

There is a critical angle $\theta$ at which for a given set of surfaces, the component of gravity pulling the object down the ramp is equal to the maximum static friction force pulling it up the ramp. Below this angle, the block will not slide. Above the angle, there will be motion down the plane. This angle, called the *slip angle*, is unique for a given system. At the slip angle,

$$\Sigma F = 0$$

$$mg \sin\theta - \mu mg \cos\theta = 0$$

$$\mu = \sin\theta / \cos\theta$$

The fraction $\sin\theta / \cos\theta$ is also equal to $\tan \theta$. So the slip angle is

$$\tan^{-1} \mu_s$$

## SAMPLE PROBLEM 9

A 4.0 kg block sits at rest on an inclined plane. You slowly raise the angle of the inclined plane until the block slips, finding that the highest you can lift the plane is 15°.

(a)  Find the coefficient of friction between the block and the plane.

(b)  At 15°, what is the normal force the plane exerts on the block?

(c)  What is the effective weight, the force causing the motion of the block down the plane?

(d)  Why is the force that is pulling the block down the plane not equal to the block's weight?

(e)  Find the frictional force acting on the block.

## SOLUTION TO PROBLEM 9

(a)  $\mu = \tan \theta$
$\mu = \tan (15°) = 0.268$

(b)  The normal force is N = $mg \cos\theta = (4.0 \text{ kg}) \left(9.8 \, ^m/_{s^2}\right) \cos 15° = 37.8\text{N}.$

(c)  The effective weight is

$$F' = mg \sin \theta = (4.0 \text{ kg})\left(9.8 \, ^m/_{s^2}\right)\sin 15.0° = \textbf{10.2 N}$$

(d)  The weight vector points vertically downward, but the ramp is inclined. Only the component of the weight parallel to the ramp causes its downward motion.

(e)  Friction on the plane is

$$f = \mu mg \cos\theta = (0.268)(4.0 \text{ kg})\left(9.8 \, ^m/_{s^2}\right)\cos 15.0° = 10.2 \text{ N}$$

## SAMPLE PROBLEM 10

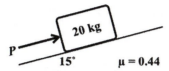

A 20.0 kg moving box is pushed up a 15° incline at constant velocity. The coefficient of friction between the underside of the box and the surface of the incline is 0.44. What push $P$ parallel to the plane is required to maintain constant velocity?

## SOLUTION TO PROBLEM 10

Constant velocity implies equilibrium and therefore forces up the plane = forces down the plane. The box is moving up the plane, friction acts down the plane, and we can write

$\Sigma F = 0$

$P - mg \sin\theta - \mu mg \cos\theta = 0$

$P = mg(\sin\theta - \mu \cos\theta)$

$P = (20 \text{ kg}) \left(9.8 \, ^m/_{s^2}\right) (\sin 15° - 0.44\cos 15°)$

134 N

# TRANSLATIONAL EQUILIBRIUM: STUDENT OBJECTIVES FOR THE AP® EXAM

- ▪ You should be able to differentiate between mass and weight.
- ▪ You should be able to discuss gravitational field strength.
- ▪ You should be able to define equilibrium and give examples.
- ▪ You should be able to state and apply Newton's first and third laws of motion to systems in equilibrium.
- ▪ You should be able to relate systems of concurrent forces and the first condition for equilibrium.

■ You should be able to discuss and calculate the normal force.
■ You should be able to differentiate between static and kinetic friction and solve for either.
■ You should be able to discuss contact forces and field forces.

## MULTIPLE-CHOICE QUESTIONS

1. A 25.0 N force applied to a 100.0 N crate moves it at constant velocity on a flat horizontal surface where the coefficient of kinetic friction between the crate and the surface is 0.25.

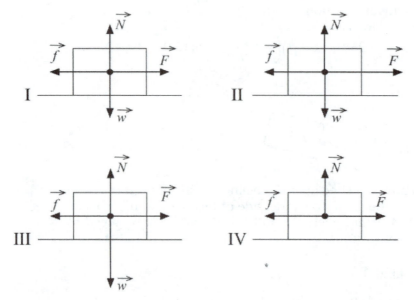

Which of the above illustrations shows the forces acting on the body correctly?
(A) I
(B) II
(C) III
(D) IV

2. While he is buckled up in the back seat of the family car, your little brother kicks very hard against the back of the driver's seat. Why does this not move the car forward?
(A) If the action force is a push against the driver's seat, the reaction force is the driver's seat pushing back. By Newton's third law, they are equal and opposite, so they cancel each other out.
(B) For someone to push against the driver's seat, to brace himself or herself, a greater force must be applied in the opposite direction.
(C) Doing so creates motion from nowhere, violating the law of inertia, which says that all objects at rest remain at rest.
(D) None of these are the reason.

3.  A 500 N crate is moved on a horizontal surface at constant speed by a 120 N force applied at an angle of 40° as shown below.

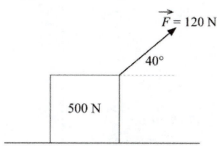

The normal force
(A) acting on the crate is equal to the weight since the normal is the reaction force to the weight of the crate on the surface.
(B) is greater than the weight because the object is moving and a force is applied to overcome the frictional force acting on the crate needed to keep the crate in equilibrium.
(C) is less than the weight of the crate because the net forces in the $y$-direction must be zero since the crate is in equilibrium in the $y$-direction and the 120 N force has a component in the $+y$ direction.
(D) is equal to the weight of the crate since the object is in translational equilibrium.

4.  A 2.00 kg box is placed at the top of a 30° inclined plane as shown below.

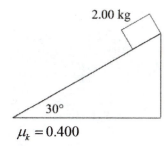

What push $P$ is required to keep the box moving down the plane at constant speed once the motion has started? The coefficient of kinetic friction, $\mu_k$, is 0.40.
(A) 6.8 N down the plane
(B) 6.8 N up the plane
(C) 3.0 N down the plane
(D) 3.0 N up the plane

5.  A 100 N block rests on a horizontal surface where the coefficient of static friction between the block and the surface is 0.64. When a 40 N force is applied to the block, the frictional force acting on the block is
(A) 64 N opposite the applied force.
(B) 40 N opposite the applied force.
(C) 24 N opposite the applied force.
(D) 0 since the block is at rest.

6. A 300 N weight hangs from three cables as shown below.

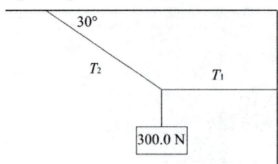

The magnitude of the tension in the horizontal cable $T_1$ is closest to
(A) 300 N.
(B) 520 N.
(C) 450 N.
(D) 600 N.

7. A physics student pulls a wooden crate across a rough horizontal surface with a constant force $\vec{F}$ at constant speed. Which of the following is a correct mathematical relationship for the free-body diagram shown below?

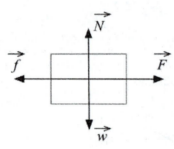

(A) $\vec{F} > \vec{f}$ and $\vec{N} < \vec{w}$
(B) $\vec{F} = \vec{f}$ and $\vec{N} > \vec{w}$
(C) $\vec{F} = \vec{f}$ and $\vec{N} = \vec{w}$
(D) $\vec{F} > \vec{f}$ and $\vec{N} = \vec{w}$

8. A 40.0 kg child stands on a bathroom scale in an elevator that is moving upward at a constant speed of 5.00 m/s. What does the scale read as the elevator moves upward?
(A) 192 N
(B) 240 N
(C) 392 N
(D) 592 N

9. An eraser is held against a whiteboard by a teacher's finger. If a student wants to construct a free-body diagram of the situation, where would the student draw the vector for the normal force?

(A) The student would draw the vector pointing up because normal force always points up.

(B) The student would draw the vector perpendicular to the surface of the board pointing in the opposite direction of the teacher's finger.

(C) The student would draw the vector perpendicular to the surface of the board pointing in the same direction as the force from the teacher's finger.

(D) The student would draw the vector pointing down because weight and normal are always equal to each other.

10. When a suitcase is rolled without friction to the right by a force $F$ applied at an angle $\theta$ above the horizontal, the floor exerts a normal force $N$. Later, the same force $F$ is applied at half the angle when a child pushes the suitcase.

While the child is pushing, the normal force is

(A) greater than $N$.

(B) equal to $N$.

(C) less than $N$.

(D) less than or greater than $N$ depending on the magnitude of the applied force $F$.

11. Student A of mass $a$ and student B of mass $b > a$ are initially at rest holding ends of a rope that hang over an ideal pulley. Initially, neither student is pulling on the rope. The floor provides a normal force $N_a$ on student A and a normal force $N_b$.

 Student B pulls the rope, causing student A to rise at a constant speed. What change, if any, occurs to $N_b$?

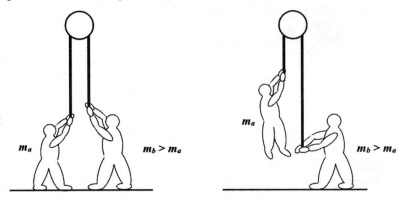

 (A) $N_b$ increases to cancel out the loss of $N_a$.
 (B) $N_b$ decreases by an amount equal to $N_a$.
 (C) $N_b$ increases because student B must apply more downward force.
 (D) $N_b$ remains the same because only student A was lifted.

12. A car is driving on a level highway at constant speed. Which of the following forces propels the car forward?
 (A) The normal force exerted by the road on the car
 (B) The normal force exerted by the car on the road
 (C) The force of friction exerted by the road on the car
 (D) The force of friction exerted by the car on the road

13. Mass $m$ is being pushed at constant speed up a ramp inclined at an angle $\theta$ above the horizontal. Between the mass and the ramp, the coefficient of static friction is $\mu$. A force $F$ acts at an angle $\phi$ relative to the horizontal. What is the magnitude of the friction force acting on this mass?

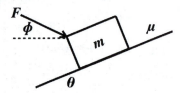

 (A) $\mu mg \cos\theta + F\sin\phi$
 (B) $\mu mg \cos(\theta + \phi) + F\sin(\theta + \phi)$
 (C) $\mu mg \cos(\theta + \phi) + \mu F\sin\phi$
 (D) $\mu mg \cos\theta + \mu F\sin(\theta + \phi)$

## Questions 14 and 15

**Directions:** For each of the questions or incomplete statements below, two of the suggested answers will be correct. For each of these questions, you must select both correct choices to earn credit. No partial credit will be awarded if only one correct choice is selected. Select the two that are best in each case.

14. A book rests on the top of a level table. The action-reaction pairs are (Select two answers.)
    (A) the force the book exerts on Earth and the gravitational force exerted by Earth on the book.
    (B) the normal force exerted on the book by the table and the gravitational force exerted by Earth on the book.
    (C) the force the book exerts on the table and the normal force exerted on the book by the table.
    (D) the normal force exerted on the book by the table and the gravitational force exerted by Earth on the table.

15. An object placed on a rough surface experiences the force shown. Which of the following statements is true? Select two answers.

    (A) Friction only exists when there is relative motion between the two surfaces.
    (B) Friction only exists when one body tends to slide past another surface because translational equilibrium applies when a stationary body experiences an external force that tries to move the body.
    (C) Friction arises from the gravitational force of attraction between the molecules on the two materials. Since friction depends on the normal between the surfaces, a larger gravitational force produces a larger normal.
    (D) Friction arises from the electromagnetic force of attraction between the molecules of the two materials that is the same as the bonding holding the materials together.

## FREE-RESPONSE PROBLEMS

1. A steel sphere of mass $m = 0.020$ kg is at rest in a 90° groove in a steel track as shown below.

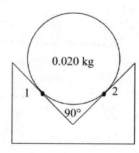

0.020 kg

1    2

90°

    (a) Show the forces acting on the sphere in a free-body diagram. Clearly label and indicate the magnitude and direction of each force.
    (b) Determine the normal forces $\vec{N}_1$ and $\vec{N}_2$ acting on the sphere.

2. As shown in the diagram, two boxes are next to each other on a horizontal surface. A student pushes one box, which in turn pushes the other box.

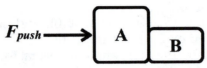

(a) Draw a free-body diagram for each box to show the forces on each box.

(b) A student makes the following comment about pushing the boxes: "The force that box A exerts on box B is greater than the force that box B exerts on box A because box A is bigger and is being pushed by my hand."
    Is the student's explanation correct? Explain your answer.

(c) The student is pushing the boxes at a constant velocity. Describe the type of friction the boxes experience.

(d) Will the magnitude of the friction force change if the student pushes at a slower constant velocity? Explain your answer.

3. A 2.00 kg box slides down an inclined plane at constant speed when the angle of the incline is 30°.

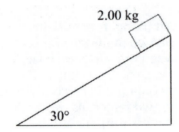

(a) Show the forces acting on the body using the tilted box below. Label the forces correctly.

(b) Write and solve the equation needed to obtain the coefficient of friction between the 2.00 kg box and the surface. Treat down the plane as positive.

(c) Mass is added to the box, increasing its total mass to 3.00 kg. As a result, the box will
    ___ increase its speed down the plane.
    ___ decrease its speed down the plane.
    ___ continue down the plane at constant speed.
    Explain your answer.

(d) If the box is pushed up the plane at constant speed, the frictional force acting on the box will
    ___ increase.
    ___ decrease.
    ___ remain the same.
    Explain your answer.

4. A student uses a string to pull a box across a table at a constant speed of 1 m/s. As shown in the diagram below, the string makes an angle $\theta$ with the horizontal.

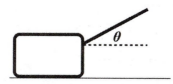

(a) Draw a free-body diagram to represent all of the forces on the box.

(b) Does the weight of the box equal the normal force in this situation? Explain your answer.

(c) The student now drops the string and picks up a stick to push the box along the table as shown. The angle $\theta$ is the same in both situations.

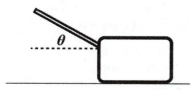

Will the magnitude of the force on the block by the applied force from the stick be greater than, less than, or equal to the applied force from tension in the string? In both cases, the box is moving at a constant speed of 1 m/s.

# Answers

## MULTIPLE-CHOICE QUESTIONS

1. **A** The 100.0 N body is in translational equilibrium, and $\Sigma \vec{F}_x = 0$. The force pulling the crate must equal the frictional force. In the $y$-direction, $\Sigma \vec{F}_y = 0$ means that the normal to the surface must be equal and opposite $\vec{W}$, the weight of the body.

   (L.O. 3.A.2.1)

2. **D** An object <u>cannot</u> exert a force on itself. Forces always come in pairs. Newton's third law tells us that the action-reaction pair always acts on different objects.

   (L.O. 3.A.3.2)

3. **C** The object is in translational equilibrium. A free-body diagram shows the direction of the forces. Since the 120 N force acts upward on the crate, $\Sigma \vec{F}_y = \vec{N} + \vec{F} \sin \theta - \vec{w}$.
   The normal will be less than the weight $\vec{w} = m\vec{g}$.

   (L.O. 2.B.1.1, 3.A.4.3)

4. **D** The object is in translational equilibrium. The force parallel to the plane is found by solving the $\sum \vec{F} = 0$ equation. Since the effective component of the weight is down the plane and friction will be up the plane, the difference of the two will give the magnitude of the push as well as indicate the needed direction. Take down the plane as the positive direction.

$$0 = m\vec{g}\sin\theta - \mu_k m\vec{g}\cos\theta \pm \vec{P}.$$

$$(2.00\text{ kg})\left(9.80\text{ }\text{m}\big/\text{s}^2\right)\sin 30° - 0.400(2.00\text{ kg})\left(9.80\text{ }\text{m}\big/\text{s}^2\right)\cos 30° \pm \vec{P}$$

$$9.80\text{ N} - 6.79\text{ N} \pm \vec{P} = 0.$$ The push is 3.0 N directed up the plane.

(L.O. 2.B.1.1, 3.A.4.3)

5. **B** The block is in equilibrium. To start the block moving, the maximum force of static friction must be overcome. The equality, $\vec{f}_{max} = \mu_s \vec{N}$, requires a maximum force of 64 N. The block remains at rest; the first condition of equilibrium applies, and $\Sigma \vec{F} = 0$. A 40 N force is applied; thus, the frictional force is 40 N.

(L.O. 2.B.1.1, 2.B.2.1, 3.A.4.3)

6. **B** Drawing a free-body diagram for the system permits you to write

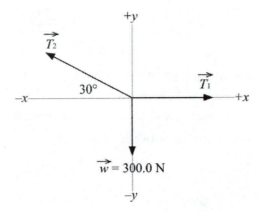

$\vec{F}_y = \vec{T}_2 \sin 30° - \vec{w}$. The tension $\vec{T}_2 = \dfrac{300.0\text{ N}}{\sin 30°} = 600.0\text{ N}$.

Writing the $\Sigma \vec{F}_x = \vec{T}_1 - \vec{T}_2 \cos 30° = 0$ equation will give the solution for $\vec{T}_1$. The tension in the horizontal cable $\vec{T}_1$ is 520 N.

(L.O. 3.A.4.3)

7. **D** Since the crate is moving at constant speed, the free-body diagram shows that $\sum \vec{F}_x = 0$; thus, the force pulling the crate and the frictional force on the crate are equal and in opposite directions. $\sum \vec{F}_y = 0$ also applies, and the normal force must equal the weight of the crate.

(L.O. 3.B.2.1)

8. **C** The elevator is moving upward at 5.00 m/s, moving at constant speed. There are no net forces acting along the $x$-direction, only in the $y$-direction. Newton's first law of motion applies $\Sigma \vec{F}_y = 0$. The bathroom scale reads the weight of the child, 392 N.

(L.O. 3.A.3.1)

9. **B** The normal force is always drawn perpendicular to the surface that applies it. $N$ represents the force of the board pushing back on the eraser. The normal force is equal in magnitude to the horizontal component of the push from the teacher's finger, setting the forces in the $x$-direction in equilibrium.

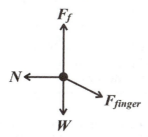

(L.O. 3.A.2.1, 3.B.2.1)

10. **C** The normal force pushing up from the floor balances the suitcase's weight and the downward component of the applied force. When the child pushes the suitcase, the downward component of the applied force, $F\sin\theta$, is reduced. Therefore, the normal force is reduced.

(L.O. 3.A.2.1)

11. **B** Initially, the floor provides a normal force on each student equal to his or her weight. When student A is no longer in contact with the floor yet is not accelerating, we know the tension force must equal student A's weight. We know this because the forces must add to zero, and the only way that will work is if the upward forces equal the downward forces. For student B, his or her weight is the same, but now there is a tension pulling upward equal to the weight of student A.

$\Sigma F = 0$

$T + N_b - m_b g = 0$

This shows that the upward forces and downward forces sum to zero.

$T = N_a$

$N_b = m_2 g - N_a$

(L.O. 3.A.3.1, 3.A.3.3, 3.A.4.2)

12. **C** The only object applying a forward force to the car is the road. Answers (A) and (B) specify only the perpendicular component of a force, while answer (D) refers to a force that acts on the road.

(L.O. 3.A.3.1, 3.A.3.3, 3.A.4.2)

13. **D** The friction force is $\mu N$. In this case, the normal force opposes both the component of the weight that is perpendicular to the ramp ($mg\cos\theta$) and the component of the applied force that is perpendicular to the ramp. The applied force makes an angle of ($\theta + \phi$) relative to the ramp, so its perpendicular component is $F\sin(\theta + \phi)$. Therefore, the normal force is $mg\cos\theta + F\sin(\theta + \phi)$.

(L.O. 3.A.3.1, 3.B.2.1)

14. **A and C** Action-reaction pairs have the same magnitude, act in opposite directions, and are applied to two different objects. The action-reaction pairs are A, the book and Earth that exert equal and opposite forces on each other whose magnitude is given by $F = G\dfrac{M_{Earth} \cdot m_{book}}{r^2}$ and C. The force the book exerts on the table and the normal force exerted on the book by the table's surface are equal. The table exerts an upward force perpendicular to the book.

(L.O. 3.A.3.3)

15. **B and D** An object that is stationary or is moving at constant speed experiences a net force $\Sigma\vec{F} = 0$. The magnitude of the frictional force will equal the applied force. The forces that arise between two bodies in contact, adhesion between unlike molecules and cohesion between like molecules, are forces of attraction and with surface irregularities are basically responsible for friction. These forces are short-range electromagnetic forces of attraction that are negligible at distances larger than a few atomic diameters.

(L.O. 3.A.3.4, 3.C.4.1, 3.C.4.2)

## FREE-RESPONSE PROBLEMS

1. (a) The diagram of the forces acting between the sphere and the sides of the groove is shown below. The normal force is a reaction to the weight on the sphere on the sides of the groove.

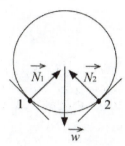

The second diagram indicates the concurrent forces and the angles these force make with respect to the x- and y-axes.

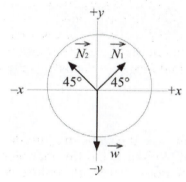

(b) Due to symmetry, the normal forces are equal in magnitude. The solution for the normal forces is determined by writing the $\Sigma \vec{F} = 0$ equation.

$$\vec{N}_1 \sin 45° + \vec{N}_2 \sin 45° = m\vec{g}$$

$$2\vec{N}_1 \sin 45° = (0.020 \text{ kg})\left(9.80 \text{ m}/_{s^2}\right)$$

$$\vec{N}_1 = \frac{0.196 \text{ N}}{0.707}$$

$$\vec{N}_1 = \vec{N}_2 = \mathbf{0.139 \text{ N}}$$

(L.O. 2.B.1.1, 3.A.2.1, 3.A.3.1, 3.A.4.1, 3.A.4.3)

2. (a) Each box must have its own free-body diagram with labeled arrows representing each force. Notice that box A experiences a normal force from box B that is equal in magnitude but opposite in direction to the normal force experienced by box B from box A. Only box A is pushed by the finger.

**Box A**        **Box B**

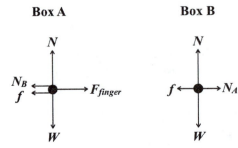

(b)   The student's statement is incorrect. The two forces represent a Newton's third law action-reaction pair. They are experiencing equal and opposite forces.

(c)   The boxes are moving, so the friction is considered kinetic friction.

(d)   Kinetic friction depends on the normal force and the coefficient of kinetic. It is independent of velocity, so changing the speed or even the force that is pushing will not change the magnitude of $f_k$.

(L.O. 3.A.2.1, 3.A.3.1, 3.A.4.2, 3.A.4.3, 3.B.2.1)

3.  (a)   The correct forces on the box are shown below.

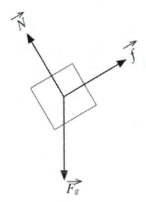

(b)   $\Sigma \vec{F} = 0$

$m\vec{g} \sin\theta = \mu_k m\vec{g} \cos\theta$

$(2.00 \text{ kg})\left(9.80 \ \frac{\text{kg}}{\text{m}^2}\right) \sin 30° = \mu_k (2.00 \text{ kg})\left(9.80 \ \frac{\text{kg}}{\text{m}^2}\right) \cos 30°$

$\mu_k = 0.577$

(c)   The correct box to check is ___√___ continue down the plane at constant speed.

In the equation in part (b), mass is a common factor in both terms. Changing the mass of the box will not change the speed of the box down the plane.

(d) The correct box to check is ___√___ remain the same.

   The frictional force is dependent on the materials in contact and the normal force but not the direction of the motion, since friction is always opposite the motion or the tendency to move in a direction. The magnitude of the force of friction has the same value up the plane as it does for the box moving down the plane.

(L.O. 3.A.2.1, 3.A.3.1, 3.A.4.2, 3.A.4.3, 3.B.2.1)

4. (a) The block is experiencing four forces: normal force, weight, the tension in the string, and friction.

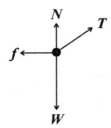

(b) Normal force does not equal the weight in this situation. The body is in equilibrium, so all of the forces in both the $x$- and $y$-directions must equal zero. Tension is in both the $x$- and $y$-directions. The $y$-component of tension is positive, so it will add to the positive normal force. Together the $y$-component of tension and the normal force must be equal and opposite to that of the weight. Otherwise, the system is not in equilibrium.

(c) The force from the stick on the block has a downward component in the $y$-direction, and for the box to be in equilibrium in the $y$-direction, the normal force must be greater than the weight. The tension in the string has an upward $y$-component, and since the block is in equilibrium, the normal force must be less than the weight. Since the weight of both blocks is the same, the normal force on the block with the stick must be greater than the normal force on the block with the string. Since the normal force with the stick is greater, the friction force is greater with the stick than with the string. And since the forces are in equilibrium in both cases, the force on the block by the stick must be greater than the tension on the block by the string.

(L.O. 3.A.2.1, 3.B.2.1)

# 3

# MOTION

## VELOCITY

In physics, we study motion by looking at how an object's position changes during some time interval. The position of an object traveling along our frame of reference's $x$-axis is described by its $x$-coordinate. The change in the object's position is its *displacement*, $\Delta x$. Initially, if the particle is located at position $x_0$ at time $t_0$ and at position $x$ at time $t$, the particle is displaced by $\Delta x = x - x_0$. Recall that displacement is a vector; remember to consider the direction. The elapsed time is $\Delta t = t - t_0$. Depending on the convenient frame of reference you've chosen, positive displacements could be, for example, to the right, down a ramp, or away from you. The *average velocity, $\overline{v}$*, of an object is the time rate of change of displacement, or

$$\overline{v} = \frac{x - x_0}{t - t_0} = \frac{\Delta x}{\Delta t}$$

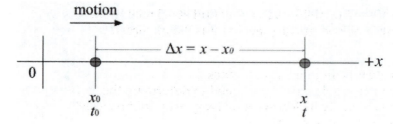

Average velocity is a vector, and its SI unit is the m/s. Average velocity measures how quickly an object's displacement changes.

## SAMPLE PROBLEM 1

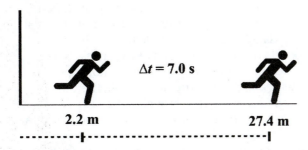

$\Delta t = 7.0$ s

2.2 m                        27.4 m

While you're watching a track meet, you notice that at one moment, a runner is 2.2 m past a light pole. 7.0 s later, that same runner is 27.4 m past the light pole. What is the runner's average velocity?

## SOLUTION TO PROBLEM 1

Average velocity compares displacement to time. The runner's displacement is 27.4 m – 2.2 m.

$$v = \frac{x - x_0}{t - t_0} - \frac{27.4\,\text{m} - 2.2\,\text{m}}{7.0\,\text{s}} - 3.6 \, ^m\!/\!_s$$

Between the two points, the runner's average velocity is 3.6 m/s.

If we pick a convenient frame of reference with the object's initial position at $x = 0$ m, and we start our timers at $t = 0$ s, the object's average velocity is

$$v = ^x\!/\!_t$$

## SAMPLE PROBLEM 2

A proton has an average velocity of $4.5 \times 10^6$ m/s at 0°. What is the displacement of the proton in a time period of 2.0 μs? One μs is a *micro second*; the prefix micro means one millionth, or $1 \times 10^{-6}$.

## SOLUTION TO PROBLEM 2

Since average velocity is expressed as $\bar{v} = ^x\!/\!_t$, displacement is $x = \bar{v}t$.

Substituting, we have

$$x = \bar{v}t = \left(4.5 \times 10^6 \, ^m\!/\!_s\right)\left(2.0 \times 10^{-6} \, \text{s}\right) = 9.0 \text{ m @ 0°}$$

The proton travels for 2.0 millionths of a second at 4.5 million meters per second. In that time period, its displacement is 9.0 meters at 0°.

## SAMPLE PROBLEM 3

A go-cart is being driven around a circular track.
(a) What two measuring tools would you need to determine the go-cart's average velocity as it travels halfway around the circle? What quantities would you measure?
(b) Describe a procedure to determine the car's average velocity around half of the track, giving enough detail so that another student could replicate the experiment.
(c) Using equations, show how your measurements could be used to determine velocity.
(d) In what way is average velocity not a good estimate of the speed of a go-cart during a race?

### SOLUTION TO PROBLEM 3

(a) To calculate average velocity, we would need to use a meter tape to measure displacement and a stopwatch to measure the time interval.

(b) Stretch the meter tape across the track at its widest point to measure the diameter. Pick two points on opposite sides of the track. Begin the stopwatch when the front of the go-cart passes one of those points and stop the stopwatch when the go-cart passes the other point. This will measure the time it takes to travel half a lap.

(c) Divide the diameter of the track by the time it took to travel half a lap.

(d) Average velocity is calculated using displacement, which is the linear distance between the starting and ending points of a motion. The displacement of a go-cart, no matter the number of laps, is never more than the diameter from the starting point. At every lap, the displacement is zero again because the car passes its starting point. A better measurement would be to compare the total distance traveled to time.

## ACCELERATION

Just as average velocity is the rate of change of displacement, *average acceleration* is the rate of change of velocity. As an equation,

$$a = \frac{v - v_0}{t - t_0} = \frac{\Delta v}{\Delta t}$$

Acceleration is a vector that has units of $m/s^2$, which is read as meters per second per second. In $m/s^2$, the unit of time shows up twice, once for velocity in m/s and again for how quickly velocity is changing.

If we again start our observations at $t_0 = 0$, $a = \dfrac{v - v_0}{t}$ and $v - v_0 = at$, solving for the final velocity yields

$$v = v_0 + at$$

which states that when something moving at initial velocity $v_0$ undergoes an acceleration $a$ for a time interval $t$, its final velocity is $v$.

If the velocity of a particle is increasing or decreasing uniformly with time, we can express the average velocity for any time interval as the *arithmetic average* of the initial velocity $v_0$ and the final velocity $v$ as

$$\overline{v} = \frac{v_0 + v}{2}$$

We can also use the following kinematic equation when we are not worried about the final velocity:

$$x = v_0 t + \frac{1}{2}at^2$$

## SAMPLE PROBLEM 4

A car getting on the highway with an initial velocity of 10.0 m/s speeds up smoothly to 28.0 m/s in 6.0 seconds.
(a)   What constant acceleration does the car experience?
(b)   What is the car's average velocity?
(c)   What is the displacement of the body over the 6.0 seconds of acceleration?

## SOLUTION TO PROBLEM 4

(a)   Since we are seeking acceleration, we will use the equation that defines it.

$$a = \frac{v - v_0}{t} = \frac{28.0 \ ^m/_s - 10.0 \ ^m/_s}{6.0 \ s} = \textbf{3.0} \ ^m/_{s^2}$$

Each second, the velocity of the body increases by 3.0 m/s.
(b)   The average velocity is simply the average of 10 m/s and 28 m/s: 19 m/s.
(c)   The car's displacement is

$$x = v_0 t + \frac{1}{2} a t^2 = \left(10 \ ^m/_s\right)(6.0 \ s) + (0.5)\left(3.0 \ ^m/_{s^2}\right)(6.0 \ s)^2 = \textbf{114 m}$$

Alternatively, $x = v_{avg} t$, so $x = (19 \ m/s)(6 \ s) = 114 \ m$.

## SAMPLE PROBLEM 5

Another car getting on the highway starts with a velocity of $12 \ ^m/_s$ and accelerates at $2.5 \ m/s^2$ for 15 s. What is the car's final velocity?

## SOLUTION TO PROBLEM 5

The final velocity is found by

$$v = v_0 + at = 12 \ ^m/_s + \left(2.5 \ ^m/_{s^2}\right)(15 \ s) = \textbf{50} \ ^m/_s$$

The equations we have can be rearranged to relate initial velocity, final velocity, acceleration, and displacement. This equation does not require time:

$$v^2 = v_0^2 + 2ax$$

When the velocity and acceleration of an object are in the same direction, the object speeds up. When velocity and acceleration are in opposite directions, the object slows down. If the velocity and acceleration are staying at a right angle, the object doesn't speed up or slow down; it moves in a circle at constant speed.

| Sign of Velocity | Sign of Acceleration | Object's Resulting Motion |
|:---:|:---:|:---:|
| + | + | Speeds Up |
| + | − | Slows Down |
| − | − | Speeds Up |
| − | + | Slows Down |

An easy way to remember this is that "Same Signs Speed Up."

Of course, in a convenient frame of reference, + and – might mean, respectively, clockwise and counterclockwise or down the ladder and up the ladder.

## KINEMATICS AND GRAPHIC RELATIONSHIPS

Instead of using sentences or equations, we can describe the motion of a particle using graphs to illustrate the particle's movement. The AP® exam uses three common graphs when discussing the motion of a particle: position vs. time, velocity vs. time, and acceleration vs. time.

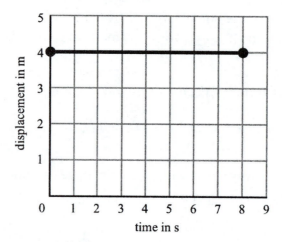

A position vs. time graph shows where an object is at many points in time. In the position vs. time graph above, the horizontal line shows that the object remains 4 meters past some reference point for the 8 seconds over which data were recorded for the time shown on the graph.

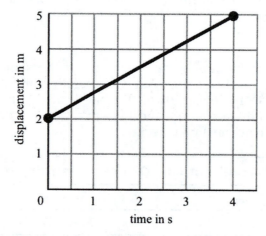

This position vs. time graph shows a line with a positive slope. This graph tells us that an observer saw an object start 2 meters from a reference point and move three more meters away over 4 seconds. Because the graph shows a constant slope, it lets us know that the object was traveling with a constant velocity. On a position vs. time graph, the velocity can be found by taking the slope of the line. Slope is rise/run, which on this graph is $\frac{\Delta x}{\Delta t}$. The average velocity shown on this graph is

$$V_{avg} = \frac{\Delta x}{\Delta t} = \frac{3\,m}{4\,s} = \textbf{0.75 m/s}$$

Below we see a velocity vs. time graph that shows how fast an object is moving at different times. The sign of the velocity—whether above or below the x-axis—tells us direction. These graphs can be used to find the displacement of an object by taking the area under the curve of the graph.

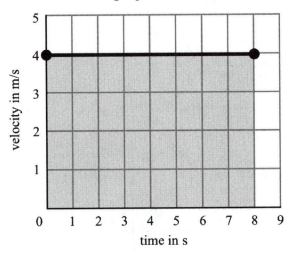

This velocity vs. time graph for constant velocity shows that displacement is given by the area beneath the curve.

## SAMPLE PROBLEM 6

The graph above shows the speed of a runner during an 8-second time interval. Calculate the displacement of this runner.

## SOLUTION TO PROBLEM 6

The area under the curve is shaded, and calculating it yields the displacement.

Area = base × height = (8 s)(4 m/s) = **32 m**

Calculating the area under a velocity vs. time graph yields displacement because we are essentially multiplying velocity and time. Calculating the slope of a velocity vs. time graph yields acceleration. Slope *m* can be calculated using the following equations:

$$slope = \frac{change\ in\ the\ vertical}{change\ in\ the\ horizontal}$$

$$m = slope = \frac{\Delta y}{\Delta x}$$

In this case, rise/run is $\frac{\Delta v}{\Delta t}$, which is acceleration.

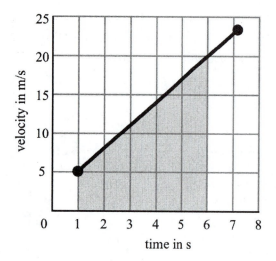

This velocity vs. time graph for uniform acceleration shows that displacement is given by the area beneath the curve.

## SAMPLE PROBLEM 7

The velocity of a car as it gradually speeds up is shown above.
(a)   Calculate the displacement of the car from $t = 1$ s until $t = 6$ s.
(b)   Calculate the car's average acceleration during that interval.

## SOLUTION TO PROBLEM 7

(a)   Displacement is the area between the velocity vs. time graph and the $x$-axis, shown by the shaded area under the curve. The displacement is conveniently found as the sum of the triangular area 1 and rectangular area 2.

$$A1 = \frac{1}{2} \text{base} \times \text{altitude} = \frac{1}{2}(6 \text{ s} - 1 \text{ s})\left(20 \text{ }^{m}\!/_{s} - 5 \text{ }^{m}\!/_{s}\right) = 37.5 \text{ m}$$

$$A2 = \text{base} \times \text{height} = (6 \text{ s} - 1 \text{ s})\left(5 \text{ }^{m}\!/_{s}\right) = 25 \text{ m}$$

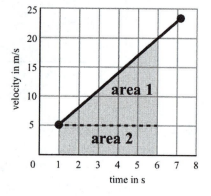

The total area is $A1 + A2 = 37.5$ m $+ 25$ m $=$ **62.5 m.**

(b)   Acceleration is the slope $m$ of a velocity vs. time graph.

$$m = \frac{\Delta v}{\Delta t} = \frac{v - v_0}{t - t_0} = \frac{20 \text{ }^{m}\!/_{s} - 5 \text{ }^{m}\!/_{s}}{6 \text{ s} - 1 \text{ s}} = \frac{15 \text{ }^{m}\!/_{s}}{5 \text{ s}} = 3 \text{ }^{m}\!/_{s^2}$$

$$\text{slope} = \frac{\text{change in the vertical}}{\text{change in the horizontal}}$$

This means that at every second, the car was moving 3 m/s faster than the second before. A straight line sloped up to the right means that the object is getting faster over time and is showing positive acceleration.

## Kinematic Graphs

| Position vs. Time | Velocity vs. Time | Acceleration vs. Time |
|---|---|---|
| shows position at points in time | shows velocity at points in time | shows acceleration at points in time |
| slope = velocity<br>+ slope shows + velocity<br>– slope shows – velocity | slope = acceleration<br>+ slope shows + acceleration<br>– slope shows – acceleration<br><br>area = displacement | Graphs on the AP® exam should be horizontal lines because acceleration on the exam will be constant. |

On the AP® exam, you will be required to describe the motion of an object using words, pictures, graphs, and mathematics.

## SAMPLE PROBLEM 8

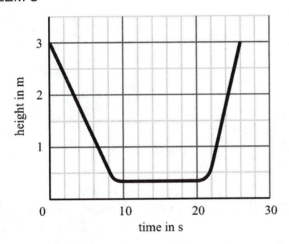

The position vs. time graph above shows a box being lowered and then raised by a forklift in a warehouse. Based on this graph, sketch a velocity vs. time graph for the box's motion. Pay attention to the magnitude and direction of the speed velocity at each part of the box's motion.

## SOLUTION TO PROBLEM 8

There are three distinct parts of this motion. The box is moving down with constant velocity between 0 s and 8 s, it is at rest from 10 s until 20 s, and it is moving upward with constant velocity from 22 s until 26 s. Between those times, the box is changing speed. Our velocity graph should show the three speeds and smooth transitions between them.

First, we find the velocities when the box is moving.

From 0 s until 8 s, $v = \dfrac{\Delta x}{\Delta t} = \dfrac{(0.6\,\text{m} - 3\,\text{m})}{(8\,\text{s} - 0\,\text{s})} = -0.45\,\text{m}/\text{s}$.

From 22 s until 26 s, $v = \dfrac{\Delta x}{\Delta t} = \dfrac{(3\,\text{m} - 0.6\,\text{m})}{(26\,\text{s} - 22\,\text{s})} = 0.6\,\text{m}/\text{s}$.

Now we connect the breaks in the graph with smooth accelerations.

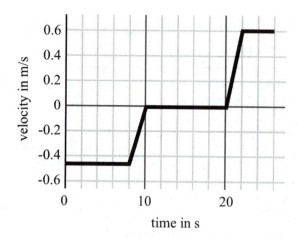

## THE BASIC EQUATIONS OF KINEMATICS

On the AP® exam, you will receive the following three kinematic equations:

$$v_x = v_{x0} + a_x t$$

$$x = x_0 + v_{x0}t + \frac{1}{2}a_x t^2$$

$$v_x^2 = v_{x0}^2 + 2a_x(x - x_0)$$

The kinematic equations on your AP® equation sheet are written to be used in the x-direction. However, they can be applied to any direction. You are the one whose physics intuition picks the convenient frame of reference.

## FREE FALL

Earth behaves as if all of its mass were concentrated at its center—the center of mass, c.m. of Earth. Due to its mass, Earth generates a *gravitational field* that extends into deep space. Within a few kilometers of Earth's surface, the gravitational field is approximately uniform. Fields are vector quantities that have a magnitude, a unit, and a direction. The magnitude and unit of the gravitational field is 9.81 m/s², with the same unit as acceleration. This *acceleration due to gravity* gets its own special symbol, *g*, and always points vertically downward toward Earth's c.m.

> ## AP® Tip
>
> On test day, use 10 m/s² for free fall acceleration. It is much faster to multiply or divide by 10. The questions and rubrics were written to accept $g = 10$ m/s² or 9.8 m/s².

When an object is under only the influence of gravitational force, it is in *free fall*. In AP® Physics, we generally ignore the effects of air resistance and assume that objects move through Earth's atmosphere unhindered by the air. Whether you drop a rock or throw it upward, it is in free fall because the only force affecting its motion is Earth's gravitational pull. The free fall model is a good approximation of how most compact, rigid objects move through the air with everyday speeds.

Even an object launched vertically upward is in free fall. Because its initial velocity is positive and its acceleration is always negative, the object's velocity will slow until it reaches its highest point. At the apex of its path, the object's velocity is changing direction from positive to negative, so it is momentarily zero. The object's acceleration is always 9.8 m/s² downward through every step of its trajectory. As it rises, peaks, and falls, the object is under the same acceleration.

The diagram below shows a ball projected vertically upward from ground level at 49.0 m/s. At the end of 1.0 s of flight, the ball is 44.1 m above the ground traveling upward at 39.2 m/s. As we continue to follow the ball up in its path, notice that the ball is slowing with time and altitude. Whether rising or falling, it has the same acceleration, $g = -9.8$ m/s². After 5.0 s, the ball reaches a maximum altitude of 122.5 m and for an instant has stopped. Then it begins to fall, accelerating at $g$ as it does.

Note the symmetry in the diagram.

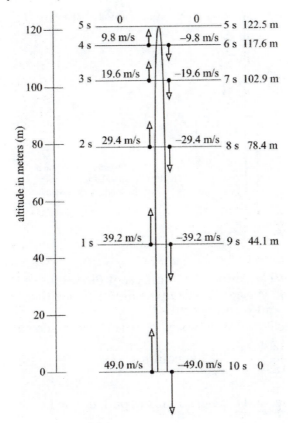

The time it takes to reach the top of its trajectory is equal to the time taken from there to the ground. This implies that the upward motions are the same as the downward motions, but in reverse.

## SAMPLE PROBLEM 9

A person standing at the edge of a cliff drops a brick from rest that impacts the ground 2.4 seconds later.
(a)    How far did the brick fall?
(b)    What is the velocity of the brick upon impact with the ground?

## SOLUTION TO PROBLEM 9

(a)    The brick is released from rest, making $v_0 = 0$. The direction of motion will be downward, so we can call that positive. The acceleration is 9.8 m/s². The distance fallen is
$y = v_0 t + \frac{1}{2} at^2 = (0)(2.4 \text{ s}) + \frac{1}{2} (9.8 \text{ m/s}^2)(2.4 \text{ s})^2 = \textbf{28.2 m}$
(b)    Since $v_0 = 0$, the velocity on impact is
$v = v_0 + at = (0) + (9.8 \text{ m/s}^2)(2.4 \text{ s}) = \textbf{23.5 m/s downward}$
In both cases, we got a positive answer which, by our frame of reference, means downward.

## SAMPLE PROBLEM 10

A person standing at the edge of a cliff 30.0 m high reaches out over the edge and throws a ball vertically upward with a velocity of 12.0 m/s.
(a)    When does the ball reach its highest point?
(b)    How high off the ground is that?
(c)    How much more time does it take the ball to fall to the ground?
(d)    How fast is the ball moving when it impacts the ground?

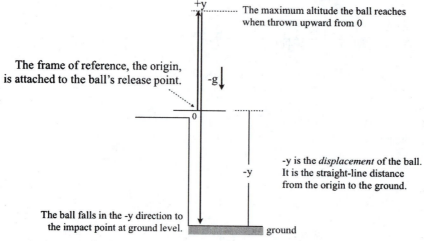

## SOLUTION TO PROBLEM 10

The initial velocity and the acceleration are in different directions, and the motion changes direction. We could pick up or down as positive. In this case, up will prove to be more convenient.
(a)    Gravity will slow a projectile's upward speed by 9.8 m/s every second.

$a = \dfrac{\Delta v}{\Delta t}$, so $\Delta t = \dfrac{\Delta v}{a} = \dfrac{-12 \, \text{m}/\text{s}}{-9.8 \, \text{m}/\text{s}^2} = \textbf{1.22 seconds}$

(b)    Initial velocity is positive, and acceleration is negative. The height above 30 m is
$\Delta x = v_0 t + 1/2 at^2 = (12 \, \text{m}/\text{s})(1.22 \text{ s}) + 1/2(-9.8 \, \text{m}/\text{s}^2)(1.22 \text{ s})^2 = 7.3 \text{ m}$
above the cliff for a total height of **37.3 m.**

(c)    At the top of its path, the ball has zero velocity.

$$\Delta x = v_0 t + \frac{1}{2} at^2, \text{ so } t = \sqrt{\left(2\Delta \frac{x}{a}\right)} = \sqrt{\left[2\left(-37.3 \text{ m}/s^2\right)\right]} = \textbf{2.8 s more}$$

(d)    The ball speeds up 9.8 m/s from rest for every second of fall.

$$v = at = (9.8 \text{ m}/s^2)(2.8 \text{ s}) = \textbf{27.4 m}/\textbf{s}$$

## AP® Tip

As long as you pick a frame of reference and continue to use it, you will find the right answer. If the motion is complicated or if you don't know its direction, choose "up" as positive and call the initial height zero.

## INVESTIGATING MOTION

Example problems often start with a phrase like "a ball is rolling 2.4 m/s across your desk" with no explanation of how you might know that. Part of being a great AP® Physics student is devising ways to determine quantities like velocity that require combining several careful measurements. Familiarity with analyzing the motion of everyday objects will help you as the course progresses into the less conveniently measured realms of astronomy and the atom.

To measure the motion of everyday objects, you first pick a frame of reference. At first, this could be as simple as using the edge of your desk for $x = 0$, and whenever you begin, the stopwatch is $t = 0$. After that, you only need two types of tools: those that tell you where the object is and those that tell you when it got there. Metersticks, digital range finders, and long measuring tapes are typical tools for measuring the distance between an object and whichever zero point you've chosen. Stopwatches measure duration, and several stopwatches can be used together as long as they begin at the same zero point in time. Clocks and calendars are useful for measuring longer time spans.

Once you have a reference frame and a set of measuring tools, you begin to collect position and time data for an object of interest. Collecting as much data as possible over several nearly identical trials will yield a collection of ordered pairs of data. Attentive physics students are quick to notice that measured data are not nearly as clean as the example problems suggest. Two people measuring the same phenomenon will have slight variations in their measurements just as two nearly identical events will never be exactly the same. Small random variations in data are expected; you can easily handle noisy data by performing more trials.

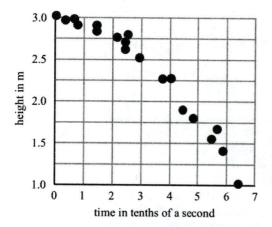

The minimum value on the y-axis is 1 m because that was the lowest data point.

The x-axis is labeled in tenths of a second, not seconds.

Random noise is typical in data taken over such a short time range.

If you have collected enough position vs. time data, plot those points on a graph with time on the x-axis and position on the y-axis. Each axis should span from your lowest measurement to your highest measurement to guarantee that all of your data are shown and that your data take up the full graph. Your axes should start at zero if that was the lowest data point you measured. For example, if your distance measurement varied from 1.5 m to 2.2 m, those should be the limits of your y-axis. Your plotted points will approximate some curve; draw the curve that passes smoothly through the clusters of data points. You are left with a graph of an object's position during some duration. The noisy data have been smoothed out, so you no longer need the plotted points.

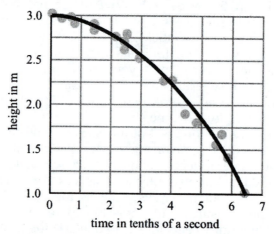

From a best-fit curve of position, you can determine velocity by taking the slope over a short time span. When calculating slopes from a graph, use the best-fit curve, not actual data points. To determine the average velocity between 0.3 s and 0.4 s in the graph above, you would use values shown in the curve:

$$v_{avg} = \text{slope} = \text{rise/run} = \frac{(2.25\,\text{m} - 2.55\,\text{m})}{(0.4\,\text{s} - 0.3\,\text{s})} = -3.0\,\text{m/s}$$

On the test's experimental design question, be clear about what measuring tools should be used and what they will measure. "Start the stopwatch when the car starts moving and stop it when the car passes the tree" is very clear. "Time the car's motion" is not. Use the simplest appropriate measuring tools; stay away from requiring highly specialized tools like the particle collider at CERN. Avoid mentioning tools you don't use in the classroom or tools that will not work for the purpose you intend, such as using a stopwatch to measure the time it takes light to cross your desk.

# PROJECTILES

In physics, we use simplified models to study the important parts of complex phenomena. Calculating the actual flight of a real object thrown near the ground is extraordinarily complex, so we use the *projectile* model to explore the essence of the motion. The projectile model treats all objects the same regardless of size, shape, and mass and ignores any effects of air resistance. In the projectile model, once the object is aloft, only the force of gravity acts on it. Projectiles may include a tossed ball, a bullet fired out of a gun, and water leaving a garden hose. The path of any projectile—its *trajectory*—is parabolic.

Projectile launches are described by their initial velocity vectors, $v_0$ at $\theta_0$. We call $y_{max}$ the *maximum altitude or maximum height* reached by the projectile and $R$ or $x_{max}$ the *maximum range* of the projectile. We call the total flight time $t_T$.

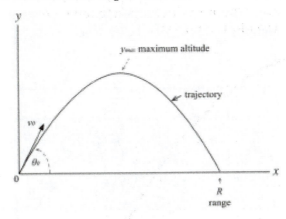

## ASSUMPTIONS OF THE PROJECTILE MODEL

The projectile continues at a constant horizontal velocity during its flight. Nothing is pulling it horizontally, so it doesn't speed up or slow down. The speed in the $x$-direction remains $v\cos\theta$.

In the $y$-direction, the projectile's velocity does change because gravity accelerates it downward at 9.8 m/s².

The velocity of the $y$-component slows down, passing from positive to negative as it changes direction, then speeds up again.

The projectile's trajectory is a symmetrical parabola. The projectile takes as much time going up as it does going down, assuming that launch and landing are at the same height.

We can treat the $x$- and $y$-components separately.

$$\cos\theta_0 = \frac{v_{0x}}{v_0} \text{ and } \sin\theta_0 = \frac{v_{0y}}{v_0}$$

$$v_{0x} = v_0 \cos\theta_0 \quad v_{0y} = v_0 \sin\theta_0$$

## SAMPLE PROBLEM 11

A projectile has an initial velocity of $v_0 = 40.0 \text{ m/s}$ at 60.0°.
Calculate initial velocity's horizontal and vertical components.

## SOLUTION TO PROBLEM 11

$$v_{0x} = v_0 \cos\theta_0 = \left(40.0 \text{ m/s}\right)\cos 60° = 20.0 \text{ m/s}$$

$$v_{0y} = v_0 \sin\theta_0 = \left(40.0 \text{ m/s}\right)\sin 60° = 34.6 \text{ m/s}$$

---

## AP® Tip

To find the instantaneous velocity of the projectile in flight at a specific time along its trajectory, we must to do <u>four</u> things:

1. Find the horizontal velocity from $v_x = v_0 \cos\theta_0$.
2. Determine the vertical velocity at that time using $v_y = v_0 \sin\theta_0 - gt$.
3. Find the speed by adding $v_x$ and $v_y$ as vectors using $v = \sqrt{v_x^2 + v_y^2}$.
4. Find the direction of the instantaneous velocity vector using $\theta = \tan^{-1}\left(\frac{v_y}{v_x}\right)$.

---

## AP® Tip

To locate the position of a projectile along its trajectory, we do <u>two</u> things:

1. Determine its distance downrange using $x = (v_0 \cos\theta_0)t$.
2. Find its altitude using $y = (v_0 \sin\theta_0)t - \frac{1}{2}gt^2$.

---

## SAMPLE PROBLEM 12

Calculate the position of the projectile from Problem 11 at $t = 2.0$ s.

## SOLUTION TO PROBLEM 12

To locate the projectile 2.0 s into flight, we do two things:

1. Find how far it has moved horizontally using

$$x = (v_0 \cos\theta_0)t = \left(40.0 \text{ m/s}\right)(\cos 60°)(2.0 \text{ s}) = 40.0 \text{ m}$$

2. Find how high off the ground it is as follows

$$y = (v_0 \sin \theta_0)t - \frac{1}{2}gt^2$$

$$= (40.0 \text{ m/s})(\sin 60°)(2.0 \text{ s}) - \frac{1}{2}(9.8 \text{ m/s}^2)(2.0 \text{ s})^2$$

$$= 49.7 \text{ m}$$

At $t = 2.0$ s into its flight, the projectile is **40.0 m** downrange and is **49.7 m** above the ground.

## SAMPLE PROBLEM 13

How high off the ground is the projectile from Problem 11 when it has traveled 20.0 m horizontally?

## SOLUTION TO PROBLEM 13

We can find out when it has traveled that far with $v_x = \frac{\Delta x}{\Delta t}$ or $\Delta t = \frac{\Delta x}{v_x}$.

$$\Delta t = \frac{\Delta x}{v} = \frac{20.0 \text{ m}}{20 \text{ m/s}} = 1 \text{ s}$$

At a time of 1 s, we can find the projectile's altitude with

$$y = v_{0y}t + \frac{1}{2}at^2 = (34.6 \text{ m/s})(1 \text{ s}) + \frac{1}{2}(-9.8 \text{ m/s}^2)(1 \text{ s})^2 = \textbf{29.7 m}$$

# MOTION: STUDENT OBJECTIVES FOR THE AP® EXAM

- ▓ You should be able to define displacement, velocity, time interval, and acceleration.
- ▓ You should be able to calculate velocity and acceleration from time intervals and displacement.
- ▓ You should be able to explain the subsequent motion of a particle that, at one instant, has zero velocity but constant acceleration.
- ▓ You should be able to discuss the subsequent motion of a particle that has negative acceleration.
- ▓ You should be able to discuss the motion of a particle that, at one instant, has negative velocity but a positive acceleration.
- ▓ You should be able to discuss the properties of a motion graph.
- ▓ You should be able to analyze the motion of a body in free fall.
- ▓ You should be able to analyze the motion of a projectile launched into two-dimensional space.

# MULTIPLE-CHOICE QUESTIONS

1. A body moves in the $x$-$y$ plane with some initial velocity $v_0$. A short time later it has a velocity $v_f$. Which situation is impossible?
   (A) Velocity and acceleration vectors are parallel.
   (B) Velocity and acceleration vectors are anti-parallel.
   (C) Velocity and acceleration vectors are perpendicular.
   (D) Velocity and acceleration are both constant (nonzero).

2.  Six graphs are shown below. Which combination could represent the motion of an object moving in one dimension with a constant nonzero acceleration?

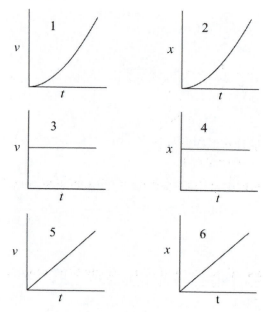

(A)  3 and 6
(B)  1 and 4
(C)  5 and 2
(D)  5 and 6

3.  A graph of position as a function of time is shown below. Rank the points A, B, C, and D for the velocity at that point from highest to lowest with negative values below positive values.

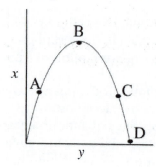

(A)  B > A = C > D
(B)  A > B > C > D
(C)  A = C > B > D
(D)  D > A = C > B

Four cars move as indicated on a $v = f(t)$ graph below. Use the graph to answer questions 4 and 5.

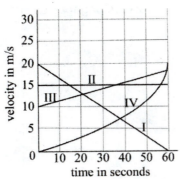

4. Which car has zero acceleration during the time interval shown?
   (A) I
   (B) II
   (C) III
   (D) IV

5. Which car has the greatest displacement during the time interval shown?
   (A) I
   (B) II
   (C) III
   (D) IV

6. A 2.0 kg ball and a 1.0 kg ball are dropped from the roof of a tall building at the same time. If air resistance is neglected, then
   (A) the 2.0 kg ball hits the ground first.
   (B) the 1.0 kg ball hits the ground first.
   (C) they strike the ground at the same time.
   (D) it will depend on the ball with greater radius.

7. Three cars travel linearly along the $x$-axis as shown in the graph below. At the end of 10.0 s, they have the same

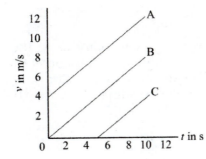

   (A) displacement.
   (B) final velocity.
   (C) acceleration.
   (D) speed.

8.  A ball is projected upward reaching a maximum height of 10.0 m. Neglecting air resistance, what is the ball's velocity when it returns to its initial height, assuming downward as negative direction.

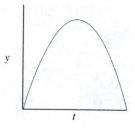

(A)  0
(B)  14.0 m/s
(C)  10.0 m/s
(D)  −14.0 m/s

9.  The motion of an object is variable over time as shown in the graph below. Rank the accelerating periods shown from highest to lowest acceleration.

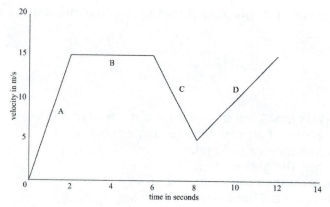

(A)  A > D > B > C
(B)  A > C > D > B
(C)  A > D > C > B
(D)  A > C > B > D

10. A body is projected upward with some velocity $v_0$. At the highest point in its path, which of the following correctly describes the sign of the body's displacement, velocity and acceleration? Ignore air resistance. Assume upward as being the positive sense.

|     | *y* | *v* | *A* |
| --- | --- | --- | --- |
| (A) | 0   | 0   | 0   |
| (B) | +   | 0   | −   |
| (C) | +   | +   | 0   |
| (D) | +   | 0   | −   |

11. A body is projected horizontally from a table that is 1.0 m above the floor with some initial velocity $v_0$. Neglecting air resistance, its acceleration after leaving the table
    (A) is tangent to the parabolic path of the body.
    (B) is directed vertically downward.
    (C) is directed horizontally.
    (D) has a horizontal and vertical component.

12. Two blocks, $m_A = 1.0$ kg and $m_B = 2.0$ kg, moving parallel to each other slide on the surface of a frictionless table whose height is 1.0 m as illustrated in the diagram below. When the blocks land on the floor, which of the following best describes their time of flight and horizontal range?

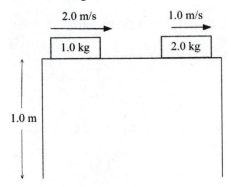

|     |              |     |              |
| --- | ------------ | --- | ------------ |
| (A) | $t_A = t_B$  | and | $x_A = x_B$  |
| (B) | $t_A > t_B$  | and | $x_A > x_B$  |
| (C) | $t_A = t_B$  | and | $x_A = x_B$  |
| (D) | $t_A > t_B$  | and | $x_A > x_B$  |

**Questions 13 to 15**
**Directions:** For each of the questions or incomplete statements below, two of the suggested answers will be correct. For each of these questions, you must select both correct choices to earn credit. No partial credit will be earned if only one correct choice is selected. Select the two that are best in each case.

13. A particle moves in one dimension as shown on the graph below. The graph indicates that the particle (select two answers)

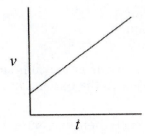

    (A) has a constant (nonzero) positive acceleration.
    (B) has a constant positive velocity.
    (C) started with some nonzero initial velocity.
    (D) started with some initial displacement from the origin (0,0).

14. You want to increase the distance that an object of mass $M$ will travel horizontally when projected from a tall building with some initial horizontal velocity $v_0$. You can do this by (Select two answers.)
    (A) increasing the height from which it is released.
    (B) decreasing the height from which it is released.
    (C) decreasing the mass of the object.
    (D) increasing the horizontal velocity.

15. A student wants to design an experiment to investigate the motion of an object. The experimental equipment the student will need for this investigation includes the following (Select two answers.):
    (A) a meterstick and a platform balance
    (B) an air track with glider, timer, and a protractor
    (C) a photogate and a meterstick
    (D) a constant velocity cart, meterstick, and timer

## FREE-RESPONSE PROBLEMS

1. A student evaluating the motion of a cart over 1.0-second intervals obtains the following data:

| $V_{avg}$ | $t$ |
|-----------|-----|
| cm/s | s |
| 1.3 | 1.0 |
| 2.5 | 2.0 |
| 3.7 | 3.0 |
| 5.1 | 4.0 |
| 6.3 | 5.0 |
| 7.6 | 6.0 |

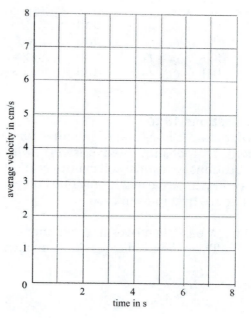

   (a) Plot the data on the graph shown on the right.
   (b) What does the shape of the best-fit curve suggest?
   (c) Using the graph, how can you determine the acceleration of the body? Explain your reasoning without calculations but in enough detail that another student could duplicate your methods and obtain the same results.

2. Two students watching a softball game are arguing about the last hit, which traveled from the batter to the outfield in a perfect parabola.

   Student A claims, "Projectiles reach their maximum height when the velocity and acceleration are zero. After they run out of acceleration, they can't get any higher."

   Student B claims, "Projectiles don't ever run out of acceleration. They reach their maximum height when the direction of the acceleration switches from up to down."

(a) What information about each student's reasoning is correct?

(b) What information about each student's reasoning is incorrect?

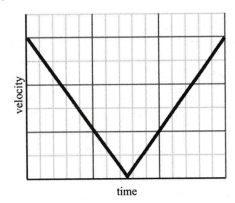

time

3. While examining free fall in the lab, the students toss a ball straight up and watch it fall back to the same position. They measure the velocity using a motion detector and use the data to sketch this graph of velocity vs. time.

(a) What, if anything, is wrong with the students' graph? Explain your answer. If something is incorrect on the graph, sketch what the proper graph would look like in your explanation.

(b) Sketch an accurate acceleration vs. time graph for the ball from the time it left the student's hand until it returned. Include numbers and units in your axes labels.

(c) Compare the initial velocity of the ball to the final velocity of the ball when it has returned to its original position. Explain your reasoning.

4. (a) A projectile follows the path shown below. Clearly indicate the vectors, drawn to scale, representing $v_x$, $v_y$, and $a$ for each of the points shown on the graph.

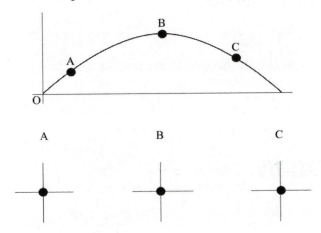

(b) For each of the following, at what angle would you launch the projectile to obtain the

i. greatest range?

ii. greatest height?

iii. longest time of flight?

5. A car initially at rest accelerates at the rate of 4.0 m/s² for 5.0 seconds. For the next 6.0 seconds, it travels at constant speed and then decelerates at 5.0 m/s² until it has a velocity of –10.0 m/s.
   (a) On the graphs provided, plot
      i.   velocity vs. time.
      ii.  acceleration vs. time.
   (b) Use your velocity vs. time graph to calculate the total displacement.

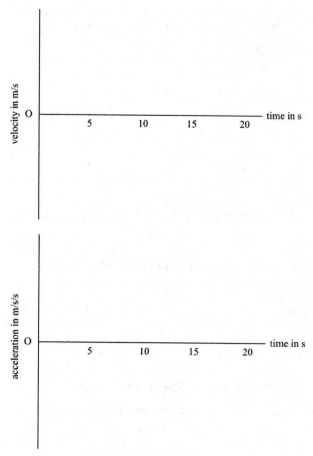

   (c) When the velocity of the car is zero, is the displacement of the car zero? Justify your answer.

# Answers

## Multiple-Choice Questions

1. **D** Average acceleration is defined as $\vec{a} = \dfrac{\Delta \vec{v}}{\Delta t}$. The speed, the direction, or both can change. Both cannot remain constant.

   (L.O. 3.A.1.1)

2.  **C** A constant displacement (graph 4) indicates zero values for both velocity and acceleration. To have a constant nonzero acceleration, velocity vs. time must be linear (graph 5). The displacement graph must be parabolic, which is graph 2.

    (L.O. 3.A.1.1)

3.  **B** The graph represents a body that moved to its highest position in a gravitational field or a body that was initially traveling along the +x-axis, then stopped and reversed its direction. Since the question asks to rank for velocity, which is a vector having both a magnitude and a direction, points C and D are negative and therefore smaller than the zero value at point B.

    (L.O. 3.A.1.1)

4.  **B** The slope of a velocity vs. time graph is the acceleration of the object. The slope for car B is zero.

    (L.O. 3.A.1.1)

5.  **B** The area under a *velocity vs. time* graph is the displacement of the car. The displacement of car I is $\frac{1}{2}bh$, which is

    $\frac{1}{2}\left(20 \, \text{m/s}\right)(60 \, \text{s}) = 600$ m. For car II, the area is

    $bh = \left(15 \, \text{m/s}\right)(60 \, \text{s}) = 900$ m. The displacement for car III is

    $\left(10 \, \text{m/s}\right)(60 \, \text{s}) = \frac{1}{2}\left(17.5 \, \text{m/s} - 10 \, \text{m/s}\right)(60 \, \text{s}) = 830$ m. For car IV, counting the blocks gives a good value for the displacement. Car IV's displacement is about 450 m.

    (L.O. 3.A.1.1)

6.  **C** Both balls are dropped from rest at the same time. Their acceleration is the gravitational acceleration, $g = -9.80 \, \text{m/s}^2$.

    (L.O. 3.A.1.1)

7.  **C** The cars have different initial velocities and move for different times, but they are parallel to each other and have the same slope. Therefore, they have the same acceleration.

    (L.O. 3.A.1.1)

8.  **D** The initial velocity of the ball is determined from substitution into $v_f^2 = v_o^2 + 2ax$. $0 = v_0^2 + 2(-9.80 \, \text{m/s}^2)(10.0 \, \text{m})$. The initial velocity is 14.0 m/s. Since the gravitational acceleration is constant, the ball

returns to the origin with the same speed, but since it is moving downward, the velocity is –14.0 m/s.

(L.O. 3.A.1.1)

9. **A** The slope of each section of the line is the acceleration for that time period.

   Both **A** and **D** show a positive acceleration. **B** shows a constant velocity and hence no acceleration, while **C** indicates a negative acceleration.

   (L.O. 3.A.1.1)

10. **B** Since air resistance is neglected, the body is under a constant gravitational acceleration, $-g$ for its entire path. At the highest point, $y_{max}$, its velocity is zero; the body will change direction, returning to the ground.

    (L.O. 3.A.1.1)

11. **B** The body is moving in two dimensions under the action of a gravitational acceleration $a = -g$, which will increase the $v_y$ component of its velocity. Provided air resistance is negligible, the horizontal velocity will not change.

    (L.O. 3.A.1.1)

12. **C** Both blocks fall the same vertical distance, and in the absence of air resistance, the times to fall 1.0 m are the same. Block A has the larger horizontal velocity as the blocks leave the table, and knowing that $x = v_x t$, block A will travel farther horizontally than block B.

    (L.O. 3.A.1.1)

13. **A** and **C** The graph of velocity vs. time shows that the graph is linear. Therefore, the slope of the line, its acceleration, is positive and constant. The $y$-intercept of the graph indicates that there is some initial velocity $v_0$.

    (L.O. 3.A.1.1)

14. **A** and **D** The range that an object projected into space will travel horizontally is determined by $x = v_x t$. Increasing the time it takes to reach the ground or increasing the horizontal velocity will increase the horizontal distance traveled.

    (L.O. 3.A.1.1)

15. **B** and **D** Since you can elevate the air track and measure the elevation, the change in the velocity can be determined by $\Delta v = v_f - v_0$. The glider starts from rest, and using the timer and the built-in

distance markings on the air track, you can determine the average velocity, $v_{avg} = \dfrac{v_f - v_0}{2}$, and thus the final velocity $v_f$.

A constant velocity cart moves at a set speed. Using the timer and a meterstick, you can determine the constant or average velocity, $v_{avg} = \dfrac{d}{t}$.

(L.O. 3.A.1.2)

# FREE-RESPONSE PROBLEMS

1. To the right is the graph with the best-fit curve:
   (a) The graph is $v_{avg}$ vs. time. Since the body is uniformly accelerated, the data for the average velocity is the approximate value for the midpoint of the time period. Points should be plotted correctly at the 0.5 s, 1.5 s, 2.5 s, etc., times.
   (b) Since the graph is linear, the acceleration is constant.
   (c) The acceleration is the change in the velocity divided by the change in time. Since the body starts from rest, the acceleration is the final velocity divided by the time. Data were obtained for the average velocity that is the sum of the initial velocity and the final velocity divided by 2. To obtain the correct value for the acceleration, multiply the slope by 2.

   (L.O. 3.A.1.1, 3.A.1.3)

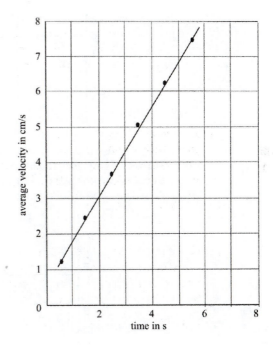

2. (a) Student A is almost correct about the velocity at the top of the path. Vertical velocity is zero at the top of a projectile's path. Student B is right that acceleration never runs out.
   (b) Student A is incorrect about acceleration running out. The acceleration from gravity is always 9.8 m/s² downward. It never runs out. Student A is also incorrect about velocity being zero since the softball has horizontal velocity. Student B is incorrect about the direction of the acceleration vector. Acceleration from gravity points downward whether the softball is rising or falling.

   (L.O. 3.A.1.1)

3. (a) The graph is drawn incorrectly. The velocity of a ball tossed straight up begins in the positive direction, reaches zero velocity at the top, and then reverses its direction back to the original position. The graph shown describes the speed of the ball, which does not account for the direction required by velocity. On a velocity graph, positive values show upward speeds and negative values show downward speeds. The line of the graph should maintain a constant slope to represent constant acceleration.

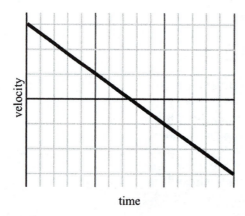

time

(b) Acceleration of an object in free fall is always 9.8 m/s² directed downward. The acceleration vs. time graph of an object in free fall is always a flat line.

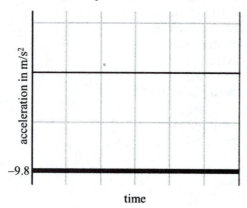

time

(c) The velocity of the ball at the beginning of the toss should have a magnitude that is equal to the ball at the end of the toss. The end velocity should have the opposite direction since it is now moving in a downward (negative) direction.

(L.O. 3.A.1.1, 3.A.1.3)

4. (a) The vectors applied to points A, B, and C are shown below.

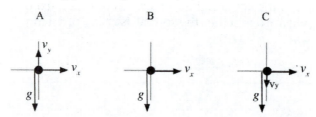

(b)   45°. The greatest range is determined using $R = \dfrac{(V_0)^2 \sin 2\theta}{g}$.

The sin of twice 45° is the sin of 90°, or 1.

i.   The angle that gives the greatest height is the angle that has no horizontal component: $\theta = 90°$.

ii.  The angle that has the longest time of flight is also 90°. Again, there is no horizontal component.

(L.O. 3.A.1.1)

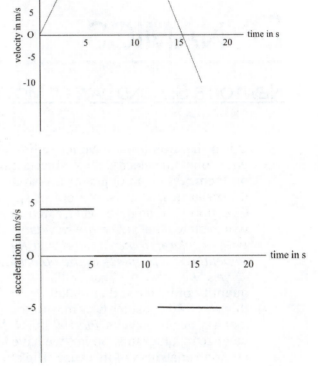

5.  (a)  To plot the graphs, the last time period is needed. Solving $v_f = v_0 + at$, $-10.0\ \mathrm{m/s} = 20.0\ \mathrm{m/s} - 5.0\ \mathrm{m/s^2}(t)$.

The time for the deceleration is 6.0 s.

(b)  We can find the area bounded by each of the three parts of this motion and add them to find the total displacement. Because the third part of the motion involved a direction change as positive velocity slowed down and became negative velocity, we will count the small triangle as negative displacement.

Area 1 is a triangle with a base of 5 s and a height of 20 m/s. $A_1 =$ **50 m**.

Area 2 is a rectangle with a base of 6 s and a height of 20 m/s. $A_2 =$ **120 m**.

Area 3 has two triangles, one with a base of 4 s and a height of 20 m/s and another with a base of 2 s and a height of –10 m/s. $A_3 = 40\ \mathrm{m} - 10\ \mathrm{m} =$ **30 m**.

The sum of all of these displacements is 50 m + 120 m + 30 m = **200 m**.

(c)  When the velocity is zero, the car begins to reverse direction, but it is 210 m from its origin. 50 m + 120 m + 40 m = 210 m.

(L.O. 3.A.1.1)

# 4

# DYNAMICS

## NEWTON'S SECOND LAW OF MOTION

Up to this point, we have looked at objects either at rest or traveling with constant velocity. Such objects have a resultant force of zero acting on them. When an object or a system experiences a nonzero net force, it accelerates. That is, the object speeds up, slows down, or changes direction. We analyze accelerated motion with Newton's second law, which states that *when the resultant force is not zero, the object moves with accelerated motion and that the acceleration, with a given force, depends on a property of the object known as its mass.*

As we saw in Chapter 2, an object's mass is a measure of the quantity of material that makes up the object. When we describe mass this way and consider how mass affects acceleration, we are specifically looking at an object's *inertial mass*. Mass is a scalar measured on a balance in kilograms. So far, we have looked at *kinematics*, the study of motion regardless of its cause. The study of motion and the forces that make motion occur is called *dynamics*. Equilibrium is a special case in dynamics when the acceleration is zero. In its broadest sense, dynamics comprises nearly the whole of mechanics.

We know from experience that an object at rest cannot start to move by itself; some other object must exert an external push or a pull on it. We also know that a net force is required to slow or to stop moving an object and that a sidewise or lateral force must be applied to an object to curve it out of a straight line. Speeding up, slowing down, and changing direction are changes in the magnitude or the direction of an object's velocity. Every time an object is accelerated, we know that an external force must be acting on it to produce the acceleration. We also know that the acceleration is always in the same direction of the net force that caused it.

Newton's second law takes the equation form $\Sigma F = ma$. It is a simple equation stating that a resultant force always makes a "ma," a mass accelerate.

**116**

## SAMPLE PROBLEM 1

A 12.0 kg mass is accelerating at 4.0 m/s². What resultant force causes the acceleration?

## SOLUTION TO PROBLEM 1

Systems accelerate because there is a resultant or net or unbalanced force causing the acceleration. By Newton's second law,

$$\Sigma F = ma = (12.0 \text{ kg})\left(4.0 \text{ }^m\!/_{s^2}\right) = \textbf{48.0 N}$$

## SAMPLE PROBLEM 2

Students attach an inflated balloon to a motion cart and let go. The deflating balloon applies a 2.0 N force to the cart and accelerates it from rest across the 10.0 m classroom. A motion sensor records a final speed of 2.4 m/s. Calculate $M$, the mass of the cart/balloon system.

## SOLUTION TO PROBLEM 2

The unbalanced force acting on $M$ is $\Sigma F = Ma$, which we can write as $F = Ma$ with the understanding that $F$ also represents the unbalanced force. Since we do not know the time interval involved during the acceleration process, we make use of the time-independent equation $v^2 = v_0^2 + 2a\Delta x$. Since the object starts from rest, the acceleration is $a = \dfrac{v^2}{2\Delta x}$. Substituting into $F = Ma$ and solving for $M$ yields $M = \dfrac{2F\Delta x}{v^2}$,

and then $M = \dfrac{2(2.0 \text{ N})(10.0 \text{ m})}{\left(2.4 \text{ }^m\!/_s\right)^2} = 6.9 \text{ kg}.$

## THE NORMAL FORCE AND THE FRICTION FORCE

Remember that in many cases, the normal force is equal to the weight of the object pressing downward on the surface; however, as we will see, the normal force can be greater or less than the weight of the object. Sometimes the normal force may have no relation to the weight at all. The key is to ensure that the normal force acts perpendicular to the surface on which the object is resting. Make sure you are paying attention to every new situation so that you know where all of the forces are acting.

Also recall that the friction force is directly proportional to the normal force between two surfaces and how much drag exists between those surfaces, a property called the coefficient of friction, $\mu$, that is unique for different pairs of surfaces. Friction works to oppose the motion of the object or the applied force and can be calculated with $f = \mu N$.

## SAMPLE PROBLEM 3

An airline passenger walks 0.9 m/s through an airport and then stands on a moving walkway. Within 0.75 s, the passenger speeds up to 2.2 m/s. What is the coefficient of friction between the commuter's feet and the moving walkway?

## SOLUTION TO PROBLEM 3

Since the commuter does not accelerate up or down, we know that he or she is in vertical equilibrium. That means that the normal force is equal to his or her weight. Because the friction force is the only horizontal force in this situation, it must be the force responsible for the acceleration. In equation form, we would say

$$\Sigma F = ma$$

$$f = ma$$

since $f = \mu N$, $N = mg$, and $a = \Delta v / \Delta t$

$$\mu mg = \frac{m \Delta v}{\Delta t}$$

$$\mu = \frac{\Delta v}{g \Delta t}$$

$$\mu = \frac{1.3 \text{ m/s}}{9.8 \text{ m/s}^2 \times 0.75\text{s}} = 0.18$$

## SAMPLE PROBLEM 4

In Problem 3, the passenger is clearly moving. Is the friction force static or kinetic?

## SOLUTION TO SAMPLE PROBLEM 4

Although the passenger is in motion, there is no motion between his or her feet and the moving walkway. That means that the friction is static. Kinetic friction occurs only between surfaces that are sliding over each other.

## SAMPLE PROBLEM 5

An 8.0 kg wooden block is pushed to the right across a horizontal surface with a force of 10.0 N. The frictional force between the block and the surface is 1.2 N.
(a)   What acceleration does the block experience?
(b)   Find the coefficient of kinetic friction for the surfaces involved.

## SOLUTION TO PROBLEM 5

First, we make a free-body diagram.

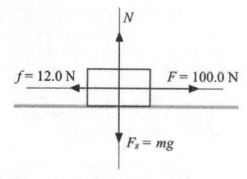

(a)   Friction always opposes the motion and is negative. From our free-body diagram, the resultant force, $\Sigma F$, is $\Sigma F = F - f$. We set this resultant force equal to $ma$ with $F - f = ma$ or $ma = F - f$. Solving for $a$ gives

$$a = \frac{F - f}{m} = \frac{10.0 \text{ N} - 1.2 \text{ N}}{8.0 \text{ kg}} = 1.1 \text{ m/s}^2$$

(b) To find the coefficient of friction, we need the normal force. Because the block isn't accelerating up or down, we know the block is in equilibrium. That means that the sum of all of the upward forces equals the sum of all of the downward forces, or

$$N = F_g = mg = (8.0 \text{ kg})(9.8 \, ^m/_{s^2}) = 78.4 \text{ N}$$

By definition, the frictional force is $f = \mu N$, and solving for the coefficient gives

$$\mu = \frac{f}{N} = \frac{1.2 \text{ N}}{78.4 \text{ N}} = 0.015.$$

Recall that coefficients of friction are dimensionless, so they have no units.

## SAMPLE PROBLEM 6

A sports vehicle travels at 50.0 mph on a horizontal, dry section of asphalt highway. If the coefficient of friction between the tires and the surface of the highway is 0.88, what is the minimum stopping distance of the car when the brakes are fully applied? Ignore air resistance.

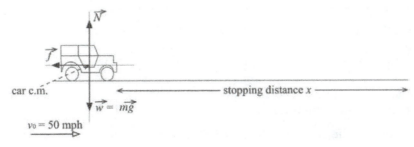

## SOLUTION TO PROBLEM 6

First, note that the initial speed of the vehicle is expressed in English units: 50.0 mph. Convert to SI as follows:

$$50.0 \frac{\text{mi}}{\text{h}} \times \frac{88 \, ^{\text{ft}}/_s}{60 \, ^{\text{mi}}/_h} \times \frac{1 \text{ m}}{3.28 \text{ ft}} = 22.4 \, ^m/_s$$

The vehicle is in equilibrium vertically, and $N = w = mg$. Horizontally, the only force acting is the force of friction, $f = \mu N = \mu(mg) = -\mu mg$. The frictional force is an unbalanced force. Applying Newton's second law: $f = ma = -\mu mg$, and mass divides out on both sides leaving the acceleration as $a = -\mu g$. Acceleration is negative since the vehicle is slowing and the brakes act in the opposite direction to the motion.

The time to stop is not given in the problem. We know the initial speed and the final speed, and we can find the acceleration. So we use the time-independent equation, $v^2 = v_0^2 + 2a\Delta x$.

Since $v = 0$,

$$0 = v_0^2 + 2a\Delta x$$

$$a = F/m = -\mu g; \text{ so}$$

$$0 = v_0^2 + 2(-\mu g)\Delta x$$

$$\text{and } x = \frac{v_0^2}{2\mu g} = \frac{\left(22.4 \, ^m/_s\right)^2}{2(0.88)\left(9.8 \, ^m/_{s^2}\right)} = 29.1 \text{ m}$$

Stopping distance is proportional to the square of the initial speed. The faster a vehicle travels, the greater the distance the vehicle requires to stop. Note that we did not need the mass of the vehicle to find the stopping distance.

## SAMPLE PROBLEM 7

A 35g toy car is initially at rest near the top of an inclined ramp that is about 1.5 m long; the ramp has marks every centimeter. The car can be released, and it is free to roll smoothly down the ramp.

(a) Design an experiment that uses only a stopwatch to determine the acceleration of the car and the net force acting on the car as it accelerates down the ramp.

(b) What quantities would you measure, and how would you measure them?

(c) Describe a procedure to determine the car's acceleration. Give enough detail so that another student could replicate the experiment.

(d) Using equations, show how your measurements could be used to determine the car's acceleration and the net force acting on it.

## SOLUTION TO PROBLEM 7

To find the car's acceleration, we need to know how far the car moves and how much time the motion takes. Since the car is a physical object and not a dimensionless point, we need to pick one spot on the car as a reference for our measurement. A good answer must describe how to measure the distance a specific part of the car travels and how to use the stopwatch to measure the time it takes:

"Use the markings on the inclined ramp to measure the distance from the front of the car to the low end of the ramp. Start the stopwatch when the car is released and stop it when the front of the car reaches the low end of the ramp."

Since we know the time and distance, we can describe the motion using

$\Delta x = v_0 t + \frac{1}{2}at^2$.

The initial velocity is zero, so

$\Delta x = \frac{1}{2}at^2$ or $a = 2\Delta x/t^2$.

We can find the net force using $\Sigma F = ma$.

## ELEVATORS

An elevator consists of a motor mechanism for lifting or lowering, a support cable, and an elevator car. The attachment point for the support cable acts as a knot when vector or free-body diagrams are made.

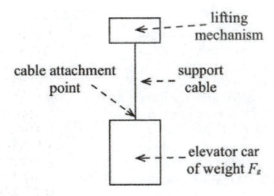

Elevator cars experience tension from the support cable. If the tension is greater than the car's weight, the elevator accelerates upward. If the tension is less than the car's weight, the elevator accelerates downward. If the elevator is traveling with constant speed, including being at rest with zero speed, we know the tension force is equal to the weight of the car. An elevator car is confined to vertical motion. Make sure you are paying attention to the signs of the acceleration and velocity. Remember, negative acceleration does not always mean that the object is slowing down!

### SAMPLE PROBLEM 8

A 2000 kg elevator is accelerated upward at 1.2 m/s². What tension exists in the support cable?

### SOLUTION TO PROBLEM 8

First, make a free-body diagram. Since the car is accelerating upward, we know the tension is greater than the weight. We draw the tension vector longer than the weight vector and draw the $a$ vector pointing upward. The symbol sigma $\Sigma$ means to sum, but since $F_g$ is directed downward, we treat it as negative and subtract it. $\Sigma F = T - F_g = ma$. We solve for the tension as follows:

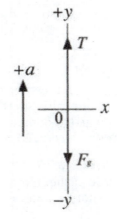

$$T = ma + mg$$
$$= m(a + g) = (2000 \text{ kg})\left(1.2 \text{ m}/{s^2} + 9.8 \text{ m}/{s^2}\right)$$
$$= 2.2 \times 10^4 \text{ N}$$

This answer is reasonable since accelerating upward requires the tension vector to be greater than the weight, $T > F_g$.

### SAMPLE PROBLEM 9

A 2000 kg elevator is being accelerated downward at 1.2m/s². What tension exists in the support cable?

## SOLUTION TO PROBLEM 9

First, make a free-body diagram. Since the car is accelerating downward, we know the tension is less than the weight. Tension is still drawn pointing upward because the support cable is pulling upward. We draw the tension vector shorter than the weight vector and show the *a* vector pointing downward. Next, we write $\Sigma F = T - F_g = m\ (-a)$. $F_g$ is directed along the $-y$-axis, making it negative. The acceleration vector is also directed downward, making it negative in this case. We solve for the tension as follows:

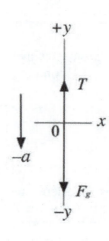

$$T = mg - mg$$
$$= m(g - a)$$
$$= (2000 \text{ kg})\left(9.8 \text{ m}/\text{s}^2 - 1.2 \text{ m}/\text{s}^2\right)$$
$$= 1.7 \times 10^4 \text{ N}$$

This answer is reasonable since accelerating downward requires the tension to be less than the weight, $T < F_g$.

## SAMPLE PROBLEM 10

What is the maximum downward acceleration an elevator car can have?

## SOLUTION TO PROBLEM 10

The maximum acceleration is the acceleration due to gravity, *g*. This can happen only if the support cable is detached from the elevator car, placing it in free fall.

## SAMPLE PROBLEM 11

What is the tension in the support cable if the 2000 kg elevator car is
(a) at rest in the elevator shaft?
(b) moving upward at constant speed?
(c) moving downward at constant speed?

## SOLUTION TO PROBLEM 11

In all three cases, the elevator car is NOT accelerating. It is *in equilibrium*. In such a state, all of the upward forces equal all of the downward forces.

$$T = F_g = mg = (2000.0 \text{ kg})\left(9.8 \text{ m}/\text{s}^2\right) = 1.96 \times 10^4 \text{ N}$$

## AP® Tip

When encountering a problem dealing with forces, ask this question: Is the system accelerating? If the system accelerates, Newton's second law is involved. If there is no acceleration, the system is in equilibrium.

## TWO–OBJECT SYSTEMS

When investigating the motion of more complicated systems composed of multiple objects, we can apply Newton's second law to the system as a whole or to any individual object in the system. As with a single-body problem, we determine the net force that is being applied to the system and use the total mass of the entire system.

### SAMPLE PROBLEM 12

Two wooden blocks, A and B, with masses of $m_A = 4.0$ kg and $m_B = 6.0$ kg, respectively, are in contact on a frictionless surface. If a horizontal force of $F = 6.0$ N pushes them to the right, what force does block A exert on block B?

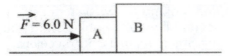

### SOLUTION TO PROBLEM 12

First, we look at blocks A and B as a single 10 kg system accelerated by a 6.0 N unbalanced force.

$$a = \frac{F}{M} = \frac{(6.0 \text{ N})}{(10.0 \text{ kg})} = 0.6 \text{ m}/_{s^2}$$

Keep in mind that the entire system accelerates at this rate.

Now we can look just at block B, which is pushed by block A with a force $F_A$.

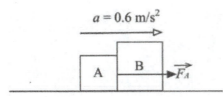

Use Newton's second law to describe what $F_A$ does to $m_B$.

$$F_A = m_B a = (6.0 \text{ kg})\left(0.6 \text{ m}/_{s^2}\right) = 3.6 \text{ N}$$

Note that since $F_A$ is perpendicular to the surface of block B, the force that block A exerts on block B is a normal force. The perpendicular normal force $F_A$ is the same magnitude as the force $F_B$ that block B exerts on block A.

*Atwood's machine* is a physics laboratory device that is used experimentally to study several areas of physics. The "ideal" Atwood's machine consists of a frictionless, massless pulley; a massless cord that does not stretch; and hanging masses that experience no air resistance. With those complications removed, the ideal Atwood's machine allows us to investigate Newton's second law.

### SAMPLE PROBLEM 13

Here is a drawing of an Atwood's machine. The masses are held at rest and then are released. Mass $m_1$ falls and $m_2$ rises. Calculate the acceleration experienced by the system and the tension in the cord.

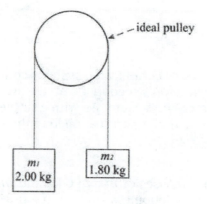

## SOLUTION TO PROBLEM 13

There are two approaches to solving problems like this: looking at individual objects in the system and looking at the system as a whole. Initially, we will apply Newton's second law to the objects individually. First, make a free-body diagram. Since $m_1$ falls, its acceleration is negative. Mass $m_2$ rises, making its acceleration positive.

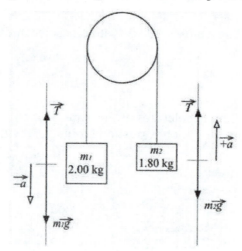

Applying Newton's second law to $m_1$, $\Sigma F = T - w_1 = T - m_1 g = m_1 (-a)$. Solving for $T$ in terms of $a$, $T = m_1 g - m_1 a$.

Next, we apply Newton's second law to $m_2$, $\Sigma F = T - w_2 = T - m_2 g = m_2 a$. Again, solving for $T$ in terms of $a$, $T = m_2 a + m_2 g$. Since $T = T$, we can write $m_2 a + m_2 g = m_1 g - m_1 a$. Transposing yields $m_1 a + m_2 a = m_1 g - m_2 g$. Factoring: $(m_1 + m_2)a = (m_1 - m_2)g$ and

$$a = \frac{(m_1 - m_2)}{(m_1 + m_2)} g$$

$$= \frac{(2.00 \text{ kg} - 1.80 \text{ kg})}{(2.00 \text{ kg} + 1.80 \text{ kg})} \left( 9.8 \ \frac{\text{m}}{\text{s}^2} \right)$$

$$= \left( \frac{0.20 \text{ kg}}{3.80 \text{ kg}} \right) \left( 9.8 \ \frac{\text{m}}{\text{s}^2} \right)$$

$$= 0.52 \ \text{m}/_{\text{s}^2}$$

From above,

$$T = m_2 a + m_2 g$$

$$= m_2(a + g) = (1.80 \text{ kg})\left(0.52 \text{ m/s}^2 + 9.8 \text{ m/s}^2\right)$$

$$= 18.6 \text{ N}$$

## ALTERNATIVE SOLUTION TO PROBLEM 13

This time we will look at the whole system of the Atwood's machine. In the case of a pulley in which one side rises and the other side falls, it is convenient to pick a rotation direction as positive. In other words, instead of using up and down as positive and negative, respectively, we can use clockwise and counterclockwise, respectively. Since this pulley will be rotating counterclockwise, we can pick that as positive acceleration. Since $m_1$ pulls the pulley counterclockwise, its weight is positive; since $m_2$ pulls the pulley clockwise, its weight is negative.

Newton's first law tells us that only external forces can accelerate a system; in this case, the "external" forces are the weight of $m_1$ and the weight of $m_2$. The tension is an internal force that does not affect the motion of the system. First, we apply Newton's second law to the whole system.

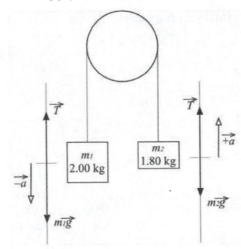

Change both accelerations to positive and remove the tensions.

$$\sum F = ma$$

$$m_1 g - m_2 g = (m_1 + m_2)a$$

$$a = \frac{(m_1 g - m_2 g)}{(m_1 + m_2)}$$

$$a = \frac{(19.6 \text{ N} - 17.6 \text{ N})}{3.8 \text{ kg}}$$

$$a = 0.52 \text{ m/s}^2$$

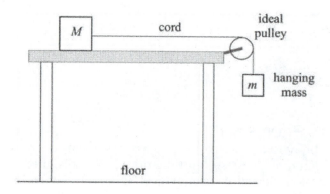

### SAMPLE PROBLEM 14

In the system shown above, block *m* is attached by a massless cord over an ideal pulley to block *M* on a frictionless table. Block *m* has a mass of 12.0 kg, and block *M* has a mass of 5.0 kg. The system is initially held in place. What acceleration will the masses experience when they are released from rest, and what tension will exist in the cord connecting them?

### SOLUTION TO PROBLEM 14

First, make a free-body diagram for both masses.

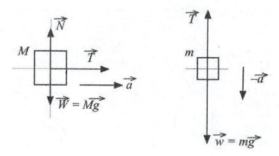

To treat this as a system, we must determine which forces are unbalanced and external. The weight of block *M* is balanced by the normal force from the table, and the tension force is an internal force. The only external unbalanced force is the weight of block *m*.

To find the acceleration, we can apply Newton's second law for the system.

$$\Sigma F = ma$$

$$mg = (M + m)a$$

Solving for the acceleration yields

$$a = \frac{m}{M+m}g = \frac{12.0 \text{ kg}}{5.0 \text{ kg} + 12.0 \text{ kg}}\left(9.8 \text{ m}/_{\text{s}^2}\right) = 6.9 \text{ m}/_{\text{s}^2}$$

Now that we know the acceleration of the system, we can use Newton's second law to look at individual objects in the system. For block *M*, the tension is the only unbalanced force. The tension is internal when we are considering the system, yet it is external when we are looking only at block *M*.

$$T = Ma = (5.0 \text{ kg})\left(6.9 \text{ m}/_{\text{s}^2}\right) = 34.6 \text{ N}$$

## SAMPLE PROBLEM 15

An inclined plane is angled at 25°, and an ideal pulley is attached to the upper end. A block of mass $M = 12.0$ kg is held at rest on the surface of the plane and is attached to a massless cord that runs over the pulley to a mass $m$. The coefficient of friction for the contact surfaces on the inclined plane is $\mu = 0.40$. When released, mass $M$ accelerates up the plane at 1.2 m/s². What is mass $m$?

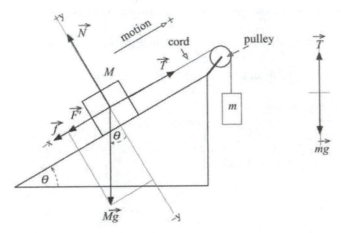

## SOLUTION TO PROBLEM 15

We first apply Newton's second law to the whole system. We begin by determining which forces are external and unbalanced. The tension force is internal, so it doesn't need to be included when we are considering the whole system. The normal force exerted where block $M$ contacts the plane is balanced by the $Mg \cos\theta$ component of block M's weight. This also means that the magnitude of the normal force must be $Mg \cos q$. The only external unbalanced forces then are the $Mg \sin\theta$ component of block M's weight directed down the ramp, the friction force $\mu F_n = \mu Mg \cos\theta$ directed down the ramp, and the weight of block $m$. Since the pulley is going to rotate clockwise, we treat forces that produce clockwise rotations as positive.

We apply Newton's second law to the whole system as follows:

$$\Sigma F = ma$$

$$\Sigma F = mg - Mg \sin\theta - \mu Mg \cos\theta = (M + m)a$$

At this point, there are several ways to solve for $m$.

$$mg - ma = Mg \sin\theta + \mu Mg \cos\theta + Ma$$

$$m = \frac{Mg \sin\theta + \mu Mg \cos\theta + Ma}{g - a}$$

$$m = \frac{12 \text{ kg} \times 9.8 \,\text{m}/\text{s}^2 \times \sin 25° + 0.4 \times 12 \text{ kg} \times 9.8 \,\text{m}/\text{s}^2 \times \cos 25° + 12 \text{ kg} \times 1.2 \,\text{m}/\text{s}^2}{9.8 \,\text{m}/\text{s}^2 - 1.2 \,\text{m}/\text{s}^2}$$

$$m = 12.4 \text{ kg}$$

## DYNAMICS: STUDENT OBJECTIVES FOR THE AP® EXAM

- You should be able to differentiate between systems in equilibrium and systems not in equilibrium.
- You should be able to explain how a particle can move if no net force is acting on it.
- You should be able to state Newton's second law of motion.
- You should be able to make and discuss free-body diagrams and to use them in problem solutions.
- You should be able to explain the reason for the direction of the force of friction.
- You should be able to define what is meant by an ideal pulley.

## MULTIPLE-CHOICE QUESTIONS

1. A 5.00 kg object is supported by a single cord that hangs from the ceiling of a room. The tension in the cord is
   (A) 5.00 kg.
   (B) 5.00 N.
   (C) 49.0 kg.
   (D) 49.0 N.

2. A 2.00 kg object is pulled along a rough horizontal surface, $m_k$ = 0.20, by a 10.0 N force acting at an angle of 40° with the horizontal. Which of the following diagrams correctly illustrates the forces acting on the object?

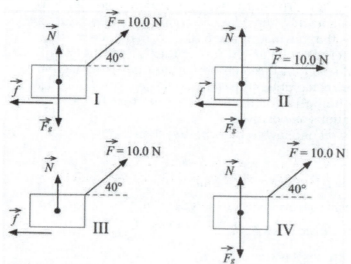

   (A) I
   (B) II
   (C) III
   (D) IV

3. Using the information in Problem 2, the acceleration of the 2.00 kg object is closest to
   (A) 1.85 m/s².
   (B) 2.50 m/s².
   (C) 2.75 m/s².
   (D) 3.80 m/s².

4. A person stands on a bathroom scale in an elevator moving downward with an acceleration of –2 m/s². The bathroom scale will read
   (A) the true weight of the person.
   (B) a weight larger than the true weight of the person.
   (C) a weight smaller than the true weight of the person.
   (D) a fractional change in the true weight, but since the true weight is not given, there is no way to determine it.

5. A car is driving on a level highway at constant speed. Which of the following forces propels the car forward?
   (A) the normal force exerted by the road on the car
   (B) the normal force exerted by the car on the road
   (C) the force of friction exerted by the road on the car
   (D) the force of friction exerted by the car on the road

6. The acceleration due to gravity on a certain planet is approximately one-fourth that on the surface of Earth. What force is required to accelerate a 25.0 kg mass on the surface of the planet?
   (A) one-fourth of the force required on the surface of Earth
   (B) the same as the force required on the surface of Earth
   (C) three-fourths of the force required on the surface of Earth
   (D) four times the force required on the surface of Earth

7. Two blocks are pushed across a frictionless surface by a constant horizontal force as shown in the diagram below.

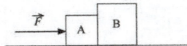

   The mass of block A is $m$, and the mass of block B is $2m$. If the blocks remain in contact as they move across the surface, block B experiences a net force of

   (A) $\dfrac{2}{3}\vec{F}$.

   (B) $\vec{F}$.

   (C) $\dfrac{3}{2}\vec{F}$.

   (D) $2\vec{F}$.

8. Two forces are applied to an object initially at rest on a frictionless surface as shown below. Rank the diagrams for the largest acceleration of the blocks to the least acceleration of the blocks.

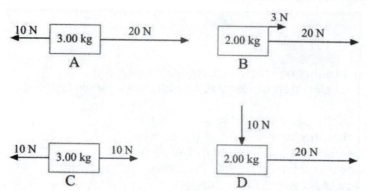

(A) B > D > A > C
(B) B > (A = C) > D
(C) C > A > B > D
(D) C > B > A > D

9. Two objects connected by a massless rod as shown in the diagram are moving with some velocity to the right on a frictionless surface.

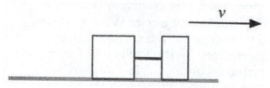

An external force is applied to the objects as shown.

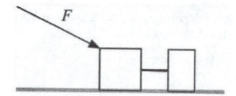

The applied force will
(A) have no effect on the velocity of center of mass of the system since it is only applied to the larger object.
(B) increase the velocity of the center of mass of the system since it will provide an acceleration to the system.
(C) only change the velocity of the larger object.
(D) maintain the initial velocity as a constant since forces must be applied to both objects to increase the velocity of the center of mass.

10. A 20.0 N force is applied to the objects in the diagram shown below. Consider the contact surfaces to be frictionless and ignore air resistance.

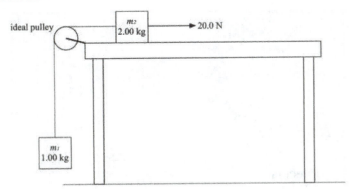

The magnitude of the acceleration of the masses indicates that

(A) $\vec{a}_{2\,kg} > \vec{a}_{1\,kg}$.

(B) $\vec{a}_{2\,kg} < \vec{a}_{1\,kg}$.

(C) $\vec{a}_{2\,kg} = \vec{a}_{1\,kg}$.

(D) the only acceleration is $\vec{a}_{2\,kg}$.

11. A 1.80 kg object experiences a force acting on it for 5.00 s. The graph representing the velocity of the object vs. time is given below.

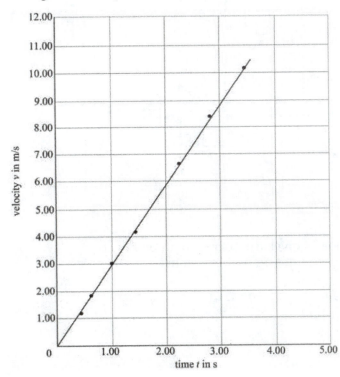

What was the magnitude of the force?

(A) 5.40 N

(B) 3.00 N

(C) 1.50 N

(D) 0.600 N

12. An object weighing 49 N is lifted by a cord as shown in the diagram below. What acceleration does the object experience?

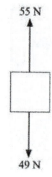

55 N

49 N

The acceleration of the object is
(A) 6.0 m/s².
(B) 5.0 m/s².
(C) 1.2 m/s².
(D) 0.83 m/s².

13. A diagram showing the forces acting in a system is illustrated below. Consider the pulley to be an ideal pulley. Ignore friction and air resistance.

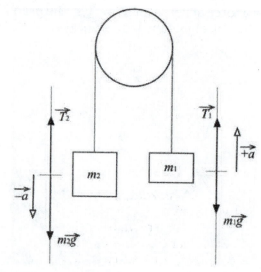

Which of the four choices gives the correct mathematical expression for the representation of the system?
(A) $T_2 - T_1 = (m)a$
(B) $T_2 - T_1 = (m_1 + m_2)a$
(C) $m_2g - m_1g = (m_1 - m_2)a$
(D) $m_2g - m_1g = (m_1 + m_2)a$

**Questions 14 and 15**
**Directions:** For each of the questions or incomplete statements below, two of the suggested answers will be correct. For each of these questions, you must select both correct choices to earn credit. No partial credit will be earned if only one correct choice is selected. Select the two that are best in each case.

14. A 2.00 kg object accelerates down a 30° inclined plane where the coefficient of friction between the object and the plane is small. Increasing the (select two answers)
    (A) angle the inclined plane makes with the horizontal will increase the effective weight down the plane, and thus, the acceleration down the plane will increase.
    (B) mass of the object will increase the acceleration since it increases the effective weight down the plane.
    (C) angle will have no effect on the acceleration down the plane since it will not change the net force on the object.
    (D) mass will not change the acceleration.

15. A student makes a free-body diagram of an object accelerating down an inclined plane as shown below.

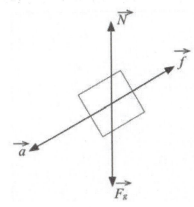

    Which of the following explains the mistake(s), if any, the student made in the diagram? Select two answers.
    (A) The normal force from the surface is drawn opposite the weight but should be perpendicular to the surface.
    (B) Friction opposes the motion of the body and should be directed down the plane.
    (C) An arrow showing the acceleration down the plane is incorrect since acceleration is not a force.
    (D) The direction of the weight vector is incorrect. It should be drawn perpendicular to the surface.

## FREE-RESPONSE PROBLEMS

1. In a laboratory experiment, the acceleration of a small cart is measured by the separation of dots burned at regular intervals onto a paraffin-coated tape. Weights are transferred from the small cart that is connected by a massless cord to a weight hanger that passes over a frictionless pulley. The surface is considered

frictionless; thus, the weight at the end of the cord is the resultant force on the system. Students obtained the following data:

| $\vec{F}$ (N) | $\vec{a}$ (m/s²) |
|---|---|
| 1.60 | 1.00 |
| 2.30 | 1.50 |
| 4.00 | 2.50 |
| 4.75 | 3.00 |
| 5.50 | 3.50 |
| 6.50 | 4.50 |
| 7.25 | 5.50 |
| 8.00 | 6.50 |

(a) Plot the graph of force versus acceleration on the grid below.

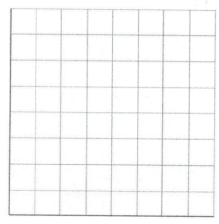

(b) What is the slope of the graph and its significance?
(c) Why did students in this experiment transfer weights from the small cart to the weight hanger attached to the cord? Explain your answer using the correct scientific terms.

2. A 1.50 kg block is placed at the top of a 38° inclined plane and released. The block accelerates down the plane at 3.80 m/s².
   (a) Draw the forces acting on the block as it moves down the plane.

   (b) Write the equation to determine the coefficient of friction between the block and the plane as the block accelerates down the plane. Calculate the coefficient of friction.

(c) Sometime later, the block is pushed up the plane by a push *P* of 16.0 N applied to the object and directed upward and parallel to the plane. The friction acting between the plane and the 1.50 kg block is

___ greater than the friction acting when the block accelerated down in the plane.

___ less than the friction acting when the block accelerated down in the plane.

___ the same as the friction acting when the block accelerated down in the plane.

Justify your answer.

3. Two metal guide rails for a 450 kg mine elevator each exert a constant frictional force of 110.0 N on the elevator car when it is moving upward with an acceleration of 2.50 m/s² as shown in the diagram below. Attached to the lower right side of the cable lifting the elevator is a counterweight of mass *M*. The pulley is an ideal pulley.

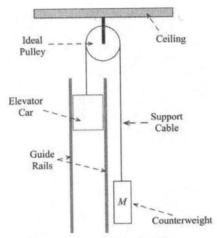

(a) What is the direction of the net force on the elevator car? Justify your answer.

(b) On the elevator cage represented below, sketch and clearly label all forces acting on the elevator during its motion.

(c) What is the tension in the supporting cable when the elevator is accelerating as described?

(d) Explain using the correct scientific terms and without writing a mathematical equation how you would determine the mass *M* of the counterweight needed to give the elevator cage the described acceleration.

4. When you and your lab partners enter the physics classroom, your teacher has laboratory equipment on the tables and asks you determine what, if any, relationship exists between a constant force and a variable mass.

(a) Design a laboratory experiment in enough detail that another student could duplicate your results and reach the same conclusion(s) about your inquiry lab.

(b) Make a sketch of your equipment and label each part of the sketch correctly.

(c) What measurements will you take, and how will you use them to answer your experimental question?

(d) Another group of students obtained data to plot the graph below. How does this graph answer the experimental question?

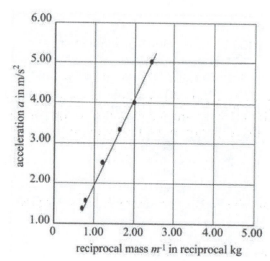

# Answers

## MULTIPLE-CHOICE QUESTIONS

1. **D** The gravitational force is the tension in the cord. $F_g = mg$. The tension is $F_g = 5.00 \text{ kg}\left(9.80 \dfrac{m}{s^2}\right) = 49.0 \text{ N}$.

   (L.O. 2.B.1.1)

2. **A** The 10.0 N force has two components, $F_x = F\cos\theta$, which will give a force in the $+x$ direction. The vertical component $F_y = F\sin\theta$ is used to determine the normal force. Since the applied force is upward, the normal force will be smaller than the weight by this value. A coefficient of friction is given in the question; therefore, a frictional force will act opposite the motion of the object.

   (L.O. 2.A.2.1)

3. **B** $\Sigma F = F_x - f = ma$ applies to the problem.

   $F_x = F\cos\theta = 10.0 \text{ N}\cos40° = 7.66 \text{ N}$.

The friction is determined from $f = \mu_k N$.
The solution to the $\Sigma F_y = 0$ equation will give the normal force.

$\Sigma F_y = 0 = N + F\sin\theta - mg$.

The normal is $2.00 \text{ kg} \left( 9.80 \text{ } \frac{m}{s^2} \right) - 10.0 \text{ N} \sin 40°$.

Solving gives the value for the normal as $19.6 \text{ N} - 6.43 \text{ N} = 13.2 \text{ N}$.
Then the frictional force is $f = 0.20(13.2 \text{ N}) = 2.64 \text{ N}$.

Finally, the acceleration is $7.66 \text{ N} - 2.64 \text{ N} = (2.00 \text{ kg})a$.

$a = 2.51 \text{ } \frac{m}{s^2}$

(L.O. 3.B.1.3)

4.  **C** The elevator is moving downward with a negative acceleration; therefore, the tension in the cable is smaller than the weight of the person. The apparent weight is less than the true weight.

    (L.O. 3.B.1.1)

5.  **D** Newton's second law applies to this problem: $F_g = (m_1 + m_2)a$.

    $1.00 \text{ kg} \left( 9.80 \text{ } \frac{m}{s^2} \right) = (1.00 \text{ kg} + 2.00 \text{ kg})a$.

    The acceleration is $a = 3.67 \text{ } \frac{m}{s^2}$.

    The absolute value of the velocity of the center of mass of the system, its speed, is found from $v_f^2 = +2ay$. Substitution

    $v_f^2 = +2\left( -3.67 \text{ } \frac{m}{s^2} \right)(-0.500 \text{ m}) \text{ } v_f = -1.81 \text{ } \frac{m}{s}$.

    (L.O. 3.A.3.1)

6.  **B** The weight of the object is one-fourth its weight on the surface of Earth, but its mass has not changed. $F = ma$ applies; the force is the same as on the surface of Earth.

    (L.O. 3.B.1.4)

7.  **A** Newton's second law $F = ma$ applies. The acceleration of the system $F = 3ma$ is $a = \dfrac{F}{3m}$. Block B experiences a force that is

    $F' = 2m(a) = 2m\left( \dfrac{F}{3m} \right) = \dfrac{2}{3}(F)$

    (L.O. 3.A.1.1)

8.  **A** Forces that act parallel to each other as in diagram B will produce the largest net force. Any force applied perpendicular to

the direction of motion on the frictionless surface as in diagram D will not change the acceleration of the object. In diagram A, the net force on the object is 10 N in the direction of motion, and in diagram C, the net force is zero.

(L.O. 3.A.3.1)

9. **B** The acceleration and, hence, the velocity of the center of mass are related to the net external force. The motion of the system is the same as if it were a point mass $M$ acted upon by the $\Sigma F_{ext}$. This is then $\Sigma F_{ext} = M_{total} a_{cm}$. Therefore, the velocity increases.

(L.O. 4.A.3.2)

10. **C** The objects move together as a system and will have the same acceleration.

(L.O. 3.B.1.4)

11. **A** The slope of the velocity as a function of time graph will give the acceleration of the object.

$$a = \frac{\left(9.00 \ ^m/_s - 3.00 \ ^m/_s\right)}{(3.00 \ s - 1.00 \ s)} = 3.00 \ ^m/_{s^2}$$

Then using $\Sigma F = ma$, the force acting on the object is

$$\Sigma F = 1.80 \ kg \left(3.00 \ ^m/_{s^2}\right) = 5.40 \ N.$$

(L.O. 3.B.1.2, 3.B.1.3)

12. **C** The acceleration of the system is determined from $T - mg = ma$. The weight of the object, not the mass, is shown. The mass is found from $49 \ N = m \left(9.80 \ ^m/_{s^2}\right)$. $m = 5.0 \ kg$.

Then $56 \ N - 49 \ N = (5.0 \ kg)a$ and $a = 1.2 \ ^m/_{s^2}$.

(L.O. 3.B.1.3)

13. **D** Newton's second law applies $\Sigma F = ma$. Since the tension on either side of the rope is the same, the tension does not enter into the correct solution. The correct substitution is

$m_2 g - m_1 g = (m_1 + m_2)a.$

(L.O. 3.B.1.3)

14. **D** Initially, the floor provides a normal force on each student equal to his or her weight. When student A is no longer in contact with the floor yet is not accelerating, we know the tension force must equal student A's weight. We know this because the forces must

add to zero, and the only way for that to work is if the upward forces equal the downward forces. For student B, his or her weight is the same, but now there is a tension pulling upward equal to the weight of student A.

$\Sigma F = 0$

$T + N_b - m_b g = 0$

This shows that the upward and downward forces sum to zero.

$T = N_a$

$N_b = m_2 g - N_a$

(L.O. 3.B.2.1)

15. **A and D** Increasing the angle of the incline increases the effective weight down the plane and thus the acceleration. Since Newton's second law applies, $\Sigma F = ma = mg\sin\theta - \mu_f mg\cos\theta$. Increasing the angle increases the value for the $\sin\theta$ while decreasing the $\cos\theta$. Increasing the mass will not change the acceleration since it is a common factor in all three terms of the equation.

$ma = mg\sin\theta - \mu_f mg\cos\theta$

(L.O. 3.A.1.1)

16. **A and C** The normal to the surface is a perpendicular force exerted on the object by the plane in response of the $\Sigma F_y = 0$ equation. The normal must cancel the $y$-component of the gravitational force relative to the plane. Acceleration is not a force. What causes the acceleration down the plane is the $x$-component of the gravitational force minus the frictional force up the plane.

$ma = mg\sin\theta - \mu_f mg\cos\theta$

(L.O. 3.B.2.1)

## FREE-RESPONSE PROBLEMS

1. (a) The graph of force versus acceleration is shown below.

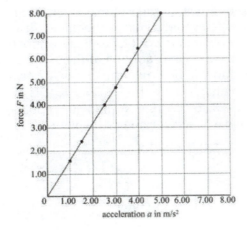

(b) The slope of the line is the total mass in the system: $\Sigma F = ma$.

$$m = \frac{(8.00\ \text{N} - 1.60\ \text{N})}{\left(5.00\ \text{m}/\text{s}^2 - 1.00\ \text{m}/\text{s}^2\right)} = 1.60\ \text{kg}$$

(c) Students transferred mass from the cart to the mass hanger to keep the total mass in the experiment a constant. In the absence of frictional forces, the only net force acting on the cart is the weight on the weight hanger since $F_g = mg$. If mass does not vary in the experiment, the acceleration is directly proportional to the net unbalanced force acting on the constant mass.

(L.O. 2.B.1.1, 4.A.2.2)


2. (a) The correct force diagram is shown below.

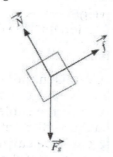

(b) The block is in equilibrium in the y-direction since it is neither lifting off the incline nor sinking into it. $\Sigma F_y = N - F_g \cos\theta$. The normal is the reaction force applied by the plane to what it perceives as the "weight" of the block. $N = mg \cos\theta$.

The motion of the block is down the plane. $\Sigma F_{net} = F_{effective} - f$.

Since the effective force down the plane is given by $F_{effective} = mg \sin\theta$, substitution into $\Sigma F_{net} = F_{effective} - f$ is as follows:

$$(1.50\text{kg})\left(3.80\ \text{m}/\text{s}^2\right) = (1.50\ \text{kg})\left(9.80\ \text{m}/\text{s}^2\right)\sin 38° - \mu_k (1.50\ \text{kg})\left(9.80\ \text{m}/\text{s}^2\right)\cos 38°$$

$$5.70\ \text{N} = 9.05\ \text{N} - \mu_k (11.6\ \text{N})$$

$$\mu_k = 0.289$$

(c) The correct line checked is __ √ __, the same as the friction acting when the block accelerated down in the plane. Frictional forces act parallel to the objects in contact and are opposite the direction of motion. Since a push is applied to the block to send it up the plane, the friction will be down the plane. (Checking to see that it moves up the plane, write the $\Sigma F_{net} = F_{effective} + f$ equation. $\Sigma F_{net} = 9.05\ \text{N} + 3.35\ \text{N} = 12.4\ \text{N}$. Since the applied force is 16.0 N, the block will accelerate up the plane.)

(L.O. 2.B.1.1, 3.A.2.1, 3.A.4.3, 3.B.1.3, 3.B.2.1)

3. (a) Since the cage is moving upward with an acceleration of 2.50 m/s², the net force on the cage is upward.
   (b) The forces on the cage are shown below; since both rails exert frictional forces, acting downward, they as well as the normal forces exerted on the two guide rails must be included in the diagram. The last two forces are the tension in the cable and the weight of the cage.

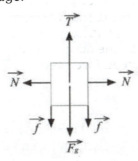

   (c) Because the mass of the counterweight is unknown, the solution is $T_{cable} - 2f - mg = ma$. Substitution into the equation is

$$T_{cable} - 2(110.0\ N) - (450.0\ kg)\left(9.80\ \frac{m}{s^2}\right) = (450.0\ kg)\left(2.50\ \frac{m}{s^2}\right).$$

   (d) Since the force on the counterweight is in the $y$-direction, applying Newton's second law would give the mass of the counterweight. The upward tension in the cable minus the weight of the counterweight would give the net force on the counterweight. Then the mass $M$ could be calculated by dividing the net force by the acceleration of the system.

   (L.O. 2.B.1.1, 3.A.2.1, 3.A.4.3, 3. B.1.1, 3.B.1.3, 3.B.1.4, 3.B.2.1)

4. (a) What relationship exists between a constant force and a variable mass?

   A known force will be applied to an object that is free to move by passing a cord connected to the object over an ideal pulley to a weight hanger at the other end of the cord.

   Using a platform balance, determine the mass of the object.

   Connect the object via the cord to the known force on the weight hanger.

   Using a meterstick, measure the distance between two motion sensors.

   Calculate the acceleration of the object using $v_f^2 = v_0^2 + 2ax$.

   Repeat the observation for at least three readings, taking the average of the readings and recording this average.

   Increase the mass of the glider by attaching an additional mass to the glider, then record the new mass. Using the same force on the weight hanger, determine the acceleration of the new mass.

(b) Several possible pieces of equipment may be used.

    i. An air track is the best choice of equipment if it is available since it has very low friction between the track and the glider.

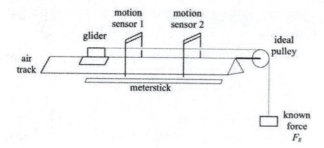

    ii. The equipment can be set up on a lab table, but friction between the object and the known net force must be measured for each trial. The frictional force, $f_k = \mu_k N$ will increase with additional weight on the block since the normal force increases. Once the increase in friction has been adjusted, the acceleration of the object can be measured for the known net force.

(c) Measurements taken in the experiment include the known weight and the mass for each trial.

The motion sensors will record the velocity of the glider as it passes through the sensor. The separation of the motion sensors will be needed to determine the acceleration using $v_f^2 = v_0^2 + 2ax$.

Knowing the acceleration will permit a calculation of the product of the mass and the acceleration. If the force calculations are equal to the known force, then experimentally we can write that the acceleration is directly proportional to the force and inversely proportional to the mass.

(d) The slope of the acceleration as a function of the reciprocal of the mass will give the value of the known net force. Increasing the mass results in a decrease in acceleration when a constant force is applied.

(L.O. 1.C.1.1, 2.B.1.1, 3.A.1.2, 3.B.1.3, 3.B.1.4, 3.B.2.1)

# 5

# WORK, ENERGY, AND POWER

## WORK

The term *work* is restricted in physics to cases in which there is a force and a displacement along the line of the force. When a force $F$ moves an object through a displacement $d$ and the directions of these two vectors are not the same, the work $W$ is

$$W = F_{\parallel}d = Fd\cos\theta,$$

where $\theta$ is the angle between the direction of the force and the displacement. The $\cos\theta$ term means that only the component of the force that is in line with the displacement does work on an object. If the force and displacement are in the same direction, this simplifies to $W = F_{\parallel}d$. Work is a scalar quantity with a unit of newton-meters, Nm, or *joules*, J.

$$1\,J = 1\,N \cdot m = 1\,\frac{kg \cdot m^2}{s^2}$$

This equation uses displacement, not distance. If a force causes no displacement in the direction of the force, no work is done. If you held two physics books in your outstretched arms without raising or lowering them, you'd be doing no work even though your arms would get very tired. A force that causes something to move in a circle also does no work because there is zero displacement in the direction of the force.

## SAMPLE PROBLEM 1

A wooden box is pulled 4.6 m across a horizontal floor at constant speed by a rope that makes an angle of 20° with the floor. The tension in the rope is 120.0 N. How much work is done on the box by the person pulling it?

**143**

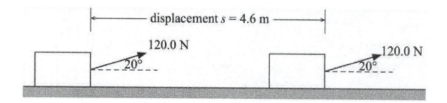

## SOLUTION TO PROBLEM 1

The force vector makes a 20° angle with respect to the displacement. So the work is

$$W = Fd\cos\theta = (120.0 \text{ N})(4.6 \text{ m})(\cos 20°) = \textbf{518.7 J}$$

---

### AP® Tip

The work done on an object is always done by some force.

---

## SAMPLE PROBLEM 2

A 600.0 N force is developed in the braking system of a car to bring it to rest over a horizontal distance of 80.0 m. What work is done by the braking system?

## SOLUTION TO PROBLEM 2

The force developed by the brakes acts in the exact opposite direction to the displacement making $\theta = 180°$ and the $\cos 180° = -1$. Notice that the brake system is the *agent* doing the work on the car. The work done will be negative.

$$W = Fd\cos\theta$$

$$= (600.0 \text{ N})(80.0 \text{ m})(\cos 180)$$

$$= (600.0 \text{ N})(80.0 \text{ m})(-1)$$

$$= -48.0 \text{ kJ}$$

## KINETIC ENERGY

In physics, we say that anything that has *energy* has the capacity to do work. Energy exists in many forms and can be transformed from one form to another. In our study of *mechanics,* our primary interest is *mechanical energy.* So we consider these:

1. Kinetic energy, $K$
2. Potential energy, $U$
3, Total mechanical energy, $E = K + U$

All moving objects have *kinetic energy.* Objects at rest do not have kinetic energy. To give an object motion requires an unbalanced force, and that force causes an acceleration or does work on the object to change its energy.

> ## AP® Tip
>
> When the force and displacement vectors are parallel, the $\cos 0° = 1$, making the work done on an object $W = F\Delta x$.

Work transforms energy. The amount of work it takes to accelerate an object from rest to some velocity is equal to the amount of kinetic energy the object gains.

$$K = \frac{1}{2}mv^2$$

> ## AP® Tip
>
> The work done by something on an object to give it speed $v$ equals the change in kinetic energy of that object.

### SAMPLE PROBLEM 3

A 0.03 kg bullet is fired horizontally from a rifle, giving it a muzzle velocity of 300.00 m/s. The rifle barrel has a length of 75.0 cm.
(a) What is the kinetic energy of the bullet as it exits the muzzle of the rifle?
(b) What work is done on the bullet by the expanding gases as it travels the length of the barrel?
(c) What acceleration does the bullet experience while in the barrel?
(d) How much time does the bullet spend in the rifle barrel?
(e) Ignoring air resistance, how much work will the bullet do when it strikes a thick tree trunk 100 m from the rifle?

### SOLUTION TO PROBLEM 3

(a) The kinetic energy of the bullet is

$$K = \frac{1}{2}mv^2 = \frac{1}{2}(0.03 \text{ kg})\left(300.0 \ \frac{\text{m}}{\text{s}}\right)^2 = \textbf{1.35 kJ}$$

(b) Expanding hot gases do work on the bullet, driving it down the barrel. The work done to give the bullet a kinetic energy of **1.35 kJ** is the same, **1.35 kJ**.

(c) The work done on the bullet is $W = Fd$. The force exerted on the bullet by the gases is $F = W/d = ma$. The acceleration of the bullet is then

$$a = \frac{W}{md} = \frac{1.35 \times 10^3 \, J}{(0.03 \text{ kg})(0.75 \text{ m})} = 60 \times 10^3 \ \text{m}/_{\text{s}^2}$$

(d) Acceleration is defined by $a = \dfrac{v - v_0}{t} = \dfrac{v - 0}{t}$. The time period is

$$t = \frac{v}{a} = \frac{\left(300.0 \; \text{m}/\text{s}\right)}{\left(60 \times 10^3 \; \text{m}/\text{s}^2\right)} = 5.0 \; \text{ms}$$

Since the bullet has 1.35 kJ of kinetic energy, it will do **1.35 kJ** of work when it hits the tree trunk in coming to rest.

Like work, kinetic energy is a scalar quantity and is measured in the same units as work.

---

## AP® Tip

Unlike work, kinetic energy is never negative.

If the object has some initial speed, $v_0$, other than zero, its final kinetic energy is

$$K = \frac{1}{2}mv^2 - \frac{1}{2}mv_0^2$$

Note that the preceding equation also may be written as

$$\Delta K = \frac{1}{2}mv^2 - \frac{1}{2}mv_0^2 = K - K_0$$

The quantity $\Delta K$ is the change in kinetic energy.

---

## THE WORK-KINETIC ENERGY THEOREM

Work serves as a connection between a force (or forces) and the energy it can produce. We call this relationship the *work-kinetic energy theorem.*

$$\Delta K = W = F_\parallel d = Fd\cos\theta$$

The work done on a moving system changes the kinetic energy of that system. We saw earlier that when the force is in the same direction as the motion, positive work is done. In relationship to energy, this tells us that we had an increase in kinetic energy. The object sped up. We also saw that a force in the opposite direction of the motion gives negative work. The object has a decrease in kinetic energy, so it slows down.

In Problem 3(e) the question is "Ignoring air resistance, how much work will the bullet do when it strikes a thick tree trunk 100 m from the rifle?" Since air resistance was ignored, the bullet strikes the tree trunk with a speed equivalent to the muzzle velocity, 300.0 m/s. The bullet hits the tree trunk and penetrates it to some depth before coming to rest. The tree trunk is the agent stopping the bullet, and the work done is found by the work-kinetic energy theorem, or

$$W = \Delta K = K - K_0$$
$$= \frac{1}{2}mv^2 - \frac{1}{2}mv_0^2$$
$$= \frac{1}{2}m(0)^2 - \frac{1}{2}(0.03 \text{ kg})\left(300.0 \text{ m/s}\right)^2$$
$$= -1.35 \text{ kJ}$$

The work the tree trunk does on the bullet is negative. The tree trunk stops the bullet. It takes away kinetic energy.

## POTENTIAL ENERGY

When an object is lifted to a position above floor or ground level, the object is then in a position where it can do work. It can fall! Any object that is in a position where it can do work is said to have *potential energy, U*, or more specifically, *gravitational potential energy, U*. The object has been given gravitational potential energy equal to

$$U = mgy = mgh$$

The joule, J, is the SI unit of potential energy.

## CONSERVATIVE FORCES

The major feature of a conservative field, like Earth's gravitational field, is that the work done in moving a mass is totally independent of the path taken from the ground to the final position. The change in gravitational potential energy, the work done, is only dependent on the vertical displacement. We say that in general, a force is conservative when the work done by that force acting on an object moving between two points is independent of the path the object takes between the points. A conservative force has the property that the *total work* done by the conservative force is zero when the object moves around any closed path and returns to its initial position. The most frequently used example of a conservative force is the force due to gravity.

## THE LAW OF CONSERVATION OF MECHANICAL ENERGY

In a conservative field, the total energy is always a constant. This means that the total initial energy in a closed system before an event will always equal the total final energy in that same system after the event, or, symbolically,

$$\Sigma E_0 = \Sigma E,$$

where $\Sigma E_0 = K_0 + U_0$ and $\Sigma E = K + U$, which means that

$$K_0 + U_0 = K + U$$

We call this mathematical statement the *law of conservation of mechanical energy*.

### SAMPLE PROBLEM 4

Consider the following diagram where a crate of mass $m$ is held at rest at point A. The crate is released and slides along the frictionless curved surface to point B. What is the velocity of the crate when it reaches point B?

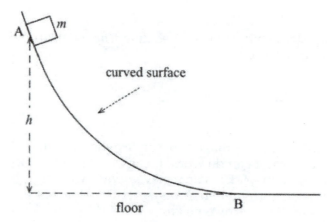

### SOLUTION TO PROBLEM 4

Since friction is ignored, the system is a conservative system. The total energy at point A is equal to the total energy at point B, or

$$K_0 + U_0 = K + U.$$

At point A, initially the crate is at rest and its kinetic energy is zero. The initial gravitational potential energy at A is $U_0 = mgh$. At ground level the gravitational potential energy becomes zero, and the kinetic energy of the crate is $K = \frac{1}{2}mv_B^2$. Writing $K_0 + U_0 = K + U$ and substituting gives $0 + mgh = \frac{1}{2}mv_B^2 + 0$. Solving for $v_B$ yields $v_B = \sqrt{2gh}$.

## AN IDEAL SPRING AS A CONSERVATIVE SYSTEM

Springs obey a rather simple law discovered by Robert Hooke in the eighteenth century. In Hooke's law, we must consider restoring forces, which are forces that want to return objects to their original position. If we stretch a spring within reason or compress it, it wants to return to its original shape. Hooke's law states that a restoring force acting in a spring is directly proportional to the amount of stretching or compressing, or

$$F = -kx$$

The negative sign (–) reminds us that $F$ is a restoring force, $k$ is the spring constant that is unique to a given spring, and $x$ is the amount of stretching or the amount of compression in the spring. In most calculations, you will ignore the negative sign. The free end of a spring in the relaxed position is called the *equilibrium position*.

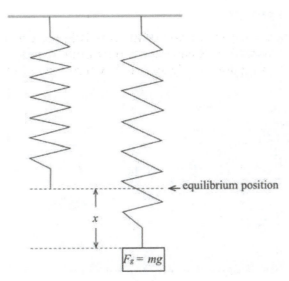

The work required to stretch a spring that follows Hooke's Law is found by

$$W = \frac{1}{2}kx^2$$

For a spring that follows Hooke's law, the restoring force is conservative. Work done to stretch or compress the spring equals the change in the spring's elastic potential energy.

$$U_s = \frac{1}{2}kx^2$$

## SAMPLE PROBLEM 5

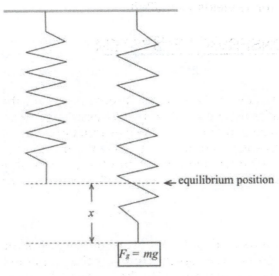

A spring is attached to a rod as shown above. The spring has a length of 0.22 m. A 0.44 kg mass is attached to the free end of the spring, stretching it to a new length of 0.28 m.
(a)    What is the spring constant of the spring?
(b)    How much work was done on the spring elongating it?

## SOLUTION TO PROBLEM 5

(a) The spring is stretched by $x = x_{loaded} - x_{relaxed} = 0.28\text{ m} - 0.22\text{ m} = 0.06\text{ m}$.
The stretching force is the weight of the load attached to the spring:

$$F = F_g = mg = (0.44\text{ kg})\left(9.8\text{ m}/_{s^2}\right) = 4.31\text{ N}$$

Hooke's law is $F = kx$

and $k = \dfrac{F}{x} = \dfrac{4.31\text{ N}}{0.06\text{ m}} = 71.87\text{ N}/_{m}$.

(b) The work done is $W = \dfrac{1}{2}kx^2 = \dfrac{1}{2}\left(71.87\text{ N}/_m\right)(0.06\text{ m})^2 = 0.13\text{ J}$.

## NON-CONSERVATIVE FORCES

A force is non-conservative if the work done by the force on an object moving between two points depends on the path taken. Friction, $f$, is such a force. Friction is *degrading force* since it takes away kinetic energy and converts it to another energy form, *thermal energy, Q.* Thermal energy or heat is a wasteful energy form, a dissipative form, and it cannot be recovered into the mechanical system. Heat escapes into the environment of the system, warming it.

The work, $W_{nc}$, done by all non-conservative forces in a system equals the change in the total mechanical energy of the system.

$$-W_{nc} = \Delta K + \Delta U$$

or

$$-W_{nc} = (K + U) - (K_0 + U_0)$$

## SAMPLE PROBLEM 6

A 2.00 kg block of wood slides across a horizontal floor with an initial velocity of 0.40 m/s and comes to rest after sliding 1.20 m as shown in the diagram below. Using energy considerations, calculate the average frictional force acting on the block.

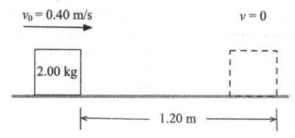

## SOLUTION TO PROBLEM 6

The block moves along the horizontal, making the change in gravitational potential energy zero.

Since friction brings the block to rest, this is a non-conservative system and $-W_{nc} = \Delta K + \Delta U = K - K_0 + 0$. The final kinetic energy is zero since the block comes to rest and $-W_{nc} = -fx = -K_0 = -\dfrac{1}{2}mv_0^2$.

The average frictional force is

$$f = \frac{mv_0^2}{2x} = \frac{(2.00 \text{ kg})\left(0.40 \text{ m}\!/\!_s\right)^2}{2(1.20 \text{ m})} = \textbf{0.13 N}$$

## WORK DONE BY A CONSTANT FORCE

Lifting a mass $m$ through a vertical displacement $h$ is an example of work done by a *constant force*. In this simple case, we can write $F =$ constant, and the work done by the lifting agent is $W = Fh$. In this example, we use the vertical displacement $h$ to show that an object was lifted up; $h$ is equal to displacement in $W = Fd$.

If we draw a graph with the force $F$ plotted along the vertical axis and the distance $h$ through which the force acts, we get a straight horizontal line as shown in the graph below.

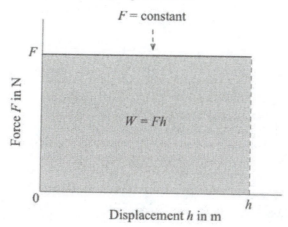

For any given $F$ and $h$, the work done is simply equal to the shaded area in the graph. This idea that *the area under a curve can represent work done* is extremely useful. It can be extended to practical cases where the force may not be constant, but varies.

## WORK DONE BY A VARIABLE FORCE

Up to this point in the treatment of work, we have assumed that an applied force is constant. We could easily use the formula given to find the work done by the force. However, in numerous situations, a force is far from being constant. One such case is in the stretching of a spring. The spring shown below is elongated through a displacement $x$ by a horizontal force $F$. The stretching force is not a constant; it varies with displacement. How then can we find the work done? We need to collect data and plot a $F$ vs. $x$ graph.

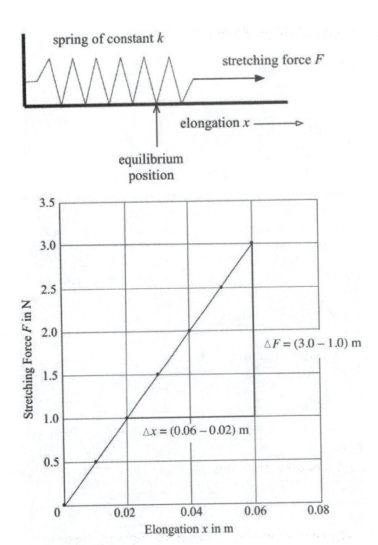

Before we do anything else, we need to find the slope of the curve. The slope is the spring constant, $k$. Earlier, when we were making a study of motion, we defined slope as

$$\text{slope} = \frac{\text{change in the vertical}}{\text{change in the horizontal}}$$

And we can write

$$\text{slope} = k = \frac{\Delta F}{\Delta x} = \frac{F - F_0}{x - x_0} = \frac{3.0 \text{ N} - 1.0 \text{ N}}{0.06 \text{ m} - 0.02 \text{ m}} = 50.0 \text{ N}\!/\!\text{m}$$

The area beneath a $F$ vs. $x$ graph is the work done by the force $F$. Note in the following graph that the shaded area represents the work done in stretching the spring by 0.06 m.

Recall that the area bounded by a triangle, which is the work done by the variable force, is one-half the base multiplied by the height, or area $= W = \frac{1}{2}(x)(F)$. And since $F = kx$, $W = \frac{1}{2}(x)(kx)$ or $W = \frac{1}{2}kx^2$.

The work done in stretching the spring by 0.06 m is then

$$W = \frac{1}{2}kx^2 = \frac{1}{2}\left(50.0 \text{ N}\!/\!\text{m}\right)(0.06 \text{ m})^2 = 0.09 \text{ J}$$

Or taking one-half the base times the height,

$$\text{area} = \frac{1}{2}(0.06 \text{ m})(3.0 \text{ N}) = 0.09 \text{ J}$$

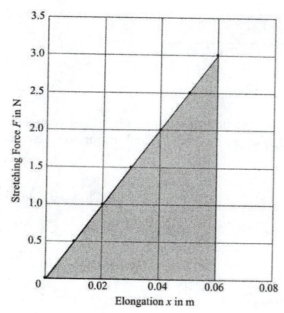

## POWER

We define power as the work done per unit time.

$$P = \frac{W}{t}$$

Power, like work, is a scalar quantity. In the SI, the unit of power is the watt (W), and we define 1 W as

$$1.00 \text{ W} = 1.00 \frac{\text{J}}{\text{s}}$$

Sometimes the kilowatt, kW, is more advantageous to use, and 1 kW = 1000 W.

Although not given on the AP® equation sheet, power also may be expressed as the product of the average force acting on an object and the average speed of the object:

$$P = Fv$$

### SAMPLE PROBLEM 7

An 1100 kg car accelerates from 5.0 m/s to a speed of 25.0 m/s in a time interval of 6.0 s. Ignoring frictional losses, what average power must the drive train produce to cause this acceleration?

### SOLUTION TO PROBLEM 7

The work done in accelerating the car is

$$W = \Delta K = K - K_0 = \frac{1}{2}mv^2 - \frac{1}{2}mv_0^2 = \frac{1}{2}m\left(v^2 - v_0^2\right)$$

Power is defined as follows:

$$P = \frac{W}{t}$$

$$= \frac{m\left(v^2 - v_0^2\right)}{2t} = \frac{(1100 \text{ kg})\left[\left(25.0 \text{ m/s}\right)^2 - \left(5.0 \text{ m/s}\right)^2\right]}{2(6.0 \text{ s})}$$

$$= 55.0 \text{ kW}$$

## SAMPLE PROBLEM 8

A 1120 W electric motor at a construction site can lift a load vertically at the rate of 0.10 m/s. What maximum mass of building materials can it lift at this constant speed?

## SOLUTION TO PROBLEM 8

In 1.0 s, the load *mg* is lifted a vertical distance of 0.10 m. The work done in 1.0 s is $W = mgh$.

By definition, $P = \dfrac{W}{t} = \dfrac{mgh}{t}$. Solving for *m* yields

$$m = \frac{Pt}{gh} = \frac{\left(1.12 \times 10^3 \text{ W}\right)(1.0 \text{ s})}{\left(9.8 \text{ m/s}^2\right)(0.10 \text{ m})} = 1.14 \times 10^3 \text{ kg}$$

## WORK, ENERGY, AND POWER: STUDENT OBJECTIVES FOR THE AP® EXAM

- You should be able to define and calculate work.
- You should be able to differentiate between total mechanical energy, kinetic energy, and gravitational potential energy.
- You should be able to state and use the work-kinetic energy theorem.
- You should be able to state and use the law of conservation of mechanical energy.
- You should be able to calculate the work done in stretching a spring.
- You should be able to explain why friction is called a dissipative force.
- You should be able to compare the effects of a conservative vs. non-conservative force.
- You should be able to explain what will happen to the kinetic energy of an object if the net work done on it is negative.

## MULTIPLE-CHOICE QUESTIONS

1. A 4.00 kg block moving with a velocity of 1.5 m/s on a flat horizontal surface enters a region where the coefficient of friction has increased. The kinetic energy of the block will
   (A) increase as the block enters the region.
   (B) decrease as the block enters the region.
   (C) remain constant in the region.
   (D) decrease then increase as the block leaves the region.

2.  A 5.00 kg object is lifted vertically a distance of 1.20 m and then carried a horizontal distance of 4.00 m. The work done by gravity on the object is
    (A) –245 J.
    (B) –196 J.
    (C) –58.8 J.
    (D) –24.0 J.

3.  A block moves from point A to B to C and back to A along the path shown.

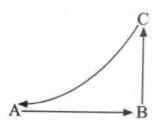

The path is conservative if the
    (A) final kinetic energy is the same as the initial kinetic energy.
    (B) final gravitational potential energy is the same as the initial gravitational potential energy.
    (C) the loss of gravitational potential energy equals the loss in kinetic energy.
    (D) the sum of the total mechanical energy remains the same.

4.  An object experiences a variable force applied over a displacement of 8.00 m as shown in the graph. What is the change in the kinetic energy as the object moves through the displacement of 8.00 m?

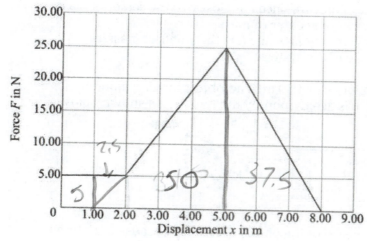

    (A) 67.5 J
    (B) 92.5 J
    (C) 130 J
    (D) 160 J

5. A 2.00 kg block slides on a frictionless horizontal surface with a velocity of 1.60 m/s when it strikes the massless bumper on a spring attached to a wall as shown below.

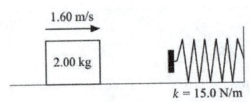

1.60 m/s

2.00 kg

$k = 15.0$ N/m

The maximum increase in the internal potential energy of the massless bumper and spring is
(A) 5.12 J.
(B) 3.20 J.
(C) 2.56 J.
(D) 1.60 J.

6. Two blocks, $m_1 = 1.00$ kg and $m_2 = 2.00$ kg, are connected by a massless 0.60 m long rod. The system is moving horizontally with a velocity of 1.50 m/s. A force of 20.0 N, as shown below, is applied to the system and moves it 3.00 m.

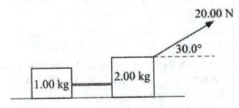

20.00 N

30.0°

1.00 kg

2.00 kg

This force will change the
(A) kinetic energy of only $m_2$ since it is closest to the center of mass.
(B) kinetic energy of the center of mass.
(C) kinetic energy of only $m_1$ since it is the smallest mass.
(D) internal energy of the system.

7. With an initial velocity of 1.50 m/s, an object moving along a horizontal surface is acted upon by the two forces shown below.

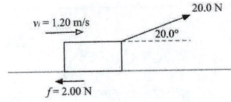

20.0 N

$v_i = 1.20$ m/s

20.0°

$f = 2.00$ N

(A) The kinetic energy of the object decreases because a frictional force acts on it, reducing its initial velocity.
(B) The kinetic energy increases because the horizontal component of the 20.0 N force is larger than the frictional friction and is in the same direction as the initial velocity.
(C) The kinetic energy is constant since the two forces acting on the object change neither the respective magnitude nor the direction of the initial velocity.
(D) The kinetic energy cannot be determined since the mass of the object is not indicated in the question.

8.  A 4.00 kg block is attached to a horizontal spring that has spring
    constant $k = 15.0$ N/m. The block is pushed to the left, compressing
    the spring by 8.00 cm. The block is then released from rest.
    Determine the maximum velocity of the block as its center of mass
    passes through the equilibrium position. (Ignore friction.)

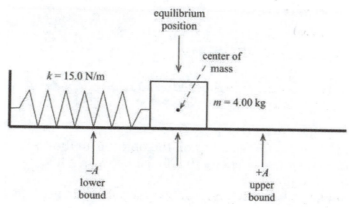

(A) 0.155 m/s
(B) 0.387 m/s
(C) 0.548 m/s
(D) 0.745 m/s

9.  A block is moved 12.0 m across a frictionless horizontal surface by
    a constant force of 30.0 N that acts at some angle $\theta$ less than 90° to
    the horizontal. If $\theta$ changes, what happens to the kinetic energy of
    the block?
    (A) It remains the same since the force and the displacement
        remain the same.
    (B) The kinetic energy will increase if $\theta$ increases toward 90°.
    (C) The kinetic energy will decrease if $\theta$ decreases toward 0°.
    (D) The kinetic energy will increase only if you increase the force
        or the displacement as you change the angle.

10. A 3.00 kg object is pushed in the $+x$ direction by a force as shown below. The velocity of the object is 1.80 m/s when it enters a rough region where the coefficient of friction between the object and the surface is $\mu_k = 0.200$.

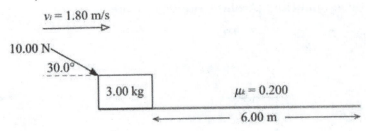

$v_i = 1.80$ m/s

10.00 N

30.0°

3.00 kg

$\mu_k = 0.200$

6.00 m

The kinetic energy of the object will
(A) increase because the net force acting on the object is in the same direction as the initial velocity the object had when it entered the rough region.
(B) remain constant since the two forces acting on the object are applied over the same displacement.
(C) decrease because the frictional force, which is larger than the 10.0 N force, is acting in the direction opposite the object's initial velocity.
(D) decrease because the net force acting on the object is in the opposite direction of the initial velocity the object had when it entered the rough region.

11. A 5.00 kg block with an initial velocity of 10.0 m/s is projected up a rough 30° incline. When the center of mass of the block has been elevated by 3.00 m, its velocity is 4.00 m/s.

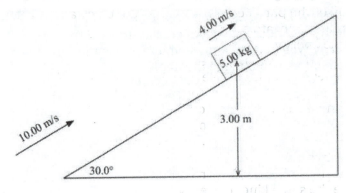

4.00 m/s

5.00 kg

3.00 m

10.00 m/s

30.0°

Which of the following best describes the work done by the frictional force?
(A) −250. J
(B) −147. J
(C) −63.0 J
(D) −40.0 J

12. Two masses, $m_1 = 2.00$ kg and $m_2 = 1.00$ kg, are connected to a massless string that passes over a frictionless pulley. What is the speed of the two masses after the larger mass has descended 1.50 m to the floor?

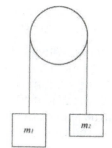

   (A) $v_1 > v_2$
   (B) $v_2 > v_1$
   (C) $v_1 = v_2$
   (D) $v_2 = 2v_1$

13. With a speed of 1.50 m/s, a 2.50 kg object slides across a frictionless tabletop that is 1.20 m above the floor. Relative to the tabletop, the energy associated with the object is
   (A) only kinetic energy.
   (B) only gravitational potential energy.
   (C) the sum of the kinetic energy and the gravitational potential energy.
   (D) the difference between the kinetic energy and the gravitational potential energy.

**Questions 14 and 15**
**Directions:** For each of the questions or incomplete statements below, two of the suggested answers will be correct. For each of these questions, you must select both correct choices to earn credit. No partial credit will be earned if only one correct choice is selected. Select the two that are best in each case.

14. The work done by the frictional force in question 11 goes into the system as (select two answers)
   (A) internal energy of the particles that compose the block and the incline as increased vibratory energy.
   (B) non-recoverable energy.
   (C) energy that leaves the system entirely.
   (D) some form of energy that is completely recoverable.

15. Two balls of the same mass are released from a height $h$. The first ball is dropped from rest, and the second ball is thrown vertically downward with some initial velocity $v_0$. Just before the balls reach the ground, (select two answers)
   (A) both balls have the same gravitational potential energy.
   (B) both balls have the same kinetic energy.
   (C) the ball that was thrown downward has the greater total energy.
   (D) the ball that was dropped has the greater total energy since it spends more time in the air.

## FREE-RESPONSE PROBLEMS

1.  (a)  You are to design an experiment to determine the amount of work done in stretching a spring. Provide enough detail so that another student could duplicate your experiment and obtain the same results.
    (b)  What measurements would you take, and how would you use these measurements to determine the amount of work done in your experiment?
    (c)  Suppose you replace your spring with one whose spring constant $k_2 = \dfrac{1}{2}k_1$. The work done in stretching the spring will

    ___ increase.
    ___ decrease.
    ___ remain constant.
    Justify your answer without calculations.
    (d)  Another student plots the following graph from data obtained in his experiment.

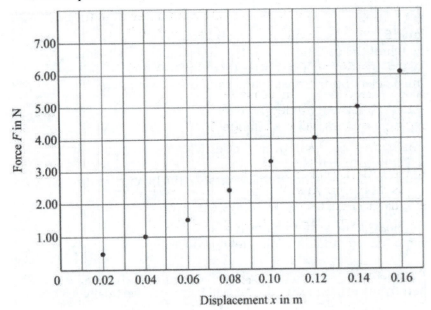

    i.  What does the plotted data indicate about the spring constant for the experiment?
    ii. Will the spring return to its original unstretched position when the load is removed?
    Explain your answer.

2.  A small, heavy block of mass $m$ as shown in the diagram rests on top of a smooth curved track whose lowest end is a height $h_2$ above a horizontal floor. When released from a height $h_1$ above the table, the block moves down the track and leaves the table horizontally, striking the floor a distance $D$ from the base of the table.

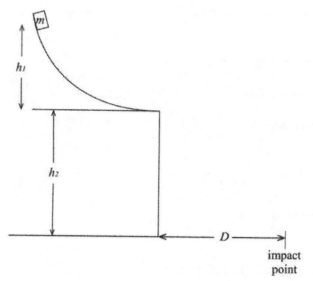

(a) Explain, without calculations, how you would determine the horizontal velocity of the block as it leaves the table.

(b) Explain, without calculations, how you would determine the horizontal distance $D$ the block travels before it strikes the floor.

(c) How will reducing the mass of the block change the distance $D$?

_____ Reducing the mass will increase $D$.
_____ Reducing the mass will decrease $D$.
_____ Reducing the mass will not change $D$.
Justify your answer.

3. (a) A 2.00 kg block $M_2$ is pulled across a frictionless surface by a 1.00 kg mass $m_1$ as shown in the figure below, which illustrates a massless, frictionless pulley in the system. Determine the velocity of the system when the 1.00 kg mass has descended to the floor.

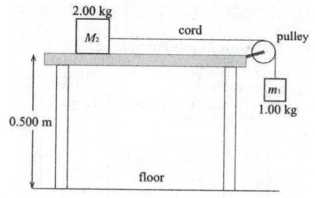

(b) The system is altered to the figure on the following page.

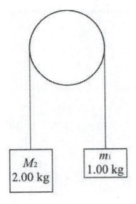

The velocity of the system will
___ increase.
___ decrease.
___ not change.
Justify your answer.

4. A 0.500 kg block slides on a frictionless curved track that is 1.50 m in height. It slides 2.00 m on a straight track where the coefficient of kinetic friction $\mu_k$ is 0.100 before it strikes a spring with a spring constant of 100.0 N/m.

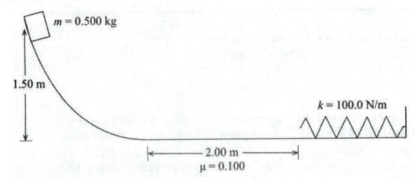

(a) Find the maximum compression of the spring.
(b) The mass rebounds returning up the curved ramp. Determine the rebound height.
(c) Complete the bar charts below for the energies associated with the mass.

| initial | | | | | compression | | | | | rebound | | | |
|---|---|---|---|---|---|---|---|---|---|---|---|---|---|
| $U_g$ | $K$ | $W_f$ | $U_s$ | | $U_g$ | $K$ | $W_f$ | $U_s$ | | $U_g$ | $K$ | $W_f$ | $U_s$ |

# Answers

## MULTIPLE-CHOICE QUESTIONS

1. **B** The kinetic energy of the block will decrease. Frictional forces act opposite the motion of an object and will do negative work on the block. Increasing the coefficient of friction will increase the frictional force.

   (L.O. 3.E.1.2)

2. **C** Gravity, a force that acts at a distance, does negative work on the 5.00 kg object as it is lifted 1.20 m in the gravitational field. It does no work on the object as it is carried horizontally 4.00 m since the force is perpendicular to the displacement. $W_g = -mgh$.

   Substitution gives $-5.00 \text{ kg}\left(9.80 \, \dfrac{\text{m}}{\text{s}^2}\right) 1.20 \text{ m} = -58.8 \text{ J}$.

   (L.O. 5.B.5.5)

3. **D** If only conservative forces do work, the total mechanical energy will be conserved. $K_0 + U_0 = K + U$. Either kinetic energy or potential energy can change, but the sum is constant.

   (L.O. 4.C.1.1, 4.C.1.2)

4. **B** The change in the kinetic energy is equal to the work done. $W = \Delta K$. The work done is the area under the force-displacement graph.

   $$5.00 \text{ N}(5.00 \text{ m} - 0 \text{ m}) + \frac{1}{2}(25.00 \text{ N} - 5.00 \text{ N})(5.00 \text{ m} - 2.00 \text{ m})$$

   $$+ \frac{1}{2}(25.0 \text{ N} - 0 \text{ N})(8.00 \text{ m} - 5.00 \text{ m})$$

   The work done is 25.0 J + 30.0 J + 37.5 J = 92.5 J.

   (L.O. 3.F.1.4)

5. **C** The increase in the internal potential energy of the spring comes from the conversion of the kinetic energy of the block when it strikes the spring and compresses beyond its equilibrium position. $\Delta K = -\Delta U_s$. The change in the kinetic energy is

   $$\Delta K = -\frac{1}{2}(2.00 \text{ kg})\left(\left(0 \, \frac{\text{m}}{\text{s}}\right)^2 - \left(1.60 \, \frac{\text{m}}{\text{s}}\right)^2\right) = -2.56 \text{ J}$$

   (L.O. 5.B.3.3)

6.  **B** The center of mass is the point in the object or a system of objects where the mass of the objects appears to be concentrated. A force applied to the center of mass will accelerate the system and change its velocity and hence its kinetic energy.

    (L.O. 4.C.2.2)

7.  **B** The net force acting on the object is in the direction of the initial velocity of the object. The work of the net force will increase the velocity and thus the kinetic energy of the object.

    (L.O. 3.E.1.2)

8.  **A** The internal energy of the spring is converted into kinetic energy of the block. The maximum velocity of the block occurs when the spring has a displacement from the equilibrium position of $x = 0$. $U_s = \Delta K$. Solving the energy equation gives

    $\frac{1}{2}kx^2 = \frac{1}{2}mv^2$. Thus, $\frac{1}{2}(15.0 \text{ N/m})(0.080 \text{ m})^2 = \frac{1}{2}(4.00 \text{ kg})v^2$. The velocity is $\pm 0.155 \text{ m/s}$.

    (L.O. 5.B.4.2)

9.  **C** The work done in moving an object by a given force is $W = |\vec{F}|(\cos\theta)d$. Increasing the angle decreases the horizontal component of the force acting in the direction of the displacement since the maximum value of the cosine of 0° is 1 and the minimum value of the cosine is 0 when the angle is 90°. This increase of the angle will decrease the amount of work done by the force compared with the original force over the 12.00 m displacement.

    (L.O. 3.E.1.1)

10. **A** The 10.0 N force acting downward at an angle of 30° has two components. The horizontal component of the force is $F_x = 10.0 \text{ N}$ cos30° = 8.66 N. Since the 10.0 N force pushes downward on the object, it increases the normal force exerted on the object by the surface. $N = F_y + F_g$. Therefore, the normal is

    $N = 10.0 \text{ N } (\sin 30°) + (3.00 \text{ kg})(9.80 \text{ m/s}^2) = 34.4 \text{ N}.$

    The net force acting on the object is

    $\Sigma F_x = 8.66 \text{ N} - 0.200 (34.4 \text{ N}) = 8.66 \text{ N} - 6.88 \text{ N} = 1.78 \text{ N}$

    The net force is in the direction of the initial velocity, and the kinetic energy will increase $\Sigma F_x \cdot x = \Delta K$.

    The kinetic energy will increase by 10.7 J.

    (L.O. 3.E.1.3)

11. **C** Conservation of energy applies. $K_0 + U_0 = K + U + W_f$. The initial gravitational potential energy is zero since the base of the incline is the reference point for this problem.

$$\frac{1}{2}(5.00 \text{ kg})\left(10.0 \text{ m}/_{s^2}\right)^2 - \frac{1}{2}(5.00 \text{ kg})\left(4.00 \text{ m}/_{s^2}\right)^2$$

$$-(5.00 \text{ kg})\left(9.80 \text{ m}/_{s^2}\right)(3.00 \text{ m}) = W_f$$

250 J – 40 J – 147 J = $W_f$. Since friction acts opposite the displacement of the object, the work done by friction is negative. The answer is –63 J.

(L.O. 4.C.1.1)

12. **C** The objects are connected with a string and move together with the same speed.

(L.O. 3.E.1.4, 5.B.4.2)

13. **A** Since the reference position is the tabletop, the object has no gravitational potential energy; it possesses only kinetic energy.

(L.O. 5.B.1.1)

14. **A** and **C** Work done by a non-conservative force either leaves the system for the surrounding environment or stays in the system in some form of an increase in internal energy that is vibratory energy that could show up as thermal energy or sound.

(L.O. 4.C.1.1, 4.C.1.2, 5.B.3.3, 5.B.4.1)

15. **A** and **C** Since the balls are released from the same height, they have the same gravitational potential energy. The ball that was projected downward with some initial velocity $v_0$ has initial kinetic energy. When they reach the ground, the object that was thrown downward has the greatest total energy. $K_0 + U_0 = K$.

(L.O. 4.C.1.1, 5.B.4.2)

## FREE-RESPONSE PROBLEMS

1. The work done in stretching a spring is the area under the curve of a force vs. displacement graph.
   (a) Measure the length of the spring.

   Hang the spring from a hook collar attached to a support.

   Add a mass to the free end of the spring and measure the elongation.

   Continue adding masses one at a time, making sure the spring does not oscillate.

   Record the elongation of the spring. Repeat until you have a minimum of six to eight readings.

(b) The measurements taken will be the mass in kg added to the spring and the force in N that each of the masses exert converted from $F_g = mg$.

The elongation of the spring measured in cm will be converted into $m$ units for each mass added to the spring.

A graph of force as a function of elongation will be plotted from the data that was recorded. The graph is linear if the spring obeys Hooke's law $F = -kx$. The slope of the line is the spring constant $k$ measured in N/m, and the area under the curve is the work done on the spring.

(c) The correct line checked is *increased*. If the spring constant is reduced such that $k_2 = \frac{1}{2}k_1$, forces applied to the spring will increase the elongation, thereby increasing the work done.

(d)  i. The graph indicates that the spring constant is not uniform over the graph.
    ii. There are two possible answers:
        No, the spring will not return to its initial position when the forces are removed. The spring is distorted.
        Yes, because the spring is non-Hookean does not mean the spring will not return to its original position. Unloading data are not given.

(L.O. 5.B.5.1, 5.B.5.2)

2. (a) By applying conservation of energy between the highest point and the tabletop, we can find the horizontal velocity of the block when it leaves the table. Since the track is smooth, no energy is "lost" to friction and the change in the gravitational potential energy of the earth-block system will equal its gain in kinetic energy. Since mass is a common factor in the equation, the horizontal velocity will be the square root of twice the acceleration due to gravity times the height $h_1$.

(b) The object leaves the table with a constant horizontal velocity. The distance traveled will be the horizontal velocity times the time to reach the floor. Since the object leaves the table with no vertical velocity, falling a distance $h_2$, the time for it to drop is found by taking the square root of twice the height, $h_2$, divided by the acceleration due to gravity.

(c) The mass of the object was eliminated in the conservation of energy equation. Changing the mass will not change the distance $D$.

(L.O. 4.C.1.1, 5.B.4.2, 5.B.5.4)

3. (a) Conservation of energy applies to the problem. Since the surface is frictionless, the gravitational potential energy will be converted into the kinetic energy of both bodies as the falling weight drops to the floor.

$$m_1gh = \frac{1}{2}(m_1 + M_2)v^2$$

Substitution gives the following:

$$(1.00 \text{ kg})\left(9.8 \text{ }^{m}/_{s^2}\right)(0.500 \text{ m}) = \frac{1}{2}(1.00 \text{ kg} + 2.00 \text{ kg})v^2$$

$$v = 1.81 \text{ }^{m}/_{s}$$

The correct line checked is the velocity will *not change*.

When the arrangement is changed to figure (b), conservation of energy still applies to the problem.

$$M_2 gh - m_1 gh = \frac{1}{2}(m_1 + M_2)v^2$$

The result is the same velocity:

$$v = 1.81 \text{ }^{m}/_{s}$$

(L.O. 4.C.1.1, 5.B.4.2, 5.B.5.4)

4.  (a)  The maximum compressions of the spring is determined from conservation of energy, $E = E_0 - W_f$.

$$U_{g0} + U_{s0} + K_0 - W_f = U_g + U_s + K$$

$$mgh + 0 + 0 - \mu_k mgx = 0 + \frac{1}{2}kx^2 + 0$$

$$(0.500 \text{ kg})\left(9.80 \text{ }^{m}/_{s^2}\right)(1.50 \text{ m}) - (0.100)(0.500 \text{ kg})\left(9.80 \text{ }^{m}/_{s^2}\right)(2.00 \text{ m})$$

$$= \frac{1}{2}\left(100.0 \text{ }^{N}/_{m}\right)(x^2)$$

$$x = 0.356 \text{ m}$$

(b)  After compression, conservation of energy still applies.

$$U_{g0} + U_{s0} + K_0 - W_f = U_g + U_s + K$$

$$0 + \frac{1}{2}kx^2 + 0 - W_f = mgh + 0 + 0$$

$$\frac{1}{2}\left(100.0 \text{ }^{N}/_{m}\right)(0.356 \text{ m})^2 - 0.100(0.500 \text{ kg})\left(9.8 \text{ }^{m}/_{s^2}\right)(2.00 \text{ m})$$

$$= 0.500 \text{ kg}\left(9.8 \text{ }^{m}/_{s^2}\right)h$$

$$h = 1.10 \text{ m}$$

(c)

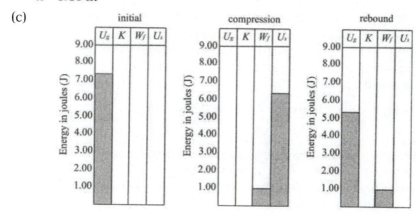

(L.O. 4.C.1.1, 4.C.1.2, 5.B.3.1, 5.B.3.2, 5.B.4.2, 5.B.5.4)

# 6

# MOMENTUM, IMPULSE, AND COLLISIONS

## LINEAR MOMENTUM

*Momentum* is a property of all moving objects. Galileo originally called the property "persistence," so momentum gets the symbol $p$. It is a vector quantity with magnitude and direction. It is defined as the product of an object's mass and velocity.

$$\vec{p} = m\vec{v}$$

The SI unit of momentum is $kg \cdot m/s$.

All moving objects have two properties: (1) momentum and (2) kinetic energy. Stationary objects do not have these properties.

### SAMPLE PROBLEM 1
Find the momentum of a proton, $m_p = 1.67 \times 10^{-27}$ kg, that has been accelerated to a velocity of $4.50 \times 10^6$ m/s in a particle accelerator.

### SOLUTION TO PROBLEM 1
Momentum is $p = mv$, and

$$p = \left(1.673 \times 10^{-27} \text{ kg}\right)\left(4.50 \times 10^6 \text{ m/}_{s}\right) = 7.53 \times 10^{-21} \text{ kg} \cdot \text{m/}_{s}$$

### SAMPLE PROBLEM 2
(a) Derive an expression relating momentum, $p$, and kinetic energy, $K$, as $p = f(K)$.
(b) Use the data from Sample Problem 1 to determine the momentum of the proton from its kinetic energy.

**168**

## SOLUTION TO PROBLEM 2

So then, solving for $p$ yields

(a)   $K = \frac{1}{2}mv^2$ and $K = \frac{1}{2}(mv)v$. Since $p = mv$, $K = \frac{1}{2}pv$.

Solving for $p$

$$p = \frac{2K}{v}$$

(b)   $p = \frac{2K}{v} = \frac{2(1.69 \times 10^{-14} \text{ J})}{(4.50 \times 10^6 \text{ m}\!/\!\text{s})} = 7.53 \times 10^{-21} \text{ kg} \cdot \text{m}\!/\!\text{s}$

## IMPULSE

Golfers, boxers, hockey players, and softball players try to "follow through" when they hit or swing; this maximizes the time they make contact. During the time $\Delta t$ that a golf club is in contact with a golf ball, a very large force is exerted on the ball by the club. Forces of this kind are called *impulsive forces*. Initially, the golf ball is at rest. Over the time interval $\Delta t$, the ball is accelerated and then separates from the club with some velocity, $v$.

We define *impulse, J,* as a vector quantity equal to the product of the force, $F$, and the time interval, $\Delta t$, during which the force acts. The direction of the impulse is the same as that of the force.

$$\vec{J} = \vec{F}\Delta t$$

The unit of impulse is the newton-second, N · s.

Impulse changes momentum. The amount of impulse delivered is equal to the amount momentum changes. The direction of vector change in momentum is the direction of the impulse vector—the same as the force vector that caused it. Every momentum change is caused by an impulse. So

$$\vec{F}\Delta t = m\Delta \vec{v}$$

## SAMPLE PROBLEM 3

A 1100 kg car traveling at 10 m/s collides with a concrete barrier and comes to rest in 0.9 second.
(a)   What force does the barrier exert on the car?
(b)   What acceleration does the car experience?

## SOLUTION TO PROBLEM 3

(a)   The car undergoes a momentum change that is caused by impulse, $m\Delta v = F\Delta t$. Since we are seeking the force, we solve for $F$:

$$F = \frac{m\Delta v}{\Delta t} = \frac{m(v - v_0)}{\Delta t} = \frac{(1100 \text{ kg})(0 - 10 \text{ m}\!/\!\text{s})}{(0.9 \text{ s})} = -12.2 \times 10^3 \text{ N}$$

(b)   Unbalanced forces cause accelerations: $F = ma$. The acceleration is

$$a = \frac{F}{m} = \frac{-12.2 \times 10^3 \text{ N}}{1100 \text{ kg}} = -11.1 \text{ m}\!/\!\text{s}^2$$

We also could have found the acceleration from

$$a = \frac{v - v_0}{t} = \frac{0 - 10 \; ^m/_s}{0.9 \; s} = -11.1 \; ^m/_{s^2}$$

## AP® Tip

The area beneath an $F$ vs. $\Delta t$ graph for a collision is the impulse generated during the collision.

A typical impact involves a normal force increasing from zero to some maximum value and then returning to zero over a short time period.

### SAMPLE PROBLEM 4

The graph below shows the force experienced by a cue ball while it impacts another pool ball. From the graph, determine the impulse generated in the collision.

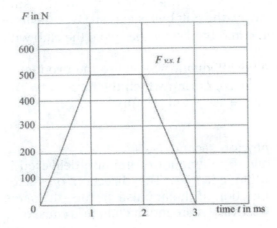

### SOLUTION TO PROBLEM 4

The impulse $J = F\Delta t$ is the area between the graph and the $x$-axis.

$$J = A = \frac{1}{2}(base)(altitude) + (base)(altitude) + \frac{1}{2}(base)(altitude)$$

$$J = \frac{1}{2}(1.00 \times 10^{-3} \; s)(500 \; N) + (1.00 \times 10^{-3} \; s)(500 \; N) + \frac{1}{2}(1.00 \times 10^{-3} \; s)(500 \; N)$$

$$= 1.00 \; N \cdot s$$

## LAW OF CONSERVATION OF LINEAR MOMENTUM

In a closed system with no external forces acting, the total initial momentum, $p_i$ of the system, is always equal to the total final momentum, $p_f$ of the system. This is a statement of one of the most important principals in all of physics, the *law of conservation of linear momentum*:

$$\Sigma \vec{p}_0 = \Sigma \vec{p}$$

## SAMPLE PROBLEM 5

A 20 g projectile is shot horizontally into a stationary 7 kg block of wood and becomes imbedded in it. Immediately after impact, the block is observed to move to the right with a velocity of 0.52 m/s. Determine the initial velocity of the projectile.

$m = 0.020$ kg

$v_0$ →

$M = 7.000$ kg
$V_0 = 0$

COLLISION
EVENT

$V$

$M + m = 7.020$ kg

## SOLUTION TO PROBLEM 5

To analyze a collision, always start by writing the law of conservation of linear momentum: $\Sigma p_0 = \Sigma p$. Before the collision, only the projectile is moving. After collision, the projectile and the block move together as a single object. The total initial momentum equals the total final momentum, $mv_0 = (M + m)V$. Solving for $V$,

$$v_0 = \frac{(M+m)V}{m} = \frac{(7.020 \text{ kg})(0.52 \text{ m/s})}{(0.020 \text{ kg})} = 182.5 \text{ m/s}$$

Note that the signs of the initial and final velocities are the same. The block + bullet combination continues moving to the right.

## COLLISIONS

When two objects strike one another, a *collision* occurs. In our model of collisions, we disregard any external forces and focus only on the forces that each object exerts on the other. During a collision, impulse changes the momentum of each object, yet the total momentum stays the same.

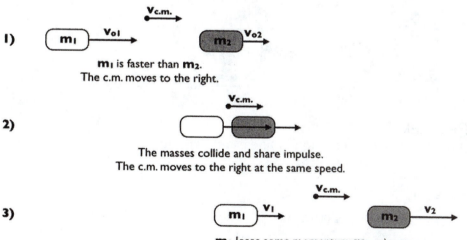

1) $m_1$ faster than $m_2$.
The c.m. moves to the right.

2) The masses collide and share impulse.
The c.m. moves to the right at the same speed.

3) $m_1$ loses some momentum; $m_2$ gains some.
The c.m. moves to the right at the same speed.

Above we see two isolated masses sliding across a flat surface (1) as they approach each other, (2) at the moment before they collide, and (3) after the collision. The white mass loses momentum in the collision, and the gray block gains momentum. The system's center of mass remains at the same velocity during the entire event. Total momentum is the same before, during, and after a collision, so the velocity of the center of mass does not change. If data show that the system's center of mass does experience a change in its velocity, then you know some outside force is acting on the whole system.

We differentiate different types of collisions based on whether total kinetic energy is lost to heat and whether the objects stick together. There are three types of collisions: totally inelastic, inelastic, and elastic. Remember that no matter what type of collision it is, momentum is always conserved.

## THE TOTALLY INELASTIC COLLISION

In a totally inelastic collision, the two objects stick together, merging their masses into one. When a bug hits the windshield of a car, it sticks to the car, and a totally inelastic collision has occurred. The sum of each object's momentum is conserved, becoming the momentum of the combined mass. The total kinetic energy is greater before the collision, as energy is converted into heat during the collision. As always, $\Sigma p_0 = \Sigma p$; conservation of linear momentum is always our starting point.

### SAMPLE PROBLEM 6

A 4.0 kg mother penguin, $m_1$, slides on the ice to the right along the $x$-axis with a velocity $v_1 = 10.0 \ \text{m/s}$. She undergoes a totally inelastic collision with her baby penguin, $m_2 = 6.0$ kg, sliding to the right with a velocity of $v_2 = 4.0 \ \text{m/s}$. Calculate the velocity of the combined mother-baby system immediately after the collision. How much kinetic energy is lost in this collision?

Before Collision        After Collision

$v_1$      $v_2$        $v$

### SOLUTION TO PROBLEM 6

Start by stating the law of conservation of linear momentum.

$\Sigma p_0 = \Sigma p$

$M_1 v_1 + m_2 v_2 = Mv.$

$(4 \ \text{kg})(10 \ \text{m/s}) + (6 \ \text{kg})(4 \ \text{m/s}) = (10 \ \text{kg})v$

$40 \ \text{kg m/s} + 24 \ \text{kg m/s} = (10 \ \text{kg})v$

$6.4 \ \text{m/s} = v$

In a totally inelastic collision, kinetic energy is not conserved and $\Sigma K_0 > \Sigma K$. Calculating both $K_0$ and $K$:

$$\Sigma K_0 = \frac{1}{2}m_1 v_1^2 + \frac{1}{2}m_2 v_2^2$$

$$= \frac{1}{2}(4.0 \text{ kg})(10.0 \text{ m/s})^2 + \frac{1}{2}(6.0 \text{ kg})(4.0 \text{ m/s})^2 = 248.0 \text{ J}$$

$$\Sigma K = \frac{1}{2}Mu^2 = \frac{1}{2}(10.0 \text{ kg})(6.4 \text{ m/s})^2 = 204.8 \text{ J}$$

By the work-kinetic energy theorem,

$$W = \Delta K = \Sigma K - \Sigma K_0 = 204.8 \text{ J} - 248.0 \text{ J} = -43.2 \text{ J}$$

During the collision, some kinetic energy is transferred to sound and heat energies.

## THE INELASTIC ELASTIC COLLISION

When colliding objects don't stick together after the collision and there is some loss of kinetic energy, the collision is considered inelastic. As always, $\Sigma p_0 = \Sigma p$ and momentum is conserved.

### SAMPLE PROBLEM 7

A 4.0 kg mass, $m_1$, travels to the right with a velocity of $v_1 = 10.0$ m/s and collides with a second mass, $m_2 = 6.0$ kg, also traveling to the right with a velocity of $v_2 = 4.0$ m/s.

    After the collision, $m_2$ is moving 5.5 m/s to the right.

(a)   Calculate the velocity of $m_1$ after the collision.
(b)   How much kinetic energy is lost in the collision?

### SOLUTION TO PROBLEM 7

(a)   Momentum before the collision must equal momentum after the collision.

$$\Sigma p_{before} = \Sigma p_{after}$$

$$m_1 v_1 + m_2 v_2 = m_1 v_1 + m_2 v_2$$

$$4.0 \text{ kg}(10 \text{ }^m/_s + 6.0 \text{ kg}(4.0 \text{ }^m/_s = 4.0 \text{ kg}(v_1) + 6.0 \text{ kg}(5.5 \text{ }^m/_s)$$

After the collision, $v_1 = 7.8 \text{ }^m/_s$

(b)   $K_{before} = \frac{1}{2} m_1 v_1^2 + \frac{1}{2} m_2 v_2^2 = \frac{1}{2} (4.0 \text{ kg})(10 \text{ }^m/_s)^2 + \frac{1}{2} (6.0 \text{ kg})(4.0 \text{ }^m/_s)^2$

$K_{before} = 248 \text{ J}$

$K_{after} = \frac{1}{2} m_1 v_1^2 + \frac{1}{2} m_2 v_2^2 = \frac{1}{2} (4.0 \text{ kg})(7.8 \text{ }^m/_s)^2 + \frac{1}{2} (6.0 \text{ kg})(5.5 \text{ }^m/_s)^2$

$K_{after} = 209 \text{ J}$

$K_{before} - K_{after} = 248 \text{ J} - 209 \text{ J} = 39 \text{ J}$

## THE PERFECTLY ELASTIC COLLISION

In perfectly elastic collisions, the amount of kinetic energy before and after the event is unaltered. No energy is lost during this collision process. Such a collision is called a perfectly elastic collision.

In nature, very few collisions are perfectly elastic. Neutrons and mono-atomic molecules such as helium approximate perfectly elastic collisions. Colliding ivory billiard balls come within several percent of being perfectly elastic. Otherwise, the collision is just an approximation.

In perfectly elastic collisions, as in all collisions, the law of conservation of linear momentum, $\Sigma p_0 = \Sigma p$, always holds true and kinetic energy is conserved, $\Sigma K_0 = \Sigma K$. ONLY in the perfectly elastic collision model does $\Sigma K_0 = \Sigma K$.

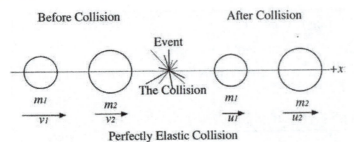

Perfectly Elastic Collision

## SAMPLE PROBLEM 8

A 4.0 kg mass, $m_1$, traveling to the right with a velocity of $v_1 = 10.0$ m/s collides elastically with a second mass $m_2 = 6.0$ kg traveling to the right with a velocity of $v_2 = 4.0$ m/s. Calculate the velocity of the masses $m_1$ and $m_2$ immediately after collision.

## SOLUTION TO PROBLEM 8

In all collisions the law of conservation of linear momentum always holds true and in a perfectly elastic collision kinetic energy is also conserved. From both of these laws we derive the following equations that only hold for perfectly elastic collisions.

$$u_1 = \frac{(m_1 - m_2)}{(m_1 + m_2)}v_1 + \frac{2m_2}{(m_1 + m_2)}v_2$$

$$u_2 = v_1 - v_2 + u_1$$

$$u_2 = \frac{2m_1}{(m_1 + m_2)}v_1 + \frac{(m_2 - m_1)}{(m_1 + m_2)}v_2$$

The velocity of mass $m_1$ immediately after collision is found by using the following equation:

$$u_1 = \left(\frac{m_1 - m_2}{m_1 + m_2}\right)v_1 + \left(\frac{2m_2}{m_1 + m_2}\right)v_2$$

$$u_1 = \left(\frac{4.0 \text{ kg} - 6.0 \text{ kg}}{10 \text{ kg}}\right)\left(10.0 \frac{m}{s}\right) + \left(\frac{12 \text{ kg}}{10 \text{ kg}}\right)\left(4.0 \frac{m}{s}\right) = 2.8 \frac{m}{s}$$

$$u_2 = \left(\frac{2m_1}{m_1 + m_2}\right)v_1 + \left(\frac{m_2 - m_1}{m_1 + m_2}\right)v_2$$

$$u_2 = \left(\frac{8.0 \text{ kg}}{10.0 \text{ kg}}\right)\left(10.0 \frac{m}{s}\right) + \left(\frac{2.0 \text{ kg}}{10.0 \text{ kg}}\right)\left(4.0 \frac{m}{s}\right) = 8.8 \frac{m}{s}$$

Calculating the total initial kinetic energy, $\Sigma K_0$:

$$\Sigma K_0 = \frac{1}{2}m_1v_1^2 + \frac{1}{2}m_2v_2^2$$

$$\Sigma K_0 = \frac{1}{2}(4.0 \text{ kg})(10.0 \text{ m/s})^2 + \frac{1}{2}(6.0 \text{ kg})(4.0 \text{ m/s})^2 = \textbf{248.0 J}$$

Calculating the total final kinetic energy, $\Sigma K$:

$$\Sigma K = \frac{1}{2}m_1u_1^2 + \frac{1}{2}m_2u_2^2$$

$$\Sigma K = \frac{1}{2}(4.0 \text{ kg})(2.8 \text{ m/s})^2 + \frac{1}{2}(6.0 \text{ kg})(8.8 \text{ m/s})^2 = \textbf{248.0 J}$$

$\Sigma K_0 = \Sigma K$ and the collision is perfectly elastic.

## AP® Tip

If asked to justify the type of collision, you must show whether the kinetic energy was conserved. Something sticking together or bouncing apart is not a proper justification.

## BALLISTIC PENDULUM

A ballistic pendulum is a device used to measure the velocities of small projectiles such as bullets. The simplest model of a ballistic pendulum consists of a block of wood hanging by vertical cords. When a bullet is fired and embeds itself in the wood block, it transfers momentum and energy to the block, causing it to swing through an arc. By measuring the mass of the block, $M$, and the mass of the bullet, $m$, and by measuring the vertical elevation, $h$, of the block, the initial velocity of the bullet can be determined.

The momentum ($mv$) of the flying bullet becomes the momentum of the combined mass ($M + m$). That mass has some kinetic energy that is converted to gravitational potential energy ($M + m$)$gh$ as the block rises.

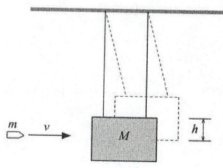

### SAMPLE PROBLEM 9

Consider the above diagram. A 10 g bullet is fired into the stationary 3990 g block of a ballistic pendulum. The bullet is captured in the block, elevating it vertically by 0.03 m.
(a)  What is the velocity, $V$, of the block-bullet system just after impact?
(b)  What is the velocity, $v$, of the bullet prior to impact with the block?

## SOLUTION TO PROBLEM 9

(a) Mechanical energy is conserved from the time the bullet impacts the block through its 0.03 m rise. Writing the law of conservation of mechanical energy:

$E_0 = E$ and at the moment just after impact

$\frac{1}{2}(m + M)V^2 = (m + M)gh.$ Solving for $V$:

$$V = \sqrt{2gh} = \sqrt{2\left(9.8\ ^m/_{s^2}\right)(0.03\ m)} = 0.77\ ^m/_s$$

(b) On impact, $\Sigma p_0 = \Sigma p$ and $mv = (m + M)V.$ Solving for the initial velocity of the bullet, $v$:

$$V = \frac{(m + M)V}{m} = \frac{(4.0\ kg)\left(0.77\ ^m/_s\right)}{0.01\ kg} = 308\ ^m/_s$$

# MOMENTUM OF THE CENTER OF MASS

In the absence of external forces, a system's *center of mass* maintains its momentum during a collision. When a collision occurs in the presence of a net external force, the system moves in such a way that its center of mass is accelerated by the net force. Internal impulses rearrange the momentum within the system while external forces affect the system as a whole by accelerating its center of mass.

Below we see an artillery shell fired from a cannon. The shell, actually the center of mass of the shell, follows a parabolic trajectory like any other projectile. At a point along the trajectory, as shown in the diagram below, the shell explodes into eight fragments.

The *cluster* of fragments forms a cloud about the center of mass, c.m. The beauty of the law of conservation of linear momentum is that the c.m. follows the original trajectory of the shell.

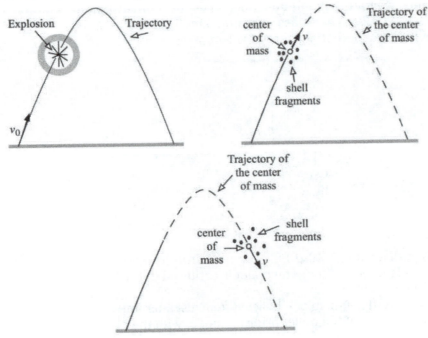

Astronauts on the International Space Station (ISS) could play three-dimensional pool, rearranging the momentum of pool balls countless times due to the impulses internal to the astronaut-balls system. All the while, the c.m. of the entire ISS and everything on it are pulled by gravity in a circle around Earth regardless of the impulse-momentum tradeoffs within.

In each system described above, the c.m.'s trajectory is subject to external forces that affect the whole system. The objects that make up the system are subject to internal forces that do not affect the path of their mutual center of mass.

## MOMENTUM, IMPULSE, AND COLLISION: STUDENT OBJECTIVES FOR THE AP® EXAM

- ▪ You should know how to determine the value of the average impulsive force if you know the change in the velocity of an object and the duration of the impulse.
- ▪ You should be able to explain how a system of two particles can have a total momentum of zero when both particles are moving.
- ▪ You should be able to identify the type of collision based on conservation of momentum and conservation of kinetic energy.

## MULTIPLE-CHOICE QUESTIONS

1. A 2.00 kg object moves under the action of a variable force as shown in the graph below.

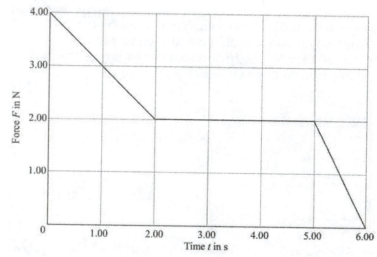

What is the change in the momentum of the object after 6.00 s?
(A) 7.00 N · s
(B) 10.0 N · s
(C) 12.0 N · s
(D) 13.0 N · s

2. A small rocket explodes into two fragments of equal mass at the highest point in its trajectory as shown.

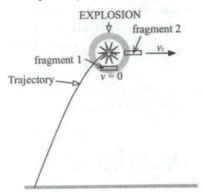

EXPLOSION

fragment 2

fragment 1

$v_x$

$v = 0$

Trajectory

Fragment 1 has zero velocity and drops as a falling object. Fragment 2 moves with horizontal velocity $\vec{v}_x$ after the explosion. After the two fragments have hit the ground, the center of mass of the system will be located
   (A) at the position of fragment 1 as it falls since the only force acting on it as it falls is the force of gravity.
   (B) at the position of fragment 2 since all of the velocity $\vec{v}_x$ remains with this fragment.
   (C) midway between the two equal massed fragments and located at the position where the rocket would have landed if there had been no explosion.
   (D) between the two equal massed fragments but closer to the landing position of the second fragment since it moved with all of the velocity $\vec{v}_x$ after the explosion.

3. Two students of $m_1 = 60.0$ kg and $m_2 = 50.0$ kg, push off each other on frictionless ice, exerting forces on each other shown below in four possible diagrams. $F_{12}$ represents the force student 1 applies to student 2, and $F_{21}$ represents the force student 2 applies to student 1.

I

$\vec{F}_{12}$          $\vec{F}_{21}$

The velocity of the center of mass is to the right since $\vec{F}_{21} > \vec{F}_{12}$.

II

$\vec{F}_{12}$          $\vec{F}_{21}$

The velocity of the center of mass is to the left since $\vec{F}_{12} > \vec{F}_{21}$.

III

$\vec{F}_{12}$          $\vec{F}_{21}$

The velocity of the center of mass is constant since $\vec{F}_{12} = \vec{F}_{21}$.

IV

$\vec{F}_{12}$          $\vec{F}_{21}$

The velocity of the center of mass is to the left since both forces are to the left.

Which diagram best represents this scenario?
(A) I
(B) II
(C) III
(D) IV

4.  A 20.0 g pellet is fired into a 1.50 kg block hanging by a vertical cord from the ceiling.

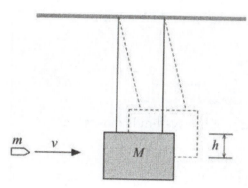

After the collision, the center of mass of the system rises 15.0 cm from its initial position.

   The initial velocity of the pellet is
   (A) 106 m/s.
   (B) 130 m/s.
   (C) 147 m/s.
   (D) 223 m/s.

5.  An object moving in the positive *x*-direction experiences a variable force as shown below in the force vs. time graph.

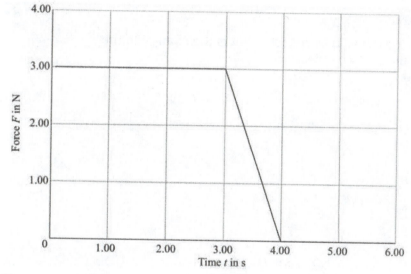

The momentum of the system will
   (A) increase since a variable but positive force acts on the object as shown by the area under the force-time graph.
   (B) decrease since the force applied to the object drops to zero as shown in the graph.
   (C) increase from 1.00 s to 3.00 s then decrease from 3.00 s to 4.00 s as shown in the force-time graph.
   (D) be undetermined since the mass of the object is not included in the problem.

6. As shown in the figure below, a 20.0 N force applied at an angle of 30° with the horizontal acts on the system for 10.0 s. The surface is flat and frictionless.

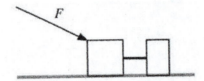

The change in the momentum of the system is closest to
(A) 200 N · s.
(B) 173 N · s.
(C) 150 N · s.
(D) 100 N · s.

7. Students performed a collision experiment measuring the recoil of two objects, $m_1$ = 4.00 kg and $m_2$ = 2.00 kg. Part of their data is shown below for the 4.00 kg object.

| x(m) | t(s) |
|------|------|
| 0.05 | 0.10 |
| 0.10 | 0.20 |
| 0.15 | 0.30 |
| 0.20 | 0.40 |
| 0.25 | 0.50 |
| 0.30 | 0.60 |

Using the data for the 4.00 kg object, what is the change in the momentum of the 2.00 kg object?

(A) 6.0 $\dfrac{\text{kg} \cdot \text{m}}{\text{s}}$

(B) 3.0 $\dfrac{\text{kg} \cdot \text{m}}{\text{s}}$

(C) 2.0 $\dfrac{\text{kg} \cdot \text{m}}{\text{s}}$

(D) 1.0 $\dfrac{\text{kg} \cdot \text{m}}{\text{s}}$

8. Two blocks of 0.500 kg and 1.00 kg on a flat horizontal surface are pushed together against a compressed spring and then released. When released, the 1.00 kg block moves to the right and the 0.500 kg block moves to the left.

    Data for the blocks at 0.100 s intervals are shown below.

| v (m/s) for 1.00 kg | v (m/s) for 0.500 kg |
|:---:|:---:|
| 1.20 | 2.40 |
| 1.00 | 2.20 |
| 0.81 | 2.00 |
| 0.61 | 1.82 |
| 0.42 | 1.62 |
| 0.21 | 1.42 |

    Evaluation of the data shows that
    (A) momentum was not conserved in the recoil since the velocities were not constant after the recoil.
    (B) momentum was conserved in the initial recoil of the two blocks, but frictional forces acting on the bodies reduced their speeds due to a frictional impulse.
    (C) momentum was conserved because momentum was transferred to the spring, increasing the internal energy and momentum of the molecules in the spring.
    (D) momentum may have been conserved, but without additional information about the spring, it cannot be determined.

9. Two blocks slide on a horizontal frictionless surface and collide as diagrammed below.

    The velocity and the kinetic energy for the 1.00 kg block are best answered by which of the following choices?

| | $v_{1.00\,kg}$ (m/s) | $K_{1.00\,kg}$ (J) | Type of Collision |
|:---:|:---:|:---:|:---|
| (A) | 1.50 | 1.50 | perfectly elastic |
| (B) | 2.00 | 2.00 | perfectly elastic |
| (C) | 1.50 | 1.50 | perfectly inelastic |
| (D) | 2.00 | 2.00 | perfectly inelastic |

10. A 2.00 kg cart moves in a straight line on a flat frictionless surface under the action of a variable force for 6.00 s. The graph of the velocity as a function of time is shown below.

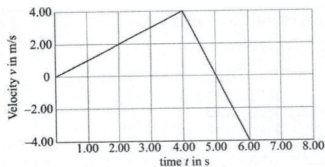

The change in the momentum of the cart during the 6.00 s interval is
(A) −16.0 N · s.
(B) −8.00 N · s.
(C) 8.00 N · s.
(D) 16.0 N · s.

11. A graph of the impulse on the 2.00 kg cart in problem 10 is best shown in which of the following graphs?

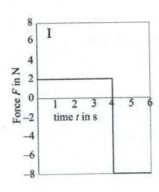

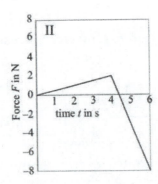

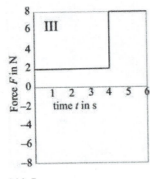

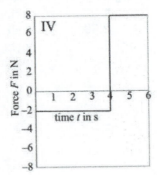

(A) I
(B) II
(C) III
(D) IV

12. A projectile of mass $M$ at the apex of its flight explodes into three fragments whose mass and velocity are shown in the illustration below.

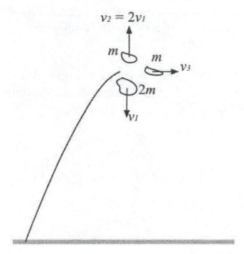

What was the velocity of the projectile just before it exploded?

(A) $2v_1$

(B) $\dfrac{v_3}{4}$

(C) $\dfrac{v_3}{3}$

(D) $\dfrac{v_2}{3}$

13. Rank the four blocks shown below for their change in momentum from the largest change to the smallest change. Each block is in motion for 10.0 s.

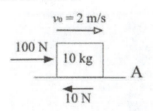

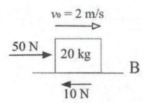

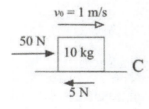

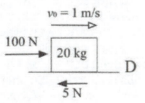

(A) D > A > C > B
(B) A > D > C > B
(C) A > B > D > C
(D) D > A > B > C

**Questions 14 and 15**
**Directions:** For each of the questions or incomplete statements below, two of the suggested answers will be correct. For each of these questions, you must select both correct choices to earn credit. No partial credit will be earned if only one correct choice is selected. Select the two that are best in each case.

14. A 0.30 kg sphere is projected toward the floor at an angle of 60° with the horizontal as shown. Which of the following statements are correct? Select two answers.

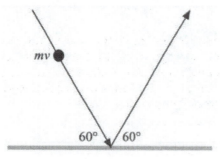

(A) Momentum is not conserved because the force between the sphere and the floor changes the vertical component of the velocity, causing it to rebound at the same angle.
(B) Momentum is conserved because the force between the sphere and the floor changes the horizontal component of the sphere so that it rebounds at the same angle.
(C) Kinetic energy is conserved since it has the same speed after the impact with the floor as it did before the impact.
(D) Kinetic energy is not conserved since the velocity changed from downward and negative to upward and positive.

15. A 1.00 kg block moving on a flat horizontal frictionless surface as shown below collides with a 2.00 kg block at rest. After the collision, they move together with the velocity shown.

After the collision, the kinetic energy will (select two answers)
(A) increase since both blocks are moving after the collision.
(B) remain constant. The 1.00 kg block moves with an initial velocity of 3.00 m/s, and afterward the composite mass, 3.00 kg, moves with a velocity of 1.00 m/s.
(C) decrease since the speed of the 3.00 kg composite mass was reduced in the collision and kinetic energy depends on $v^2$ not $v$.
(D) decrease since work was done in forming the composite mass and some undetermined amount of the initial kinetic energy was transferred into internal energy in the system.

## FREE-RESPONSE PROBLEMS

1. The series of objects shown below have been constructed by bending uniform, heavy gauge wire. Locate the center of mass of each by using symmetry and graphical methods as opposed to using mathematical equations. Explain your reasoning.

2. A 2.00 kg object, $m_A$, is attached to a cord. In the diagram below, the distance, $L$, from the attachment point of the cord to the center of mass of $m_A$ is 1.20 m. A second mass, $m_B = 1.00$ kg, is attached to a second cord with the same dimensions. Mass $m_B$ is displaced by 60.0° with the vertical and is then released from rest.

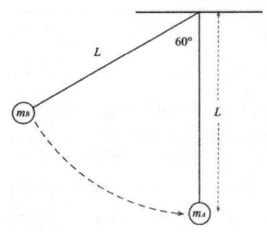

(a) Using a diagram, show how you will determine the height, $h$, that $m_B$ drops before its collision with the stationary mass, $m_A$.

(b) If the collision of the bodies is totally inelastic, what is the speed immediately after collision?

(c) Explain how you would determine the maximum vertical displacement of the composite mass after collision.

3. As shown below, a 0.030 kg projectile moving with a velocity of 300.0 m/s strikes a 2.00 kg block at rest on a post. The projectile passes through the block, emerging on the other side with a velocity of 100.0 m/s.

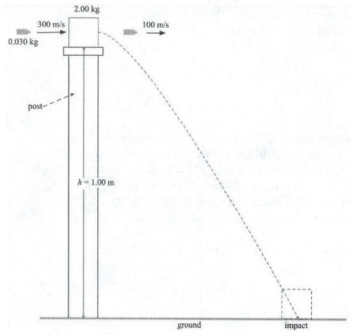

(a) What is the velocity of the block after the impact?
(b) How far from the base of the post will the block be when it impacts the ground?
(c) What is the fractional change in the kinetic energy of the block and the projectile?
(d) What are three possible reasons for the fractional change in the kinetic energy?

4. A firecracker shot into the air with some speed $v$ and at some angle $\theta$ relative to the horizontal explodes into several fragments at the apex of its trajectory. (Ignore air resistance.)
(a) Consider the time period immediately before and immediately after the explosion. Is the horizontal momentum of the center of mass conserved? Explain your answer.
(b) For the same time period, is the vertical momentum of the center of mass conserved? Explain your answer.
(c) What is the path of the center of mass?

# Answers

## MULTIPLE-CHOICE QUESTIONS

1. **D** The change in the momentum is the area under the graph of force versus time. This is $\frac{1}{2}bh + bh + \frac{1}{2}bh$. Substitution into the sum of the areas gives the following:

$$\frac{1}{2}(4.00 \text{ N} - 2.00 \text{ N})(2.00 \text{ s}) + 2.00 \text{ N}(5.00 \text{ s} - 0) +$$

$$\frac{1}{2}(2.00 \text{ N} - 0)(1.00 \text{ s}) = 13.0 \text{ N} \cdot \text{s}.$$

(L.O. 4.B.2.1)

2. **C** The center of mass of the system is located at the position where the projectile would have landed if it had not exploded. The explosion supplied an internal force only; no external force was applied in the horizontal direction. (The gravitational force was applied to the fragments—and hence the center of mass to change the vertical velocity; in the absence of friction, this results in the same vertical velocity.) In the horizontal direction, momentum is conserved. $MV_x = \frac{M}{2}(0) + \frac{M}{2}(2v_x)$. The second fragment will move horizontally a distance of $d = 2v_x(t)$. The center of mass is at the midpoint between the two fragments.

(L.O. 4.A.1.1)

3. **C** Action-reaction forces apply to the students as they push off each other on frictionless ice. These forces are equal in magnitude and opposite in direction. There will be no change in the center of mass of the system. As the students recoil, the velocity of the center of mass will remain in its original position between the two students, and since the students were initially at rest, the velocity of the center of mass will be constant—zero.

(L.O. 5.D.3.1)

4. **B** The collision is inelastic. Kinetic energy is not conserved in the collision, but after the collision, the initial kinetic energy of the block and the pellet is converted into gravitational potential energy of the system. $U_i + K_i = U_f + K_f$.

$$\frac{1}{2}(m + M)V^2 = (m + M)gh$$

Since the mass is a common factor, substitution gives

$V^2 = 2\left(9.80 \text{ m}/_{s^2}\right)(0.150 \text{ m})$ and the velocity of the center of mass of the system is $1.71 \text{ m}/_s$. Solving for momentum $mv = (m + M) V$. The velocity of the pellet is

$$v = \frac{(0.020 \text{ kg} + 1.50 \text{ kg})}{0.020 \text{ kg}}\left(1.71 \text{ m}/_s\right) = 130 \text{ m}/_s$$

(L.O. 5.D.2.5)

5. **A** The change in momentum is given as the area under the force-time graph. The force is variable but always positive; thus, the momentum will increase.

   (L.O. 4.B.2.2)

6. **B** Since the surface is frictionless, the net force acting on the system is $F = 20.0$ N$(\cos30°) = 17.3$ N. The change in the momentum is given by $Ft = \Delta p$. The change in the momentum is 173 N · s.

   (L.O. 4.B.2.1)

7. **C** The velocity is determined from the change in $x$ divided by the change in $t$. The velocity is $\dfrac{0.30 \text{ m} - 0.10 \text{ m}}{0.60 \text{ s} - 0.20 \text{ s}} = 0.50 \text{ m}/\text{s}$. The momentum of the 4.00 kg object is $2.0 \dfrac{\text{kg} \cdot \text{m}}{\text{s}}$. Since momentum is conserved, the momentum of the 2.00 kg object is also $2.0 \dfrac{\text{kg} \cdot \text{m}}{\text{s}}$.

   (L.O. 4.B.1.1)

8. **B** Momentum is conserved. After the initial recoil, both blocks move on a surface where friction acts. Friction applies an impulse to the blocks, reducing their momentum as indicated with the velocities in the data chart.

   (L.O. 5.D.2.4)

9. **B** Conservation of momentum $\vec{p}_i = \vec{p}_f$ will give the velocity of the 1.00 kg object after the collision.

   $$(2.00 \text{ kg})\left(1.50 \text{ m}/\text{s}\right) = (2.00 \text{ kg})\left(0.500 \text{ m}/\text{s}\right) + (1.00 \text{ kg})v$$

   The velocity of the 1.00 kg object is 2.00 m/s. The kinetic energy is found as follows:

   $$\frac{1}{2}(2.00 \text{ kg})\left(1.50 \text{ m}/\text{s}\right)^2$$
   $$= \frac{1}{2}(2.00 \text{ kg})\left(0.500 \text{ m}/\text{s}\right)^2 + \frac{1}{2}(1.00 \text{ kg})\left(2.00 \text{ m}/\text{s}\right)^2$$
   $$2.25 \text{ J} = 0.25 \text{ J} + 2.00 \text{ J} .$$

   The kinetic energy of the 1.00 kg object is 2.00 J, and since the initial kinetic energy equals the final kinetic energy, the collision is elastic.

   (L.O. 5.D.1.5)

10. **B** The change in momentum is $m\Delta v$, which is

$$(2.00 \text{ kg})\left(-4.00 \text{ m/s} - 0 \text{ m/s}\right) = -8.00 \text{ N} \cdot \text{s}$$

(L.O. 3.D.2.1)

11. **A** The shape of the graph can be determined using $F\Delta t = m\Delta v$ or taking the slope of the velocity-time graph $a = \dfrac{\Delta v}{\Delta t}$ and then using $\Sigma F = ma$. In the first method,

$$F(4.00 \text{ s}) = 2.00 \text{ kg}\left(4.00 \text{ m/s} - 0\right) = 2.00 \text{ N}$$

The next section of the graph is

$$F(1.00 \text{ s}) = 2.00 \text{ kg}\left(0 - 4.00 \text{ m/s}\right) = -8.00 \text{ N}, \text{ and the last section is}$$

$$F(1.00 \text{ s}) = 2.00 \text{ kg}\left(-4.00 \text{ m/s} - 0\right) = -8.00 \text{ N}$$

Using the slope requires taking the slope of the line from $t = 0.00$ s to 4.00 s and the line from 4.00 s to 6.00 s, and this gives from 0.00 s to 4.00 s:

$$a = \frac{4.00 \text{ m/s} - 0}{4.00 \text{ s}} = 1.00 \text{ m/s}^2 \text{ then using}$$

$$\Sigma F = 2.00 \text{ kg}\left(1.00 \text{ m/s}^2\right) = 2.00 \text{ N}.$$

Then the slope of the line from 4.00 s to 6.00 s is as follows:

$$a = \frac{\left(-4.00 \text{ m/s} - 4.00 \text{ m/s}\right)}{2.00 \text{ s}} = -4.00 \text{ m/s}^2 \text{ and}$$

$$\Sigma F = 2.00 \text{ kg}\left(-4.00 \text{ m/s}^2\right) = -8.00 \text{ N}.$$

(L.O. 4.B.1.2)

12. **B** The momentum has $v_x$ and $v_y$ components; looking at the information in the illustration, the change in the momentum $\Delta p_y = 0$ since $m(2v_1) - (2m)v_1 = 0$. The solution arises from $\Delta p_x = 0$. $MV = mv_3$, and since $M = m + 2m + m = 4m$, the initial velocity of the projectile is $V = \dfrac{mv_3}{4m} = \dfrac{v_3}{4}$.

(L.O. 3.D.2.1)

13. **A** Since the ranking asks for the change in the momentum, neither the mass nor the initial velocity is needed for the ranking. $F\Delta t = \Delta p$. The forces are applied for the same amount of time; thus, the largest net force will produce the largest change in the momentum.

(L.O. 3.D.1.1)

14. **A and C** In the collision with the floor, the force the floor applies is upward; it changes the vertical component of the initial velocity from $-\vec{v}_y$ to $+\vec{v}_y$ as

$$\left|\vec{F}_y\right|\Delta t = \left(mv\sin 60° - \left(-mv\sin 60°\right)\right) = 2mv\sin 60°$$ and makes no

change to the $+\vec{v}_x$ component. The sphere leaves the floor at the same angle with the horizontal; thus, the same speed and therefore momentum is not conserved here. Since the magnitude of the velocity components remained the same, the kinetic energy, which is a scalar, is conserved.

(L.O. 5.D.1.1)

15. **C and D** The collision was inelastic. Momentum but not kinetic energy is conserved in the collision. Kinetic energy after the collision depends on $\frac{1}{2}(m_1 + m_2)V^2$, not on $v$. (Although not part of the question, these values can be calculated. The initial kinetic energy was 4.5 J, and the final kinetic energy was 1.5 J.)

(L.O. 5.D.2.3)

## FREE-RESPONSE PROBLEMS

1. The center of mass of a uniform segment is located at its midpoint. Symmetry and graphical methods are shown with the dotted lines, and the intersection of the dotted lines locates the c.m. of the shape.

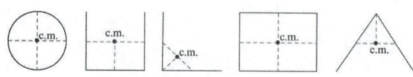

(L.O. 1.A.5.2)

2. (a) The mass $m_B$ was displaced through an angle of 60° forming a right triangle of altitude $L - h$ and hypotenuse $L$. The vertical drop $h$ can then be found by writing $\cos 60° = \dfrac{L-h}{L}$ $L - h = L\cos 60°$, and then $h$ becomes $h = L - L\cos 60° = L(1\ \cos 60°)$.

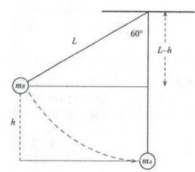

(b) By using the law of conservation of energy, the velocity of $m_B$ can be determined just before the collision.

$$U_0 + K_0 = U + K$$

$$m_B gh + 0 = 0 + \frac{1}{2} m_B v_B^2 \text{ and } v_B^2 = 2gh$$

From part (a), $h = L(1 - \cos 60°) = (1.20 \text{ m})(1 - \cos 60°) = 0.60 \text{ m}$.

Solving for $v$: $v = \sqrt{2\left(9.80 \text{ m}/_{s^2}\right)(0.60 \text{ m})} = \textbf{3.43 m}/_\textbf{s}$.

The law of conservation of linear momentum will allow us to calculate the velocity of the composite mass after collision.

$$m_B v_B = (m_A + m_B)u$$

Solving for the final velocity, $u$:

$$u = \frac{m_B v_B}{(m_A + m_B)} = \frac{(1.00 \text{ kg})\left(3.43 \text{ m}/_s\right)}{3.00 \text{ kg}} = \textbf{1.14 m}/_\textbf{s}$$

(c) Use conservation of mechanical energy to find the height to which the final system is elevated. The final gravitational potential energy of the system is related to the initial kinetic energy by total initial energy of the system = total final energy of the system.

Before collision is the kinetic energy of $m_B$, and after collision, the system has only gravitational potential energy when the system comes to rest.

$$\frac{1}{2} m_B v_B^2 = (m_A + m_B)gH$$

And then $H = \dfrac{m_B v_B^2}{2(m_A + m_B)g}$.

(L.O. 3.A.1.1, 4.C.1.1, 5.A. 2.1, 5.B 4.2, 5.D.1.5)

3. (a) Conservation of momentum occurs.

$$(0.030 \text{ kg})\left(300.0 \frac{\text{m}}{\text{s}}\right) = (0.030 \text{ kg})\left(100.0 \frac{\text{m}}{\text{s}}\right) + (2.00 \text{ kg})V_{block}$$

The velocity of the block is $\textbf{3.00} \dfrac{\textbf{m}}{\textbf{s}}$.

(b) The block drops from the post with an initial vertical velocity of zero. The time it takes the block to drop to the ground is given by $y = \frac{1}{2}gt^2$. $-1.00 \text{ m} = \frac{1}{2}\left(-9.80 \frac{\text{m}}{\text{s}^2}\right)t^2$. The time $t = 0.452$ s. The distance the block is from the base of the post is determined from $x = v_x t$.

$$x = \left(3.00 \frac{\text{m}}{\text{s}}\right)(0.452 \text{ s}) = \textbf{1.36 m}$$

(c) The fractional change in the kinetic energy is

$$\frac{1}{2}(0.030 \text{ kg})\left(300.0 \ \frac{m}{s}\right)^2 \neq \frac{1}{2}(0.030 \text{ kg})\left(100.0 \ \frac{m}{s}\right)^2 + \frac{1}{2}(2.00 \text{ kg})\left(3.00 \ \frac{m}{s}\right)^2$$

$$13.5 \times 10^2 \text{ J} \neq 1.50 \times 10^2 + 9.00 \text{ J} \ .$$

$$\frac{159 \text{ J}}{13.5 \times 10^2 \text{ J}} = 0.118 \text{ or } 11.8\% \text{ of the initial kinetic energy}$$

remains in the system.

(d) Possible explanations for the reduction in the kinetic energy include the following:

- The projectile did work on the block as it bored through the block.
- The projectile transferred kinetic energy to the block, giving it a horizontal velocity of 3.00 m/s.
- The internal energy of the block and the surrounding air are increased.
- Energy was transferred to the surroundings in the production of sound and heat energy.

(L.O. 5.B.3.1, 5.B.4.1, 5.D.1.5, 5.D.2.3)

4. (a) Horizontal momentum is conserved. Before and after the collision, the c.m. of the system is moving to the right with a velocity of $v \cos\theta$. Some pieces fly backwards, and some fly forwards, but the total horizontal momentum is unchanged.

(b) Vertical momentum is conserved. Before and after the collision, the c.m. follows the parabolic path of a projectile.

(c) In the absence of friction, the center of mass of the system follows the parabolic path the firecracker would have followed if it had not exploded. The explosion supplied an internal force only; no external force was applied to the center of mass of the system in the explosion. The gravitational force will be applied to the fragments (and hence the center of mass) to change the vertical velocity in the downward part of the trajectory. In the absence of friction, this causes the fragments to have the same vertical speed when they return to the ground. The horizontal velocity is constant.

(L.O. 3.D.1.1, 5.D.3.1)

# CIRCULAR MOTION AND ROTATIONAL MOTION

## CENTRIPETAL ACCELERATION

When an object is moving in a circular path with constant speed, its velocity is constantly changing because the direction of the tangential velocity is constantly changing. For this reason, the object experiences a constant acceleration. Remember that acceleration causes the object to speed up, slow down, or change direction. The acceleration causing the object to move in a circle produces a change in direction but not speed. The acceleration is always directed toward the center of the circle. Such an acceleration is called a *centripetal acceleration*. *Centripetal* means "center-seeking."

---

### AP® Tip

When you establish a frame of reference for a problem, choose a convenient rotation direction as positive. Usually, the direction of rotation is a good choice. If you're unsure, treat counterclockwise as positive by default.

---

Centripetal acceleration, $a_c$, is determined from the radius of curvature of the path and the speed as

$$\text{centripetal acceleration} = \frac{(\text{tangential speed})^2}{\text{radius of circular path}}$$

$$a_c = \frac{v^2}{r}$$

The SI units of centripetal acceleration are m/s², the same as any other acceleration.

## CENTRIPETAL FORCE

Recall that an unbalanced net force causes a mass to experience acceleration. The unbalanced net force that causes centripetal acceleration is called *centripetal force* and is directed toward the center of the circular path.

> ## AP® Tip
>
> Any body that travels along a curved path experiences a centripetal acceleration.

Centripetal force is

$$F_c = ma_c = \frac{mv^2}{r} = m\omega^2 r$$

The SI unit of centripetal force, as with all forces, is the newton, N.

Any object moving in a curved path does so because it experiences centripetal force. Something supplies the centripetal force. A ball tied to a string that is twirled overhead in a flat circle experiences centripetal force; so does the diesel engine rounding a curve on railroad tracks; so does Earth in its orbit about the sun. The tension in the string supplies the centripetal force on the ball. The rails provide centripetal force to the diesel engine. The force of gravity is the agent providing the centripetal force on Earth.

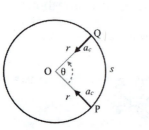  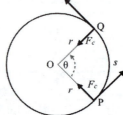

Diagram 1    Diagram 2

In diagram 1, a particle, as seen from above, travels with uniform speed *v* counterclockwise in a horizontal circle of radius *r*. Regardless of its location, the particle experiences a centripetal acceleration $a_c$ directed along the radius toward the center of the circle, O.

In diagram 2, the particle experiences a centripetal force $F_c$ to the center of the circle. At point P, the particle experiences a tangential velocity *v*. Of course, a tangent touches a curve at one point and one point only. The tangential velocity vector is always perpendicular to the centripetal acceleration vector and the centripetal force vector.

If the centripetal force were suddenly removed at point Q, no outside force would act on the particle and it would obey Newton's first law of motion, flying away in a straight line with speed $v$ at some direction.

When you are asked to draw a free-body diagram of something moving in a circular path, you do not label a separate vector arrow as centripetal force, $F_c$. The forces should be labeled as the type of force they are. The sum of all of the forces acting on the object, $\Sigma F$, is what causes the acceleration. For example, a ball attached to a string swinging horizontally in a circle would have the free-body diagram shown below.

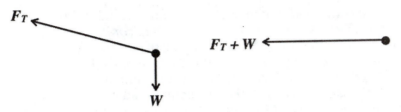

Tension and weight are the two forces acting on the ball as it spins. The sum of tension and weight is a vector pointing toward the center of the circular path.

## SAMPLE PROBLEM 1

A 2000 kg car rounds a flat curve of radius 60 m with a velocity of 9 m/s.
(a)  What centripetal force acts on the car?
(b)  What provides the centripetal force?

## SOLUTION TO PROBLEM 1

(a)  Centripetal force:

$$F_c = \frac{mv^2}{r} = \frac{(2000 \text{ kg})\left(9 \text{ m/s}\right)^2}{60 \text{ m}} = 2.7 \times 10^3 \text{ N}$$

(b)  The agent providing the force is the friction between the road surface and the tires.

## SAMPLE PROBLEM 2

A 500 g wooden ball is attached to a piece of string rated with a maximum tension of 8 N. The ball is tied to a 1 m piece of the string and is whirled overhead in a horizontal circle. What maximum speed can the ball have?

## SOLUTION TO PROBLEM 2

The centripetal force cannot exceed 8 N. Since $F_c = \dfrac{mv^2}{r}$, then $v^2 = \dfrac{F_c r}{m}$, and taking the square root of both sides gives

$$v = \sqrt{\frac{F_c r}{m}} = \sqrt{\frac{(8 \text{ N})(1 \text{ m})}{0.5 \text{ kg}}} = 4 \text{ m/s}$$

# TORQUE

In previous chapters, we assumed that forces were applied at an object's center of mass. Unbalanced forces applied at the c.m. accelerate objects in straight line, translational motion. If forces are applied at areas other than the c.m., we must consider the point of application of each force as well as the magnitude and direction of each force.

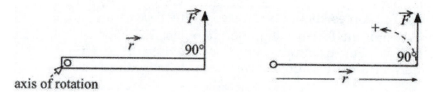

axis of rotation

The diagram above shows a rod that is attached to an external axle about which it can rotate freely. We call that point an *axis of rotation*. A force *F* is applied to the opposite end of the rod. The distance from the axis of rotation to the point of application of the force is called the *moment arm, r.* The force applied will cause the rod to rotate around the axis. A counterclockwise, CCW, rotation is considered to be a positive rotation (+), and a clockwise, CW, rotation is considered to be a negative rotation (–).

We define the torque $\vec{\tau}$, a vector, exerted by a force on an object as the measure of how well a force can cause an object to rotate around an axis. Torque is defined as

$$\tau = Fr\sin\theta$$

The angle $\theta$ is the angle between the moment arm and the applied force. When $\theta$ is 90°, when the moment arm and applied force are perpendicular, the sin90° = 1 and then

$$\tau = Fr$$

If a force is applied directly in line with the rod, it doesn't cause a torque since sin 0° = 0.

If more than one force causes torque on an object, we add the torques to find the *net torque, $\Sigma\tau$.* The magnitude of net torque is the difference between the sum of all clockwise torques and the sum of all counterclockwise torques.

When the sum of the torques applied to an object is zero, $\Sigma\tau = 0$, the object is said to be in rotational equilibrium. The sum of the torques that would cause the object to rotate counter clockwise must equal the sum of the torques that would cause the object to rotate clockwise.

The SI unit of torque is the Newton meter, Nm. Torque is not a unit of energy or work, so Nm in reference to torque is never a Joule.

## SAMPLE PROBLEM 3

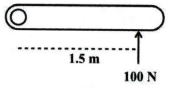

1.5 m

100 N

A force of 100.0 N is applied at the end of a 1.5 m rod and at a 90° angle. Ignore the mass of the rod. If the rod has a length $r = 1.5$ m, what is the torque acting on the rod?

## SOLUTION TO PROBLEM 3

Note that the force is applied upward, making the torque CCW (+). The torque about the axis of rotation is then

$$\tau_0 = Fr = +(100.0 \text{ N})(1.5 \text{ m}) = +150.0 \text{ N} \cdot \text{m}$$

The system will rotate CCW.

## SAMPLE PROBLEM 4

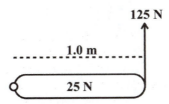

Calculate the torque on a 1.0 m long uniform rod that weighs 25 N and is pivoted at one end by a 125 N force applied 1.0 m from the point of rotation.

## SOLUTION TO PROBLEM 4

Both the 125 N force and the rod's weight apply a torque. The weight force is applied in the center of the rod at a distance of 0.5 m. The weight vector tends to make the system rotate CW (–). The applied force on the system tends to make the system rotate CCW (+). The total torque involved is then

$$\Sigma \tau_0 = \tau_w + \tau_F = -w(0.5r) + Fr = -(25.0 \text{ N})(0.5 \text{ m}) + (125.0 \text{ N})(1.0 \text{ m})$$

$$\Sigma \tau = +112.5 \text{ N} \cdot \text{m}$$

The system will rotate CCW.

## SAMPLE PROBLEM 5

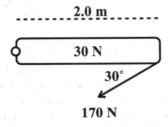

Calculate the torque on a 2.0 m long 30 N rod pivoted at one end by a 170 N force applied at an angle of 30° relative to the rod.

## SOLUTION TO PROBLEM 5

Both the weight vector and the applied force tend to make the system rotate CW about the axis of rotation. The total torque is

$$\Sigma \tau_0 = \tau_w + \tau_F$$
$$= -w(0.5r) + (-F)r \sin 30°$$
$$= (-30.0 \text{ N})(0.5 \times 2.0 \text{ m}) + (-170 \text{ N})(2.0 \text{ m})(\sin 30°)$$

$$\Sigma \tau_0 = -200.0 \text{ N} \cdot \text{m}$$

The negative sign implies that the system will rotate CW.

## TORQUE AND EQUILIBRIUM

When there is no resultant torque acting in a system, $\Sigma\tau = 0$, the system is in a state of rotational equilibrium. We have already addressed the first condition for equilibrium where the sums of the $x$- and $y$-components add to zero: $\Sigma x = 0$ and $\Sigma y = 0$. We must now state the second condition for equilibrium and stipulate that there is no resultant torque in the system, or $\Sigma\tau = 0$.

### SAMPLE PROBLEM 6

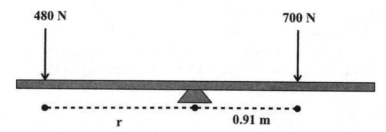

Where should a 480 N girl sit on a seesaw to balance her 700 N dad sitting 0.91 m from the pivot point?

### SOLUTION TO PROBLEM 6

The term *balance* implies equilibrium. The CCW torque and the CW torque will be equal.

$$(480 \text{ N})r = (700 \text{ N})(0.91 \text{ m})$$

$$r = \frac{(700 \text{ N})(0.91 \text{ m})}{(480 \text{ N})} = \textbf{1.33 m} \text{ from the pivot point}$$

## CENTER OF MASS

### SAMPLE PROBLEM 7

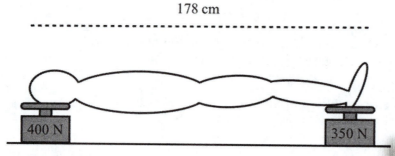

This student, holding perfectly still, is lying flat across two scales. Where is the student's center of mass?

### SOLUTION TO PROBLEM 7

The student's weight is the sum of the scale readings, 750 N.

Because the student is not rotating, we can choose any convenient point as the rotation point; for now, we will use the top of the student's head. The CW torque balances the CCW torque. The c.m. is some distance $r$ from the student's head.

$$(750 \text{ N})(r) = (350 \text{ N})(178 \text{ cm})$$

$$r = \frac{(350 \text{ N})(178 \text{ cm})}{(750 \text{ N})} = \textbf{83 cm} \text{ from the student's head}$$

## ANGULAR MOTION

Angles are commonly measured and expressed in degrees, where one full turn or one complete *revolution* is 360°. In the study of physics, radians are a more suitable unit in some cases.

$$\text{angle in radians} = \frac{\text{arc length}}{\text{radius}}$$

$$\theta = \frac{s}{r}$$

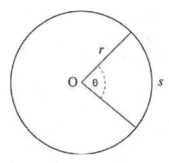

Since the circumference of a circle is $2\pi r$, there are $2\pi$ rad in one complete revolution, rev.

$$1 \text{ rev} = 360° = 2\pi \text{ rad}$$

and

$$1 \text{ rad} = 57.3°$$

## ANGULAR VELOCITY

The *angular velocity*, $\omega$, defines how rapidly an object is turning, spinning, or rotating.

$$\text{angular velocity} = \frac{\text{angular displacement}}{\text{time}}$$

$$\omega = \frac{\theta}{t}$$

In physics, angular velocity, $\omega$, is expressed in radians per second, rad/s. Another common unit for $\omega$ is revolutions per minute, RPM.

$$1 \frac{\text{rev}}{\text{s}} = 2\pi \frac{\text{rad}}{\text{s}}$$

The linear velocity, $v$, of a particle that travels in a circle of radius $r$ with uniform angular velocity, $\omega$, is related by

linear velocity = angular velocity × radius of the circle

or

$$v = \omega r$$

The time required to make one complete revolution is the period, $T$, and

$$T = \frac{2\pi}{\omega}$$

Since $f = \dfrac{1}{T}$, the angular frequency, $\omega$, is $\omega = 2\pi f$.

## ANGULAR ACCELERATION

A rotating body whose angular velocity changes from $\omega_0$ to $\omega$ in a time interval $t$ undergoes an *angular acceleration, $\alpha$.*

$$\text{angular acceleration} = \frac{\text{change in angular velocity}}{\text{time interval}}$$

or

$$\alpha = \frac{\omega - \omega_0}{t}$$

The unit for angular acceleration is rad/s².

Remember from Chapter 3 that if velocity and acceleration have the same signs, the object will speed up. If an object's angular velocity and angular acceleration have the same signs, the object will spin more rapidly. If $\omega$ and $\alpha$ have opposite signs, angular velocity will decrease.

The kinematic equations from Chapter 3 can be related to angular equations.

| | |
|---|---|
| $v_x = v_{x0} + a_x t$ | $\omega = \omega_0 + \alpha t$ |
| $x = x_0 + v_{x0}t + \dfrac{1}{2}a_x t^2$ | $\theta = \theta_0 + \omega_0 t + \dfrac{1}{2}\alpha t^2$ |
| $v_x^2 = v_{x0}^2 + 2a_x(x - x_0)$ | $\omega^2 = \omega_0^2 + 2\alpha\theta$ |

### SAMPLE PROBLEM 8

A wheel of radius $r = 0.30$ m spins at the rate of 900 rpm.
(a)   What is the angular velocity of all points on the wheel?
(b)   If the wheel slows uniformly to 60 rpm in 15 s, what angular acceleration does the wheel experience?

### SOLUTION TO PROBLEM 8

(a)   900 rpm = 900 revolutions per minute. To find the angular velocity, we need to convert revolutions into radians and minutes into seconds.

$$\omega = 900 \ \frac{\text{rev}}{\text{min}} \times \frac{2\pi \ \text{rad}}{1 \ \text{rev}} \times \frac{1 \ \text{min}}{60 \ \text{s}} = 94.2 \ \text{rad}/\text{s}$$

This angular velocity is shared by all points on the wheel.

(b)   First, we need to find the final angular velocity:

$$\omega = 60 \ \frac{\text{rev}}{\text{min}} \times \frac{2\pi \ \text{rad}}{1 \ \text{rev}} \times \frac{1 \ \text{min}}{60 \ \text{s}} = 6.28 \ \text{rad}/\text{s}$$

Angular acceleration is the rate of change of angular speed. Define angular acceleration and set the initial angular velocity as $\omega_0$.

$$\alpha = \frac{\omega - \omega_0}{t} = \frac{6.28 \ \text{rad}/\text{s} - 94.2 \ \text{rad}/\text{s}}{15 \ \text{s}} = -5.9 \ \text{rad}/\text{s}^2$$

## SAMPLE PROBLEM 9

A small pulley attached to the shaft of an electric motor has a radius $r = 0.05$ m and is turning at $\omega_0 = 5$ rad/s and speeds up to $\omega = 8$ rad/s in 2.5 s.

(a)   What acceleration does the pulley experience?
(b)   What is the angular displacement during this time period?
(c)   How many revolutions is this?

## SOLUTION TO PROBLEM 9

(a)   Both the initial and final angular velocities are in rad/s. Then

$$\alpha = \frac{\omega - \omega_0}{t} = \frac{8 \ \text{rad}/\text{s} - 5 \ \text{rad}/\text{s}}{2.5 \ \text{s}} = 1.2 \ \text{rad}/\text{s}^2$$

(b)   Since we know the time period,

$$\theta = \omega_0 t + \frac{1}{2}\alpha t^2 = \left(5 \ \frac{\text{rad}}{\text{s}}\right)(2.5 \ \text{s}) + \frac{1}{2}\left(1.2 \ \frac{\text{rad}}{\text{s}^2}\right)(2.5 \ \text{s})^2 = 16.2 \ \text{rad}$$

(c)   To find the number of revolutions, we use a conversion factor:

$$n = 16.2 \ \text{rad} \times \frac{1 \ \text{rev}}{2\pi \ \text{rad}} = 2.6 \ \text{rev}$$

## ANGULAR AND TANGENTIAL RELATIONSHIPS

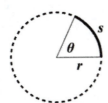

When a wheel rotates through an angle $\theta$, the motion of a point on the wheel's edge can be described in terms of the distance, $s$, it has moved, its tangential speed, $v$, and its tangential acceleration, $a$. These quantities are related to the angular displacement, $\theta$, angular velocity, $\omega$, and angular acceleration, $\alpha$, by the following relationships:

$$s = \theta r \qquad\qquad v = \omega r \qquad\qquad a = \alpha r$$

## SAMPLE PROBLEM 10

A bicycle wheel mounted on a test frame has a diameter of 1.2 m and spins at the rate of 4.0 rad/s.

(a) What is the linear or tangential speed of a particle on the circumference of the wheel?

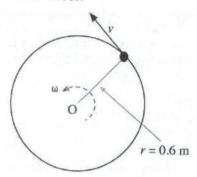

$r = 0.6$ m

(b) What is the linear or tangential speed of a particle 0.3 m from the center of the wheel?

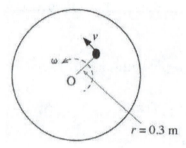

$r = 0.3$ m

## SOLUTION TO PROBLEM 10

(a) The relationship between the angular velocity of a wheel and the linear velocity at any point on the wheel is

$$v = \omega r = \left(4.0 \ \frac{\text{rad}}{\text{s}}\right)(0.6 \ \text{m}) = 2.4 \ \frac{\text{m}}{\text{s}}$$

If the particle were free to leave the circumference of the wheel, it would fly off in a straight-line tangent to the circumference at 2.4 m/s.

(b) The linear speed at 0.3 m from the center of the wheel is

$$v = \omega r = \left(4.0 \ \frac{\text{rad}}{\text{s}}\right)(0.3 \ \text{m}) = 1.2 \ \frac{\text{m}}{\text{s}}$$

Radians do not carry a unit.

## MOMENT OF INERTIA

All rotating bodies have a property called a *moment of inertia*. The moment of inertia, *I*, of an object is a measure of the rotational inertia of the body. It depends on the size, shape, and mass distribution of the object with respect to an axis of rotation. Moment of inertia is a scalar quantity with units of kg · m². The moments of inertia of several bodies are given below. On the AP® exam, if you need a moment of inertia, it will be provided for you.

| Object | Moment of Inertia | Shape |
|---|---|---|
| Hoop of mass $M$ and radius $R$. The axis of rotation is through the geometric center. | $I = MR^2$ | |
| Solid disk or cylinder of mass $M$ and radius $R$. The axis of rotation is through the geometric center. | $I = \frac{1}{2}MR^2$ | |
| Solid sphere of mass $M$ and radius $R$. The axis of rotation is through the center of mass. | $I = \frac{2}{5}MR^2$ | |
| Hollow sphere of mass $M$ and radius $R$. The axis of rotation is through the center of mass. | $I = \frac{2}{3}MR^2$ | |
| Thin rod of mass $M$ and length $L$. The axis of rotation is through the center of mass and is perpendicular to the length. | $I = \frac{1}{12}ML^2$ | |

## SAMPLE PROBLEM 11

Calculate the moment of inertia about an axis through the center of a 25 kg solid sphere whose diameter is 0.30 m.

## SOLUTION TO PROBLEM 11

$$I = \frac{2}{5}MR^2 = 0.4(25 \text{ kg})(0.15 \text{ m})^2 = \textbf{0.23 kg} \cdot \textbf{m}^2$$

# THE ROTATIONAL SECOND LAW OF MOTION

Just as unbalanced forces cause acceleration, unbalanced torques cause angular acceleration. The angular acceleration produced by a given net torque depends on the mass as well as the moment of inertia.

Torque = moment of inertia × angular acceleration

Often called the rotational second law of motion, a resultant torque, $\vec{\tau}$, acting on a rigid body of moment of inertia, $I$, about an axis produces an angular acceleration, $\vec{\alpha}$.

$$\vec{\tau} = I\vec{\alpha}$$

Earlier, we defined torque as the product of the force $\vec{F}$ applied at a distance $\vec{R}$ from the axis of rotation. Recall that we called $\vec{R}$ the moment arm.

When $\vec{F}$ and $\vec{R}$ are perpendicular,

$$\vec{\tau} = \vec{R}\vec{F} \text{ and } I\alpha = RF$$

When a ball rolls, friction is the force applying the torque that causes the ball to rotate. If there were no friction to apply the torque to the ball, it would merely slide across the floor. On the AP® exam, pay attention to descriptions of objects in motion. If they are sliding, there is no rotational motion. If they are rolling, you must consider rotational motion.

## SAMPLE PROBLEM 12

A motor spins a 1.2 kg pulley 1200 times per minute. The pulley's radius is 12 cm. When the motor turns off, the pulley takes 40 s to stop spinning. How much torque is acting on the pulley while it is stopping?

## SOLUTION TO PROBLEM 12

First, we need to determine the angular acceleration in rad/sec.

$$\omega_0 = 1200 \text{ rev}/\text{min} \times 1 \text{ min}/60 \text{ s} \times 2\pi \text{ rad}/1 \text{ rev} = 126 \text{ rad}/\text{s}$$

Angular acceleration $\alpha = \dfrac{\omega - \omega_0}{t} = \dfrac{0 - 126 \text{ rad}/\text{s}}{40 \text{ s}} = -3.2 \text{ rad}/\text{s}^2$

The moment of inertia of a disk is $I = \dfrac{1}{2}MR^2$

$\Sigma\tau = I\alpha$

$$\tau = \frac{1}{2}\alpha MR^2 = (0.5)\left(-3.2 \frac{\text{rad}}{\text{s}^2}\right)(1.2 \text{ kg})(0.12 \text{ m})^2 = -0.028 \text{ N} \cdot \text{m}$$

The torque is negative because the acceleration is negative.

## SAMPLE PROBLEM 13

A 0.6 kg mass hangs at rest from the end of a cord wrapped several times around a pulley with a 0.15 m radius. When released from rest, the mass falls 2.2 m in 6.0 s. Determine the value of the moment of inertia of the pulley.

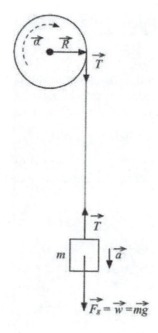

## SOLUTION TO PROBLEM 13

First, we calculate the acceleration of the falling mass.

We know how far the mass falls and how much the time it takes to fall: $y = -2.2$ m and $t = 6$ s.

$$y = v_0 t + \frac{1}{2} at^2 = 0 + \frac{1}{2} at^2$$

Solving for $a$, $a = \dfrac{2y}{t^2} = \dfrac{2(-2.2 \text{ m})}{(6 \text{ s})^2} = -0.12 \text{ m}/_{s^2}$.

Now that we know the linear acceleration, we can find the angular acceleration with $a = \alpha R$.

$$\alpha = \frac{a}{R} = \frac{\left(-0.12 \text{ m}/_{s^2}\right)}{0.15 \text{ m}} = -0.8 \text{ rad}/_{s^2}$$

The torque acting on the pulley is $\vec{\tau} = I\vec{\alpha}$, and the force acting on the mass is $\vec{F} = m\vec{a}$. The negative sign means that the pulley accelerates in the clockwise direction.

From Newton's second law, the unbalanced force acting on mass $m$ is $\vec{F}_{net} m\vec{a}$ and $T - mg = ma$. Solving for $T$,

$$T = ma + mg = m(a + g) = (0.6 \text{ kg})\left(-0.12 \text{ m}/_{s^2} + 9.8 \text{ m}/_{s^2}\right) = 5.8 \text{ N}$$

Now we write $\tau = I\alpha$ for the pulley.

$$\tau = RF = RT = Ia$$

Solving for $I$,

$$I = \frac{RT}{\alpha} = \frac{(0.15 \text{ m})(-5.8 \text{ N})}{\left(-0.8 \text{ rad}/_{s^2}\right)} = 1.1 \text{ kg} \cdot \text{m}^2$$

## ROTATIONAL ENERGY

All moving objects have kinetic energy. Rotating objects have *rotational kinetic energy*, $K_{rot}$, and we define it as $K_{rot} = \frac{1}{2} I\omega^2$. Rotational kinetic energy is measured in Joules.

## SAMPLE PROBLEM 14

A 12 kg solid steel disk has a radius of 0.06 meter. The disk spins with a velocity of 10 rad/s. What is the rotational kinetic energy of the disk?

## SOLUTION TO PROBLEM 14

A solid disk has $I = \frac{1}{2} MR^2$, and rotational kinetic energy is

$$K_{rot} = \frac{1}{2} I\omega^2$$

$$K_{rot} = \frac{1}{2}I\omega^2$$

$$= \frac{1}{2}\left(\frac{1}{2}MR^2\right)\omega^2$$

$$= \frac{1}{4}MR^2\omega^2$$

$$= (0.25)(12 \text{ kg})(0.06 \text{ m})^2\left(10\ \frac{\text{rad}}{\text{s}}\right)^2$$

$$K_{rot} = 1.08 \text{ J}$$

## SAMPLE PROBLEM 15

A mass $M$ is free to slide without friction across a horizontal tabletop. This mass is connected by a light string to a mass $m$ that hangs over the edge of the table. The connecting string passes over a frictionless pulley in the shape of a disk with radius $R$ and mass $m_p$. Find the velocity of the falling mass as it strikes the floor. The mass $m$, starting from rest, falls a distance $h$ to the floor.

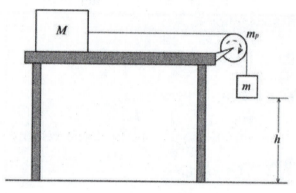

## SOLUTION TO PROBLEM 15

Because the system is frictionless, we know that mechanical energy is conserved.

$$U_0 + K_0 = U + K$$

$$mgh + 0 = 0 + \frac{1}{2}mv^2 + \frac{1}{2}Mv^2 + \frac{1}{2}I\omega^2$$

The pulley's moment of inertia is $I = \frac{1}{2}m_pR^2$, and $\omega = v/R$.

$$mgh = \frac{1}{2}mv^2 + \frac{1}{2}Mv^2 + \frac{1}{2}\left(\frac{1}{2}m_pR^2\right)\left(\frac{v}{R}\right)^2$$

Solving for $v$ yields

$$v = \sqrt{\frac{2mgh}{m + M + m_p/2}}$$

Objects that roll have rotational motion as well as linear motion. The total kinetic energy of a rolling object is

$$K_{total} = K_{rot} + K = \frac{1}{2}I\omega^2 + \frac{1}{2}Mv^2$$

## SAMPLE PROBLEM 16

A solid 10 kg sphere rolls without slipping across a horizontal surface at 15 m/s and then rolls up an inclined plane tilted at 30°. If friction losses are negligible, at what height, $h$, above the floor will the ball come to rest?

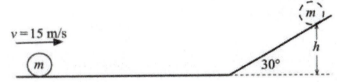

## SOLUTION TO PROBLEM 16

The rotational and translational kinetic energies of the rolling sphere at the bottom of the incline will be totally changed into gravitational potential energy when it stops.

$$U_0 + K_0 = U + K$$

$$0 + K_{rot} + K = mgh + 0 + 0$$

$$\frac{1}{2}I\omega^2 + \frac{1}{2}mv^2 = mgh$$

The sphere's moment of inertia is $I = \frac{2}{5}mR^2$, and the linear velocity of the sphere at its periphery is $v = \omega R$. Solving for angular velocity, we get $\omega = v/R$.

$$\frac{1}{2}\left(\frac{2}{5}mR^2\right)\left(\frac{v}{R}\right)^2 + \frac{1}{2}mv^2 = mgh$$

Solving for $h$,

$$\frac{1}{5}v^2 + \frac{1}{2}v^2 = gh$$

And

$$h = \frac{0.7v^2}{g} = \frac{0.7\left(15 \; ^m/_s\right)^2}{9.8 \; ^m/_{s^2}} = \textbf{16 m}$$

Notice that the mass $m$, the radius $R$, and the angular velocity $\omega$ did not enter into the calculations.

## ANGULAR MOMENTUM

Any rigid body that rotates has *angular momentum*, $\vec{L}$, and its angular momentum is $\vec{L} = I\vec{\omega}$. Bodies that travel along a curved path also have angular momentum about a given point $\vec{L} = m\vec{v}\vec{r}$.

The unit of angular momentum is the $kg \cdot m^2/s$.

> **AP® Tip**
>
> The angular momentum is a vector quantity. Bodies rotating or orbiting counterclockwise have a positive (+) angular momentum, and when rotations are negative (–), the rotation is clockwise.

The quantity $\vec{\tau}\Delta t$ is called angular impulse, and $I\Delta\vec{\omega}$ is the change in the angular momentum, $\Delta\vec{L}$. Just as linear impulse changes linear momentum, angular impulse changes angular momentum.

$$\vec{\tau}\Delta t = I\Delta\vec{\omega} = \Delta\vec{L}$$

## SAMPLE PROBLEM 17

A 4 kg hollow sphere with a mass of 4 kg and a radius of 4 cm is set into motion about an axis through the center. If the sphere has an angular speed of 20 rad/s, what is its angular momentum?

## SOLUTION TO PROBLEM 17

The moment of inertia of a hollow sphere is $\frac{2}{3}MR^2$. Its angular momentum is

$$L = I\omega$$

$$= \left(\frac{2}{3}MR^2\right)\omega^2$$

$$= \frac{2(4\text{ kg})(0.04\text{ m})^2\left(20\ \text{rad}/_s\right)^2}{3}$$

$$= 1.7\text{ kg} \cdot \frac{\text{m}^2}{\text{s}}$$

## SAMPLE PROBLEM 18

An 8 kg grinding wheel with a 0.12 m radius is mounted on the shaft of an electric motor and initially is at rest. The motor is turned on, and after 20.0 s, the grinding wheel reaches its maximum speed of 1800 rpm. What torque does the motor apply to the wheel?

## SOLUTION TO PROBLEM 18

First, we convert 1800 rpm to rad/s.

$$\omega = 1800\text{ rpm} \times \frac{1\text{ min}}{60\text{ s}} \times \frac{2\pi\text{ rad}}{1\text{ rev}} = 188.5\ \text{rad}/_s$$

Since the wheel starts from rest, $\omega_0 = 0$.

The impulse-momentum relationship tells us that $\vec{\tau}\Delta t = I\Delta\vec{\omega}$.

A solid disk has $I = \frac{1}{2}MR^2$. In substituting and solving for the torque,

$$\tau = \frac{I\Delta\omega}{\Delta t}$$

$$= \frac{\frac{1}{2}MR^2(\omega - \omega_0)}{t} = \frac{(0.5)(8\text{ kg})(0.12\text{ m})^2\left(188.5\ \text{rad}/\text{s} - 0\right)}{(20\text{ s})}$$

$$= 0.5\text{ N}\cdot\text{m}$$

## THE LAW OF CONSERVATION OF ANGULAR MOMENTUM

If no external torque acts on an object, the angular momentum of an object rotating about a fixed axis is constant.

In a system where there is no external torque, the total angular momentum before any event is always equal to the total angular momentum after the event. We call this statement the *law of conservation of angular momentum*.

$$\Sigma\vec{L} = \Sigma\vec{L}_0$$

$$I_0\omega_0 = I\omega$$

The product of the moment of inertia and the angular velocity, $I\omega$, is the conserved quantity, not the angular velocity $\omega$. Changing a system's moment of inertia causes its angular velocity to change. When a figure skater begins to spin with open arms, her moment of inertia is largely due to the wide distribution of her mass. As the skater pulls her arms in and decreases her moment of inertia, her angular velocity will increase. At all times during her spin, her angular momentum is the same.

### SAMPLE PROBLEM 19

A neutron star is formed when a star, such as our sun, collapses in on itself. Before collapse, the mass of the star is $M_0$ and its radius is $R_0$. After collapse, the neutron star has mass $M_0$ and radius $(1 \times 10^{-5})\ R_0$. The mass does not change, but the radius shrinks by a factor of one hundred thousand. Before collapse, the star rotated at 1 revolution every 25 days. Determine the rotation rate of the neutron star in rev/s.

### SOLUTION TO PROBLEM 19

Angular momentum must be conserved, and the total angular momentum before collapse must equal the total angular momentum after collapse. We write the law of conservation of angular momentum as $I_0\omega_0 = I\omega$.

$$\frac{1}{2}M_0R_0^2\omega_0 = \frac{1}{2}M_0\left(1\times10^{-5}R_0\right)^2\omega$$

Note that the one-half $M_0$ and $R_0^2$ appear on both sides of the equation and divide out.

$$\omega = \frac{\omega_0}{\left(1\times10^{-5}\right)^2} = \frac{\dfrac{1\text{ rev}}{25\text{ d}}\times\dfrac{1\text{ d}}{24\text{ h}}\times\dfrac{1\text{ h}}{3600\text{ s}}}{\left(1\times10^{-10}\right)} = 4630\ \frac{\text{rev}}{\text{s}}$$

## CIRCULAR MOTION AND ROTATIONAL MOTION: STUDENT OBJECTIVES FOR THE AP® EXAM

■ You should be able to discuss why a particle moving in a horizontal circle at constant speed experiences a centripetal acceleration and a centripetal force.

■ You should be able to explain the relationship between angular and linear descriptions of rotational motion.

■ You should be able to explain the meaning of the term *moment of inertia*.

■ You should be able to explain why an ice skater spins faster when she pulls her arms in toward her body than when she extends her arms.

## MULTIPLE-CHOICE QUESTIONS

1.  A cylinder is rotating clockwise about a frictionless axle when two forces are applied to the rim of the cylinder as shown below.

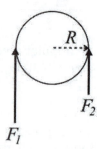

The cylinder will rotate with

(A) increasing angular speed in the clockwise direction since $\left| \vec{F_1} \right| > \left| \vec{F_2} \right|$.

(B) decreasing angular speed in the clockwise direction since the net force acting on the cylinder is $\left| \vec{F}_{net} \right| = \left| \vec{F_1} \right| - \left| \vec{F_2} \right|$.

(C) decreasing angular speed in the clockwise direction since $\left| \vec{\tau_1} \right| > \left| \vec{\tau_2} \right|$.

(D) increasing angular speed in the clockwise direction since $\left| \vec{\tau_1} \right| > \left| \vec{\tau_2} \right|$.

2.  A graph of the torque applied to a cylinder as a function of time is shown below.

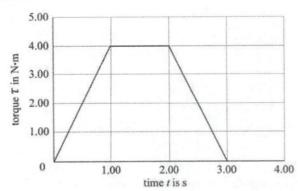

What is the change in the angular momentum of the cylinder?

(A) $8.0 \dfrac{\text{kg} \cdot \text{m}}{\text{s}}$

(B) $6.0 \dfrac{\text{kg} \cdot \text{m}}{\text{s}}$

(C) $4.0 \dfrac{\text{kg} \cdot \text{m}}{\text{s}}$

(D) $2.0 \dfrac{\text{kg} \cdot \text{m}}{\text{s}}$

3.  A disk with a moment of inertia $I_1$ is rotating with an angular velocity of $\omega_0$ when a small lump of clay drops onto the edge of the disk. The angular velocity of the disk will
    (A) increase because the falling clay strikes the disk with an initial kinetic energy that will be given to the disk, increasing its rotational speed.
    (B) decrease because the lump of clay with no initial angular velocity makes an inelastic collision with the disk, causing the composite mass to slow.
    (C) remain constant since the mass of the disk is much larger than the lump of clay.
    (D) increase as the small lump of clay makes contact with the disk but will decrease as the lump of clay spreads out on the disk.

4. A plate with a moment of inertia of 0.300 kg · m² that is initially at rest experiences two torques applied to the rim of the plate as shown in the diagram below for 10.0 s.

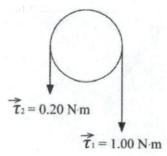

$\vec{\tau_2} = 0.20$ N·m

$\vec{\tau_1} = 1.00$ N·m

What is the change in the angular momentum of the plate?

(A) $10.0 \ \dfrac{\text{kg} \cdot \text{m}^2}{\text{s}}$

(B) $8.00 \ \dfrac{\text{kg} \cdot \text{m}^2}{\text{s}}$

(C) $5.00 \ \dfrac{\text{kg} \cdot \text{m}^2}{\text{s}}$

(D) $3.00 \ \dfrac{\text{kg} \cdot \text{m}^2}{\text{s}}$

5. Various forces are applied to four identical rods as shown below. Rank the forces from the greatest change in the angular velocity to the least change in angular velocity.

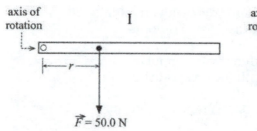

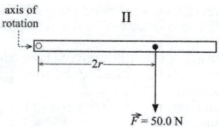

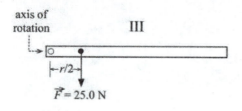

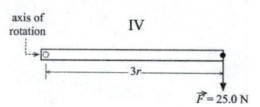

(A) II > IV > I > III
(B) IV > II > I > III
(C) II > I > IV > III
(D) I > II > IV > III

6. A skater with a moment of inertia of 2.40 kg · m² rotating with an angular velocity $\omega_0$ pulls her arms in toward her body, changing her moment of inertia to 1.50 kg · m².
   (A) Because angular momentum is conserved, her angular speed will increase; her kinetic energy will also increase because she does work in moving her arms closer to her body.
   (B) Only her angular speed will increase since conservation of angular momentum applies.
   (C) Her angular speed and kinetic energy will increase because the force applied by her arms in bringing them closer to her body provides a torque, increasing her angular momentum and thus her kinetic energy.
   (D) Her angular speed and her rotational kinetic energy will remain the same since no net torque was applied to the system.

7. A counterclockwise rotating platform with a diameter of 1.00 m shown in the diagram below slows from 4.00 rad/s to 2.00 rad/s in 1.00s.

   Which of the acceleration vectors shown best illustrates the magnitude and the direction of the acceleration of the point on the rim when its angular velocity is 2.00 rad/s ?

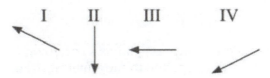

   (A) I
   (B) II
   (C) III
   (D) IV

8. A satellite moving in an elliptical orbit around Earth is shown in the diagram below. Which statement best describes the torque on the satellite about Earth while in orbit?

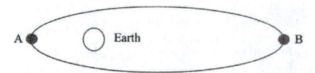

(A) The torque on the satellite is greatest at point B because it is farthest from Earth at that point and a larger torque is needed to cause it to move in toward perigee.

(B) The torque on the satellite is greatest at point A since the satellite is at perigee.

(C) The torque at A is larger than the torque at point B since the angular velocity at A is greater than the angular velocity at point B.

(D) The torque at A and B are both zero since angular momentum in the system is conserved.

9. In Problem 8, consider the satellite to be a GPS satellite. At closest approach, perigee, its distance from the center of Earth is 20,200 km and its orbital velocity is 13,900 km/h. At apogee, its distance is 26,560 km. What is its orbital velocity at this position?

(A) 10,570 km/h

(B) 12,500 km/h

(C) 13,900 km/h

(D) 18,280 km/h

10. Various forces are exerted on four identical cylinders as shown. Rank the cylinders from the greatest change to the smallest change in angular velocity.

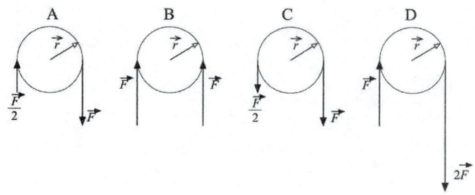

(A) D > A > C > B

(B) D > ( A = C) > B

(C) B > C > A > D

(D) (A = C) > D > B

11. At a playground, a 60.0 kg student stands on the outer rim of a merry-go-round that has a moment of inertia of 500.0 kg · m². The initial angular velocity of the merry-go-round is 3.00 rad/s when the student is 2.50 m from the center. What is the angular velocity of the system when the student walks to a position that is 1.50 m from the center of the merry-go-round? (The student may be considered as a point mass with a moment of inertia given by $I = mr^2$.)
    (A) 1.77 rad/s
    (B) 4.13 rad/s
    (C) 5.25 rad/s
    (D) 8.33 rad/s

12. A hockey puck is attached to a string that passes through a hole in a frictionless table and initially is moving in a circle of radius $r$. A student slowly increases the force on the opposite end of the string beneath the hole in the table keeping that portion of the string vertical. As a result,

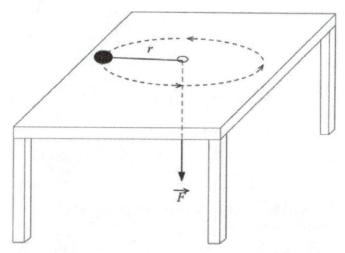

    (A) the angular momentum of the puck is conserved and the kinetic energy of the puck increases.
    (B) angular momentum is conserved and kinetic energy decreases.
    (C) kinetic energy is conserved and angular momentum increases.
    (D) kinetic energy is conserved and angular momentum decreases.

13. A disk rotating with some initial angular velocity $\omega_0$ was engaged by a second smaller disk that dropped vertically onto the larger rotating disk, reducing the speed of the system to $\omega_f$. To bring the system back to its initial angular velocity will require
    (A) a force applied in the direction of rotation for a period of time.
    (B) a torque applied in the direction of rotation for a period of time.
    (C) an angular impulse opposite the direction of rotation.
    (D) a linear impulse opposite the direction of rotation.

14. A comet is in an elliptical orbit around the sun. Rank the positions shown from the largest to the smallest centripetal force acting on the comet.

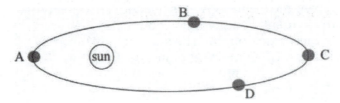

(A) C > D > B > A
(B) A > B > D > C
(C) A = B = C = D
(D) A > D > B > C

15. A small block slides without friction through a loop-the-loop as shown in the diagram below.

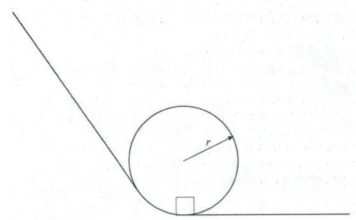

Which of the force diagrams best illustrates the forces on the block at the bottom of the loop-the-loop?

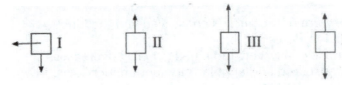

(A) I
(B) II
(C) III
(D) IV

## FREE-RESPONSE PROBLEMS

1.  A cylinder and a hoop are released from rest at the top of a long inclined plane and roll down the plane without slipping.

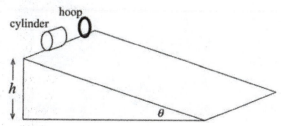

    (a) Which of the two objects reaches the bottom of the inclined plane first?
        ___ The hoop
        ___ The cylinder
        ___ They both reach the bottom at the same time.
        Justify your answer without a mathematical solution.
    (b) Since the objects roll without slipping,
        ___ the friction acting on them is up the plane.
        ___ the friction acting on them is down the pane.
        ___ there is no friction since they roll without slipping.
    (c) If the mass of each is doubled and the radius of each is reduced by half, how does this affect the velocity when the objects reach the end of the incline?
        Justify your answer with a mathematical solution in enough detail to answer your question. You do not need to solve for the velocity at the bottom of the incline.

2.  Your teacher has set up an Atwood's machine in the room and tells your laboratory group to design an experiment to determine the change in the angular momentum of the cylinder.

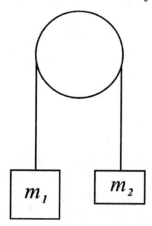

    (a) Design an experiment to determine the change in the angular momentum of the cylinder in sufficient detail that another student could duplicate your experiment and obtain the same result.
    (b) What measurements will you take and how will you use them to determine the change in the angular momentum?

(c) If you plot a graph, what will you plot and how will this help you answer your experimental question?

(d) Another group of students obtained data to produce the following graph. What information can you obtain from the analysis of this graph?

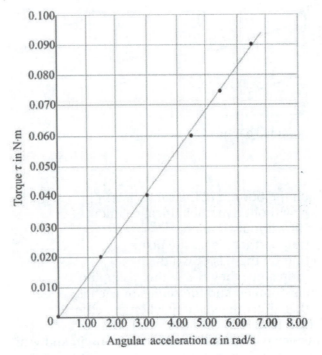

3. A small disk, mass $M = 1.00$ kg, slides on a flat frictionless table connected by a massless rod 0.500 m long that is attached by a ring to a pin in the table.

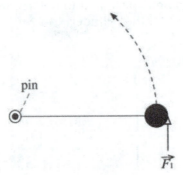

A force applied tangentially to the disk for a short period of time and then removed causes the disk to rotate in a counterclockwise direction with a tangential velocity $v_t = 2.00$ m/s.

(a) Shortly after the force is removed, a small piece of modeling clay, $m = 0.100$ kg, drops onto the center of the disk. The velocity of the disk will
___ increase.
___ decrease.
___ remain constant.
Justify your answer in a well-written statement using the correct scientific terms.

(b) Write but do not solve the mathematical equation related to your answer. Include all steps needed to show your final answer. (Both the disk and the modeling clay may be considered to be point masses $I = mr^2$ for this part of the question.)

(c) Sometime after the modeling clay has fallen on the disk, a second force is applied to the system as shown.

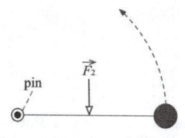

later

When the force is applied initially, the velocity of the system
___ increases.
___ decreases.
___ remains constant.
Justify your answer in a well-written statement using the correct scientific terms.

4. A large wheel and axle, as shown in the diagram below, experiences two forces

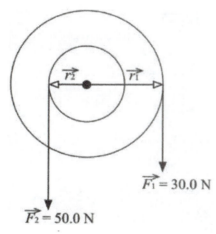

$\vec{F_2} = 50.0$ N

$\vec{F_1} = 30.0$ N

$\vec{F_1} = 30.00$ N is applied to the outer radius $\vec{r_1} = 0.500$ m, and
$\vec{F_2} = 50.0$ N is applied to the inner radius $\vec{r_2} = 0.250$ m.

(a) The net torque acting on the wheel and axle produces
___ clockwise rotation.
___ counterclockwise rotation.
___ no rotation.
Justify your answer.

(b) Several students wanted to determine experimentally the moment of inertia of the wheel and axle. They attached a weight hanger to the cord wrapped around the outer rim of the wheel and axle shown above and measured the distance from the bottom of the outer wheel to the floor. Keeping this distance constant in their experiment, they used an electronic timer to measure the time it took various masses to descend to the floor. The torque was then calculated for each descending mass to produce the following data.

| Torque (N · m) | Time (s) |
|---|---|
| 1.00 | 5.00 |
| 2.00 | 4.00 |
| 2.75 | 3.50 |
| 3.25 | 3.00 |
| 3.75 | 2.50 |
| 4.50 | 2.00 |
| 5.00 | 1.50 |
| 5.50 | 1.00 |

Using the data, plot a net torque vs. time graph. How would you use this graph to determine the change in the angular momentum of the wheel and axle?

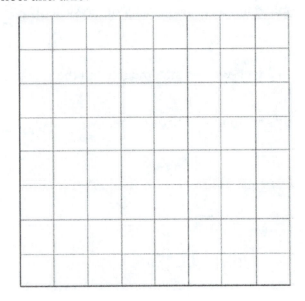

# Answers

## MULTIPLE-CHOICE QUESTIONS

1.  **D** The net torque on the system is $\left|\vec{\tau}_{net}\right| = \left|\vec{\tau}_{1} \cdot r\right| - \left|\vec{\tau}_{2} \cdot r\right|$. The cylinder will increase its angular speed in the clockwise direction because $\left|\vec{\tau}_{1}\right| > \left|\vec{\tau}_{2}\right|$.

    (L.O. 3.F.2.1)

2.  **A** The change is the angular momentum $\Delta L$ is the area under a torque as a function of time graph. The total area is $\frac{1}{2}bh + bh + \frac{1}{2}bh$.

The area is

$$\frac{1}{2}(4.0\ \text{N}\cdot\text{m})(1.0\ \text{s})+(4.0\ \text{N}\cdot\text{m})(1.0\ \text{s})+\frac{1}{2}(4.0\ \text{N}\cdot\text{m})(1.0\ \text{s})$$

$$=8.0\ \frac{\text{kg}\cdot\text{m}}{\text{s}}.$$

(L.O. 3.F.3.2)

3. **B** This is an inelastic collision. The small lump of clay has no initial angular velocity, but when it strikes the disk, conservation of momentum applies.

$(I_{\text{disk}})(\omega_{\text{disk}}) = (I_{\text{disk}} + I_{\text{clay}})(\omega_f)$

So in order for momentum to be conserved, the angular velocity must decrease because the moment of inertia of the system has increased.

(L.O. 3.F.3.1)

4. **B** The change in the angular momentum is given by $\vec{\tau}_{\text{net}} = \frac{\Delta L}{t}$.

The net torque is 1.00 N · m − 0.200 N · m = 0.800 N · m. Thus, the change in the angular momentum is $\Delta L = (0.800\ \text{N}\cdot\text{m})(10.0\ \text{s}) =$ 8.00 N · m · s. The answer is also expressed as $8.00\ \frac{\text{kg}\cdot\text{m}^2}{\text{s}}$.

(L.O. 4.D.3.1)

5. **A** $\tau = rF\sin\theta$ and $\tau = I\alpha$ apply. The greatest change in the angular velocity is produced by the torque that produces the largest angular acceleration. Since the rods are identical and are hinged at the left end, their moment of inertia is the same, $I = \frac{1}{3}mL^2$.

(L.O. 3.F.2.1)

6. **A** Conservation of angular momentum applies. Thus, her speed will increase since she moved her arms closer to her body, reducing her moment of inertia, which results in an increase in angular velocity. Work must be done to bring her arms closer to her body. This results in an increase in rotational kinetic energy.

(L.O. 5.E.1.1)

7. **D** The angular acceleration of the platform is found from $\omega_f = \omega_0 + \alpha t$. Substituting into the equation gives $2.00\ \text{rad}/\text{s} = 4.00\ \text{rad}/\text{s} + \alpha(1.00\ \text{s}) = -2.00\ \text{rad}/\text{s}^2$. The platform is rotating counterclockwise initially. Thus, the tangential acceleration of the point is $a_t = \alpha r = 2.00\ \text{rad}/\text{s}^2(0.500\ \text{m}) = 1.00\ \text{m}/\text{s}^2$ pointing downward (clockwise direction).

The centripetal acceleration is

$$a_c = v^2/r = \omega^2 r \cdot a_c = \left(2.00 \text{ rad}/s\right)^2 (0.50 \text{ m}) = 2.00 \text{ rad}/s^2 \text{ and inward.}$$

The resultant acceleration is $a = \sqrt{\left(a_c^2 + a_t^2\right)} = 2.24 \text{ rad}/s^2$ pointing downward and to the left into the third quadrant.

(L.O. 3.B.1.3, 3.B.2.1)

8.  **D** No torque is applied to the system; therefore, angular momentum is conserved.

    (L.O. 5.E.1.1)

9.  **A** Conservation of angular momentum applies. $I_P\omega_P = I_a\omega_a$. The GPS satellite may be considered as a point mass at this distance, and its moment of inertia is $I = mr^2$. The orbital velocity is $v = \omega r$;

    therefore, $mr_p^2 \left(\dfrac{v_p}{r_p}\right) = mr_a^2 \left(\dfrac{v_a}{r_a}\right) = v_p r_p = v_a r_a.$

    $(20\ 200 \text{ km})\left(13\ 900 \text{ km}/h\right) = (26\ 560 \text{ km})v_a$

    giving the orbital speed at apogee as $10\ 570 \text{ km}/h$.

    (L.O. 5.E.1.2)

10. **A** The net torque on the cylinders is determined by $\tau = rF\sin\theta$ and how the applied force will cause the cylinder to rotate either clockwise or counterclockwise. In the case of B, the net torque on the cylinder is zero since equal torques cause opposite and balanced rotation and the cylinder is at rest or is rotating at constant speed. D will cause the largest change in speed since the net torque is $\Sigma\tau = 3rF$, followed by A and then C.

    (L.O. 3.F.2.1)

11. **B** Conservation of angular momentum is required to solve this question. The correct equation is $\left(I_{\text{platform}} + I_{\text{child}_0}\right)\omega_0 = \left(I_{\text{platform}} + I_{\text{child}f}\right)\omega_f.$ The initial moment of inertia of the child considered as a point mass is found by substituting into $I = mr^2$. Then $I_{\text{child}_0} = 60.0 \text{ kg}(2.50 \text{ m})^2 = 375.0 \text{ kg} \cdot \text{m}^2$. The final moment of inertia of the child is $I_{\text{child}_0} = 60.0 \text{ kg}(1.50 \text{ m})^2 = 135.0 \text{ kg} \cdot \text{m}^2$. Substitution into the first equation given by $\left(I_{\text{platform}} + I_{\text{child}_0}\right)\omega_0 = \left(I_{\text{platform}} + I_{\text{child}f}\right)\omega_f$ is

    $\left(500.0 \text{ kg} \cdot \text{m}^2 + 375.0 \text{ kg} \cdot \text{m}^2\right)3.00 \text{ rad}/s$

    $= \left(500.0 \text{ kg} \cdot \text{m}^2 + 135.0 \text{ kg} \cdot \text{m}^2\right)\omega_f$

    $= 4.13 \text{ rad}/s.$

    (L.O. 5.E.1.2)

12. **A** As the student pulls the cord through the table, the radius decreases, but since no torque is applied to the hockey puck, conservation of angular momentum applies, $L_0 = L_f$. As the puck moves closer to the hole, its moment of inertia decreases since $r$ decreases. The kinetic energy will also increase since

$$K_{rotation} = \frac{1}{2} I \omega^2$$

(L.O. 5.E.1.1)

13. **B** The interaction was an inelastic collision. To return the system to its original angular velocity, a torque must be applied to the rotating disk. $\Sigma \tau = \dfrac{\Delta L}{\Delta t}$

(L.O. 4.D.1.1)

14. **B** The centripetal force $F_c = m\,v^2\!/_r$ arises from the gravitational force between the sun and the comet. The force is greater at point A since this position is closest to the sun.

    Since the orbit is elliptical and not circular, the force will vary in magnitude, so answer C is not a choice. The correct ranking is A > B > D > C.

(L.O. 3.A.3.4)

15. **C** The small block is in circular motion, and the centripetal force acting on it is directed into the center of the circle. At the lowest point, the normal force acting on the body is equal to the sum of the centripetal force and the gravitational force. Thus, diagram III best illustrates the size of the normal force and the gravitational force.

(L.O. 3.A.2.1)

## FREE-RESPONSE PROBLEMS

1. (a) The correct line checked is __√__ the cylinder.
   Justification
   Two possible answers:

   The cylinder has a smaller moment of inertia $I = \frac{1}{2} mr^2$. More of the gravitational potential energy will be transferred into translational kinetic energy and less into rotational kinetic energy, and the cylinder will reach the bottom with the higher linear velocity.

   The hoop, because its mass is concentrated on the rim, has a larger moment of inertia $I = mr^2$. More of the gravitational potential energy will be transferred into rotational kinetic

energy and less into translational kinetic energy, and the hoop
will reach the bottom with the lower linear velocity.

(b) The correct line checked is __√__ the friction acting on them is
up the plane.

To cause either object to roll down the plane, the friction
must provide the torque on the outer surface at the contact
point with the plane. This causes rotation of the object in a
clockwise direction as it rolls down the plane.

(c) Changing the mass and the radius of either or both objects will
not change the linear velocity as they roll down the plane.

$$U_0 + K_{0_T} + K_{0_i} = U_f + K_{f_T} + K_{f_R}$$

$$mgh = \frac{1}{2}mv^2 + \frac{1}{2}I\omega^2$$

Then knowing the moment of inertia and $v = \omega r$,
substitution into a solution for the cylinder is

$mgh = \frac{1}{2}mv^2 + \frac{1}{2}\left(\frac{1}{2}mr^2\right)\frac{v^2}{r^2}$. The mass m and the radius $r$ are

common factors in all terms and can be eliminated, yielding

$gh = \frac{1}{2}v^2 + \frac{1}{2}\left(\frac{1}{2}\right)v^2$. The solution for the hoop proceeds in the

same manner.

(L.O. 4.C.1.1, 4.C.1.2, 5.A.2.1)

2. (a) The net torque acting on the cylinder for a period of time will
give the change in the angular momentum of the cylinder.
Measure and record the radius of the cylinder. Measure and
record the distance from the descending mass attached to the
cord wrapped around the cylinder to the floor so that the
height will be constant in the experiment. Balance the system
with sufficient weight $F_g = w = mg$ on the mass hangers so that
the weights will rise and descend at constant speed. The
difference, if any, will indicate the frictional force on the
system.

Increase the weight on the descending mass hanger so that
the system will accelerate. Record this force.

Measure the time needed for the descending weight to
reach the floor. Repeat the observation at least three times.

Increase the force on the descending mass hanger, making
at least three observations for this force. Record the force and
the time for each descent. Record the average time, making
sure the times were consistent.

Repeat the above step for five to eight additional readings.

(b) Since we determined the net force acting on the system in part
(a), we can calculate each torque and the net torque acting on
the system. Each torque is determined by $|\vec{\tau}| = \vec{r}|\vec{F}|\sin\theta$, and
the net torque acting on the system is $\Sigma|\vec{\tau}| = |\vec{\tau}_2| - |\vec{\tau}_1|$.
Use the data to plot a graph of $\Sigma|\vec{\tau}|$ as a function of time.

(c) The plot of $\Sigma|\vec{\tau}|$ as a function of time will give a linear graph. The area under the curve is the change in the angular momentum $\Delta\vec{L}$.

(d) This plot of $|\vec{\tau}|$ versus $\vec{\alpha}$ is linear, and the slope of this line will determine the moment of inertia of the system. Since the angular acceleration is given, if you knew the time of the descent, you could determine the change in the angular velocity. Then the change in the angular momentum could be found: $\Delta L = I\Delta\omega$.

(L.O. 4.D.1.1, 4.D.1.2, 4.D.2.2, 4.D.3.1, 4.D.3.1)

3.  (a) The correct line checked is __√__ decrease.
    The small piece of putty makes an inelastic collision with the disk. It has a moment of inertia that must be added to the moment of inertia of the disk. Since there is an increase in the moment of inertia, there must be a corresponding decrease in the velocity.

    (b) Solution $\vec{L}_0 = \vec{L}_f$.

    $I_0\omega_0 = I_f\omega_f$

    The moment of inertia of a point mass is given as $I = mr^2$

    $Mr^2\omega_0 = (M + m)\,r^2\omega_f$

    The angular velocity is related to the linear velocity by $\omega = \dfrac{v_t}{r}$.

    Then

    $Mr^2\dfrac{v_0}{r} = (M + m)r^2\dfrac{v_f}{r}$

    and finally

    $Mrv_0 = (M + m)rv_f$

    (c) The correct line checked is __√__ decrease.
    The force produces a torque that is opposite the rotation of the system, causing it to slow. (If the force was applied long enough, it might cause the system to stop and perhaps change direction of rotation.)

    (L.O. 3.F.2.1, 3.F.3.1, 3.F.3.2, 4.D.3.1)

4.  (a) The correct line checked is __√__ clockwise rotation. The net torque acting on the system is negative. The solution gives

    $\Sigma\vec{\tau} = (50.0 \text{ N})(0.250 \text{ m}) - 30.0(0.500 \text{ m})$

    $= 12.5 \text{ N} \cdot \text{m} - 15.0 \text{ N} \cdot \text{m}$

    $= -2.50 \text{ N} \cdot \text{m}$

    The correct graph is shown on the next page.

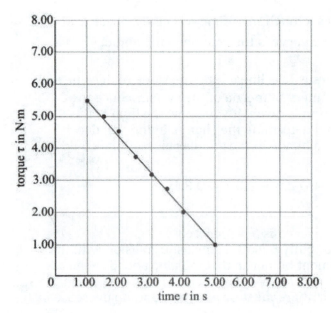

The area beneath the curve will give the change in the angular momentum of the system.

(L.O. 4.D.1.1, 4.D.3.1)

# 8

# GRAVITATION

## KEPLER'S LAWS OF PLANETARY MOTION

Years before the invention of the telescope, Johannes Kepler waged a mathematical war on Mars using the best astronomical data available. The results of his relentless analysis of planetary orbits are Kepler's three laws of planetary motion.

Kepler's first law (1609): *The planets move in elliptical orbits with the sun at one of the foci.*

The *eccentricities* of the planetary orbits are so small that they can almost be considered circles.

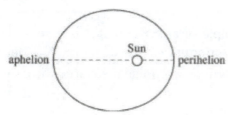

Kepler's second law (1609): *The straight line joining the sun and any planet sweeps out equal areas, A, in equal intervals of time, T.*

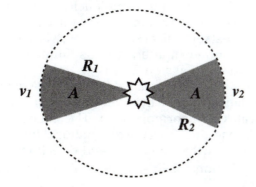

Earth's orbit is slightly elliptical. When Earth is farthest from the sun at $R_1$, it moves with a slower speed $v_1$. At a closer radius $R_2$, it moves with faster speed $v_2$. In the same time interval though, a planet sweeps out the same area regardless of its orbital speed. Planets orbit a sun with constant angular momentum. With velocity $v$ and radius $R$, the product $vR$ is constant.

Kepler's third law (1619): *The ratio of the square of the period, T, of one full orbit about the Sun and the cube of the radius, R, of the orbit is a constant.*

This law goes even further, giving a specific way to calculate the speed of an orbiting body. Mathematically, the third law can be expressed as $\dfrac{T^2}{R^3} = \text{constant}$ in which $T$ is the time for the orbiting body to complete one lap and $R$ is the distance between the objects.

## NEWTON'S UNIVERSAL LAW OF GRAVITATION

Isaac Newton discovered that *every particle in the universe attracts every other particle with a force that is directly proportional to their masses and is inversely proportional to the inverse square of the distance between them,* or

$$\vec{F} \propto \frac{m_1 m_2}{R^2}$$

This is significant because it says that all masses—astronomical and everyday—cause gravitational attraction.

To make the proportionality an equation, we introduce the constant, $G$, and call it the *universal gravitational constant*. The universal law of gravitation is then written:

$$\vec{F} = G\frac{m_1 m_2}{R^2}$$

This relationship is an example of an *inverse square law* since the force drops off as the inverse of the square of the distance, $\dfrac{1}{R^2}$, between interacting bodies, $m_1$ and $m_2$. Henry Cavendish measured the universal gravitational constant as

$$G = 6.67 \times 10^{-11} \text{ N}\cdot\text{m}^2\!\big/\!\text{kg}^2$$

Gravity is the weakest of the three fundamental forces in nature. At short distances, the electro-weak force and the strong nuclear force are overwhelmingly stronger. However, gravity dominates over the great distances and great masses of the astronomical scale. Gravity is always attracting all matter toward all other matter, yet its influence is exceedingly small between everyday objects on Earth.

Anything with mass generates a gravitational field, a $g$-field. A small amount of mass generates a relatively weak field, and a very massive object, such as a planet, generates an enormous field. The interaction between these $g$-fields and other masses causes the gravitational force of attraction. $\vec{g}$, the $g$-field, is a vector with magnitude and direction and describes the intensity of the gravitational field at some point in space.

## SAMPLE PROBLEM 1

Calculate the gravitational attraction between two identical 100 kg spheres of spent uranium that have radii of 10.8 cm when their surfaces are 2.78 m apart.

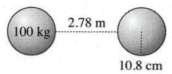

## SOLUTION TO PROBLEM 1

We calculate the distance between the centers with

0.108 m + 2.78 m + 0.108 m = 3.0 m

The gravitational force is

$$F = \frac{Gm_1m_2}{R^2} = G\left(\frac{m}{R}\right)^2$$

$$= \left(6.67 \times 10^{-11} \text{ N} \cdot \text{m}^2 \Big/ \text{kg}^2\right)\left(\frac{100 \text{ kg}}{3.0 \text{ m}}\right)^2$$

$$= 7.4 \times 10^{-8} \text{ N}$$

This gravitational force is on the order of the force an amoeba generates pushing off the surface of a tiny grain of sand in a drop of water.

## SAMPLE PROBLEM 2

Calculate the gravitational attraction between Jupiter and its moon Io. Jupiter's mass is $1.9 \times 10^{27}$ kg. Io's mass is $4.8 \times 10^{22}$ kg. Their centers are $4.22 \times 10^8$ m apart.

## SOLUTION TO PROBLEM 2

$$F_g = \frac{Gmm}{r^2}$$

$$F_g = \frac{\left(6.67 \times 10^{-11} \text{ Nm}^2 \Big/ \text{kg}^2\right)\left(1.9 \times 10^{27} \text{ kg}\right)\left(4.8 \times 10^{22} \text{ kg}\right)}{\left(4.22 \times 10^8 \text{ m}\right)^2}$$

$$F_g = 6.36 \times 10^{22} \text{ N}$$

# THE ACCELERATION DUE TO GRAVITY

The gravitational force acting on a body is its weight, $\vec{w} = m\vec{g}$. Since weight is a gravitational force, we can write

$$F = w = G\frac{mM}{R^2} = m\left(\frac{GM}{R^2}\right)$$

The term after the $m$ is $g$, the acceleration due to gravity at the surface.

$$\vec{g}_P = G\frac{M_P}{R_P^2}$$

$\vec{g}_p$ is the acceleration due to gravity on the planet's surface, $M_p$ is the planet's mass, and $R_p$ is the planet's radius.

## SAMPLE PROBLEM 3

Mars has a mass of $6.387 \times 10^{23}$ kg and a radius of $3.332 \times 10^6$ m. To two decimal places, determine the acceleration due to gravity on Mars.

## SOLUTION TO PROBLEM 3

$$g_{Mars} = G\frac{M_{Mars}}{R^2_{Mars}}$$

$$= \frac{\left(6.673\times10^{-11}\ N\cdot m^2\big/_{kg^2}\right)\left(6.387\times10^{23}\ kg\right)}{\left(3.332\times10^6\ m\right)^2}$$

$$= 3.83\ ^m\big/_{s^2}$$

Mars is considerably less massive and has a smaller radius than Earth does. At its surface, Mars has 39% of the acceleration due to gravity as Earth does.

## SAMPLE PROBLEM 4

A 100 kg object weighs 980 N on Earth. What would the object weigh on Mars?

## SOLUTION TO PROBLEM 4

From the above problem, $g_{Mars} = 3.83$ m/s$^2$. The weight on the surface of Mars is

$$w_{Mars} = mg_{Mars} = \left(100\ kg\right)\left(3.83\ ^m\big/_{s^2}\right) = 383\ N$$

## SAMPLE PROBLEM 5

We know a model is good when its predictions match our observations. Part (a) is about the moon's observed acceleration, while part (b) looks at Newton's model. The moon circles Earth with a 3.846 m orbit every 27.32 days.
(a)  What is the centripetal acceleration of the moon in its orbit?
(b)  Earth's mass is $5.97 \times 10^{24}$ kg. What is the centripetal acceleration due to gravity at any point along the orbit of the moon?

## SOLUTION TO PROBLEM 5

(a)  Convert the period, $T$, into seconds.

$$27.32\ day \times \frac{8.64\times10^4\ s}{1\ day} = 2.36\times10^6\ s$$

Next, find the orbital speed of the moon.

$$v = \frac{C}{T} = \frac{2\pi R}{T} = \frac{2\pi\left(3.846\times10^8\ m\right)}{\left(2.36\times10^6\ s\right)} = 1.02\times10^3\ ^m\big/_s$$

The centripetal acceleration experienced by the moon is $a_c = \frac{v^2}{R}$

and $a_c = \dfrac{\left(1.02\times10^3\ ^m\big/_s\right)^2}{\left(3.846\times10^8\ m\right)} = 0.0027\ ^m\big/_{s^2}$

(b) The acceleration due to gravity at a point $3.846 \times 10^8$ m from the center of the Earth is

$$g = G \frac{M_{Earth}}{R^2}$$

$$= \frac{\left(6.73 \times 10^{-11} \ N \cdot m^2/_{s^2}\right)\left(5.97 \times 10^{24} \ kg\right)}{\left(3.846 \times 10^8 \ m\right)^2}$$

$$= 0.0027 \ m/_{s^2}$$

## SAMPLE PROBLEM 6

Calculate $g$ on a planet with three times Earth's mass and double Earth's radius.

## SOLUTION TO PROBLEM 6

Most years ratio questions like this appear on the AP® exam. We can solve them by considering what will happen when we change each of the terms in an equation.

$$g_p = \frac{Gm_p}{R_p^2}, \text{ so on Earth, } g_E = \frac{Gm_E}{R_E^2}.$$

On the other planet, $g = \frac{G3m_E}{(2R_E)^2} = \frac{3}{4} \frac{Gm_E}{R_E^2}$, which is 3/4 of $g_E$.

That means that $g$ on the other planet is 3/4 of $g$ on Earth, or **7.35 m/s²**.

# CIRCULAR ORBIT SPEED

When a space vehicle is launched atop a rocket from the surface of Earth to orbit, the initial liftoff direction is vertically upward. As the rocket gains altitude, control jets and fins slowly make it turn toward a horizontal trajectory. At the proper point, the satellite separates from the rocket. With too low of an initial speed, the vehicle will follow a nearly parabolic trajectory and will strike Earth. With just the right speed, the satellite will follow a circular orbit of radius $R$. We call this speed *the circular orbit speed*. At higher speeds, the satellite will go into an elliptical orbit or will completely escape Earth.

A satellite moving in a circular orbit about a central body of mass $M$ and orbiting a distance $R$ from the center of mass of the body will have an orbital speed given by

$$v = \sqrt{\frac{GM}{R}}$$

The equation holds for any satellite moving in a circular orbit around any astronomical body.

## SAMPLE PROBLEM 7

A 500 kg satellite is placed in a circular orbit 300 km above the surface of Earth.
(a) What is its orbital speed?
(b) What is its orbital period, $T$?

## SOLUTION TO PROBLEM 7

(a)  The orbital speed is independent of the mass of the orbiting satellite. The radius of the orbit, $R$, is the distance from the center of the c.m. of Earth to the c.m. of the satellite:

$$R = R_E + h = 6{,}378 \text{ km} + 300 \text{ km} = 6{,}678 \text{ km} \times \frac{1{,}000 \text{ m}}{1 \text{ km}} = 6.678 \times 10^6 \text{ m}$$

$$V_{orbit} = \sqrt{\frac{GM_{Earth}}{R}}$$

$$= \sqrt{\frac{\left(6.67 \times 10^{-11} \text{ N} \cdot \text{m}^2 \middle/ \text{kg}^2\right)\left(5.97 \times 10^{24} \text{ kg}\right)}{\left(6.678 \times 10^6 \text{ m}\right)}}$$

$$= 7.72 \times 10^3 \text{ m}/\text{s}$$

(b)  By definition, speed is defined as $v = \frac{s}{T}$. The distance traveled in one orbit is the circumference $C = 2\pi R$. The period then is

$$T = \frac{2\pi R}{v} = \frac{2\pi\left(6.678 \times 10^6 \text{ m}\right)}{7.72 \times 10^3 \text{ m}/\text{s}} = 5.43 \times 10^3 \text{ s} = 90.6 \text{ min}$$

(b)  Alternative solution: Using Kepler's third law $\frac{T^2}{R^3} = \frac{4\pi^2}{GM}$ and solving for $T$,

$$T = \sqrt{\frac{4\pi^2 R^3}{GM_E}} = 2\pi R\sqrt{\frac{R}{GM}}$$

$$= 2\pi\left(6.678 \times 10^6 \text{ m}\right)\sqrt{\frac{\left(6.678 \times 10^6 \text{ m}\right)}{G\left(5.97 \times 10^{24} \text{ kg}\right)}}$$

$$= 5.43 \times 10^3 \text{ s} = 90.6 \text{ min}$$

Note that both solutions used the same equation but were approached differently.

## SAMPLE PROBLEM 8

A weather satellite circles Earth in a *geosynchronous orbit* with a period of exactly one day. In this way, the satellite is over the same spot all of the time.

(a)  Find the altitude above Earth's surface of a geosynchronous orbit.
(b)  Calculate the altitude in terms of Earth's radius.

## SOLUTION TO PROBLEM 8

A satellite covers a distance of $2\pi R$ in a time period $T$. First, convert one day into seconds: $1 \text{ d} = (24 \text{ h})(60 \text{ min}/1 \text{ h})(60 \text{ s}/1 \text{ min}) = 8.64 \times 10^4 \text{ s}$.

Next, write Kepler's third law: $T^2 = \left(\frac{4\pi^2}{GM_E}\right)R^3$. Solving for $R$ gives

$$R = \sqrt[3]{\frac{GMT^2}{4\pi^2}}$$

$$= \sqrt[3]{\frac{\left(6.672\times10^{-11}\ \text{N}\cdot\text{m}^2\big/\text{kg}^2\right)\left(5.97\times10^{24}\ \text{kg}\right)\left(8.64\times10^4\ \text{s}\right)^2}{4\pi^2}}$$

$R = 4.23\times10^7\ \text{m}$

Next, subtract the radius of Earth to find the altitude above Earth, $h$.

$h = 4.23\times10^7\ \text{m} - 0.64\times10^7\ \text{m} = \mathbf{3.59\times10^7\ m}$

The satellite orbits 22,310 miles above the surface of Earth.

(b) To look at how many Earth radii that distance represents, divide the distance by Earth's radius.

$$\frac{h}{R_E} = \frac{3.59\times10^7\ \text{m}}{6.4\times10^6\ \text{m}} = 5.6 \text{ Earth radii from the surface}$$

## ESCAPE VELOCITY

For a space vehicle to escape from the gravitational field of Earth and never return, it must have a velocity greater than the circular orbit speed. The minimum *escape velocity* from the surface of any planet or moon is $v_{escape} = \sqrt{\dfrac{2GM}{R}}$.

The escape velocity of a space vehicle is independent of the mass $m$ of the vehicle. Note that planets with a greater mass or smaller radius require a greater escape velocity.

### SAMPLE PROBLEM 9

Find the escape velocity from Earth.

### SOLUTION TO PROBLEM 9

$$v_{escape} = \sqrt{\frac{2GM_{Earth}}{R_{Earth}}}$$

$$= \sqrt{2\frac{\left(6.67\times10^{-11}\ \text{N}\cdot\text{m}^2\big/\text{kg}^2\right)\left(5.97\times10^{24}\ \text{kg}\right)}{\left(6.378\times10^6\text{m}\right)}}$$

$$= \mathbf{1.12\times10^4\ m/s}$$

## GRAVITATIONAL POTENTIAL ENERGY

Doing work to lift a mass against gravity increases an object's gravitational potential energy, $U$. We know that $U$ depends on the object's mass, the $g$-field, and the distance between the object and a reference point. Essentially,

$U = mgr$

We know for a planet of mass $M$ and radius $r$ that $g = \dfrac{GM}{r^2}$

So the gravitational potential energy of a small mass $m$ in orbit would be

$$U = \dfrac{-GmM}{r}$$

There is a negative sign because a negative amount of work is done to bring the object closer to the Earth. When Earth is used as a reference point, gravitational potential energy is a negative number. Objects have zero gravitational potential energy when the distance between them is infinite; they have even lower potential energy as they get closer.

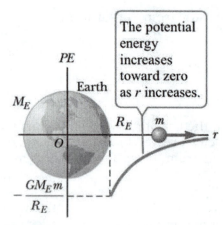

This describes an interaction between a mass and the field it's in. If we instead consider the gravitational energy per unit mass, or $U/m$, we are describing properties of the space around a massive object whether or not an orbiting body is there.

$$\dfrac{U}{m} = \dfrac{-GM}{r}$$

At great distances, the *gravitational potential energy* approaches its maximum of zero. Close to a massive body, the gravitational potential energy approaches $-\infty$.

$U/m$ has the units Joules per kilogram and answers this question: How much energy would a kilogram of mass have at this point? This property is called the *gravitational potential.*

The gravitational potential surrounding a massive object varies with distance, and mapping it can tell us about the object's gravitational field. We can map the gravitational potential around a planet by marking the regions over which the potential is the same, the *equipotential lines.*

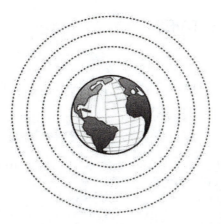

The gravitational field is perpendicular to the equipotential lines. Freely falling objects fall to regions of lower potential.

## SAMPLE PROBLEM 10

A 2000 kg satellite that is $0.23 \times 10^6$ m above the surface of Earth is in a circular orbit. Calculate the potential energy of the satellite.

## SOLUTION TO PROBLEM 10

We need the distance from the center of Earth to the satellite.

$$R = h + R_E = 0.23 \times 10^6 \text{ m} + 6.4 \times 10^6 \text{ m} = 6.63 \times 10^6 \text{ m}$$

The satellite's potential energy is

$$U = -\frac{GmM}{R}$$

$$= -\frac{\left(6.672 \times 10^{-11} \text{ N} \cdot \text{m}^2 / \text{kg}^2\right)\left(2 \times 10^3 \text{ kg}\right)\left(5.97 \times 10^{24} \text{ kg}\right)}{6.63 \times 10^6 \text{ m}}$$

$$U = -1.21 \times 10^{11} \text{ J}$$

## WEIGHTLESSNESS

Astronauts orbiting Earth in an artificial satellite feel *apparent weightlessness* that is similar to the sensation experienced by a person in a freely falling elevator. When the elevator accelerates downward with uniform acceleration *a*, the apparent weight of the person in the elevator is $w = m(g - a)$. When the elevator accelerates downward, the passenger feels lighter.

If the elevator is in *free fall*, the downward acceleration is $a = g$, and $w = m(g - g) = 0$. Thus, a person feels apparent weightlessness in a freely falling elevator. We feel weightless if we and our surroundings are together in freefall. The international space station orbits Earth because Earth's gravitational field continuously pulls on it and everything inside. A satellite is in a continuous state of free fall. Because the force of gravity acts on the space station, an astronaut inside experiences apparent weightlessness.

# GRAVITATION: STUDENT OBJECTIVES FOR THE AP® EXAM

- You should be able to explain the differences between mass and weight.
- You should be able to state Newton's universal law of gravitation.
- You should be able to calculate gravitational forces between two bodies.
- You should be able to determine the orbital speed and angular momentum of a satellite in a circular orbit.
- You should understand the concept of weightlessness.
- You should be able to explain why the moon does not crash into Earth despite the large gravitational force acting on it.

## MULTIPLE-CHOICE QUESTIONS

1.  A radius vector, *r*, is a straight line that runs from the center of mass of one body to the center of mass of a second body, as in the diagram below depicting Earth and a satellite. The radius vector changes length as the satellite moves toward perigee (closest distance to Earth) because of gravitational forces between Earth and the satellite.

    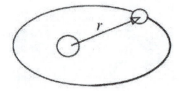

    The velocity of the center of mass as the satellite approaches perigee will
    (A) increase.
    (B) decrease.
    (C) remain constant.
    (D) be zero.

2.  A 10.0 kg body is lifted 2.00 m above the surface of Earth. The work done by the gravitational force is
    (A) –20.0 J.
    (B) 20.0 J.
    (C) –196 J.
    (D) 196 J.

3.  Earth moves in an elliptical orbit around the sun as shown in the diagram below.

    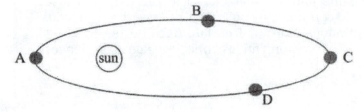

Rank the positions from highest to lowest in terms of the force acting on Earth due to the sun.
(A) A > C > B > D
(B) A > B > D > C
(C) C > D > B > A
(D) C > A > D > B

4. A body is moved from one isoline of equal gravitational potential to another. Rank the positions of gravitational potential energy per unit mass from greatest to least.

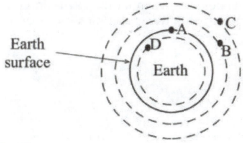

(A) A > B > C > D
(B) D > A > B > C
(C) C > B > A > D
(D) D > C > B > A

5. If you could place an object at the center of Earth, which of the following statements would be true for the object at that location?
(A) The mass and the weight of the object would be the same as the values on the surface of Earth.
(B) The mass would be zero, and the weight would be the same as on the surface.
(C) The mass and the weight would both be zero.
(D) The weight would be zero, and the mass would be the same as the value on the surface of Earth.

6. Two spheres lie along a line as shown below. Sphere $M$ is considerably more massive than sphere $m$. At which of the marked points is the sum of the gravitational forces from both spheres zero?

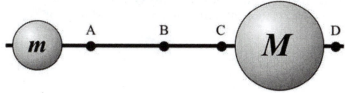

(A) A
(B) B
(C) C
(D) D

7. The gravitational force between two point masses A and B is $\vec{F}$ when they are $r$ meters apart. When the mass of A is increased by a factor

of 4, the mass of B is halved, and the distance of separation is increased by a factor of 3, what is the new force acting between them?

(A) $\dfrac{2}{9}\vec{F_1}$

(B) $\dfrac{2}{3}\vec{F_1}$

(C) $2\vec{F_1}$

(D) $3\vec{F_1}$

8. Which of the following graphs best represents the gravitational force as a function of distance between two point masses $m_1$ and $m_2$ as they are moved farther apart?

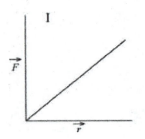 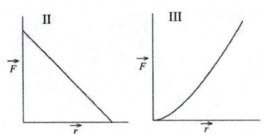 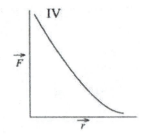

(A) I
(B) II
(C) III
(D) IV

9. A satellite is in circular motion a distance of $3R_E$ above the surface of Earth. To what is the acceleration experienced by the satellite at that location closest?

(A) 9.8 m/s²

(B) 1.1 m/s²

(C) 0.61 m/s²

(D) 0

10. Earth's radius is 6.4 × 10¹¹ kg. What is the change in the gravitational potential energy of the Earth satellite system when a 20,000 kg satellite is boosted from a distance of $2R_E$ to a distance of $3R_E$ above the surface of Earth?

(A) −1.03 × 10¹¹J
(B) 1.03 ×10¹¹J
(C) −2.08× 10¹¹J
(D) 2.08 × 10¹¹J

11. What is the gravitational acceleration on the surface of a planet that has three times the mass and twice the radius of Earth?

(A) $\dfrac{3}{4}g$

(B) $g$

(C) $\dfrac{4}{3}g$

(D) $2g$

12. A small moon is in a circular orbit around its planet as shown below.

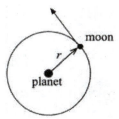

Which of the vectors displayed below gives the correct direction for the velocity and the acceleration of the moon at the point indicated in its orbit?

(A) I
(B) II
(C) III
(D) IV

13. A 600 N astronaut travels from Earth to the moon. The mass of the astronaut as measured on the moon is
(A) zero.
(B) one-sixth the mass of the astronaut on Earth.
(C) equal to the mass of the astronaut on Earth.
(D) six times the mass of the astronaut on Earth

14. Two satellites of different masses are in the same circular orbit around Earth. Which of the following statements is correct?
(A) The magnitude of the gravitational force is the same for both satellites since they are in the same orbit, and both satellites have the same period.
(B) The magnitude of the gravitational force is zero for the smaller satellite because it in the same orbit as the larger satellite. Thus, the gravitational force exerted on the larger satellite gives it a larger period.
(C) The magnitude of the gravitational force depends on the masses of the satellites, and the orbital periods are the same.
(D) The magnitude of the gravitational force is zero for both satellites, and they have the same period.

15. Which of the illustrations shows the correct relationship between the gravitational field strength vectors and the isolines of gravitational potential energy per unit mass?

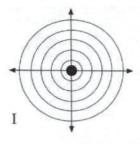 I

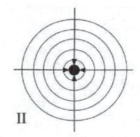 II

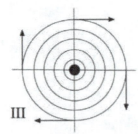 III

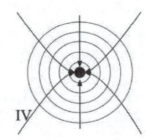 IV

(A) I
(B) II
(C) III
(D) IV

## FREE-RESPONSE PROBLEMS

1. (a) You are given an unknown mass with a hook on it and asked to determine its gravitational mass. Using any of the following materials—a spring of known spring constant, a spring stand, a ruler, and a stopwatch—design an experiment to determine the object's gravitational mass. Explain what quantities you will measure and how you would make your measurements.

   (b) Using the same equipment in part (a), design an experiment to determine the inertial mass of the object. Explain what quantities you will measure and how you would make your measurements.

   (c) Distinguish between your two experiments by clearly explaining how the experiment outlined in (a) determines inertial mass and the experiment outlined in (b) determines gravitational mass.

2. (a) A satellite moves around a central body as shown below.

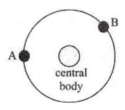

The force on the satellite is
___ greatest at point A.
___ greatest at point B.
___ the same at points A and B.
Justify your answer without a mathematical solution.

(b) The orbit is now changed to the diagram shown below.

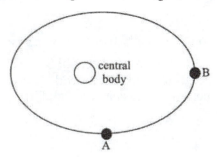

The force on the satellite is
___ greater at point A.
___ greater at point B.
___ the same at points A and B.
Justify your answer without a mathematical solution.

(c) The orbit is now changed to the diagram shown below.

The velocity of the satellite is
___ greatest at point A.
___ greatest at point B.
___ the same at points A and B.
Justify your answer.

3. A 100.0 kg body is taken from the surface of Earth to a satellite in orbit at an altitude of 1.50 $R_E$ above the surface of Earth.
   (a) Determine the weight of the body on the surface of Earth.
   (b) Determine the weight of the body at the altitude 1.50 $R_E$ above the surface of Earth.
   (c) What is the fractional change in the weight of the body?
   (d) What is the fractional change in the mass of the body?

4. A 2000.0 kg satellite is inserted into an orbit $7.00 \times 10^6$ m above the surface of Earth.
   (a) What is the gravitational force exerted between the satellite and Earth?
   (b) Determine the orbital velocity of the satellite in this orbit.
   (c) What is the change in the potential energy and the work done by the force of gravity in lifting the satellite to an orbit that is $12.0 \times 10^6$ m above the surface of Earth?

# Answers

## MULTIPLE-CHOICE QUESTIONS

1. **A** The external force applied to the satellite from $F = G\dfrac{m_s M_E}{r^2}$ increases as the satellite moves inward toward Earth. Since $\Sigma F = ma$, the satellite in its elliptical path increases in speed. There is an increase in the velocity of the center of mass as the satellite approaches perigee.

   (L.O. 4.A.3.1)

2. **C** The work done is the negative of the change in the gravitational potential energy of the Earth-object (10.0 kg body) system.

   (L.O. 5.B.5.5)

3. **B** The force acting between Earth and the sun is given by $F = G\dfrac{M_s M_E}{r^2}$. When Earth is closest to the sun, the gravitational force is the greatest; at aphelion, the farthest point in the orbit, the force is the least. The correct ranking is A >B > D > C.

   (L.O. 3.A.4.2)

4. **C** Gravitational potential energy is measured to a reference point. The surface of Earth is set at zero gravitational potential energy. Point D is shown below the surface of Earth and therefore is negative. The isoline showing a position farthest from the surface has the greatest gravitational potential energy per unit mass.

   (L.O. 5.B.5.5)

5. **D** Mass is the measure of inertia of the body; the mass will not change as we move the body from location to location. The weight of the body is determined by $F = w = mg$, and if the body is located at the center of Earth, it will experience equal pulls in all directions, making the net gravitational force on it $\Sigma F = 0$; thus, both its acceleration and weight will be zero.

(L.O. 2.B.1.1)

6. **B** The net force on $M_2$ is the vector sum of the forces between $M_2$ and the other two spheres.

$$F_{12} = \left(6.67 \times 10^{-11} \frac{\text{N} \cdot \text{m}^2}{\text{kg}^2}\right)\left(\frac{10.0 \text{ kg} \cdot 20.0 \text{ kg}}{(1.00 \text{ m})^2}\right) = 1.33 \times 10^{-8} \text{ N}$$

directed toward $M_1$. The force between $M_2$ and $M_3$ is

$$F_{23} = \left(6.67 \times 10^{-11} \frac{\text{N} \cdot \text{m}^2}{\text{kg}^2}\right)\left(\frac{20.0 \text{ kg} \cdot 30.0 \text{ kg}}{(1.00 \text{ m})^2}\right) = 4.00 \times 10^{-8} \text{ N}$$

directed toward $M_3$.

The net force is $2.67 \times 10^{-8}$ N directed toward $M_3$.

(L.O. 3.C.1.1)

7. **A** The original force between the two bodies is $F = G\dfrac{M_A M_B}{r^2}$. The

new force is $F_2 = G\dfrac{4M_A\left(\frac{1}{2}M_B\right)}{(3r)^2}$. Setting up a ratio $\dfrac{F_1}{F_2} =$

$\dfrac{9r^2}{4M_A\left(\frac{1}{2}M_B\right)} \cdot \dfrac{M_A M_B}{r^2}$ and solving for the new force gives $\dfrac{2}{9}F_1$.

(L.O. 3.C.1.1)

8. **D** The force between two point masses is an inverse square law. Increasing the distance by a factor of 2 reduces the force by one-fourth, increasing $r$ by a factor of 3 drops the force to one-ninth of its initial value.

(L.O. 3.C.1.1)

9. **C** The acceleration due to gravity at the location of $4R_E$ from the center of Earth is given by

$$\vec{g} = \left(6.67 \times 10^{-11} \frac{\text{N} \cdot \text{m}^2}{\text{kg}^2}\right)\left(\frac{5.98 \times 10^{24} \text{ kg}}{(4 \cdot 6.38 \times 10^6 \text{ m})^2}\right) = 0.61 \frac{\text{m}}{\text{s}^2}.$$

(L.O. 2.B.2.1)

10. **B** The change in the altitude of the satellite is measured as the distance $3R_E$ from the center of Earth to $4R_E$ from the center of Earth. Substitution into the equation

$$PE_{4R_E} - PE_{3R_E} = -GM_E m_s \left( \frac{1}{4R_E} - \frac{1}{3R_E} \right)$$

$$PE_{4R_E} - PE_{3R_E} =$$
$$-\left( 6.67 \times 10^{-11} \ \frac{N \cdot m^2}{kg^2} \right) (5.98 \times 10^{24} \ kg)(2.00 \times 10^4 \ kg) \left( \frac{1}{4(6.38 \times 10^6 \ m)} - \frac{1}{3(6.38 \times 10^6 \ m)} \right)$$

$$PE_{4R_E} - PE_{3R_E} = -7.98 \times 10^{18} \left( 3.94 \times 10^{-8} - 5.23 \times 10^{-8} \right) \ J =$$
$$1.03 \times 10^{11} \ J$$

(L.O. 5.B.5.5)

11. **A** The acceleration due to gravity on the planet can be determined from the ratio of the acceleration in terms of "$g$" to the acceleration $g$ on the surface of Earth.

$$\frac{g_{planet}}{g_{Earth}} = \frac{G \left( \dfrac{3M}{(2R)^2} \right)}{G \left( \dfrac{M}{R^2} \right)} = \frac{\left( \dfrac{3}{4} \right)}{\left( \dfrac{1}{1} \right)} = \frac{3}{4} g$$

(L.O. 2.B.2.1)

12. **A** The acceleration of the moon is directed toward the planet along the radial line connecting the centers of mass of the moon and the planet. The velocity of the moon is tangential to its orbit.

(L.O. 3.B.2.1)

13. **C** The mass of the astronaut is the same as the mass on Earth. The lunar value for $g_{moon}$ as determined from $g_{moon} = G \dfrac{M_{moon}}{r_{moon}^2}$ gives

$$g_{moon} = 6.67 \times 10^{-11} \ \frac{N \cdot m^2}{kg^2} \frac{(7.36 \times 10^{22} \ kg)}{(1.74 \times 10^6 \ m^2)} \qquad g_{moon} = 1.62 \ \frac{m}{s^2} \ . \ \text{The}$$

weight of the astronaut is one-sixth the weight on Earth.

(L.O. 2.B.2.1, 2.B.2.2)

14. **C** Newton's law of gravitation $F = G\dfrac{M_s M_E}{r^2}$ shows that the force is dependent on the product of the masses. Thus, the satellite of larger mass will experience the larger force. Kepler's third law also applies. $T^2 = \dfrac{4\pi^2}{GM} r^3$ indicates that the periods are the same since they are in the same orbit.

(L.O. 3.A.3.4, 3.A.4.1)

15. **B** The direction of the gravitational field strength vector (acceleration) is into the center of mass of the body creating the field. The isolines of gravitational potential are perpendicular to the direction of the gravitational field strength vector. Work is done in moving a body in a gravitational field between two isolines. No work is done when the mass is moved along an isoline.

(L.O. 5.B.5.4)

## FREE-RESPONSE PROBLEMS

1. (a) Hang the spring from the spring stand and measure its unstretched length. Hang the mass from the spring and measure the spring's new length.
   (b) With the mass hanging at rest from the spring, give the mass a small displacement and then release it. Once the spring is bouncing steadily, record the time for ten full up-and-down bounces oscillations.
   (c) Part (a) uses Hooke's law to measure the weight of the hanging mass. This experiment specifically looks at the interaction between the force of gravity and mass. The spring's stretch is caused by the mass's heaviness.

      Part (b) uses simple harmonic motion to measure the object's inertial mass. During one period of a mass bouncing spring, the force applied by the spring changes in both magnitude and direction. The object's resistance to these changes is caused by the mass's inertia.

(d) The laboratory setup might look like this or some other variation of a known force that produces acceleration of a body.

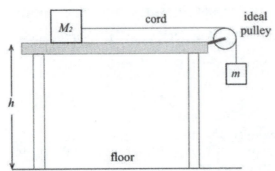

Adjust the system so that the system moves with constant velocity. This is the force needed to overcome friction between the mass on the table and the table. Measure the height of the table and record it. Add known weights to the weight hanger and record this force in Newtons. Determine the time for the hanging weight to reach the floor. From the height of the table and the time, the acceleration can be determined from

$$y = \frac{1}{2}at^2.$$

(e) Plot a graph of $F$ vs. $a$. Since the graph is linear, the slope of the line will be the inertial mass in kg.

Gravitational mass can be determined by comparing the weight of a body from a spring scale using Hooke's law.

The inertial mass and the gravitational mass can be compared. They are proportional and independent of each other.

(L.O. 1.C.1.1, 1.C.3.1, 2.B 1.1)

2. (a) The correct line checked is ___ √ ___ the same at points **A** and **B**.
   The orbit is circular and the radius is constant; thus, the force is constant.

   (b) The correct line checked is ___ √ ___ greatest at point **A**.
   The orbit is now elliptical, and the radius will vary from point to point. The satellite is closer to the central body at point **A**; therefore, the force is greater at **A** than at **B**.

   (c) The correct line checked is ___ √ ___ greatest at point **A**. The force on the body $F = G\dfrac{M \cdot m}{r^2}$ produces a larger centripetal force $F = \dfrac{m \cdot v^2}{r_A}$ at point **A** since the distance from the central body is much smaller at **A** than at **B**.

(L.O. 3.B.1.1, 3.C.1.2)

3.  (a) The weight of the body on the surface of Earth is determined from $F = w = mg$. On the surface of Earth, the weight is

$$100.0 \text{ kg}\left(9.80 \text{ m}/_{s^2}\right) = \textbf{980.0 N}.$$

(b) At the elevation of 1.50 $R_E$, the 100.0 kg body is 2.50 $R_E$ from the center of Earth.

At that location, the acceleration due to gravity is determined from $g = G\dfrac{M_E}{r_E^2}$

Substitution into the equation gives

$$g = \left(6.67 \times 10^{-11} \text{ }\frac{\text{N} \cdot \text{m}^2}{\text{kg}^2}\right)\frac{5.98 \times 10^{24} \text{ kg}}{\left(1.60 \times 10^7 \text{ m}\right)^2} = \textbf{1.56 m}/_{s^2}$$

(c) The weight of the body at that location is

$$W_{2.50\,R_E} = 100.0 \text{ kg}\left(1.56 \text{ m}/_{s^2}\right) = \textbf{156.0 N}.$$

The fractional change is determined by comparing the value of $g$ at the two locations.

$$W_{2.50\,R_E}/W_{R_E} = \left(156\,N/980\,N\right) \times 100 = 15.9\%$$

$$100\% - 15.9\,\% = 84.1\%$$

Thus, there is a 84.1% loss of weight at that height.

(d) The mass of the body is constant. It is **100.0 kg** both on the surface of Earth and at the elevation of 1.5 $R_E$ above the surface of Earth. Therefore, the fractional change is 0%.

(L.O. 2.B.1.1, 2.B.2.1)

4.  (a) The gravitational force acting between Earth and the satellite is determined from $F = G\dfrac{M \cdot m}{r^2}$. Substitution into the equation is

$$\vec{F} = \left(6.67 \times 10^{-11} \text{ }\frac{\text{N} \cdot \text{m}^2}{\text{kg}^2}\right)\left(\frac{5.98 \times 10^{24} \text{ kg} \cdot 2000 \text{ kg}}{\left(6.38 \times 10^6 \text{ m} + 7.00 \times 10^6 \text{ m}\right)^2}\right)$$

The force is **4450 N**.

(b) The orbital velocity is determined from $F_g = F_c$. Substitution into the centripetal force equation, $\vec{F} = \dfrac{m \cdot v^2}{r}$ , gives the

orbital velocity $4450 \text{ N} = \dfrac{2000.0 \text{ kg} \cdot v^2}{1.34 \times 10^7 \text{ m}}$

The velocity $v = \sqrt{\dfrac{5.96 \times 10^{10} \text{ N} \cdot \text{m}}{2000.0 \text{ kg}}}$; thus, $v = \textbf{5.46} \times \textbf{10}^3 \dfrac{\textbf{m}}{\textbf{s}}$

(c)  The change in the gravitational potential energy is

$$\Delta U = -\frac{GM_E m}{r_f} - \left( -\frac{GM_E m}{r_f} \right)$$

$$\Delta U = -GM_E m \left( -\frac{1}{r_f} + \frac{1}{r_i} \right)$$

$$\Delta U = 7.98 \times 10^{17}\ \text{N} \cdot \text{m}^2 \left( -\frac{1}{1.84 \times 10^7\ \text{m}} + \frac{1}{1.34 \times 10^7\ \text{m}} \right)$$

The change in the potential energy of the satellite is $1.62 \times 10^{10}$ J.

The work done by the gravitational force is $-\Delta U = -1.62 \times 10^{10}$ J.

(L.O. 3.B.1.4, 3.C.1.1, 3.C.1.2, 4.C.1.1, 5.B.4.2)

# 9

# OSCILLATORY MOTION

## PERIODIC MOTION

There is a common and important type of motion called *periodic motion,* or *oscillatory* motion, that repeats itself at regular intervals. In its simplest form bouncing back and forth in a set rhythm, it is straightforward to analyze. This kind of motion appears in many areas of physics. Periodic motion can be seen in the swinging pendulum of a grandfather clock. When we watch the pendulum, we see that its speed and acceleration are not constant. The pendulum moves back and forth along a fixed path, repeating in a fixed set of motions and returning to each position and velocity after a definite period of time.

The repeating motions of an oscillating object are usually caused by the object being stretched or bent from its normal position and then released. Besides being called oscillatory or periodic motion, it is also called *harmonic* motion. As the object moves side to side or up and down, it experiences varying forces. Because the force changes, the acceleration also varies.

The AP® exam commonly uses the example of a block attached to a spring on a frictionless tabletop. The spring is fastened to a wall, as shown in the diagram on the next page. If we pulled the block away and let go, it would begin to move back and forth on the surface. If we pulled the block to point +A, it would oscillate between the two points −A and +A and never go any further than those two points. −A and +A are the maximum amplitude of the block-spring system.

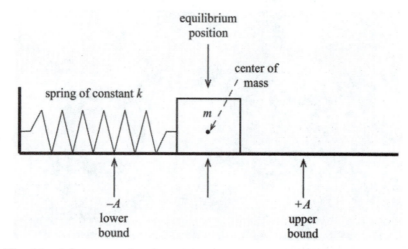

The block bounces back and forth because of the forces applied by the spring. When the block moves right, the spring pulls it left, and when the block moves left, the spring pushes it to the right. The spring exerts a *restoring force* on the block that does just what its name implies. The force tends to restore the block to its original position, the *equilibrium position*. The restoring force is given to us by Hooke's law. When we pulled the block, the spring applied a restoring force in proportion to how far the spring had been stretched.

$F = -kx$

When the block is released, the restoring force, which is the net force, produces an acceleration $a$:

$F = ma = -kx$

$$a = -\frac{k}{m}x$$

Because the acceleration is caused by the restoring force, the acceleration $a$ is proportional to the displacement $x$ but opposite in direction.

At the maximum displacement, either $-A$ or $+A$, the block experiences the most force, so it also experiences its highest acceleration in these positions. As the block begins to move, its velocity increases, but the force and the acceleration decrease. At the equilibrium position, the velocity will be at its maximum, whereas force and acceleration will be zero. Because of its momentum, the block continues past the equilibrium position, but the retarding force immediately returns and begins to increase until the block reaches the other amplitude position, where it stops momentarily and begins its trip all over again. At all times during this motion, the net force—and hence the acceleration—is proportional to the displacement and directed toward the equilibrium position.

This type of oscillatory motion, when the acceleration is linearly proportional to the displacement and is always directed toward the equilibrium position, is called *simple harmonic motion,* SHM. Simple harmonic motion is always motion along a straight line, the acceleration and velocity constantly changing as the oscillating block moves through its series of positions. The example of the grandfather clock pendulum also is an example of SHM and will be discussed in more detail later in this chapter.

## AP® Tip

An object can only be in simple harmonic motion when there is a restoring force.

## AMPLITUDE, PERIOD, AND FREQUENCY

We can describe oscillatory motion by its amplitude, period, and frequency. These terms will also be used in a discussion of waves. As shown in the previous block-spring example, the *amplitude*, *A*, of oscillatory motion is the maximum displacement from the equilibrium position. ±*A* are the boundaries or highest limits of the motion. Amplitude is expressed in meters.

The *period*, *T*, of an oscillating body is the time for one complete back-and-forth motion, or a complete oscillation. Period is expressed in seconds, *s*. The period of the moon moving around Earth is four weeks. The period of a pendulum is the time for it to swing all the way back and forth. When determining the period, make sure you include the entire cycle.

On the AP® exam equation sheet for both the Physics 1 and Physics 2 exams, you will be given the formulas for the period of a mass-spring system and the period of a pendulum.

The period for a mass-spring system like the block attached to the spring is

$$T = 2\pi\sqrt{\frac{m}{k}}$$

The period of a simple pendulum is

$$T = 2\pi\sqrt{\frac{L}{g}}$$

The *frequency*, *f*, of the oscillatory motion is the number of complete oscillations per second. Frequency tells us how many times a system repeated its motion in one second. The frequency is the reciprocal of the period $f = \frac{1}{T}$.

Frequency is expressed in hertz, Hz, and $1 \text{ Hz} = \frac{1}{s} = s^{-1}$.

On the AP® exam, you may be asked to design a lab in which part of the procedure requires you to determine the period or frequency of an object. For a fast motion, it can be difficult and inaccurate to measure just one period. A better lab technique is to time multiple periods. For example, to find the period of a block bouncing on a spring, time how long it takes for 20 complete bounces; then divide the time by 20. This gives a more accurate measurement of the period, which can then be used to find the frequency.

The *angular frequency*, *ω*, is defined as $\omega = 2\pi f = \frac{2\pi}{T}$.

This equation will be written on the AP® equation sheet as

$$T = \frac{2\pi}{\omega} = \frac{1}{f}$$

Angular frequency is expressed in rad/s. It also is expressed in Hz, and $1\,\text{Hz} = 1\frac{\text{rad}}{\text{s}} = \text{s}^{-1}$.

## LOCATION OF A HARMONIC OSCILLATOR

When an oscillator like a block-spring system or a pendulum is displaced and then released, it undergoes SHM. If we could place a pen on the underside of the block and pull a long sheet of paper underneath the block as it moved back and forth, the pen would trace a pattern that looks like a sine or cosine graph. A graph of something undergoing SHM is shown below.

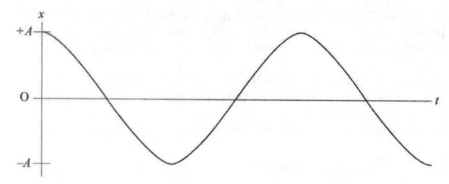

At any time $t$ after the block in our example is released, its position $x$ can be found using

$$x = A\cos\omega t$$

On the AP® equation sheet for AP® Physics 1, the equation appears as

$$x = A\cos\omega t(2\pi ft)$$

On the equation sheet for AP® Physics 2, it is slightly modified to

$$x = A\cos(\omega t) = x = A\cos\omega t(2\pi ft)$$

Note that the equation is a cosine function. The graph of the cosine function starts at $+A$ and oscillates through $-A$ back to $+A$ for a complete cycle or oscillation.

## THE ACCELERATION OF A HARMONIC OSCILLATOR

The varying acceleration of an oscillator is caused by the restoring force. For the block-spring example, the restoring force is described using Hooke's law. Because $\Sigma F = ma$ and and $\Sigma F = -kx$, $a = -\frac{k}{m}x$. The maximum acceleration, $a_{max}$, occurs at the amplitude positions, $a_{max} \infty \pm A$.

## SAMPLE PROBLEM 1

A 0.40 kg mass is attached to a vertical spring of $k = 10.0$ N/m. The mass is displaced 0.04 m vertically downward from equilibrium and is released from rest. The mass executes SHM. Neglecting friction and air resistance, find the period of motion.

## SOLUTION TO PROBLEM 1

The period of the oscillator is $T = \dfrac{2\pi}{\omega} = \dfrac{2\pi \text{ rad}}{5.0 \text{ rad}/_{\text{s}}} = \mathbf{1.26\ s}$

# THE ENERGY OF A HARMONIC OSCILLATOR

The total energy, $E$, of a mechanical system is the sum of its kinetic energy, $K$, and potential energy, $U$, or $E = K + U$. This means that as something oscillates, its energy will change from potential energy to kinetic energy and back as it travels.

The kinetic energy is zero at the amplitude positions since the object is momentarily at rest. As the object moves toward the equilibrium position, its velocity increases. The gain in kinetic energy comes from the loss of potential energy. At the equilibrium position, $x = 0$, the velocity and kinetic energy of the body is a maximum and is $K = \dfrac{1}{2}mv^2$.

When the oscillator's kinetic energy is at its maximum, the potential energy is at a minimum. The elastic potential energy, $U = \dfrac{1}{2}kx^2$, of the system is zero at the equilibrium position since $x = 0$. At the amplitude positions, $U$ is a maximum.

The total energy of a simple harmonic oscillator is $E = \dfrac{1}{2}kA^2$. The velocity of the oscillator at any position, $x$, is given by $v = \pm\omega\sqrt{A^2 - x^2}$. Therefore, $v_{max} = \pm\vec{\omega}A$.

## SAMPLE PROBLEM 2

A 0.50 kg block is connected to a horizontal spring, $k = 20.0$ N/m. The block is pulled to the right and then released from rest where it begins to oscillate on a horizontal frictionless surface with amplitude 0.03 m.
(a)  Where is the acceleration at a maximum? Where is the acceleration zero?
(b)  Where is the velocity at a maximum? Find the maximum velocity.
(c)  What is the total energy?
(d)  What are the period and frequency of the system?

## SOLUTION TO PROBLEM 2

(a)  The spring's pulling force increases as the block is pulled back. Acceleration is proportional to force. The acceleration is zero at equilibrium. Maximum acceleration occurs at $\pm A$.
(b)  Maximum velocity occurs at equilibrium where $x = 0$.

   The energy of the system is constant, so
   $K_{max} = U_{max}$ ½ $mv^2$ = ½$kx^2$

$$v^2 = \frac{(20.0)(0.03)^2}{.50}$$

$v = 0.19$ m/s

(c)  The total energy of the system is found by

$$E = \frac{1}{2}kA^2 = \frac{1}{2}(20.0 \ \text{N/m})(0.03 \ \text{m})^2 = \textbf{0.009 J}$$

(d)  $T = 2\pi\sqrt{m/k}$

$T = 19.9$ s

Period is the reciprocal of the frequency of vibration.

$T = 1/f$

$f = 0.05$ s$^{-1}$

## Sample Problem 3

Use the information about the mass oscillating on a spring in Sample Problem 2.

(a)  Determine the velocity of the block when its displacement is 0.02 m and it is moving to the right.

(b)  Find the acceleration of the block when its displacement is 0.02 m and it is moving to the right.

(c)  What are the kinetic energy, elastic potential energy, and total energy of the oscillator when it is at the 0.02 m position to the right of equilibrium?

## Solution to Problem 3

(a)  $v = \pm\omega\sqrt{A^2 - x^2} = \pm(6.325 \ \text{s}^{-1})\sqrt{(0.03 \ \text{m})^2 - (0.02 \ \text{m})^2} = \pm0.141 \ \dfrac{\text{m}}{\text{s}}$

Since the block travels to the right, the velocity is positive and

$v = \textbf{0.141} \ \dfrac{\text{m}}{\text{s}}$

(b)  $a = \pm\omega^2 x = \pm(6.325 \ \text{s}^{-1})^2 (0.02 \ \text{m}) = \pm0.800 \ \dfrac{\text{m}}{\text{s}^2}$

(c)  Since the block moves to the right, the elastic force acts in the opposite direction, slowing it. Thus, $a = -\textbf{0.800} \ \dfrac{\text{m}}{\text{s}^2}$.

(d)  Kinetic energy:  $K = \dfrac{1}{2}mv^2 = \dfrac{1}{2}(0.50 \ \text{kg})\left(0.141 \ \dfrac{\text{m}}{\text{s}}\right)^2 = \textbf{0.005 J}$

Potential energy:  $U = \dfrac{1}{2}kx^2 = \dfrac{1}{2}(20.0 \ \text{N/m})(0.02 \ \text{m})^2 = \textbf{0.004 J}$

Total energy:  $E = K + U = 0.005 \ \text{J} + 0.004 \ \text{J} = \textbf{0.009 J}$

## The Simple Pendulum

Another type of object that experiences SHM is a simple pendulum. It behaves similarly to the block-spring system. A simple pendulum is an idealized body that consists of a light, inextensible cord, one end of which is attached to a fixed support O, and a small mass, called a *pendulum bob,* which is attached to the other end. When at rest, the bob hangs straight down and is considered to be at the equilibrium position. When pulled aside to the amplitude position and released from rest, it begins to swing back and forth. It travels an arc length *x* through the equilibrium position to the amplitude position on the other side. If the amplitude is made very small, the motion of the pendulum is simple harmonic.

### Simple Pendulum

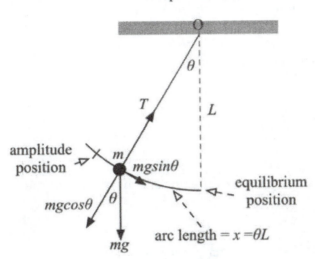

Just like the block-spring system, when the pendulum is at its maximum amplitude, it experiences its maximum restoring force and acceleration. It also has its maximum potential energy at this point. However, unlike the block-spring system, the potential energy experienced here is gravitational potential energy, $U = mgh$. The height *h* is the vertical distance above the equilibrium position.

When the pendulum bob is released and begins to move toward the equilibrium position, the restoring force and acceleration decrease and the velocity increases. The kinetic energy also increases, reaching a maximum at the equilibrium position.

The forces acting on the bob are its weight, *mg*, and the tension, *T*, in the cord. The weight both pulls the bob down and helps it swing. To show the two functions of the weight, we break the weight vector into its components. The radial component, $mg\cos\theta$, along with the tension *T*, supplies the centripetal acceleration that keeps the bob moving along a circular arc. The tangential component, $mg\sin\theta$, is the restoring force pulling the bob back toward the equilibrium position. The restoring force is

$$F = -mg\sin\theta$$

The period for a simple pendulum is $T = 2\pi\sqrt{\dfrac{L}{g}}$ .

The period of the pendulum will change as you vary the length and gravity. It is not affected by the mass of the bob. AP® exam questions may ask you how the pendulum is affected when you change any of these variables.

### SAMPLE PROBLEM 4

Calculate the length of a simple pendulum with a period of 4.00 seconds.

## Solution to Problem 4

By definition, $T = 2\pi\sqrt{\dfrac{L}{g}}$, and solving for the length yields

$$L = \frac{T^2 g}{4\pi^2} = \frac{(4.00 \text{ s})^2 \left(9.80 \text{ m/s}^2\right)}{4\pi^2} = 3.97 \text{ m}$$

## Sample Problem 5

Find an expression for the angular frequency of a simple pendulum.

## Solution to Problem 5

Period is $T = 2\pi\sqrt{\dfrac{L}{g}}$, and frequency is defined as $f = \dfrac{1}{T}$.

For the simple pendulum, $f = \dfrac{1}{2\pi\sqrt{\dfrac{L}{g}}} = \dfrac{1}{2\pi}\sqrt{\dfrac{g}{L}}$.

Cross-multiplying yields $2\pi f = \sqrt{\dfrac{g}{L}}$.

Angular frequency is defined by $\omega = 2\pi f$; the angular frequency of the simple pendulum is then $\omega = \sqrt{\dfrac{g}{L}}$.

# Oscillatory Motion: Student Objectives for the AP® Exam

- You should be able to define the term *simple harmonic motion* and explain how it relates to periodic motion.
- You should be able to relate the terms *period* and *frequency*.
- You should understand the variables that affect the period of a mass-spring system and a simple pendulum.
- You should be able to explain the relationship between the kinetic, potential, and total energies for a harmonic oscillator, especially in a mass-spring system and a simple pendulum.
- You should be able to explain the source of the restoring force for the simple pendulum and mass-spring systems.
- You should be able to plan and collect data to analyze the motion of a system in simple harmonic motion.
- You should understand the relationship between the angular frequency, the ordinary frequency, and the period of an oscillator.

## MULTIPLE-CHOICE QUESTIONS

1. A bob of weight *w* on a simple pendulum is pulled back to give it a small amplitude. The pendulum and its free-body diagram are shown below. Which force or force component causes the pendulum to return to equilibrium?

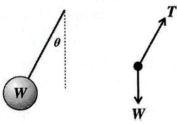

(A) T
(B) *w*
(C) *w*sin$\theta$
(D) *w*cos$\theta$

2. Which of the following energy vs. time graphs would represent the total mechanical energy of a mass spring system as it completes one oscillation?

(A)

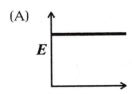

(B)

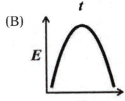

(C)

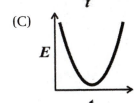

(D)

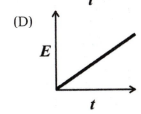

3.  A body of mass $m$, which is attached to an ideal spring whose spring constant is $k$, is set into motion on a horizontal frictionless surface with an amplitude of vibration of $A$. Which of the following statements is correct?

    (A) Increasing the amplitude of vibration by a factor of 2 will increase the period of vibration since the mass must travel a greater distance in making a complete vibration.

    (B) Increasing the mass on the spring by a factor of 2 will keep the amplitude the same but will change the period of the motion by a factor of $\sqrt{2}$.

    (C) Increasing the spring constant $k$ by a factor of 2 will increase the amplitude by a factor of $\sqrt{2}$ but will not change the period of vibration.

    (D) Increasing the spring constant by a factor of 2 will increase both the amplitude and period of motion by a factor of $\sqrt{2}$.

4.  A mass of 0.20 kg is attached to a spring whose spring constant is $k = 25.0 \text{ N/m}$. The spring is compressed 0.40 m on the frictionless surface as shown below and then released from rest.

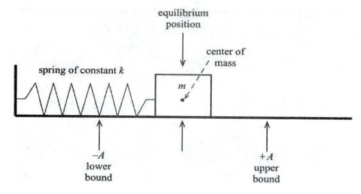

    As the mass moves through the equilibrium position, its kinetic energy is

    (A) 0.20 J.

    (B) 0.50 J.

    (C) 2.0 J.

    (D) 5.0 J.

5. Several simple pendulums are shown in the diagram below. Each pendulum is set into motion by releasing the masses from the same angle with the vertical.

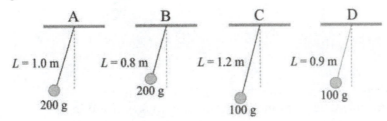

Rank the pendulums from the pendulum with the highest frequency to the one with the lowest frequency.
(A) A > B > C > D
(B) A > C > D > B
(C) B > A > D > C
(D) B > D > A > C

6. A block of mass $m = 0.250$ kg is attached to a spring, $k = 20.0$ N/m and is undergoing simple harmonic motion (SHM) on a frictionless surface. The mass oscillates under the action of an elastic restoring force. Determine the acceleration of the mass when its displacement is −0.150 m.
(A) $12.0 \text{ m/s}^2$
(B) $-12.0 \text{ m/s}^2$
(C) $5.0 \text{ m/s}^2$
(D) $-5.0 \text{ m/s}^2$

7. A body attached to a spring oscillates on a frictionless horizontal surface. A graph of the amplitude as a function of time is shown below.

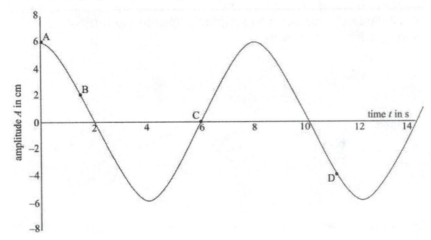

Rank the potential energy of the spring from largest to smallest for the points shown in the graph.
(A) A > D > B > C
(B) A > B > C > D
(C) A > C > D > B
(D) A > B > D > C

8.  A simple pendulum of mass $m$ and length $L$ on the surface of Earth oscillates with a period $T$. The original mass is replaced with a mass of $2m$ while keeping its length as $L$. Its new period is
    (A) $T_2 = T_1$
    (B) $T_2 = \sqrt{2}T_1$
    (C) $T_2 = 2T_1$
    (D) $T_2 = 4T_1$

9.  A variable force acts on a mass attached to a spring on a frictionless horizontal surface, elongating the spring as shown in the graph below.

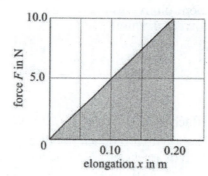

    When the spring is set into SHM, the maximum internal energy in the spring is
    (A) 0.20 J at the maximum amplitude ±$A$.
    (B) 0.20 J at the equilibrium.
    (C) 1.0 J at the maximum amplitude ±$A$.
    (D) 1.0 J at the equilibrium.

10. A spring supported vertically from a clamp is elongated 0.100 m when a 0.200 kg body is hung on the free end. If this mass is replaced by a 0.400 kg body and set into simple harmonic motion (SHM), what is the period of vibration?
    (A) 0.634 s
    (B) 0.897 s
    (C) 1.10 s
    (D) 1.58 s

11. Several students perform an experiment using a 0.150 kg pendulum bob attached to a string and obtain the following data:

| Length of the string (m) | Time for 50.0 vibrations (s) |
|---|---|
| 1.40 | 119 |
| 1.20 | 110 |
| 1.00 | 99.9 |
| 0.90 | 95.0 |
| 0.70 | 83.9 |
| 0.50 | 70.9 |

They want to determine an experimental value for the acceleration due to the gravitational force in the classroom, using information from the slope of the line. To do this, they should plot the data using which of the graphs shown below?

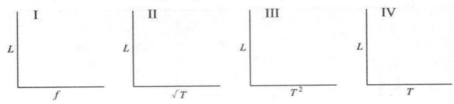

(A) I
(B) II
(C) III
(D) IV

12. A 0.300 kg block attached to a spring moves on a flat frictionless surface under the action of an elastic restoring force. The spring constant for the spring is 25.0 N/m. Which of the following statements is correct?

(A) The force acting on the spring is proportional to the displacement of the spring and is directed to the equilibrium position; therefore, the acceleration is variable and points to the equilibrium position as the displacement changes.

(B) The force acting on the spring is inversely proportional to the displacement of the spring and is directed to the equilibrium position; therefore, the acceleration is variable and points to the equilibrium position.

(C) The force is constant; the block moves with constant acceleration in the direction of the elastic restoring force.

(D) Since the force points to the equilibrium and is proportional to the displacement, the magnitude of the acceleration is variable but will point toward the maximum displacement.

13. A mass on a spring oscillates between $x = \pm A$. The elastic potential energy graph as a function of time is best shown in which of the graphs shown below?

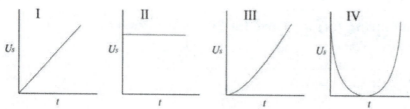

(A) I
(B) II
(C) III
(D) IV

14. A certain spring undergoes an elongation/compression about the equilibrium position as shown below.

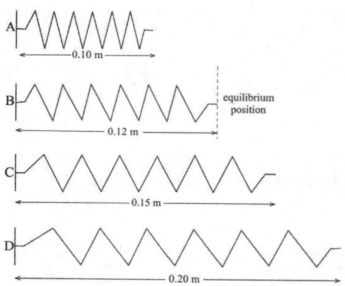

Rank the positions of the system from the one that has the most elastic potential energy to the one that has the least.
(A) D > C > A > B
(B) A > B > C > D
(C) D > C > B > A
(D) B > A > C > D

15. The simple harmonic motion of a 2.00 kg mass oscillating on a spring of spring constant $k$ is shown in the graph below.

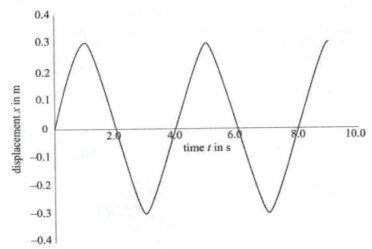

What is the value of the spring constant in N/m?

(A) $\dfrac{4}{\pi}$

(B) $\dfrac{\pi^2}{2}$

(C) $\dfrac{8}{\pi^2}$

(D) $\dfrac{\pi^2}{4}$

16. An object swings on the end of a cord as a simple pendulum with period $T$. Another object oscillates up and down on the end of a vertical spring, also with period $T$. If the masses of both objects are doubled, what are the new values for the periods?

|  | Pendulum | Mass on the Spring |
| --- | --- | --- |
| (A) | $\dfrac{T}{\sqrt{2}}$ | $T\sqrt{2}$ |
| (B) | $T$ | $T\sqrt{2}$ |
| (C) | $T\sqrt{2}$ | $\dfrac{T}{\sqrt{2}}$ |
| (D) | $T$ | $T$ |

17. A student observes that the motion of a body of mass, $m$, that is attached to a spring of constant $k_i$ and is moving on a horizontal frictionless surface has a maximum velocity $v_i$ as it passes through the equilibrium position. If the student replaces the spring with another spring with a spring constant of $4k_i$, the maximum velocity of the mass will
(A) increase by a factor of 4 while keeping the same amplitude.
(B) remain the same but the amplitude will increase by a factor of 2.
(C) increase by a factor of 2 while keeping the same amplitude.
(D) remain the same but the amplitude will increase by a factor of 4.

# FREE-RESPONSE PROBLEMS

1.  A 0.60 kg body oscillates on a horizontal frictionless surface as a simple harmonic oscillator attached to a spring with a constant $k$ as shown in the graph below.

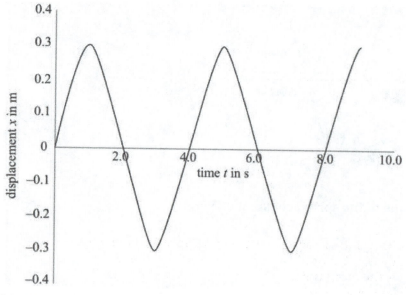

(a)  Determine the amplitude of vibration for this body. Explain your answer.
(b)  What is the period of motion of the spring-mass system? Explain how you determined your answer.
(c)  Determine the spring constant $k$ for the system.
(d)  Sketch the velocity of the mass on the graph below for at least two cycles of motion.

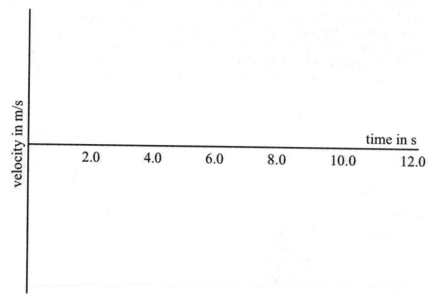

(e) Sketch the acceleration of the spring-mass system on the graph below for at least two cycles of motion.

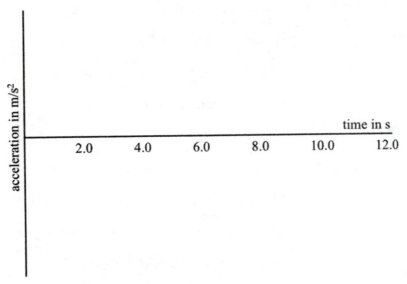

2. (a) If the mass in Problem 1 is doubled to 2 *m*, how will it change the amplitude of vibration?
   ___ Increase the amplitude.
   ___ Decrease the amplitude.
   ___ Produce no change in the amplitude.
   Justify your answer.
   (b) If the mass in Problem 1 is doubled to 2 *m*, how will it change the period of vibration?
   ___ Increase the period.
   ___ Decrease the period.
   ___ Produce no change in the period.
   Justify your answer without calculations.
   (c) In terms of your answers to parts (a) and (b), how will this change the graph in Problem 1? Make a sketch of the position of the mass as a function of time for the motion of the new spring-mass system for at least two cycles on the graph given below.

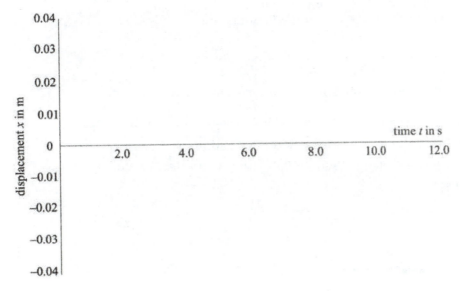

3. (a) Design an experiment to determine the spring constant of aspring-mass system undergoing simple harmonic motion (SHM) in enough detail that another student can repeat your experiment and obtain the same results.
   (b) What measurements will you take, and how will you use them to determine the spring constant?
   (c) If you plot a graph, what will you plot and how will you use the graph in the experiment?
   (d) What assumption(s) did you make in the experiment, and how might it affect your experiment?

4. A 0.100 kg mass vibrates on the end of a spring with a period of vibration of 1.50 s and an amplitude of vibration of 0.08 m. (Friction is negligible.)
   (a) Determine the frequency of motion of the mass-spring system.
   (b) What is the spring constant of the spring?
   (c) Determine the maximum velocity of the system and indicate the position at which this occurs.
   (d) What is the maximum acceleration of the system? Indicate where this occurs.

# Answers

## MULTIPLE-CHOICE QUESTIONS

1. **C** The restoring force depends only on the mass of the bob. The force is the tangential component of the weight of the bob because that is the component of the weight that causes the pendulum to swing. So the restoring force is $mg\sin\theta$.

   (L.O. 3.B.1.3)

2. **A** Total mechanical energy for and system in SHM is constant. While choice B shows a kinetic energy graph and choice C shows a potential energy graph, the total mechanical energy is a combination of both, $E = K + U$, and will not change.

   (L.O. 3.B.3.1)

3. **B** The period of motion of a spring mass system is given by $T = 2\pi\sqrt{\dfrac{m}{k}}$. Increasing the mass of the body attached to the spring by a factor of 2 will increase the period of vibration by a factor of $\sqrt{2}$. It will not increase the amplitude of vibration of the system.

   (L.O. 3.B.3.1)

4. **C** The surface is frictionless and the conservation of energy applies to the system and $\Delta U_s = \Delta K \cdot \frac{1}{2} kx^2 = \Delta K$. Substituting into the equation gives $\frac{1}{2}\left(23.0\ \text{N}/_\text{m}\right)(0.40\ \text{m})^2 = 2.0$.

(L.O. 5.B.4.2)

5. **D** The frequency of a simple pendulum is determined from the length of the pendulum, not its mass because $f = \dfrac{1}{2\pi} \sqrt{\dfrac{g}{L}}$. A shorter pendulum will have a shorter period and a higher frequency.

(L.O. 3.B.3.1)

6. **A** The acceleration of the mass is determined from $\Sigma F = ma$. The force acting on the mass due to the elastic restoring force is $\left|\vec{F}_s\right| = k\left|\vec{x}\right|$. Since the displacement is $-0.150$ m, the acceleration will point toward equilibrium and is positive. Its value is determined from $k\left|\vec{x}\right| = ma$. The acceleration is $\left(20.0\ \text{N}/_\text{m}\right)(0.15\ \text{m}) = 0.250\ \text{kg}(a)$. The acceleration is $12.0\ \text{m/s}^2$.

(L.O. 3.B.3.4)

7. **A** The internal energy of a spring is given by $\dfrac{1}{2} kx^2$. The position indicating the greatest amplitude will have the greatest internal energy. This is point A, followed by D, then B. The energy at point C is all kinetic energy since the body is passing through equilibrium.

(L.O. 5.B.3.1)

8. **A** The period of a simple pendulum is independent of its mass.

(L.O. 5.B.3.1)

9. **C** The work done on the spring in elongating it goes into the spring as elastic potential energy and is the area under the curve. $W = \dfrac{1}{2} bh$ gives $\dfrac{1}{2}(10.0\ \text{N})(0.20\ \text{m}) = 1.0\ \text{J}$. This occurs at $x = \pm A$. At the equilibrium position, the energy of the system is 1.0 J and is kinetic energy.

(L.O. 5.B.3.1)

10. **B** The period of vibration for a mass-spring system is given by $T = 2\pi \sqrt{\dfrac{m}{k}}$.

The spring constant $k$ is determined from $\left|\vec{F}_s\right| = k\left|\vec{x}\right|$.

$$k = \frac{(0.200 \text{ kg})\left(9.80\,{}^{m}\!/_{s^2}\right)}{0.100 \text{ m}} = 19.6\,{}^{N}\!/_{m}.$$ Then the period of vibration

becomes $T = 2\pi\sqrt{\dfrac{0.400 \text{ kg}}{19.6\,{}^{N}\!/_{m}}} = 0.897$ s.

(L.O. 3.B.3.3)

11. **C** The slope of length $L$ as a function of $T^2$ will give the students information they can use to determine the experimental value for $g$. $g_{exp} = 4\pi^2\dfrac{L}{T^2}$.

12. **A** The force acting on the spring is $\left|\vec{F}_s\right| = k\left|\vec{x}\right|$. This force causes the mass on the spring to undergo a variable acceleration given by $\Sigma F = ma$. The mass will undergo simple harmonic motion (SHM) under this force. The acceleration will vary in magnitude and will always point toward the equilibrium position.

(L.O. 3.B.3.4)

13. **D** The spring oscillates between the extremes of the amplitude of vibration. As it does, the internal energy of the spring changes from $\dfrac{1}{2}kA^2$ to zero to $\dfrac{1}{2}kA^2$.

(L.O. 3.B.4.1)

14. **A** The elastic potential energy of the spring is determined from $\dfrac{1}{2}kx^2$. The spring with the greatest elongation or compression from equilibrium has the greatest internal energy. The ranking is $D > C > A > B$.

(L.O. 5.B.4.1)

15. **B** From the graph, the period of oscillation of the 2.0 kg mass is 4.0 s. Using $T = 2\pi\sqrt{\dfrac{m}{k}}$ and substituting, $4.0 \text{ s} = 2\pi\sqrt{\dfrac{2.0 \text{ kg}}{k}}$ and solving, $k = \dfrac{\pi^2}{2}$.

(L.O. 3.B.3.3)

16. **B** The necessary equations are $T = 2\pi\sqrt{\dfrac{L}{g}}$ and $T = 2\pi\sqrt{\dfrac{m}{k}}$. The simple pendulum is independent of the mass of the bob, and increasing the mass of the bob by a factor of 2 will not change the period of vibration. Increasing the mass on the spring by a factor of 2 will increase the period by a factor of $\sqrt{2}$.

(L.O. 3.B.3.1)

17. **C** Increasing the spring constant will increase the maximum velocity of the mass as it moves through the equilibrium position, not the amplitude of vibration. The energy in the system is constant since the mass moves on a flat frictionless surface.

Thus, $\dfrac{1}{2}k_i x^2 = \dfrac{1}{2}mv_i^2$. Increasing $k_i$ to $4k_i$ gives $4k_i = v_{new}^2$ and will increase the maximum velocity by a factor of 2. The velocity of the mass as it passes through the equilibrium is now $2v_i$.

Also, $v_{max} = \vec{\omega}A$ and $\omega = \sqrt{\dfrac{k}{m}}$; therefore, $v_{max}$ will increase by a factor of 2.

(L.O. 5.B.4.2)

## FREE-RESPONSE PROBLEMS

1.  (a) The amplitude of motion can be read directly from the graph of displacement vs. time. The peak positions give the amplitude. From the graph, this is $|A| = \pm$**0.30 m**.

    (b) The period of motion can also be read directly from the graph. One complete cycle is completed in **4.0 s**.

    (c) The spring constant can be determined from $T = 2\pi\sqrt{\dfrac{m}{k}}$.

    Substitution gives $4.0\text{ s} = 2\pi\sqrt{\dfrac{0.60\text{ kg}}{k}}$. Squaring and solving will give $(4.0\text{ s})^2 = 4\pi^2\left(\dfrac{0.60\text{ kg}}{k}\right)$.

(d)  $k = 1.5 \, \text{N/m}$. The velocity as a function of time is shown below.

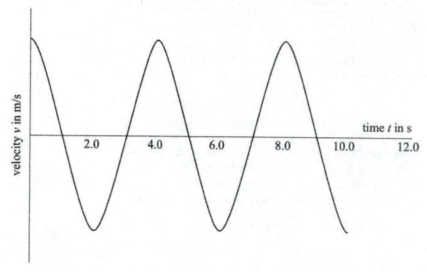

(e)  The acceleration as a function of time is shown below.

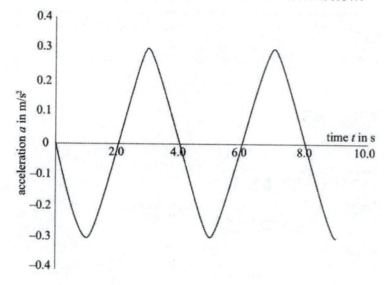

(L.O. 3.B.3.1, 3.B.3.3)

2.  (a)  The correct line checked is ___√___ Produce no change in the amplitude.

    The amplitude of vibration is independent of the mass on the spring.

    (b)  The correct line checked is ___√___ Increase the period.

    The period of vibration is proportional to the square root of the mass. Therefore, doubling the mass will change the period of vibration by $T_2 = \sqrt{2}\,(T_i)$.

    The new period is then 5.66 s.

(c) The position as a function of time for the new mass on the spring is shown below.

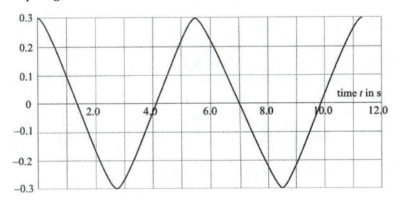

(L.O. 3.B.3.1, 3.B.3.3)

3. (a) Hang a spring on a vertical support and attach a mass hanger to one end of the spring.

Place a mass on the end of the spring and record the mass in the data chart.

Indicate the number of complete vibrations you will record for the oscillations.

Pull the mass downward a few centimeters and release.

Determine the time for the number of complete vibrations you chose for your experiment.

Record this time and then calculate the time for one complete vibration.

Repeat the reading for at least three trials and average the time for one complete vibration.

Increase the mass on the spring and repeat the above procedure for at least six mass readings.

(b) Data to be recorded will be the mass on the spring, the time for one complete vibration, and the period squared. The period is related to the spring constant $T = 2\pi\sqrt{\dfrac{m}{k}}$. The spring constant may be determined from substitution into the equation.

(c) A more consistent value for the spring constant can be obtained using the data from the slope of the mass as a function of period squared graph and multiplying the slope by $4\pi^2$.

(d) Assuming that the spring is massless will cause an error in the experiment. When the spring-mass system vibrates, one-third of the mass of the spring is actually vibrating during the oscillation. (The derivation for this requires calculus, which is not part of the scope of this course, but you can look it up on the Internet or in a calculus-based physics text.) The spring constant will have a smaller experimental value if its mass is ignored. Using a spring-mass system such as an experimental Hooke's law apparatus requires that you use the mass of the pan attached to the spring as well as the mass placed on the pan.

The equation $T = 2\pi\sqrt{\dfrac{m_{effective}}{k}}$ is the correct form, making the

effective mass equal to $m_{pan} + m_{supported} + \dfrac{1}{3}m_{spring}$.

Counting too few vibrations might increase the error in the experiment if they are miscounted. A larger error exists when 9 or 10 vibrations are completed rather than 30.

(L.O. 3.B.3.1, 3.B.3.2, 3.B.3.3)

4.  (a)  Period and frequency are reciprocals of each other. $f = \dfrac{1}{T}$. The frequency is

$$f = \frac{1}{1.50\ s}; f = \frac{1}{1.50\ s} = \mathbf{0.667\ Hz}.$$

(b)  The spring constant $k$ is determined from $T = 2\pi\sqrt{\dfrac{m}{k}}$.

$$1.50\ s = 2\pi\sqrt{\frac{0.100\ kg}{k}}$$

and $k = \mathbf{1.75\ N/m}$.

(c)  Since friction is negligible, the maximum internal energy of the spring $U_s = \dfrac{1}{2}kA^2$ will be equal to the maximum kinetic energy as the mass moves through the equilibrium position.

$\dfrac{1}{2}kA^2 = \dfrac{1}{2}mv_{max}^2$. Substitution gives

$$\left(1.75\ \frac{N}{m}\right)(0.080\ m)^2 = (0.100\ kg)v_{max}^2$$

$$v_{max} = \mathbf{0.335\ m/s}$$

(d)  The maximum acceleration occurs at the end points and is calculated from $|\vec{a}| = \dfrac{k}{m}x$.

The acceleration is $\left(\dfrac{1.75\ N/m}{0.100\ kg}\right)(0.080\ m) = \mathbf{1.40\ m/s^2}$

(L.O. 3.B.3.1, 3.B.3.3, 5.B.3.2, 5.B.4.1, 5.B.4.2)

# 10

# FLUIDS

## FLUIDS IN GENERAL

Matter exists in three phases: the solid phase, the liquid phase, and the gaseous phase. The particles in a solid have strong attractions for one another and give the solid a shape and a volume. The particles of a liquid are moving in a way that reduces the attractions between them, allowing them to flow. Liquids have the volume and shape of their container. The molecules of a gas are a thousand times farther apart with little or no attractions on each other. Gases have no shape or volume. The individual molecules move freely until they fill their container.

Because liquids and gases flow, we call them *fluids*. The study of fluids at rest is called *hydrostatics*. The study of fluids in motion is called *hydrodynamics*.

## DENSITY

The mass density or *density*, ρ, of a substance is defined as its mass per unit volume.

$$\rho = \frac{m}{V}$$

The SI density unit is the kilogram per cubic meter, $kg/m^3$. Copper, for example, has a density $\rho = 8890 \ kg/m^3$. Water has a density of 1000 $kg/m^3$. Density is a scalar quantity.

## SPECIFIC GRAVITY

Another quantity commonly used to describe densities is specific gravity, *SG*, which is described as the ratio of the density of a material to that of water, $\rho_w = 1000 \text{ kg/m}^3$. Specific gravity, like density, is a scalar quantity.

$$SG = \frac{\rho_{material}}{\rho_w}$$

### SAMPLE PROBLEM 1

Find the *SG* of
(a) lead: $\rho_{Pb} = 11{,}300 \text{ kg/m}^3$
(b) ice: $\rho_{ice} = 917 \text{ kg/m}^3$
(c) What is the density of ethyl alcohol whose *SG* is 0.81?

### SOLUTION TO PROBLEM 1

(a) $SG = \dfrac{\rho_{Pb}}{\rho_w} = \dfrac{11{,}300 \text{ kg/m}^3}{1000 \text{ kg/m}^3} = \textbf{11.3}$. Notice that no units are

associated with specific gravity. The answer is a pure number and means that lead is 11.3 times more dense than an equal volume of water.

(b) $SG = \dfrac{\rho_{ice}}{\rho_w} = \dfrac{917 \text{ kg/m}^3}{1000 \text{ kg/m}^3} = \textbf{0.917}$. Ice is 0.917 times as dense as

an equal volume of water.

(c) To find a density from *SG*, we rearrange terms and get
$\rho_{alcohol} = SG \times \rho_w = 0.81 \times 1000 \text{ kg/m}^3 = \textbf{810 kg/m}^3$.
To find *SG* from density, we divide by 1000 kg/m³, and to find a density from *SG*, we multiply by 1000 kg/m³.

## PRESSURE

When a force acts perpendicular to a surface, the *pressure, P,* exerted is the ratio of the magnitude of the normal force, *F,* to the area, *A,* of the surface. Pressure measures how hard the fluid pushes on a surface per unit area.

$P = F/A$

The SI unit of pressure is the *pascal*, Pa. The pascal is defined as a newton per square meter, $1 \text{ Pa} = 1 \text{ N/m}^2$. Pressure is a scalar; it has only magnitude.

If a fluid is placed in various containers, the pressure depends only on the depth of the fluid. That is, at any given depth, the force exerted in a fluid is the same in all directions regardless of the shape or size of the containers. The diagram below shows three bottles full of water. At lower heights, the pressure is greater. At $h_1$ in all three bottles, the pressure is the same: $P_1$. At $h_3$ in all three bottles, the pressure is the same: $P_3$.

The forces the fluids exert on the walls of its container (and the forces the walls exert on the fluid) always act perpendicular to the walls.

An external pressure exerted on a fluid is transmitted uniformly and undiminished throughout the fluid.

That does not mean that the pressures in a fluid are the same everywhere, because the weight of the fluid exerts pressure that increases with increasing depth. The pressure at a depth $h$ in a fluid having a density $\rho$ due to the weight of the fluid above is

$P = \rho g h$

A fluid in an open container has the atmosphere exerting an external pressure on it. Then the total pressure at that fluid depth is found by

$P = P_{atm} + \rho g h$

When using water, another term for gauge pressure is *hydrostatic pressure.* The term hydrostatic pressure refers to the pressure of the water only.

$P_{hydrostatic} = \rho g h$

## SAMPLE PROBLEM 2

A 1.4 m long metal cylinder has a cross section of 35 cm² and a mass of 90 kg. It stands upright on a concrete floor. What pressure does the cylinder exert on the floor?

## SOLUTION TO PROBLEM 2

Pressure is defined as force per unit area. The force the cylinder exerts is its weight, *mg*. The area, *A*, must be in SI units. To convert cm² to m², we multiply by 10⁻⁴.

$$P = \frac{F}{A} = \frac{w}{A} = \frac{mg}{A} = \frac{(90 \text{ kg})\left(9.8 \, {}^{m}\!/_{s^2}\right)}{\left(35 \times 10^{-4} \, m^2\right)} = 252 \text{ kPa}$$

## SAMPLE PROBLEM 3

Fresh water has a density of 1000 kg/m³, which is equivalent to 1 kg/L or 1g/cm³.

(a)  What hydrostatic pressure exists at a depth of 3.0 m in a freshwater lake?

(b)  What hydrostatic pressure exists at a depth of 30.0 m in a freshwater lake?

(c)  At what depth will the hydrostatic pressure be 75.5 kPa?

## SOLUTION TO PROBLEM 3

(a)  $p = \rho g h = \left(1000 \ {}^{kg}\!\!\diagup_{m^3}\right)\left(9.8 \ {}^{m}\!\!\diagup_{s^2}\right)(3.0 \ m) = \textbf{29.4 kPa}$

(b)  $p = \rho g h = \left(1000 \ {}^{kg}\!\!\diagup_{m^3}\right)\left(9.8 \ {}^{m}\!\!\diagup_{s^2}\right)(30 \ m) = \textbf{294 kPa}$

(c)  $h = \dfrac{p}{\rho g} = \dfrac{75.5 \times 10^3 \ {}^{N}\!\!\diagup_{m^2}}{\left(1000 \ {}^{kg}\!\!\diagup_{m^3}\right)\left(9.8 \ {}^{m}\!\!\diagup_{s^2}\right)} = \textbf{7.7 m}$

## GAUGE PRESSURE

Pressure gauges measure the difference in pressure between an unknown pressure and atmospheric pressure, $P_{atm}$. What they measure is called *gauge pressure*, $P_{gauge}$, and the true pressure is called the *absolute pressure*.

Absolute pressure = gauge pressure + atmospheric pressure

$P_{abs} = P_{gauge} + P_{atm}$

Atmospheric pressure at sea level is 101.3 kPa. By definition, 1 standard atmosphere at sea level and 0°C is 101,300 Pa. 1 atm = 101,300 Pa.

A common device for measuring gauge pressure is the open-tube manometer. The manometer is a U-shaped tube, as shown below, containing a dense liquid such as mercury.

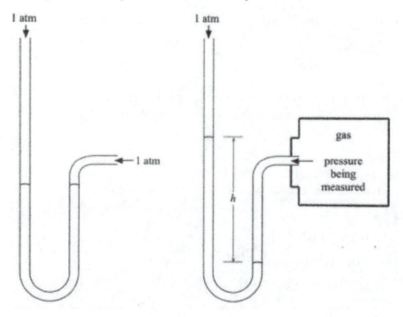

The element mercury has a density $\rho = 13,600 \ kg/m^3$. When both ends of the tube are open, the liquid seeks its own level because 1 atmosphere of pressure is exerted. When one end of the manometer is attached to a pressurized tank of a gas, the mercury rises in the open end until the pressures are equalized. The difference between the two levels of the mercury in the manometer is a measure of the gauge pressure.

## SAMPLE PROBLEM 4

A pressurized tank containing carbon dioxide is attached to a mercury manometer. The difference in the mercury levels is 36 cm. Calculate the absolute pressure of the carbon dioxide inside the tank.

## SOLUTION TO PROBLEM 4

The gauge reads a difference of 0.36 m. This is the gauge pressure. Atmospheric pressure is 1.00 atmosphere or 0.76 m of Hg. Absolute pressure is equal to the gauge pressure plus the atmospheric pressure.

Absolute pressure = 0.36 m + 0.76 m = 1.12 m of Hg

The pressure in the tank is equivalent to the pressure at the bottom of a column of mercury 1.12 m in height.

# ARCHIMEDES' PRINCIPLE

Buoyant force is the net upward force a fluid exerts on an object within it. Buoyancy occurs because the bottom of an object experiences more force per unit area than the top since the bottom is at higher pressure. The force pushing up against the bottom is greater than the force pushing down on the top.

Archimedes' principle: The buoyant force on a submerged object is equal to the weight of the fluid the object displaces.

$F_B = \rho Vg$ is the volume of the fluid displaced, not the volume of the object in the fluid; $\rho$ is the density of the fluid.

If an object is less dense than the fluid in which the object is placed, the object will float. The upward force from the fluid, $\rho g V$, balances the weight, $mg$, and the object will float in equilibrium.

For floating: buoyant force from fluid = weight of the object, so $\rho_f V_f g = m_{obj} g$.

A useful consequence of that relationship is that for a floating object, $\rho_f / \rho_o = V_o / V_f$. The ratio of the densities is the same as the fraction of the object that will sink in the fluid. Cork is 25% as dense as water, so a piece of cork will float with 25% of its volume underwater and 75% above the surface.

## SAMPLE PROBLEM 5

The specific gravity of ice is 0.917. What fraction of an iceberg is visible above the surface of the water in which it floats?

## SOLUTION TO PROBLEM 5

Ice is 91.7% as dense as water, so 91.7% of the ice will be below the surface, leaving **8.3% above the surface**.

If an object is more dense than the fluid, the object will sink. Weight will pull the object down with $W = mg$, and buoyancy will push it up with $F = F_B = \rho Vg$. The net force acting on a solitary object submerged in fluid whether it is sinking or floating is buoyancy (up) – weight (down) $= \Sigma F = \rho Vg - mg$.

### SAMPLE PROBLEM 6

The liquid element mercury has a density of $13{,}600 \text{ kg/m}^3$. Will steel float in mercury? The mass density of steel is $\rho_{steel} = 7860 \text{ kg/m}^3$.

### SOLUTION TO PROBLEM 6

Yes, it will float since $SG_{steel} < SG_{Hg}$.

### SAMPLE PROBLEM 7

A block of aluminum, $SG_{Al} = 2.700$, has a mass of 25 g.
(a)  Find the volume of the block.
(b)  Calculate the buoyant force acting on the block.
(c)  Determine the tension, $T$, in the cord that suspends the block when it is totally submerged in water.

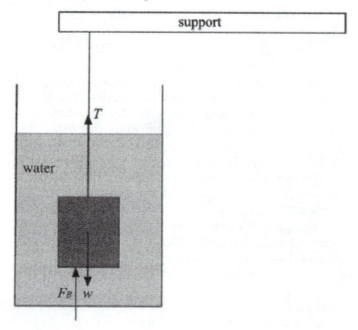

### SOLUTION TO PROBLEM 7

(a)  For a problem like this, grams and cubic centimeters are more convenient units.

From $\rho = \dfrac{m}{V}$, we write

$$V = \frac{m}{\rho}$$

$$V = \frac{(25 \text{ g})}{\left(2.7 \dfrac{g}{cm^3}\right)} = 9.26 \text{ cm}^3$$

(b)  The block displaces 9.26 cm³ of water when submerged. The buoyant force, $F_B$, acting on the block is equal to the weight of the water.

9.26 cm³ of water has a mass of 9.26 grams, or 0.00926 kg.

$$W = mg = (0.00926 \text{ kg})\left(9.8 \text{ m}/_{s^2}\right) = 0.091 \text{ N}$$

(c)   Weight pulls the aluminum down while tension and buoyancy lift it up.

$$W = F_b + T, \text{ so } T = W - F_b$$

$$T = (0.245 \text{ N}) - (0.091 \text{ N}) = \textbf{0.154 N}$$

## FLUID FLOW

In our study of the dynamics of fluids, we assume that all fluids in motion exhibit *streamline flow*. Streamline flow is the motion of a fluid in which all particles of the fluid follow the same path.

The *rate of flow, R,* is defined as the volume of fluid that passes a certain cross-sectional area per unit time. It is also called the *rate of discharge*.

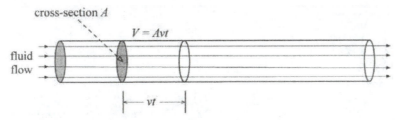

Consider a fluid flowing through the pipe diagrammed above with an average speed *v*. Over a time interval, *t*, each particle in the fluid travels a distance *vt*. The volume, *V*, flowing through a cross section is *V = Avt*. Then the *volume rate of flow*, the volume per unit time, is found by

$$V/t = Av$$

The SI unit of *Av* is expressed in $\text{m}^3/\text{s}$.

Regardless of the size of the pipe, the volume flowing past one point every second must equal the volume flowing past another point every second. The same volume flows in one end of the pipe as flows out. Variations in the pipe cross section will result in change in speed, yet the volume rate of flow *Av* remains constant.

$$A_1v_1 = A_2v_2$$

We call this relationship the equation of fluid flow continuity. A liquid flows faster through a narrow pipe section and slower through a wider cross section.

### SAMPLE PROBLEM 8

Home heating oil flows through a pipe 10 cm in diameter with a velocity of 3 m/s. What is the flow rate?

### SOLUTION TO PROBLEM 8

The area of the pipe is $A = \pi r^2 = \dfrac{1}{4}\pi D^2$. The flow rate is

$$R = Av = \left(\frac{1}{4}\pi D^2\right)v = \frac{\pi(0.10 \text{ m})^2\left(3 \text{ m}/_{s}\right)}{4} = \textbf{0.024 m}^3/_{\textbf{s}}$$

## SAMPLE PROBLEM 9

The velocity of benzene in a pipe having a radius of 4 cm is 0.44 m/s. Find the velocity of the liquid in a 1 cm radius pipe that connects with it. Both pipes have the benzene flowing to full capacity.

## SOLUTION TO PROBLEM 9

$R = A_1 v_1 = A_2 v_2$ = constant. Radii are given, and area is $A = \pi r^2$. $A_1 v_1 = A_2 v_2$ and $\pi r_1^2 v_1 = \pi r_2^2 v_2$. Solving for $v_2$ gives

$$v_2 = \frac{r_1^2 v_1}{r_2^2} = \frac{(0.04 \text{ m})^2 \left(0.44 \text{ }^{m}/_{s}\right)}{(0.01 \text{ m})^2} = 7.04 \text{ }^{m}/_{s}$$

## SAMPLE PROBLEM 10

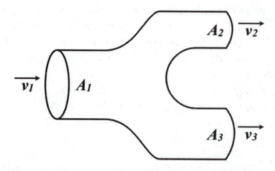

A fluid flows into this manifold through $A_1$ and leaves through $A_2$ and $A_3$. Express $v_3$ in terms of all given areas and velocities.

## SOLUTION TO PROBLEM 10

First, we find $v_3$, the velocity through $A_3$.

The total volume per second entering the manifold through $A_1$ equals the combined volume leaving per second through $A_2$ and $A_3$.

$A_1 v_1 = A_2 v_2 + A_3 v_3$

$v_3 = (A_1 v_1 - A_2 v_2)/A_3$

## BERNOULLI'S THEOREM

In the diagram on the next page, fluid flows smoothly through a pipe that changes height and thickness. The volume of liquid passing through cross section $A_1$ with same speed $v_1$ must equal the volume passing in the same time through $A_2$ with speed $v_2$.

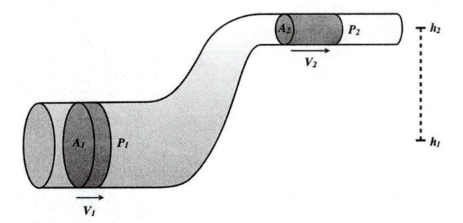

The energy per unit volume of the fluid changes as it flows. The potential energy increases as the fluid rises; the kinetic energy decreases as the fluid slows. Still, the total energy per unit volume is the same everywhere. Energy is conserved as the fluid flows between the beginning and end of the pipe.

Daniel Bernoulli was the first to relate pressure, density, fluid velocity, and fluid depth to describe the behavior of fluid flow. These relationships are shown in the equation

$$P_1 + h_1 \rho g + \frac{v_1^2 \rho}{2} = P_2 + h_2 \rho g + \frac{v_2^2 \rho}{2}$$

Bernoulli's theorem: At any two points along a streamline in an ideal fluid in steady flow, the sum of the pressure, the potential energy per unit volume, and the kinetic energy per unit volume are constant.

Two important things are immediately apparent from the equation. Since $A_1 v_1 = A_2 v_2$, it follows that the speed of flow in a pipe is greater where there is a constriction. The speed in the middle is greater than at the ends. The pressure is least where the speed is greatest.

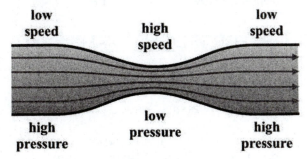

## TORRICELLI'S THEOREM

In the diagram on the next page, liquid is contained in a wide tank with an open top and a small hole near the bottom, a distance $h$ below the water level. The water at the top is barely moving, the atmospheric pressure against the top of the water is the same as the outside pressure experienced by the water that is leaving the hole, and the height of the hole is zero.

$$p_1 + \frac{1}{2}\rho v_1^2 + \rho g h_1 = p_2 + \frac{1}{2}\rho v_2^2 + \rho g h_2$$

$P_1 = P_2$, so they cancel out. There is no velocity at the top, $v_1 = 0$, so the $\frac{1}{2}\rho v_1^2$ term is zero. We can assume $h_2 = 0$, so the $\rho g h_2$ term is removed leaving $\rho g h_1 = \frac{1}{2}\rho v_2^2$; thus rearranged, the velocity of the outflow from the opening is then $v = \sqrt{2gh}$. As the liquid escapes, it gains kinetic energy at the expense of the potential energy of the remaining liquid.

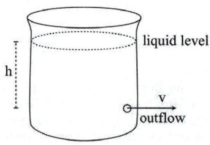

## SAMPLE PROBLEM 11

A cylindrical water tank has an orifice of 4 cm². The opening on the wall is at the bottom, 3 m below the surface of the water as diagrammed above. Assuming that no energy is wasted as the water exits the orifice, what is the rate of discharge?

## SOLUTION TO PROBLEM 11

Rate of discharge = $Av$. The speed of the water as it exits is $v = \sqrt{2gh}$. Then

$$v = \sqrt{2(10\ \frac{m^2}{s})(3\ m)} = 7.75\ \frac{m}{s}$$

The rate of discharge = $Av = (0.0004\ m^2)\left(7.75\ \frac{m}{s}\right) = 0.0031\ \frac{m}{s}$

# FLUIDS: STUDENT OBJECTIVES FOR THE AP® EXAM

■ You should be able to define density and specific gravity and relate them.

■ You should be able to define absolute pressure, gauge pressure, and atmospheric pressure.

■ You should be able to explain why pressure in a fluid depends only on depth beneath the surface and not on the shape of the container.

■ You should be able to express Archimedes' principle and to design an experiment to determine the volume of an irregular solid.

■ You should be able to use Bernoulli's equation to explain the relation between pressure, velocity, and height of a flowing fluid.

## MULTIPLE-CHOICE QUESTIONS

1. Bernoulli's equation $p + \rho gh + \frac{1}{2}\rho v^2 = \text{constant}$ is a statement of the

   (A) conservation of energy for a moving fluid.
   (B) equation of continuity for a moving fluid.
   (C) conservation of mass for a moving fluid.
   (D) conservation of momentum for a moving fluid.

2. Water flows through a horizontal pipe from a region where the diameter of the pipe is 5.0 cm into a region where the diameter of the pipe is 3.0 cm. What is the velocity $v_2$ in terms of $v_1$?

   (A) $v_2 = \frac{5}{3}v_1$

   (B) $v_2 = \sqrt{\frac{5}{3}}\ v_1$

   (C) $v_2 = \frac{9}{25}v_1$

   (D) $v_2 = \frac{25}{9}v_1$

3. A horizontal pipe, diagrammed below, carries water, $\rho = 1000\,\text{kg}/\text{m}^3$, from a region of cross-sectional area 12.0 cm² and pressure 2.15 × 10⁵ Pa where the velocity of the water is 5.00 m/s into a region where the cross-sectional area is 6.00 cm². What is the pressure in this region?

   (A) 1.78 × 10⁵ Pa
   (B) 2.13 × 10⁵ Pa
   (C) 2.28 × 10⁵ Pa
   (D) 2.52 × 10⁵ Pa

4. An ideal fluid moves through a horizontal pipe from a region of cross-sectional area $A_1$ through a constriction into a region of cross-sectional area $A_2$ as shown below. Which of the following statements is correct?

   (A) Friction acts parallel to the flow of the fluid supplying the pressure differential in the two regions.
   (B) Friction acts anti-parallel to the flow slowing the fluid when it enters $A_2$.
   (C) Friction depends on the area of the pipe and is therefore different in the two sections.
   (D) Since the fluid is ideal, there is no friction between the layers of the fluid.

5. A wooden block is placed in a beaker of water as shown below. It is observed that 40% of its volume is above the water level.

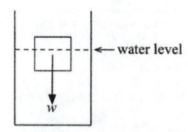

Which of the following vector arrows best illustrates the correct magnitude and direction for the buoyant force acting on the block?

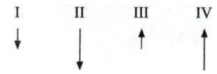

(A) I
(B) II
(C) III
(D) IV

6. Four cylinders contain a fluid of a given density. Rank the containers from greatest to least for pressure exerted on the bottom of the cylinders.

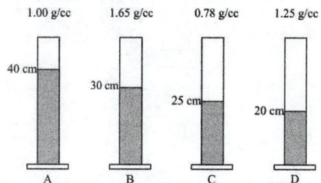

(A) B > D > A > C
(B) A > B > C > D
(C) B > A > C > D
(D) B > A > D > C

7. A pipe carries water to the third floor in a building. Which bar chart best describes the conservation of energy using Bernoulli's equation?

(A)
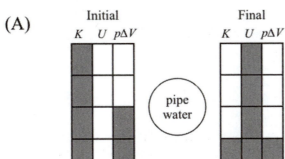

(B)
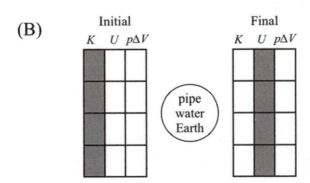

(C)
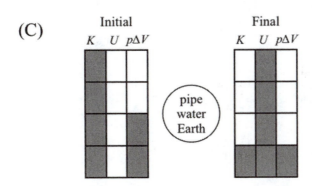

(D)
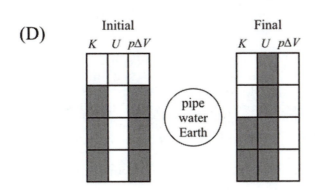

Use the following diagram to answer **Questions 8 and 9**.

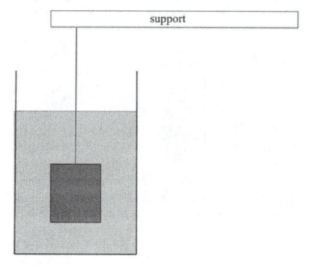

8. A 3.31 N aluminum block whose volume is 1.25 × 10⁻⁴ m³ is suspended from a cord as shown in the diagram above and submerged in water.

   The buoyant force acting on the block is
   (A) 2.34 N directed downward.
   (B) 2.34 N directed upward.
   (C) 0.968 N directed downward.
   (D) 0.968 N directed upward.

9. The tension in the string holding the aluminum block is
   (A) 0.968 N.
   (B) 2.34 N.
   (C) 3.31 N.
   (D) 4.28 N.

10. Bernoulli's equation can be stated simply as

    $$p + \frac{1}{2}\rho v^2 + \rho g h = \text{constant}.$$

    The units for each term in the equation are units of
    (A) energy.
    (B) force.
    (C) momentum.
    (D) pressure.

11. A fluid, because it has mass and density, must obey the same conservation laws that apply to solids. The work done on a fluid must equal the total changes in gravitational potential energy of the fluid-Earth system and kinetic energy of the moving fluid. This work is the
    (A) sum of the work of the input force and the work of the resistive output force.
    (B) sum of the work of the input force and the negative work done by the resistive output force.
    (C) work done by the input force.
    (D) work done by the resistive output force.

12. Three differently shaped inner connected glass tubes contain a liquid. Compare the pressure at the bottom of each tube.

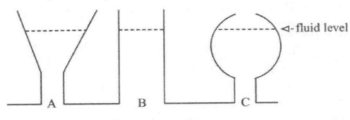

(A) B has the greatest pressure since it contains the most liquid.
(B) A has the greatest pressure since it has the smallest cross-sectional area.
(C) C has the greatest pressure since it is spherical.
(D) All three have the same pressure at the bottom of the tubes.

**Questions 13 to 15**
**Directions:** For each of the questions or incomplete statements below, two of the suggested answers will be correct. For each of these questions, you must select both correct choices to earn credit. No partial credit will be earned if only one correct choice is selected. Select the two that are best in each case.

13. A horizontal pipe diagrammed above contains water moving from $A_1$ to $A_2$ without backing up and requires the (select two answers)
    (A) speed $v_2$ in the constricted area $A_2$ to be greater than its speed in $A_1$.
    (B) speed $v_2$ in the constricted area $A_2$ to be smaller than its speed in $A_1$.
    (C) pressure $p_2$ in the constricted area $A_2$ to be larger than the pressure in $A_1$.
    (D) pressure $p_2$ in the constricted area $A_2$ to be smaller than the pressure in $A_1$.

14. An open tank with three equally spaced holes with the same diameter as shown in the diagram discharges water through the holes. The water strikes the floor some distance from the base of the tank. Select two answers.
    (A) The velocity of the emerging water is greatest when it comes from the bottom hole.
    (B) Water coming from the center hole has the greatest range.
    (C) Water coming from the top and bottom holes has the greatest range.
    (D) The rate of flow is greatest for the hole that is in the center of the tank.

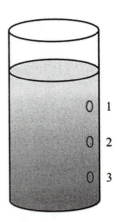

15. You are watering your garden with a hose and need to reach a section, but you do not have sufficient length of hose to move to that region. You can water this section by (select two answers)
    (A) increasing the size of the opening of the hose to obtain a larger volume of water.
    (B) increasing the rate of flow by opening the tap connected to the hose to produce a greater flow rate of water.
    (C) reducing the size of the opening to get a greater speed, thereby increasing the range of the water stream.
    (D) elevating the hose so that gravitational potential energy gives more energy to the water issuing from the hose.

## Free-Response Problems

1.  (a) The density of pure water is 1000 kg/m³. Seawater, however, can have different densities based on its salt content. Students add varying amounts of salt and some dye to several samples of pure water, creating the samples below that represent seawater.

| Red | Pure water |
| --- | --- |
| Blue | 1024 kg/m³ |
| Yellow | 1026 kg/m³ |

   If the different solutions are poured separately and slowly into a graduated cylinder, what happens to the different colors? Explain your answer.
    (b) Students want to find the density of an unknown sample of seawater. In addition to a large beaker full of the seawater, the students are given a balance, a ruler, several hooked masses of known density, and a spring scale. Describe a procedure that would allow the students to determine the density of the liquid. List any measurements you would take and how they would be used.
    (c) Another student hangs a mass from a spring scale over a beaker of fluid and slowly lowers the mass. The student first records the reading on the spring scale before the mass enters the fluid and again for every centimeter of depth. The cylinder has a cross-sectional area of $4.91 \times 10^{-5}$ m².

| Scale reading | Depth of cylinder submerged |
| --- | --- |
| N | $10^{-2}$ m |
| 3.80 | 0 |
| 3.45 | 1.0 |
| 3.00 | 2.0 |
| 2.65 | 3.0 |
| 2.30 | 4.0 |
| 1.90 | 5.0 |
| 1.50 | 6.0 |

i.  Plot the data on the graph provided below.

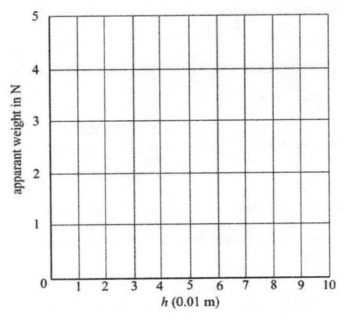

ii.  Calculate the slope of the best-fit curve.
iii. Using the slope of the line, how can you determine the density of the cylinder used in this experiment?

2.  A water tank open at the top is filled with water to a depth $d$. A small hole opens at the base of the tank at height $h$ above the ground, and a stream of water emerges.

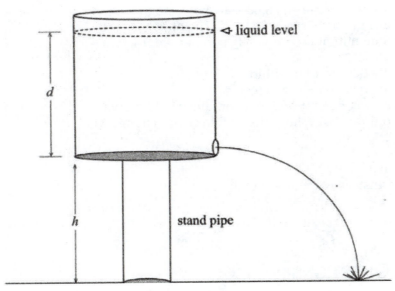

(a) Without calculations or measurements, explain how you can determine the range of the water stream after leaving the opening in the tank.

(b) If you know the diameter of the hole in the tank, explain, without calculations or measurements, how you can determine the rate of flow of the water from the hole.

3.  Water flows through a horizontal pipe from a region where the pipe diameter is 6.0 cm into a region where the diameter is 3.0 cm. See the diagram below.

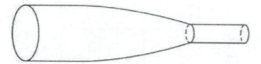

(a)  Check which of the three possible answers is correct for determining the speed of the water in the narrower section of the pipe.
___ increases
___ decreases
___ remains constant
Justify your answer qualitatively without equations or calculations.

(b)  Check which of the three possible answers is correct for determining the pressure in the narrower section of the pipe.
___ increases
___ decreases
___ remains constant
Justify your answer qualitatively without equations or calculations.

(c)  Check which of the three possible answers is correct for determining the volume rate of flow going from the larger pipe section to the smaller pipe section.
___ increases
___ decreases
___ remains constant
Justify your answer.

4.  When an object whose volume is $3.00 \times 10^{-4} \text{m}^3$ is hung from a spring scale in air, the scale reads 25.50 N. The object is then lowered into a fluid until it is completely submerged; the scale reading is now 21.50 N.
(a)  Determine the density of the fluid.
(b)  Determine the density of the object.
(c)  If the body is later submerged in a container holding water, compared to the initial values, the spring scale reading will
___ increase.
___ decrease.
___ remain constant.
Justify your answer without calculations.

# Answers

## MULTIPLE-CHOICE QUESTIONS

1.  **A** The statement $p + \rho g h + \frac{1}{2}\rho v^2 = \text{constant}$ is a statement of conservation of energy as applied to an ideal fluid.

    (L.O. 5.B.10.4)

2. **D** The equation of continuity is $A_1 v_1 = A_2 v_2$. Entering the constriction, the flow rate must be the same. Since the area decreases, the speed must increase.

Substituting into the equation: $\pi \frac{D_1^2}{4} v_1 = \pi \frac{D_2^2}{4} v_2$ gives

$$\pi \frac{(5.0 \text{ cm})^2}{4} v_1 = \pi \frac{(3.0 \text{ cm})^2}{4} v_2 \text{ and } v_2 = \frac{25}{9} v_1$$

(L.O. 5.F.1.1)

3. **A** Bernoulli's equation applies to the problem

$$p_1 + \frac{1}{2} \rho v_1^2 + \rho g h_1 = p_2 + \frac{1}{2} \rho v_2^2 + \rho g h_2$$

The pipe is horizontal, so the terms containing $\rho g h$ will not be included.

The equation of continuity $A_1 v_1 = A_2 v_2$ must be used to determine the velocity in the constriction.

$(12.0 \text{ cm}^2)(5.00 \text{ m/s}) = (6.00 \text{ cm}^2)(v_2)$. The velocity $v_2 = 10.0 \text{ m/s}$.

Solving for the pressure $p_2$ using Bernoulli's equation gives

$$2.15 \times 10^5 \text{ Pa} + \frac{1}{2}\left(1000.0 \text{ kg/m}^3\right)\left(5.00 \text{ m/s}\right)^2$$

$$= p_2 + \frac{1}{2}\left(1000.0 \text{ kg/m}^3\right)\left(10.0 \text{ m/s}\right)^2$$

and

$$P_2 = 1.78 \times 10^5 \text{ Pa}$$

(L.O. 5.B.10.3)

4. **D** The fluid flow is streamlined and nonviscous; thus, there is no friction between the layers.

(L.O. 3.C.4.2, 5.B.10.3)

5. **D** The buoyant force arises from pressure differences between the upper and lower surfaces of the object that can also be shown to be the weight of the displaced fluid. Since the block floats, $F_B = w$, its direction is upward.

(L.O. 3.C.4.2)

6. **D** Pressure is the measure of $\frac{F}{A}$ exerted on the bottom of the container. The gauge pressure $\rho g h + p_{atm} = p_{abs}$.

The gauge pressure of cylinder A is

$$\left(1000.0 \text{ kg/m}^3\right)(0.400 \text{ m})\left(9.80 \text{ m/s}^2\right) = 3920 \text{ Pa}$$

Substitution into the equation will give the pressures for the other cylinders as follows:

cylinder B = 4850 Pa, cylinder C = 1910 Pa, and cylinder D = 2450 Pa

The correct ranking is B > A > D > C.

(L.O. 5.C.4.2)

7. **C** The system involved must include all of the items in the problem. The system includes the water, the pipe, and Earth. Bernoulli's equation is a statement of conservation of energy $p + \frac{1}{2}\rho v^2 + \rho gh = \text{constant}$. The bar chart must show that work is done on the ends of the pipe and show the correct changes in kinetic energy and gravitational potential energy. Pressure is $p = \frac{F}{A}$, and from that, $F = pA$. Work is defined as $W = F\Delta x$. Volume is $V = A\Delta x$. The work is then $W = pA\Delta x = p\Delta V$.

(L.O. 5.C.4.2)

8. **D** The buoyant force on the aluminum block is equal to the weight of the water displaced.

$$F_B = \rho_{water} V_{water} g$$

$$F_B = \left(1000.0 \ \frac{kg}{m^3}\right)(1.25 \times 10^{-4} \ m^3)\left(9.80 \ \frac{m}{s^2}\right)$$

The buoyant force is 0.968 N and is directed upward.

(L.O. 3.C.4.2)

9. **B** The tension in the string is determined from the first condition for equilibrium $\Sigma F_y = 0$ equation.

$T + F_B - W = 0$. The tension is 2.34 N.

(L.O. 3.A.2.1, 3.A.3.1, 3.C.4.2)

10. **D** The units in Bernoulli's equation are pressure units. Pressure p is measured in Pa, which is $\frac{N}{m^2}$.

The term $\frac{1}{2}\rho v^2$ unit-wise is $\left(\frac{kg}{m^3}\right)\left(\frac{m}{s}\right)^2$, which is $\frac{N}{m^2}$ or Pa.

$\rho gh$ is also in Pa units, Pa $= \left(\frac{kg}{m^3}\right)\left(\frac{m}{s^2}\right)(m)$

(L.O. 5.B.10.1, 5.B.1.2)

11. **B** Bernoulli's equation is a statement of conservation of energy. The net work done on the moving fluid is $W = (p_1 - p_2)V$. The input force and the flow of liquid are in the same direction; therefore, work is

positive. The output force is opposite the flow of liquid and therefore negative. The work goes into changing the kinetic energy of the moving fluid and gravitational potential energy of the fluid-Earth system.

(L.O. 5.B.10.4)

12. **D** The weight of the liquid in the tubes exerts a downward force determined by the amount of fluid in the tube. Pressure is determined by depth only. Since the fluid level in the tubes is the same, the fluid pressure from $\rho g h$ is the same.

(L.O. 5.C.4.1)

13. **A and D** The equation of continuity requires $A_1 v_1 = A_2 v_2$. If the area is reduced, the speed must increase. Bernoulli's equation tells us that where the fluid speed is increased, the pressure is decreased.

(L.O. 5.B.10.2)

14. **A and B** Bernoulli's equation is $p_1 + \frac{1}{2}\rho v_1^2 + \rho g h_1 = p_2 + \frac{1}{2}\rho v_2^2 + \rho g h_2$. Because the tank is open, $p_1 = p_2$. The water level drops slowly, so the velocity at the top is zero. Also, each hole is the reference point for measuring the change in the gravitation potential energy of the water-Earth system. Thus, the equation simplifies to $v = \sqrt{2gh}$. The bottom hole has the greatest change in $h$. The range ($x=vt$) is determined by the speed and the time $t = \sqrt{\frac{2h}{g}}$ it takes the water to fall to the ground from each hole. The bottom hole has the least time to drop and will not have as large a range as the water from the middle hole. In fact, water from the top and bottom holes will have the same range.

(L.O. 5.B.10.2)

15. **B and C** Reducing the size of the opening, from the equation of continuity, $A_1 v_1 = A_2 v_2$ will increase the speed of the water issuing from the opening. The horizontal distance is therefore increased. $v_x t = x$

(L.O. 5.F.1.1)

## FREE-RESPONSE PROBLEMS

1. (a) Smaller densities should float on top of larger densities. So the colors should not mix, arranging themselves with orange on the bottom, then purple, yellow, blue, and red on the top.

(b) Students should use the spring scale to find the weight of one of the blocks. Students should use the ruler to find the volume (l × w × h) of the block. The block still suspended from the spring scale should be submerged in the water, and the new spring scale reading should be taken.

$F = ma$

$F_B - W = ma$

$F_B - W = 0$, so $F_B = W$

$\rho_{liquid} V_{cube} g = W_{cube}$

$\rho_{liquid} = W_{cube}/V_{cube} g$

(c) i.   The graph of the data is shown below.

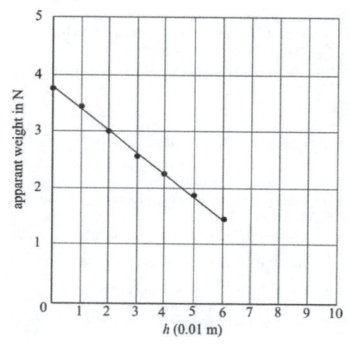

ii.   The slope of the line is $\dfrac{(1.51\ \text{N} - 3.02\ \text{N})}{(0.06\ \text{m} - 0.02\ \text{m})} = -37.75\ \text{N}/_{\text{m}}$.
The negative slope shows that the apparent weight decreases as the object is lowered into the water.

iii.   The density of the body is the mass per unit volume, and

$$\rho = \frac{\text{slope of the line}}{gA}$$

$$= \frac{\left(37.75\ \text{N}/_{\text{m}}\right)}{\left(9.8\ \text{m}/_{\text{s}^2}\right)\left(4.91 \times 10^{-4}\ \text{m}^2\right)}$$

$$= 7850\ \text{kg}/_{\text{m}^3}$$

(L.O. 1.E.1.1, 1.E.1.21)

2. (a) Bernoulli's equation can be used to determine the speed of the water as it leaves the hole. Since the tank is open to the air, the pressure at the top of the tank and at the exit hole is the same. The water level drops slowly; therefore, the speed is zero. The exit point is the reference point relative to the top of the tank used to determine the change in the gravitational potential energy of the water-Earth system.

   The velocity can be calculated from the square root of twice the product of the gravitational acceleration, $g$, times $d$, the height from the top of the container to the opening.

   When the water exits the hole, its velocity is horizontal with no vertical component.

   The time it takes the water to reach the ground is the square root of twice the drop from the tank to the ground divided by the acceleration due to gravity. The time of fall is the same time for the water to reach its range. Range is horizontal velocity multiplied by the time.

   (b) The rate of flow of the water is the velocity of the water times the cross-sectional area of the hole.

   (L.O. 1.E.1.1, 5.B.10.1, 5.B.10.3, 5.B.10.4, 5.F.1.1)

3. (a) The correct box checked is the speed of the water in the narrower section of the pipe increases.

   The equation of continuity shows that the product of the area and the velocity is a constant. The water flows from a larger area with some speed, and when it enters the smaller area, the speed must increase.

   (b) The correct box checked is the pressure in the narrower section of the pipe decreases.

   Bernoulli's equation indicates that the pressure in the smaller section will decrease because the velocity has increased. Where velocity is high, pressure is low.

   (c) The correct box checked is the volume rate of flow going from the larger pipe section to the smaller pipe section increases. Mass is conserved, and the flow rate is steady. The mass that flows through the larger section in a time interval $\Delta t$ must equal the mass that flows in the smaller section in the same time interval $\Delta t$. The rate of flow is the product of the area and the velocity.

   (L.O. 5.B.10.1, 5.B.10.2, 5.B.10.3, 5.B.10.4, 5.F.1.1)

4. (a) The buoyant force acting on the body is found from $F_B = W_{air} - W_{apparent}$. Solving for the buoyant force,

   $F_B = 25.50 \text{ N} - 21.50 \text{ N} = 4.00 \text{ N}.$

   The buoyant force is equal to the weight of the fluid displaced by the object, $F_B = \rho g V$.

   Solving for the mass density,

   $$\rho = \frac{F_B}{gV} = \frac{4.00 \text{ N}}{\left(9.8 \text{ m}/\text{s}^2\right)\left(3.00 \times 10^{-4} \text{ m}^3\right)} = 1360 \text{ kg}/\text{m}^3$$

(b) The density of the body can be found from $\dfrac{\rho_{obj}}{\rho_{fluid}} = \dfrac{W_{air}}{F_B}$. Solving for the mass density of the object,

$$\rho_{obj} = \frac{W_{air}\,\rho_{fluid}}{F_B} = \frac{(25.50\ \text{N})\left(1360\ \text{kg}/\text{m}^3\right)}{(4.00\ \text{N})} = 8670\ \text{kg}/\text{m}^3.$$ The density of the object can also be calculated from $\rho_{obj} = \dfrac{m}{V}$ and

$$\rho_{obj} = \frac{m}{V} = \frac{W/g}{V} = \frac{W}{gV} = \frac{(25.50\ \text{N})}{\left(9.8\ \text{m}/\text{s}^2\right)\left(3.00 \times 10^{-4}\ \text{m}^3\right)} = 8670\ \text{kg}/\text{m}^3.$$

(c) The correct answer is that the spring scale reading will increase. Water is less dense than the initial fluid; therefore, the weight of the displaced water is less and the buoyant force will be smaller. The apparent weight will be larger.

(L.O. 3.C.4.2)

# 11

# THERMAL PHYSICS

## PHYSICS 2

## THE ZEROTH LAW OF THERMODYNAMICS

When objects having different temperatures are placed in an isolated enclosure, all objects eventually come to the same temperature.

## TEMPERATURE SCALES

Temperature is a property that measures the relative degree of hotness or coldness and tells us the direction that heat will flow. In AP® Physics, temperature is measured in °C or Kelvin. Heat is energy that flows from regions of high temperature to regions of low temperature. The measurement of heat in Joules and the measurement of temperature in C° or Kelvin are entirely different procedures.

> ### AP® Tip
>
> A temperature of a material and a temperature difference are two different things. The temperature of a material is a measurement made in a lab. A temperature difference is a calculation to be done after temperatures are measured.

Temperature is directly related to the average kinetic energy of the particles (atoms, molecules, or ions) contained in an object. As the average kinetic energy of an object's particles decreases, the object's temperature decreases. Higher kinetic energy implies a higher temperature.

There is a minimum energy that the particles must have; a decrease below this energy is not possible. This minimum possible energy defines the minimum possible temperature: absolute zero, −273.15°C. This is a statement of the *third law of thermodynamics: the lowest possible temperature in the universe is absolute zero*. The Kelvin temperature scale is an absolute scale that is based on this fact. To convert from the Celsius to the Kelvin scale, we use K = C + 273.15.

## AP® Tip

The degree symbol is not used when we express Kelvin temperatures. The temperature is expressed in kelvins, K.

## KINETIC THEORY OF GASES

The average kinetic energy, $K$, of a gas is directly proportional to the Kelvin temperature of the gas.

$$\bar{K} = \left(\frac{1}{2}mv^2\right)_{average} = \frac{3}{2}k_BT,$$ where $k$ is the Boltzmann constant and is

$k_B = 1.38 \times 10^{-23}$ J/K

Every particle in a gas moves with its own speed, which would be hard to define, so we take the average of those speeds. The term $(v^2)_{average}$ is called the mean square speed of the molecules of a gas. When we take the square root of this term, we obtain the root-mean-square speed of gas molecules (loosely defined as the average particle speed).

$$v_{rms} = \sqrt{\frac{3k_BT}{m}}$$

### SAMPLE PROBLEM 1

Calculate the rms (root-mean-square) speed of oxygen molecules at 100°C.

### SOLUTION TO PROBLEM 1

To calculate the rms speed, we first need to find the mass of a single oxygen molecule, $O_2$. We know from the periodic table that a whole mole of $O_2$ has a mass of 32 g. A mole of something is *Avagadro's number* of that item or $6.02 \times 10^{23}$ items/mol. By dividing the molecular weight by Avagadro's number, we can calculate the mass of the oxygen molecule.

$$m = \frac{M}{N_0} = \frac{32 \times 10^{-3} \text{ kg/mol}}{6.02 \times 10^{23} \text{ mol}^{-1}} = 5.32 \times 10^{-26} \text{ kg}$$

$$\frac{1}{2}mv_{rms}^2 = \frac{3}{2}k_BT$$

$$v_{rms} = \sqrt{\frac{3k_BT}{m}} = \sqrt{\frac{3(1.38 \times 10^{-23} \text{ J/K})(100 + 273)K}{5.32 \times 10^{-26} \text{ kg}}} = \textbf{540 m/s}$$

### SAMPLE PROBLEM 2

A free electron has an average kinetic energy of $1.60 \times 10^{-20}$ J. In Celsius degrees, what is its temperature?

## SOLUTION TO PROBLEM 2

The average kinetic energy of the electron is $\bar{K} = \frac{3}{2}k_B T$. Solving for the absolute temperature, we have

$$T = \frac{2\bar{K}}{3k_B} = \frac{2(1.60 \times 10^{-20} \text{ J})}{3(1.38 \times 10^{-23} \text{ J}/_K)} = 772.95 \text{ K}$$

The Celsius temperature is

$$C = T - 273.15 = (772.95 - 273.15)^{\circ}C, \text{ or } T_C = \textbf{500}^{\circ}\textbf{C}$$

## AP® Tip

Answers should have a reasonable number of significant figures. Do your rounding off at the end of the calculations.

# THE GAS LAWS

The *state* of a sample of gas describes the sample's volume and temperature, the number of moles in the sample, and the pressure the sample is under. The pressure of a gas determines the force the gas exerts on the walls of its container. Pressure measures the average change in the momentum when molecules collide with the walls of the container. The pressure also exists inside the system, not just at the walls of the container.

Early experiments on confined gases revealed several basic but important relationships concerning the state variables pressure, $P$, the volume, $V$, the absolute temperature, $T$, and the number of moles, $n$.

Boyle's law: $P_1 V_1 = P_2 V_2$

Charles's law: $\dfrac{V_1}{T_1} = \dfrac{V_2}{T_2}$

Combined gas law: $\dfrac{P_1 V_1}{T_1} = \dfrac{P_2 V_2}{T_2}$

Combining these equations gives the ideal gas law, $PV = nRT$.

In all cases, the volume, $V$, of the gas is expressed in m³, pressure, $P$, in Pa, is absolute and not gauge pressure, the absolute temperature, $T$, in kelvins, $n$ is the number of moles of gas, and $R$ is the ideal gas constant 8.31 J/mol K. The quantity $(PV)/(nRT)$ is constant for a sample of gas. That is

$$\frac{PV}{nRT}\bigg|_{initial} = \frac{PV}{nRT}\bigg|_{final}$$

## SAMPLE PROBLEM 3

You are given a graduated beaker of gas that is sealed with a cylinder of known area, A. The beaker is clearly marked with volumes. The cylinder is free to move up and down, and it holds a thermometer. You have various blocks of different known masses as well as baths of hot and cold water.

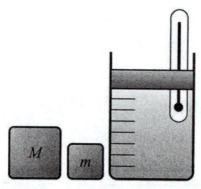

(a) Describe a procedure that would allow you to gather enough data to determine the relationship between a gas's pressure and its volume.
(b) How would you measure the pressure?
(c) Sketch a graph without any specific numbers that would show the relationship between pressure and volume.
(d) What specific feature or features of this graph demonstrate the relationship between pressure and volume?

## SOLUTION TO PROBLEM 3

(a) Place the smaller mass on the cylinder and record the volume from the cylinder. Repeat with the larger mass on the cylinder and then repeat with both masses on the cylinder.
(b) To measure pressure, divide the weight, $mg$, $Mg$, or $(M + m)g$, on top of the cylinder by the area of the cylinder.
(c)

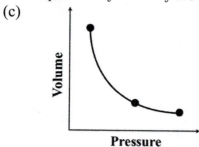

(d) As the pressure on the gas increases, the volume decreases. This is the same inverse relationship that is in Boyle's law.

## SAMPLE PROBLEM 4

(a) Describe a procedure that would allow you to gather enough data to determine the relationship between the temperature of a gas and its volume.
(b) Sketch a graph without any specific numbers that would show the relationship between temperature and volume.
(c) What specific feature or features of this graph demonstrate the relationship between temperature and volume?

## SOLUTION TO PROBLEM 4

(a)  Place the beaker in the ice bath or the water bath and record the temperature and volume at regular time intervals.

(b)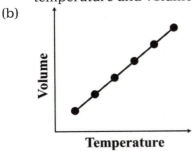

(c)  As the temperature of the gas increases, so does the volume. This is the same directly proportional relationship that is in Charles's law.

# FIRST LAW OF THERMODYNAMICS AND WORK

Thermodynamics is a branch of physics and engineering that deals with the conversion of thermal energy into useful work.

As we study thermodynamics, we use the model of a sample of gas contained within a cylinder. On top of the gas is a movable piston. The gas in the cylinder has a temperature, a pressure, and a volume. Work is done when the piston moves, and heat can be transferred to or from the gas via the cylinder.

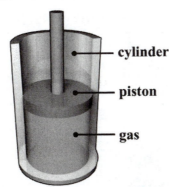

The first law of thermodynamics is a statement of the law of conservation of energy. When thermal energy (heat), $Q$, is transferred to a system, the internal energy, $U$, of the system increases and/or the system does work. This law can be expressed as

$$\Delta U = Q + W$$

$U$ in thermodynamics measures the internal energy of a gas, not its mechanical potential energy. Potential energy is an interaction between an object and its environment, whereas internal energy is a property of an object regardless of its environment.

Work is done by a moving piston when it applies a force to a gas and moves a distance. That work is equivalent to $W = P\Delta V$. Only when the gas's volume changes is work done.

When $\Delta Q$ is positive, heat is added to a system; when $W$ is positive, work is done on the system. When heat is extracted from a system and work is done by the system, both $\Delta Q$ and $W$ must be taken as negative.

| U | Q | W |
|---|---|---|
| When $\Delta U$ is positive (+), the temperature increases. | When $Q$ is positive (+), heat enters the system. | When $W$ is positive (+), work is done on the system. Usually seen as compression. |
| When $\Delta U$ is negative (–), the temperature decreases. | When $Q$ is negative (–), heat leaves the system. | When $W$ is negative (–), work is done by the system. Usually seen as expansion. |

## SAMPLE PROBLEM 5

Consider the following three systems. What is the change in the internal energy in each system?
(a)  System 1 absorbs 2.21 kJ as heat and at the same instant does 600 J of work.
(b)  System 2 absorbs 1170 J as heat, and at the same time, 500 J of work is done on it.
(c)  System 3 absorbs 400 J as heat and does 120 J of work.

## SOLUTION TO PROBLEM 5

(a)  Write the first law of thermodynamics
$$\Delta U = \Delta Q + W$$
$$\Delta U = 2.21 \times 10^3 \text{ J} + (-600 \text{ J}) = \textbf{1.61 kJ}$$
(b)  Solving for the change in internal energy
$$\Delta U = \Delta Q + W$$
$$\Delta U = 1170 \text{ J} + 500 \text{ J} = \textbf{1.67 kJ}$$
(c)  $\Delta U = 400 \text{ J} + (-120 \text{ J}) = \textbf{280 J}$

We can use the piston model to help with a first law question. When 400 J of energy are absorbed but only 120 J of energy are used to do work, the rest of the work increases the internal energy by 280 J.

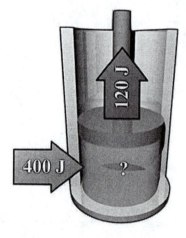

# THERMODYNAMIC PROCESSES AND *PV* DIAGRAMS

In a thermodynamic process, something causes the state to change. Heat could be input or removed. Work could be done on or by the gas. We show the changes of a thermodynamic process on a *pressure-volume diagram*, or *PV* diagram.

In the *PV* diagram below, a gas expands from state A to state B. While it expands, the gas works on its environment. The work done by the gas is the area under the curve from A to B.

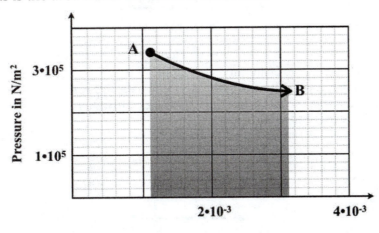

During an *isothermal* process, the temperature of the system remains constant, $\Delta T = 0$. There is no change in the internal energy of the gas and $W = -Q$.

The system does an amount of work exactly equal to the quantity of thermal energy transferred to it.

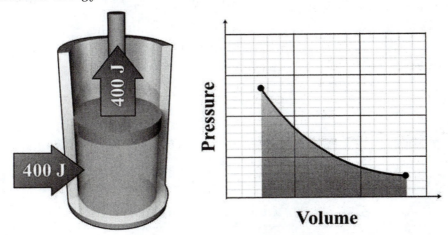

During an *isovolumetric,* which is also called *isochoric,* process, $\Delta V = 0$. Because the volume of the system does not change, no work is done. $W = 0$. All of thermal energy entering the system as heat goes into changing the internal energy of the gas. So $\Delta U = Q$.

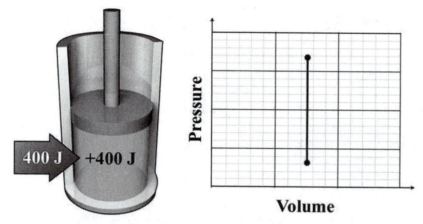

During an *isobaric* process, the pressure of the system remains constant, and $\Delta P = 0$. When heat is transferred in an isobaric process, the system changes volume to maintain constant pressure. Since this process is not isothermal, the temperature changes, which causes a change in the internal energy of the gas confined.

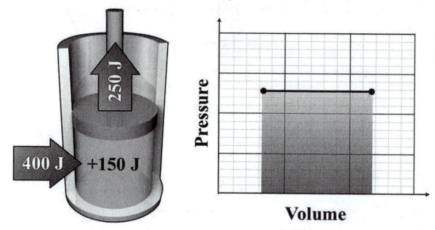

During an *adiabatic* process, no heat is transferred in or out of the system, and $Q = 0$. The work done on or by such a system changes the internal energy. Therefore, $\Delta U = W$.

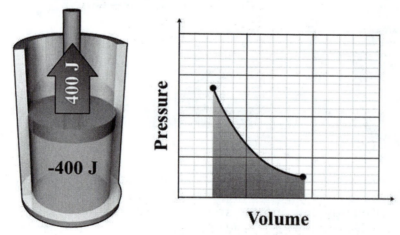

The following chart summarizes the four processes just discussed with the quantities that remain constant during these processes.

| Isothermal | $\Delta T = 0$ <br> $\Delta U = 0$ |
|---|---|
| Isochoric | $\Delta V = 0 \; W = 0$ |
| Isobaric | $\Delta P = 0$ <br> $W = P\Delta V$ <br> $\Delta U = Q + (P\Delta V)$ |
| Adiabatic | $\Delta Q = 0$ <br> $\Delta U = W$ |

In a thermodynamic *cycle*, a gas is returned to its original state after going through several intermediate steps. On a *PV* diagram, a cycle is represented as a closed loop. For a cycle, $\Delta U$ is zero. The work done by the gas will be positive if the cycle moves in a clockwise direction and negative if it moves counterclockwise. The work is found by finding the area enclosed in the cycle. The following *PV* diagram is an example of such a process. Arrows are included on the lines to indicate the order of the processes.

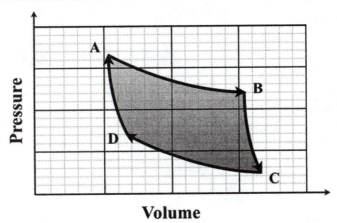

**Volume**

## SAMPLE PROBLEM 6

A 1.00 kmol sample of helium is taken through the cyclic process as shown. Path BC is isothermal. The pressure at point A is $1.20 \times 10^5$ Pa, the volume at point A is 40.0 m³, and the pressure at point B is $2.40 \times 10^5$ Pa. What are the temperatures at points A and B, and what is the volume of the gas at point C?

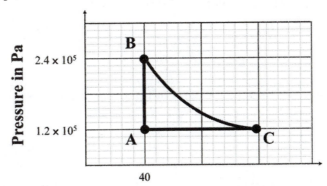

**Volume in m³**

## Solution to Problem 6

Applying the ideal gas law, $PV = nRT$, and solving for the temperature, $T_A$, we write

$$T_A = \frac{P_A V_A}{nR}$$

$$T_A = \frac{(1.20 \times 10^5 \text{ Pa})(40.0 \text{ m}^3)}{(1.00 \text{ kmol})(8.314 \times 10^3 \text{ J}/\text{kmol K})} = 577 \text{ K}$$

Because the process AB is isochoric, $T_B = \left(\frac{p_B}{p_A}\right) T_A$. Notice that $p_B = 2p_A$, making $T_B = 2T_A$ and $T_B = \textbf{1150 K}$.

For the volume of the gas at point $C$, we use $V_C = \left(\frac{T_C}{T_A}\right) V_A$. The process C is isothermal, so $T_C = T_B = 2T_A$. The process CA is isobaric, $\frac{V_C}{T_C} = \frac{V_A}{T_A}$. So $V_C = 2V_A$ and then $V_C = \textbf{80.0 m}^3$.

## Heat Transfer

Heat transfer involves a transfer of internal energy from one location to another due to the temperature difference between them. The three principal ways thermal energy can be transferred are conduction, convection, and radiation.

Conduction is heat flow through a material substance by means of molecular collisions transferring kinetic energy from one particle to another. The rate at which heat is conducted from the higher temperature face to the lower temperature face is determined from

$$H = \frac{\Delta Q}{\Delta t} = \frac{kA\Delta T}{L} \qquad H = \frac{\Delta Q}{\Delta t}$$

$H$ is the time rate of heat flow.

$L$ is the length of the conductor, $A$ is the cross-sectional area of the conductor, $k$ is the thermal conductivity of the substance measured in W/m·K, and $\Delta T$ is the temperature difference in C°.

## Sample Problem 7

A metal plate 4.0 cm thick has a cross-sectional area of 0.50 m². The inner face of the plate is at a temperature of 100°C and the outer plate is at 120°C. For the metal, $k = 75.0 \frac{\text{W}}{\text{m} \cdot \text{K}}$. What quantity of heat leaks through the plate each second?

## Solution to Problem 7

The quantity of heat flowing from the outer plate to the inner in a given time period is

$$H = \frac{\Delta Q}{\Delta t} = \frac{kA\Delta T}{L}$$

Substituting

$$H = \frac{\Delta Q}{\Delta t} = \frac{\left(75.0 \ \text{W}/\text{m}\cdot\text{K}\right)\left(0.5 \ \text{m}^2\right)\left(120°\text{C} - 100°\text{C}\right)}{0.04 \ \text{m}} = 19 \ \text{kJ}/\text{s}$$

Convection is the heat transfer in a liquid or gas by the movement of a substance due to differences in density. Warmer fluids tend to be less dense, so they rise above cooler fluids.

Radiation is heat transfer by electromagnetic radiation. Bodies hotter than their environment radiate infrared radiation that cooler bodies absorb. The surface molecules absorb this energy and gain kinetic energy as a result. In a restaurant, downward facing heat lamps shine infrared radiation on food to keep it warm.

## HEAT ENGINES

A heat engine is a device that converts thermal energy into useful work. As shown in the diagram below, a heat engine extracts heat, $Q_{hot}$, from a hot source, a reservoir, at absolute temperature $T_{hot}$ and converts a quantity of this energy into useful work. The remaining heat, $Q_{cold}$, is dumped into a heat sink, cold reservoir, at temperature $T_{cold}$. Heat flows into the heat engine, and work is done by the heat engine. This is like a car in which gasoline is burned for heat and work is done. Burning fuel in a car also makes everything under the hood get hot, which isn't useful for moving the car; wasted heat is a by-product of all heat engines.

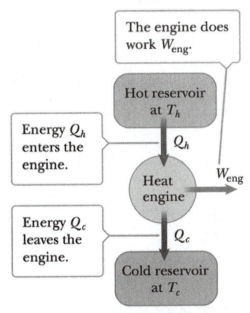

To see how well engines convert heat into useful work, we measure the efficiency of the engine. Efficiency measures the percentage of $Q_{hot}$ that actually becomes useful work. The efficiency, $e$, of heat engines producing mechanical energy is considerably less than 100%. The theoretical efficiency of a heat engine is defined as

$$e = \frac{Q_{hot} - Q_{cold}}{Q_{hot}} = \frac{W}{Q_{hot}}$$

In 1824, Nicolas Carnot described a cycle for a heat engine that can have the highest possible efficiency. No other heat engine can get more work out of an input of heat than the Carnot engine. He showed that the efficiency of such a cycle is

$$e = \frac{T_{hot} - T_{cold}}{T_{hot}}$$

As shown in the Carnot cycle below, Carnot's four-step cycle beginning at point A is isothermal expansion, adiabatic expansion, isothermal compression, and adiabatic compression.

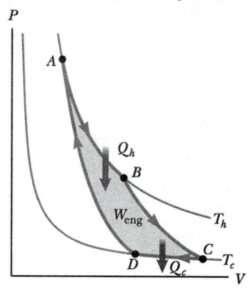

## SAMPLE PROBLEM 8

A Carnot heat engine extracts heat from a source at 327°C, does some external work, and dumps the remaining energy into a heat sink at 127°C. If 600 kJ of heat is taken from the heat source, how much heat is dumped into the heat sink? How much work does the engine do?

## SOLUTION TO PROBLEM 8

The efficiency of the heat engine is $e = \dfrac{Q_{hot} - Q_{cold}}{Q_{hot}} = \dfrac{T_{hot} - T_{cold}}{T_{hot}}$

$$\frac{600\ J - Q_{cold}}{600\ J} = \frac{(327 + 273\ K) - (127 + 273\ K)}{(327 + 273\ K)}$$

$Q_{cold} = \textbf{400 J}$

$W = \Delta Q = 600\ J - 400\ J = \textbf{200 J}$

# THE SECOND LAW OF THERMODYNAMICS AND ENTROPY

No physical system or process is ideal. In any natural process, some energy is lost to friction. Real processes are irreversible; that is, there is always more wasted heat after a process than there was before. The entropy, $S$, of an isolated system is related to the amount of waste heat generated.

---

## AP® Tip

$\Delta S = 0$ for a reversible cycle, and $\Delta S > 0$ for an irreversible cycle.

---

The second law of thermodynamics is a statement about which processes can and cannot occur in nature. The second law can be stated several ways, and all of them are equivalent.

Heat flows spontaneously from a hot body to a cold body and never flows spontaneously from a cold body to a hot body.

Like internal energy, pressure, and temperature, entropy, $S$, is a thermodynamic quantity and is a function of the state of the system. The change in entropy, $\Delta S$, during a reversible exchange of heat, $Q$, at Kelvin temperature, $T$, is

$$\Delta S = \frac{Q}{T}$$

$\Delta S$ is always positive for any natural process. The SI unit of entropy is J/K.

Addition of heat to a system increases its entropy. From a molecular point of view, additional heat increases molecular disorder. Entropy is a measure of the disorder of a system.

The second law can be stated as follows: A *natural process always moves from a state of relative order to a state of disorder. Natural processes tend to move toward a state of greater disorder.* As the disorder of the universe increases in natural processes, the energy available to do work decreases.

The most probable state of a gas is one in which the molecules are disordered. Since entropy is related to the molecular disorder, when a system increases entropy, its molecular disorder increases and the system passes from a less probable state to a more probable state. Similarly, when entropy decreases, the molecular disorder decreases and the system passes from a more probable state to a less probable state.

## THERMAL PHYSICS: STUDENT OBJECTIVES FOR THE AP® EXAM

- You should understand that the exchange of energy between two objects because of their temperature difference is called heat.
- You should know that heat can be transferred by conduction, convection, and radiation, and you should be able to solve for the rate at which energy transfer occurs.
- You should understand that $P$, $V$, $T$, and $n$ amount of gas in a container are related to one another by an equation of state, and you should be able to solve a problem involving any of these parameters.
- You should know and be able to apply the kinetic theory of gases—in particular the root-mean-square (rms) speed of a molecule.

■ You should know the equation for the work done in a thermodynamic process and be able to apply it to work done on/by a gas.

■ You should understand that work done on a gas is positive and that work done by a gas is negative in the first law of thermodynamics.

■ You should know that work is the area beneath the curve in a $PV$ diagram.

■ You should understand and be able to apply the first law of thermos-dynamics to the various processes.

■ You should be able to solve and draw $PV$ diagrams to evaluate work done, heat transferred, and changes in the internal energy.

■ You should understand the second law of thermodynamics.

## MULTIPLE-CHOICE QUESTIONS

1. Four identical containers hold 1 mol of each of the following gases at 373 K: $H_2$, He, $O_2$, and $CO_2$.

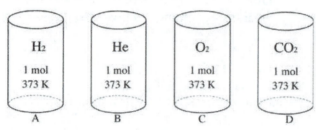

Rank the containers from highest to lowest root-mean-square speed, $v_{rms}$.

(A) D > C > A > B
(B) A = B = C = D
(C) A > B > C > D
(D) D > C > B > A

2. Every time a molecule of a gas strikes the walls of a container, there is a change in momentum for the molecule. This occurs because the
(A) walls of the container exert a force on the molecule.
(B) action-reaction forces exist between the molecule and the walls of the container.
(C) pressure exerted on the molecule is equal to the pressure exerted on the wall by the molecule.
(D) temperature of the gas has increased, increasing the velocity, and thus increasing the momentum.

3. What is the change in the momentum for a molecule of an ideal gas in the rigid container after it has struck the left wall?

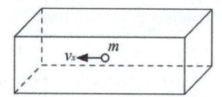

(A) $m\Delta v_x$
(B) $+2mv_x$
(C) $-2mv_x$
(D) $-m\Delta v_x$

4. What is the work done by the system in the *PV* diagram shown below as the gas is taken along the path ABCA?

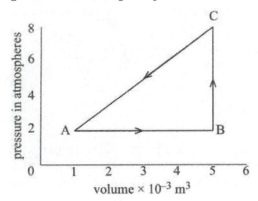

(A) −880 J
(B) +320 J
(C) +880 J
(D) −1200 J

5. In the *PV* diagram of Problem 4, what is the change in the internal energy of the gas as it moves along the path ABCA?
(A) 0 J
(B) 320 J
(C) 880 J
(D) 1200 J

6. A gas is taken through the process ABCDA as shown in the diagram below.

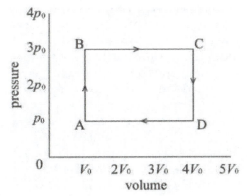

Rank the temperature of the gas at the positions from highest to lowest.
(A) A > B > C > D
(B) C > B > D > A
(C) B > A > D > C
(D) C > D > B > A

7. Heat from a reservoir is added to a gas in a container with a movable piston in a reversible isothermal process. The change in entropy of the system and the environment is best represented by which of the following?

|      | $\Delta S$ gas | $\Delta S$ reservoir | $\Delta S$ net |
|------|------|------|------|
| (A)  | +    | –    | 0    |
| (B)  | –    | +    | 0    |
| (C)  | +    | +    | +    |
| (D)  | –    | –    | –    |

8. A movable piston, with an area of $4.00 \times 10^{-3}$ m$^2$, takes a gas from 1.50 atmospheres of pressure and a volume of $3.00 \times 10^{-3}$ m$^3$ to a volume of $1.00 \times 10^{-3}$ m$^3$ during an isothermal compression. What is the magnitude of the force exerted on the piston by the molecules in the gas when the volume has reached $1.00 \times 10^{-3}$ m$^3$?
   (A) 460 N
   (B) 780 N
   (C) 1000 N
   (D) 1820 N

9. A gas in a closed container goes through an isochoric process. Thermal energy is added to the system. As a result,
   (A) work is done by the gas in expanding and the internal energy of the gas decreases.
   (B) work is done on the gas and the internal energy of the gas increases.
   (C) there is an increase in the internal energy of the system since the work done on the system in the isochoric process is zero.
   (D) there is a decrease in the internal energy of the system since the work done on the system in the isochoric process is zero.

10. One mole of an ideal gas at room temperature is in a container with a movable piston. The gas undergoes an isothermal compression when 800 J of work is done on the gas. The entropy of the gas
    (A) increases because energy is added to the gas as the gas is compressed.
    (B) remains constant since the process is an isothermal process.
    (C) decreases because the process is isothermal and the gas occupied a smaller volume.
    (D) remains constant since $Q = -\Delta W$ and then $\Delta U = 0$.

11. Students are given data from an experiment to support the graph shown below of the volume of constant mass of an ideal gas confined in a container as a function of the absolute temperature of the gas in kelvins.

| V(× 10⁻³ m³) | T (K) |
|---|---|
| 12.50 | 500.0 |
| 11.25 | 450.0 |
| 10.00 | 400.0 |
| 9.00 | 360.0 |
| 8.00 | 320.0 |
| 6.00 | 240.0 |

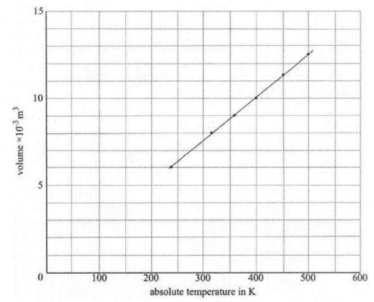

Analysis of the graph indicates that
(A) the graph cannot be extrapolated beyond the data points since not enough information is given in the data.
(B) when the volume of the gas is zero, extrapolation of the graph produces a negative value for the temperature.
(C) when the volume of the gas approaches zero, the temperature cannot be lower than 240 K because the pressure cannot be reduced to zero in the container.
(D) extrapolation of the graph indicates that when the volume of the ideal gas is zero, the temperature of the ideal gas also is zero.

12. What must happen for two bodies to have the same temperature?
(A) They must be in physical contact if energy is to transit from one body to another.
(B) They must have the same phase; otherwise, they have different heat contents and therefore are not in thermal equilibrium.
(C) They must be in thermal contact and have the same mass to cause the interactions on the microscopic level to produce an equilibrium temperature.
(D) They must be in thermal equilibrium because interactions at the microscopic level transferred energy and momentum from the body that had the higher temperature to the body with the lower temperature until they reached the same temperature.

13. For a thermodynamic system to be in thermal equilibrium, what must be true?
    (A) The pressure must be constant throughout the entire system.
    (B) The values of all of the state variables must be constant throughout the system.
    (C) The volume must be constant throughout the entire system.
    (D) The temperature must be constant throughout the entire system.

**Questions 14 and 15**
**Directions:** For each of the questions or incomplete statements below, two of the suggested answers will be correct. For each of these questions, you must select both correct choices to earn credit. No partial credit will be earned if only one correct choice is selected. Select the two that are best in each case.

14. The translational kinetic energy of a given molecule of an ideal gas is directly proportional to the (select two answers)
    (A) pressure.
    (B) kelvin temperature.
    (C) speed of the molecule.
    (D) momentum of the molecule.

15. Several students performed an experiment using a syringe form of a Boyle's law apparatus and obtained the following data.

| Absolute Pressure | Volume |
|---|---|
| ($10^5$ Pa) | ($10^{-3}$ m$^3$) |
| 1.11 | 10.0 |
| 1.20 | 9.2 |
| 1.30 | 8.5 |
| 1.40 | 7.9 |
| 1.50 | 6.2 |
| 1.80 | 5.6 |

Which of the graphs shown below is the correct graph that will permit the students to determine the work done in the compression? (Select two answers.)

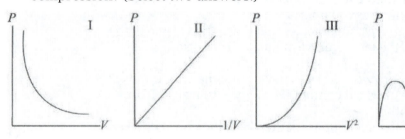

(A) I
(B) II
(C) III
(D) IV

# FREE-RESPONSE PROBLEMS

1.  Two samples of gas, A and B, are in two separately sealed
    containers. The containers are rigid and cannot change volume.
    Gas A is warmer than room temperature, and gas B is cooler than
    room temperature.
    (a) The graph below shows the speed distribution for the
        molecules of each gas. Identify which curve, dotted or solid,
        describes each sample.
        Justify your answer.

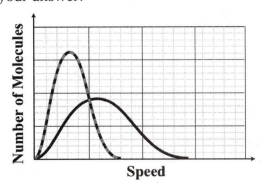

    i.   Gas A
    ii.  Gas B
    (b) The gases are allowed to reach thermal equilibrium with their
        surroundings. Describe in words other than "positive" or
        "negative" any heat flow, changes in internal energy, or work
        done as the gases reach thermal equilibrium with their
        surroundings.

|       | Heat Flow | Change in Internal Energy | Work |
|-------|-----------|---------------------------|------|
| Gas A |           |                           |      |
| Gas B |           |                           |      |

    (c) i.   Describe the entropy change, if there is any, in gas A.
        ii.  Describe the entropy change, if there is any, in gas B.
        iii. Describe the entropy change, if there is any, in the system.

2.  One mole of He at a pressure $p_0$, volume $V_0$, and temperature $T_0$
    in a closed container is taken through the following processes. It
    is heated at constant volume until its pressure is doubled; then it
    is heated at constant pressure until its volume is doubled. Next,
    it is compressed at constant volume until it reaches its initial
    pressure $p_0$ and finally is returned to its initial state conditions of
    $p_0$, $V_0$, and $T_0$.
    (a) Draw the process on a $PV$ diagram.
        i.   Indicate on your diagram where thermal energy $Q$ flows
             into the system.
        ii.  Indicate on your diagram where thermal energy $Q$ flows
             out of the system.

(b) Calculate the net amount of work done in the process in terms of $p_0$ and $V_0$.
(c) What net energy $Q$ is added to the system? Justify your answer without calculations.

3. An ideal gas is confined in a rigid container at some temperature above 0 K. A gas molecule approaches the left-hand wall of the container as shown, making an elastic collision with the wall.

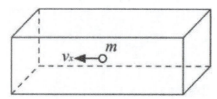

To change its momentum, a force must be applied over a short period of time.
(a) Increasing the temperature of the gas will
  ___ increase the pressure on the walls of the container.
  ___ decrease the pressure on the walls of the container.
  ___ produce no change in the pressure on the walls of the container.
  Justify your answer without calculations.
(b) The same gas is placed in a container with a movable piston and goes through an isothermal expansion. As the gas does work the piston, the entropy of the molecules will
  ___ increase.
  ___ decrease.
  ___ remain constant.
  Justify your answer without calculations.
(c) Two very large reservoirs are shown in the diagram below.

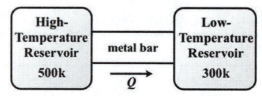

The reservoirs are very large so that their temperature remains fairly constant while the metal bar is in contact with them, transferring thermal energy. The flow of thermal energy through the bar is from the hot reservoir to the cold reservoir. The net entropy of the system
  ___ increases.
  ___ decreases.
  ___ remains constant.
  Justify your answer without calculations.

4. A closed steel cylinder contains 4.0 g of He at a gauge pressure of $9.1 \times 10^5$ Pa and a temperature of 67°C.
(a) Determine the root-mean-square speed, $v_{rms}$, for the gas.
(b) Calculate the volume of the cylinder.
(c) Sometime later, the cylinder is checked, and the temperature of the gas has dropped to 27°C and the gauge pressure is

measured at $5.2 \times 10^5$ Pa. How many grams of He have leaked from the container?

(d) Sketch the process on a $PV$ diagram.

# Answers

## MULTIPLE-CHOICE QUESTIONS

1.  **C** The root-mean-square speed (rms speed) is given by $v = \sqrt{\dfrac{3RT}{M}}$.

    The speed of the gases in the containers at the same temperature is therefore proportional to their molar mass. The order is from the lightest to the heaviest molecule: $H_2$, He, $O_2$, $CO_2$. This is the correct ranking in terms of mass and therefore the correct order for rms speed.

    (L.O. 7.A.2.1, 7.A.2.2)

2.  **B** Every time a molecule makes a collision with a wall of the container, action-reaction forces arise between the wall and the molecule. The walls are considered to be rigid, and the molecules may be considered to be hard, small spheres so that the collisions are elastic and the momentum change results in a change of direction—from into the wall to rebounding from the wall in the opposite direction.

    (L.O. 7.A.1.1)

3.  **B** The molecules are moving to the left in the container with $p_i = -mv_x$. They strike the left-hand wall and rebound to the right with $p_f = mv_x$. The change in the momentum is $\Delta p = p_f - p_i$, $\Delta p = mv_x - (-mv_x) = 2mv_x$.

    (L.O. 5.A. 2.1)

4.  **D** The work done is the area under the curve, $W = \dfrac{1}{2}bh$. The work done is

    $$\frac{1}{2}\left(5.0 \times 10^{-3}\ m^3 - 1.0 \times 10^{-3}\ m^3\right)\left(8.0\ atm - 2.0\ atm\right)\left(1.013 \times 10^5 \frac{Pa}{atm}\right) = 1200\ J$$

    Since work is done by the system, the work is negative, $W = -1200$ J.

    (L.O. 5.B.5.5, 5.B.5.6)

5. **A** The process ABCA is a cyclic process; the system is returned to its original state conditions of $p$, $T$, and $V$. There is no change in the temperature; therefore, the change in the internal energy $\Delta U$ is zero.

   (L.O. 5.B.7.1)

6. **D** The gas is confined in a container where $p$, $V$, and $T$ can change. The internal energy (absolute temperature) can be determined using $\dfrac{pV}{T} = $ constant. At A, the state conditions are

   $\dfrac{p_0V_0}{T_0}$. At B, $\dfrac{p_0V_0}{T_0} = \dfrac{3p_0V_0}{T_B}$; thus, the temperature at B is $3T_0$.

   At C, $\dfrac{3p_0V_0}{3T_0} = \dfrac{3p_0 4V_0}{T_C}$. The temperature at C is $12T_0$. Finally,

   at D, $\dfrac{3p_0 4V_0}{12T_0} = \dfrac{p_0 4V_0}{12T_0}$, the temperature is $4T_0$. The correct ranking

   is C > D > B > A.

   (L.O. 5.B.7.3)

7. **A** The entropy of the gas increases as the entropy of the reservoir decreases because the reservoir transfers thermal energy into the gas. Since the process is isothermal, both the gas and the reservoir are at the same temperature; the internal energy is the same, and the net change in entropy $\Delta S_{net}$ is 0.

   (L.O. 5.B.4.1, 7.B.2.1)

8. **D** Since it is an isothermal compression, the temperature does not change and $P_1 V_1 = P_2 V_2$. Substitution into the equation to

   determine $P_2 = \dfrac{1.50\,\text{atm}\left(1.013\times10^5\,\text{Pa}/\text{atm}\right)\left(3.00\times10^{-3}\,\text{m}^3\right)}{1.00\times10^{-3}\,\text{m}^3}$ gives a

   pressure of $P_2 = 4.56\times10^5\,\text{N}/\text{m}^2$.

   Then since $P = \dfrac{F}{A}$, force is

   $F = \left(4.56\times10^5\,\text{N}/\text{m}^2\right)\left(4.00\times10^{-3}\,\text{m}^2\right) = 1820\,\text{N}$

   (L.O. 7.A.1.2)

9. **C** The first law of thermodynamics applies. Since it is an isochoric process, no change in the volume occurs and thus no work is done. Therefore, $Q = \Delta U$. Thermal energy is added to the system; therefore, the internal energy of the system must increase.

   (L.O. 5.B.4.1)

10. **C** The first law of thermodynamics applies. Since this is an isothermal compression in which the temperature of the system remains constant, when 800 J of work is done on the gas in the container, 800 J of thermal energy, $Q$, must leave the system. The second law of thermodynamics applies since $Q$ is negative and the gas occupies a smaller volume with less disorder. Thus, the entropy decreases.

(L.O. 7.B.2.1)

11. **D** The equation of state, $pV = nRT$, for the ideal gas in the container gives a relationship of $\dfrac{p_1 V_1}{T_1} = \dfrac{p_2 V_2}{T_2}$. Extrapolation of the data points indicate that when the ideal gas reaches a volume of zero, the molecules in the gas have no energy of external vibration (still have internal energy); thus, the temperature of the gas is zero K. Since the translational kinetic energy is zero, there are no collisions with the walls of the container and the pressure also is zero.

(L.O. 7.A.3.1)

12. **D** If there is no exchange of energy between the bodies, they must be in thermal equilibrium and are therefore at the same temperature. Thermal energy will flow from the object with the higher temperature to the body with the lower temperature by molecular interaction, until both objects are in thermal equilibrium. Since thermal energy is transferred by conduction, convection, and radiation, objects may be in thermal contact but not in thermal equilibrium. They do not need to be in the same phase.

(L.O. 4.C.3.1, 5.B.6.1)

13. **B** The internal energy of any isolated system must remain constant, so $\Delta U = 0$. If the system goes through a cyclic process, the change in the internal energy will be zero and $p$, $V$, $T$, and $n$ must return to their initial values.

(L.O. 5.A.2.1)

14. **A and B** The average translational kinetic energy of a molecule is directly proportional to the both the pressure of a gas in a container and the kelvin temperature of the gas. Not all molecules in the container are moving in the same direction with the same speed or momentum.

(L.O. 7.A.2.1, 7.A.2.2)

15. **A** and **B** The work done is the area under a pressure vs. volume curve. Both graphs will show the work done. In A, work is estimated since the graph is hyperbolic; in B, the graph has been linearized, making it easier to determine the work done.

(L.O. 5.B.7.2)

## FREE-RESPONSE PROBLEMS

1. (a) i.  Gas A is the solid line. Gases at higher temperatures will have higher average speeds.
      ii.  Gas B is the dotted line. Gases at lower temperatures will have lower average speeds.

   (b)

|  | Heat Flow | Changes in Internal Energy | Work |
|---|---|---|---|
| Gas A | Heat will flow out of gas A. | Gas A will lose internal energy. | The gas will do no work. |
| Gas B | Heat will flow into gas B. | Gas B will gain internal energy. | The gas will do no work. |

   (c) i.  Gas A gets colder, and its molecules move more slowly. It loses entropy.
      ii.  Gas B gets warmer, and its molecules move more quickly. It gains more entropy than gas A lost.
     iii.  Gas A loses entropy, and gas B gains more than A lost. The net change in entropy is positive.

   (L.O. 5.B.5.6, 5.B.7.2, 5.B.7.3)

2. (a) The correct graph drawn for the information is shown below.

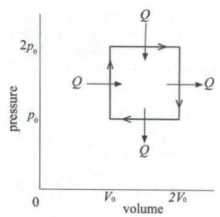

   (b) The work done on the gas in the process is the area bounded by the cycle. The work is equal to the base times the height. The work is $p_0 V_0$.

(c) The process is a cyclic process in which the gas returns to its original state conditions of $p_0$, $V_0$, and $T_0$. Therefore, since the gas returns to its original temperature, the change in the internal energy is zero. The net energy $Q$ removed from the system is equal to the network $W$, which is the area bounded by the cycle, $p_0 V_0$.

(L.O. 5.B.4.1, 5.B.7.1, 5.B.7.2, 5.B.7.3)

3. (a) The correct line checked is
    ___√___increase.
    If the temperature of the gas in the container increases, the kinetic energy and velocity will also increase. By conservation of momentum, the force also must increase. Increasing the force that each gas molecule exerts on the walls of the container will increase the pressure.
   (b) The correct line checked is
    ___√___ increase.
    An isothermal expansion of the gas requires that thermal energy must be added to the container to maintain a constant temperature for the gas, which now occupies a larger volume. The system is in a higher disordered state since the volume has increased.
   (c) The correct line checked is
    ___√___ increases.
    The entropy of the large, hot reservoir decreases, and the entropy of the large, cold reservoir increases. Because the hot reservoir loses heat, its change in entropy is negative and is smaller than the increase in the entropy of the cold reservoir.

    Because $\Delta S = QT$, the entropy of the metal bar does not change since the same amount of thermal energy enters the bar as leaves the bar. The process is an irreversible process. The net entropy change in the system will increase. In all irreversible processes, the entropy will increase. This is a general statement of the second law of thermodynamics.

(L.O. 7.A.1.1, 7.A.2.1, 7.B.1.1, 7.B.2.1)

4. (a) The root-mean-square speed for the gas is found by using
$$V_{rms} = \sqrt{\frac{3RT}{M}}$$

$$V_{rms} = \sqrt{\frac{3(8.31 \text{ J/mol} \cdot \text{K})(67 + 273 \text{ K})}{4 \times 10^{-3} \text{ kg/mol}}} = 1456 \text{ m/s}$$

   (b) Next, we find the absolute pressure of the gas
    $pV = nRT$
    $p_{abs} = p_{gauge} + p_{atm}$
    $p_{abs} = 9.1 \times 10^5 \text{ Pa} + 1.01 \times 10^5 \text{Pa} = 10.11 \times 10^5 \text{Pa}$

Now we solve for the volume:

$$V = \frac{mRT}{Mp}$$

$$V = \frac{\left(4 \times 10^{-3}\ \text{kg}\right)(8.31\ \text{J/mol K})(340\ \text{K})}{\left(4 \times 10^{-3}\ \text{kg/mol}\right)\left(10.11 \times 10^{5}\ \text{Pa}\right)} = \textbf{2.79} \times \textbf{10}^{-3}\ \textbf{m}^3$$

(c) The gauge pressure changes, changing the absolute pressure. So

$p_{abs} = 5.2 \times 10^5$ Pa + $1.01 \times 10^5$ Pa = $6.21 \times 10^5$ Pa

$pV = nRT$

Solving for the number of moles gives $n = pV/RT$. So

$$n = \frac{\left(6.21 \times 10^5\ \text{Pa}\right)\left(2.79 \times 10^{-3}\ \text{m}^3\right)}{(8.31\ \text{J/mol K})(300\ \text{K})} = 0.69\ \text{mol}$$

The number of moles of helium lost is

$\Delta n = 1.0$ mol – 0.69 mol = 0.31 mol, and the mass is

$m = \Delta n M = (0.31\ \text{mol})(4.0\ \text{g/mol}) = \textbf{1.24 g}$.

(d) A sketch is shown.

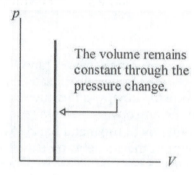

The volume remains constant through the pressure change.

(L.O. 5.B.7.2, 7.A.2.1)

# 12

# ELECTROSTATICS

## ELECTRICAL CHARGE

The basic components of all matter are the electron, the proton, and the neutron. The electron carries a negative charge, and the proton carries a positive charge. The neutron is electrically neutral. Charge is quantized, which means that every electron in nature carries the same charge as all other electrons and every proton has the same charge as every other proton. All electrons are identical to all other electrons just as all protons are identical to all other protons.

The SI unit of charge is the coulomb, C. The charge of the electron is $-1.60 \times 10^{-19}$ C, and the charge of the proton is $+1.60 \times 10^{-19}$ C. We call the magnitude of this charge the elemental charge, $e$. The electron carries the charge $-e$; the proton, $+e$.

## SAMPLE PROBLEM 1

A metal sphere is given a charge of 1.0 C. How many electrons were removed from the sphere?

## SOLUTION TO PROBLEM 1

The elemental charge is $e = 1.60 \times 10^{-19}$ C. So

$$1.0 \, C \times \frac{1e}{1.60 \times 10^{-19} \, C}$$

$$1 \, C = 6.25 \times 10^{18} e$$

As given on the exam formula sheet, the masses of the elementary particles are

$$m_e = 9.109 \times 10^{-31} \, kg$$

$$m_p = 1.673 \times 10^{-31} \, kg$$

$$m_n = 1.675 \times 10^{-31} \, kg$$

**323**

The law of conservation of electrical charge states that the total charge existing in the universe is constant. In a system, the total charge before and after changes remains the same.

The usual units of an electrical charge range from several micro coulombs, $\mu C$, to several nanocoulombs, nC, with values as follows

$$1 \ \mu C = 1 \times 10^{-6} \ C \text{ and } 1 \ nC = 1 \times 10^{-9} \ C$$

## CONDUCTORS AND INSULATORS

All metal atoms can easily lose or gain valence electrons. In some metals, electrons are free to move from atom to atom, so they're called conductors. Electricity flows easily through a conductor.

Nonmetals have no free electrons. They tend to prevent the flow of electricity. We call them insulators.

## ELECTROSTATIC FORCES OF ATTRACTION AND REPULSION

Electrostatic means electricity at rest. Electrostatic forces are non-contact forces, which means charged objects do not have to be touching for the force to be felt. Like charges repel each other, and unlike charges attract.

---

**AP® Tip**

Attractions and repulsions are forces. Forces are vectors with both magnitude and direction.

---

## COULOMB'S LAW

The magnitude of the electrostatic force $F$ between charges $q_1$ and $q_2$ that are separated by distance $R$ is given by Coulomb's law

$$F = \frac{1}{4\pi\varepsilon_o} \frac{q_1 q_2}{R^2}$$

where $\varepsilon_o$ is a constant called the permittivity of free space and its value is $\varepsilon_o = 8.85 \times 10^{-12} \ \dfrac{C^2}{N \cdot m^2}$. The constant $\dfrac{1}{4\pi\varepsilon_0}$ is called Coulomb's constant, and its symbol is $k$.

$k = 9 \times 10^9 \ Nm^2/C^2$. Coulomb's law now takes the form

$$F = k\frac{q_1 q_2}{R^2} \text{ (Physics 1 equation)}$$

## SAMPLE PROBLEM 2

Two dry socks held 5 cm apart have charges of +30 nC and +80 nC.

(a) How many electrons were removed from each sock?
(b) What is the magnitude of the electrical force between the socks?
(c) Is the force attractive or repulsive?
(d) Which sock applies more electrical force?

## SOLUTION TO PROBLEM 2

(a) For the 30 nC sock, $30 \times 10^{-9}$ C $\times \dfrac{6.25 \times 10^{18} e^-}{C} = 1.9 \times 10^{11}$ electrons.

For the 80 nC sock, $80 \times 10^{-9}$ C $\times \dfrac{6.25 \times 10^{18} e^-}{C} = 5.0 \times 10^{11}$ electrons.

(b) $F = k \dfrac{q_1 q_2}{r^2} = 9 \times 10^9$ N $\dfrac{m^2}{C^2} \dfrac{30 \times 10^{-9} \text{ C} \times 80 \times 10^{-9} \text{ C}}{(0.05m)^2} = 0.0086$ N

(c) The force is repulsive since both socks are positively charged.
(d) Both socks apply an equal repulsive force of 0.0086 N.

## SUPERPOSITION

When a number of charges in a system act on a particular charge, each exerts an electrostatic force on that charge as well as on each other. These electric forces are all calculated separately, one at a time, and are then added as vectors. This is called the superposition principle.

## SAMPLE PROBLEM 3

Two fixed point charges, $q_1 = +4$ nC and $q_2 = +6$ nC, are 10 cm apart. What is the force acting on a test charge, $q_0 = +1$ nC, placed midway between $q_1$ and $q_2$?

## SOLUTION TO PROBLEM 3

Point charge $q_1$ exerts a force of electrostatic repulsion, $F_{1,0}$, to the right on the test charge $q_0$, and charge $q_2$ exerts electrostatic force, $F_{2,0}$, to the left of test charge $q_0$ as shown.

Use Coulomb's law to determine $F_{1,0}$. Remember not to place the signs with their charges. The signs determine the direction of the force.

$$F_{10} = k \frac{q_1 q_0}{R^2}$$

$$= \frac{\left(9 \times 10^9 \text{ N} \cdot \text{m}^2/\text{C}^2\right)\left(4 \times 10^{-9} \text{ C}\right)\left(1 \times 10^{-9} \text{ C}\right)}{(0.05 \text{ m})^2}$$

$$= 1.44 \times 10^{-5} \text{N at } 0°$$

We use the same method to find $F_{2,0}$.

$$F_{20} = k \frac{q_1 q_0}{R^2}$$

$$= \frac{\left(9 \times 10^9 \text{ N} \cdot \text{m}^2/\text{C}^2\right)\left(6 \times 10^{-9} \text{ C}\right)\left(1 \times 10^{-9} \text{ C}\right)}{(0.05 \text{ m})^2}$$

$$= 2.16 \times 10^{-5} \text{N at } 180°$$

So $F_{2,0}$ is directed to the left.

Treating right as positive, we subtract the leftward force from the rightward force.

$$F = F_{1,0} - F_{2,0} = 1.44 \times 10^{-5} \text{ N} - 2.16 \times 10^{-5} \text{ N} = -7.2 \times 10^{-6} \text{ N}$$

The negative sign tells us that the resultant is to the left. The resultant force acting on $q_0$ is therefore **$7.2 \times 10^{-6}$ N** to the left.

### SAMPLE PROBLEM 4

Point charges $q_1 = -2$ nC and $q_2 = -5$ nC are 50.0 cm apart as shown. Where between the charges should a test charge $q_0$ be placed so that the resultant force acting on it is zero?

### SOLUTION TO PROBLEM 4

$F_{10} = F_{20}$ and

$$k\frac{q_1 q_0}{x^2} = k\frac{q_2 q_0}{(0.5-x)^2}$$

Here, $k$ and $q_0$ are common to both sides, so they divide out, leaving

$$\frac{q_1}{x^2} = \frac{q_2}{(0.5-x)^2}$$

Substituting for the charge on both sides of the equation gives

$$\frac{2 \text{ nC}}{x^2} = \frac{5 \text{ nC}}{(0.5-x)^2}$$

The nC units are common to both sides and divide out. So

$$\frac{2}{x^2} = \frac{5}{(0.5-x)^2}$$

Taking the square root of both sides yields $\sqrt{2}/x = \sqrt{5}/(0.5-x)$. Cross multiplying gives $1.414 (0.5 - x) = 2.2311$, or $x = $ **0.194 m**.

## THE ELECTRIC FIELD (PHYSICS 2 ONLY)

The field model of charge was developed to explain electrical interactions before atoms and their components were discovered. A field is not a physical thing, but the model's predictions are so good that we still use it. Fields are properties of the space around an object. Just like a mass that generates a gravitational field, a charged particle creates its own electric field, $E$. Fields apply forces proportional to the field strength and proportional to the property affected, such as charge or mass.

The electric field, $E$, is a vector with magnitude and direction. To test for an $E$-field, we place a positive test charge $q_0$ in the field and measure the electrostatic force $F$ acting on it. If $q_0$ is repelled, the charge creating the field is positive, and if $q_0$ is attracted, the charge creating the field is negative. Just as $g$ measures the weight per kilogram, we define the $E$-field as

$$E = \frac{F}{q_0}$$

The magnitude of the *E*-field at a point in space around a point charge can be calculated using Coulomb's law where *R* represents the distance from the charge setting up the field and the point in question:

$$E = \frac{1}{4\pi\varepsilon_0} \cdot \frac{q}{r^2} \text{ or } E = k\frac{q}{r^2}$$

The SI unit of the electric field is the newton per coulomb, N/C.

# ELECTRIC FIELD LINES

The electric model uses electric lines of force to visualize the field. To help us "see" an electric field, the concept of electric field lines is frequently used. The direction of the field lines indicates the direction a positive point charge would take if placed at various locations in the electric field. The electric field is directed away from a positive charge and directed toward a negative charge.

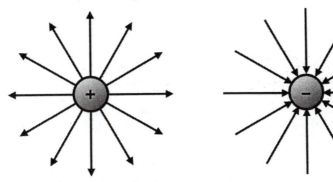

toward a positive charge       away from a negative charge

The direction of a line of force at any point in an electric field is the same as the direction of the resultant *E*-field vector at that point. We sketch the electric field following these three rules.

1. Lines never cross one another.
2. Lines in a grouping of point charges begin on a positive charge and end on a negative charge.
3. The number of lines leaving a positive charge or ending on a negative charge is proportional to the magnitude of the charges.

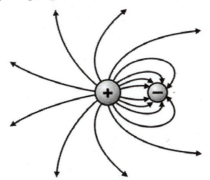

electric field lines associated with point charges +2*q* and −*q* illustrating electrostatic attraction

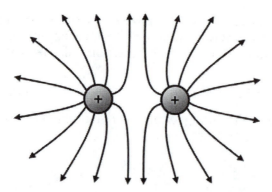

electric field lines associated with point charges +q and +q
illustrating electrostatic repulsion

## SAMPLE PROBLEM 5

A charge $q = +2$ nC is placed in a uniform electric field. In the field, $q$ experiences a force $F = 4 \times 10^{-4}$ N. What is the magnitude of the electric field intensity?

## SOLUTION TO PROBLEM 5

Using the definition of the $E$-field, we write

$$E = \frac{F}{q} = \frac{4 \times 10^{-4} \text{ N}}{2 \times 10^{-9} \text{ C}} = 2 \times 10^{5} \text{ N}\!\!\big/\!\!_{C}$$

## SAMPLE PROBLEM 6

Determine the magnitude of the electric field 2.0 cm from a point charge $q = +10$ nC.

## SOLUTION TO PROBLEM 6

This time we use the modification of Coulomb's law and substitute

$$E = k \frac{q}{R^2} = \frac{\left(9 \times 10^{9} \text{ N} \cdot \text{m}^2/\text{C}^2\right)\left(10 \times 10^{-9} \text{ C}\right)}{(0.02 \text{ m})^2} = 2.25 \times 10^{5} \text{ N}\!\!\big/\!\!_{C}$$

## SAMPLE PROBLEM 7

Two fixed point charges, $q_1 = +8 \mu$C and $q_2 = -2 \mu$C, are 10 cm apart as shown. What is the electric field intensity at the midpoint of the line connecting $q_1$ and $q_2$?

$q_1$ 〇 ———————————— $E_1$ → $E_2$ → ———————————— 〇 $q_2$

|← ———— 0.05 m ————→|← ———— 0.05 m ————→|

## SOLUTION TO PROBLEM 7

First, determine the electric field contributions due to each point charge.

$$E_1 = k \frac{q_1}{R^2} = \frac{\left(9 \times 10^{9} \text{ N} \cdot \text{m}^2\!\!\big/\!\!_{C^2}\right)\left(8 \times 10^{-6} \text{ C}\right)}{(0.05 \text{ m})^2} = 29 \times 10^{6} \frac{\text{N}}{\text{C}} \text{ to the right}$$

$$E_2 = k \frac{q_2}{R^2} = \frac{\left(9 \times 10^{9} \text{ N} \cdot \text{m}^2\!\!\big/\!\!_{C^2}\right)\left(2 \times 10^{-6} \text{ C}\right)}{(0.05 \text{ m})^2} = 7.2 \times 10^{6} \frac{\text{N}}{\text{C}} \text{ to the right}$$

Based on the diagram, we know both forces are to the right. We can dispense with the positive and negative notations in this case because we already know the resultant direction. Applying the superposition principle gives

$$E = E_1 + E_2 = 29 \times 10^6 \; \text{N}/\text{C} + 7.2 \times 10^6 \; \text{N}/\text{C} = \mathbf{3.6 \times 10^7 \; N}/\text{C} \textbf{ to the right}$$

The electric field measures the force per unit charge, $E = F/q$, just as the gravitational field measures the force per unit mass.

$g = F_g/m$ OR $F_g = mg$ and

$F = qE$

## SAMPLE PROBLEM 8

Two charged, parallel metal plates have a gap of 4.0 mm, and the $E$-field between the plates is 6000 N/C directed downward. An electron is released from rest from the negative plate.
(a)   What force does the electron experience?
(b)   Calculate the acceleration the electron experiences in the $E$-field.

## SOLUTION TO PROBLEM 8

(a)   The plates are not considered point charges in this problem, so we cannot use $F = k\dfrac{q_1 q_2}{R^2}$.

We use the above equation instead and substitute:

$$F = Eq = Ee = \left( 6000 \; \text{N}/\text{C} \right)\left( 1.60 \times 10^{-19} \; \text{C} \right) = \mathbf{9.6 \times 10^{-16} \; N}$$

Because the field is directed downward, the upper plate is positive and the force acting on the electron is upward. The force acting on the electron is therefore **9.6 × 10⁻¹⁶ N upward**. The electrical force on an electron accelerates against the field lines.
(b)   The acceleration is found from Newton's second law:

$$F = ma \text{ and } a = \frac{F}{m} = \frac{9.6 \times 10^{-16} \; \text{N}}{9.11 \times 10^{-31} \; \text{kg}} = \mathbf{1.1 \times 10^{15} \; m}/\text{s}^2 \textbf{ upward}$$

# ELECTRIC POTENTIAL ENERGY (PHYSICS 2 ONLY)

The total energy of a charge in an electric field is conserved. The electrical potential energy, $U$, of two point charges separated by distance $R$ is

$$U = k\frac{q_1 q_2}{R}$$

## SAMPLE PROBLEM 9

Two point charges, $q_A = +12 \; \mu\text{C}$ and $q_B = -22 \; \mu\text{C}$, are brought from infinity to a distance of 0.40 m of each other. How much work was done to assemble this system?

## SOLUTION TO PROBLEM 9

First, find the electrical potential energy.

$$U = k\frac{q_A q_B}{R}$$

$$= \frac{\left(9\times10^9 \text{ N}\cdot\text{m}^2\big/\text{C}^2\right)\left(+12\times10^{-6} \text{ C}\right)\left(-22\times10^{-6} \text{ C}\right)}{(0.40 \text{ m})}$$

$$= -5.9 \text{ J}$$

Because the particles attract, their fields do −5.9 J work in assembling the system. To separate the particles to infinity, some outside agent would have to do +5.9 J of work.

Note: Since $U$ is not a vector quantity and cannot be represented by a directional vector, the negative sign must be included in the calculation of $U$.

## ELECTRIC POTENTIAL (PHYSICS 2 ONLY)

Because the electrical force is a conservative force, an electrical potential energy, $U$, is associated with it. The change in electrical potential energy between two points, $\Delta U$, is the work, $W$, done moving a charged particle between these two point in an electric field. In equation form, the work done is $\Delta U = W = Fd = Eq_0 d$.

So $U = q_0 Ed$.

We define the change in electric potential, $\Delta V$, as

$$V = \frac{\Delta U}{q_0} = \frac{W}{q_0} = Ed$$

The SI unit of electric potential is the volt, $V$, and is defined as a joule per coulomb (1 $V = 1$ $J/C$).

The electric potential due to a point charge can also be calculated by

$$V = \frac{1}{4\pi\varepsilon_0} \cdot \frac{q}{r}$$

Voltage, or electric potential, is a scalar quantity.

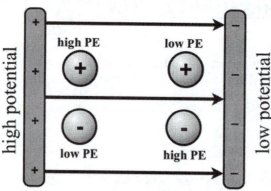

The diagram above indicates the difference between potential and potential energy. The space around the plates can be at high or low potential. If there is a charge around to experience the potential, the charge will experience electrical force toward where it would have lower potential energy. Areas of high potential are located in regions of positive charge, while areas of low potential are located in regions of negative charge. Potential is not determined by charges experiencing the electric field but rather by the charges generating the field. Potential energy is an interaction between a charge and the potential at that location. Positive charges are repelled from an area of high potential, so they have high potential energy in those areas. Negative charges are attracted to areas of high potential, so they have low potential energy in those areas.

Since both electric potential and electric potential energy are scalar quantities, if there is more than one charge, their resulting potential or potential energy can be found by adding the individual potentials or energies without having to use vector addition. With these scalar quantities, sign does matter and must be included in the mathematical equations, unlike their vector counterparts electrical force and electric field.

## SAMPLE PROBLEM 10

How much work is done moving a small body with a charge of +25 $\mu$C from point A to point B through a potential difference of 40.0 V?

## SOLUTION TO PROBLEM 10

Solving $\Delta V = W/q_0$ for work done gives

$$W = \Delta V q_0 = (40.0 \text{ V})(25 \times 10^{-6} \text{ C}) = \textbf{0.001 J}.$$

## SAMPLE PROBLEM 11

Find the electric potential 2.0 cm from a point charge $q = -12 \ \mu$C.

## SOLUTION TO PROBLEM 11

The electric potential is

$$V = k\frac{q}{R} = \frac{\left(9 \times 10^9 \text{ N} \cdot \text{m}^2\middle/\text{C}^2\right)\left(-12 \times 10^{-6} \text{ C}\right)}{(0.02 \text{ m})} = \textbf{-5.4} \times \textbf{10}^6 \textbf{ V}$$

Note: Since $V$ is not a vector quantity and cannot be represented by a directional vector, the negative sign must be included in the calculation of $V$.

## SAMPLE PROBLEM 12

A point charge of $q_1 = +2.0 \ \mu$C is placed at the origin of a frame of reference, and a second point charge of $q_2 = -8.4 \ \mu$C is placed at the position $x = 80.0$ cm as shown. Calculate the potential midway between these point charges.

## SOLUTION TO PROBLEM 12

To calculate the potential, we add both of the individual $kq/r$ contributions.

$$V = k\sum\left(\frac{q_i}{r_i}\right) = \left(9 \times 10^9 \text{ N} \cdot \text{m}^2\middle/\text{C}^2\right)\left[\left(\frac{2.0 \times 10^{-6} \text{ C}}{0.4 \text{ m}}\right) + \left(\frac{-8.4 \times 10^{-6} \text{ C}}{0.4 \text{ m}}\right)\right]$$

Simplifying the algebra, we have

$$V = \frac{\left(9 \times 10^9 \ \text{N} \cdot \text{m}^2/\text{C}^2\right)}{0.4 \ \text{m}}\left[(2.0) + (-8.4)\right] \times 10^{-6} \ \frac{\text{C}}{\text{m}} = -1.44 \times 10^5 \ \text{V}$$

Since $V$ is a scalar quantity, the potentials were added algebraically.

## Sample Problem 13

In air, a metal sphere of radius $R = 10.0$ cm is given an electrical charge of $q = +100$ nC.
(a) What is the electrical potential at the surface of the sphere?
(b) What is the electrical potential at a point 20.0 cm from the surface of the sphere?
(c) Determine the maximum electric field intensity for the sphere.

## Solution to Problem 13

(a) The electric potential at the surface of the sphere is

$$V = k\frac{q}{R} = \frac{\left(9 \times 10^9 \ \text{N} \cdot \text{m}^2/\text{C}^2\right)\left(100 \times 10^{-9} \ \text{C}\right)}{(0.10 \ \text{m})} = 9 \times 10^9 \ \text{V}$$

(b) The electric potential at a point 20.0 cm from the surface of the sphere is

$$V = k\frac{q}{R} = \frac{\left(9 \times 10^9 \ \text{N} \cdot \text{m}^2/\text{C}^2\right)\left(100 \times 10^{-9} \ \text{C}\right)}{(0.10 \ \text{m} + 0.20 \ \text{m})} = 3 \times 10^3 \ \text{V}$$

The electric field is strongest at the surface of the sphere, so we use $R = 0.1$ m.

$$E = k\frac{q}{R^2} = \frac{\left(9 \times 10^9 \ \text{N} \cdot \text{m}^2/\text{C}^2\right)\left(100 \times 10^{-9}\text{C}\right)}{(0.10 \ \text{m})^2} = 9 \times 10^4 \ \text{N}/\text{C}$$

Notice that a relationship exists between $E$ and $V$.

$$V = \frac{kq}{R} \quad \text{and} \quad E = \frac{kq}{R^2}$$

$$VR = kq = ER^2$$

Therefore,

$$V = ER.$$

So reworking part (c), we then have

$$E = \frac{V}{R} = \frac{9 \times 10^3 \ \text{V}}{0.1 \ \text{m}} = 9 \times 10^4 \ \text{V}/\text{m}$$

## SAMPLE PROBLEM 14

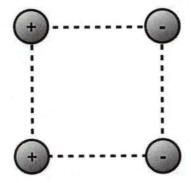

Four charges of equal magnitude are arranged at the corners of a square as shown above.

(a)   Where within the square is the electric field zero?
(b)   Where within the square could a positive charge be placed so that it experiences no net force?
(c)   Where within the square is the electric potential zero?
(d)   Where within the square could a positive charge have zero potential energy?
(e)   Use a solid line to indicate a path that a positive charge could take through the arrangement that would require no net work on the particle to move it.

## SOLUTION TO PROBLEM 14

(a)   Electric field emanates from a positive charge and ends on a negative charge. No point in this arrangement has a zero electric field.
(b)   Since no point in this arrangement has a field of zero, there cannot be a location at which a positive charge feels no force.
(c)   The electric potential will be zero, right in the middle. $V = k\Sigma Q/r$. In the middle, all of the $r$ values are the same. Two charges are positive, and two are negative, so their sum will be zero.
(d)   The potential energy is $qV$, and the voltage is zero in the middle of the square.
(e)   The work to move a charge along an equipotential is zero. Here is one equipotential curve:

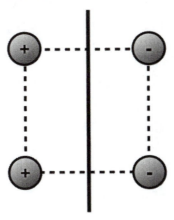

## EQUIPOTENTIAL SURFACES (PHYSICS 2 ONLY)

A collection of points that all have the same electrical potential is called an equipotential surface. The equipotential surfaces of a spherical surface are concentric spheres perpendicular to the electric field lines as shown. For a region in space where an electric field exists, the equipotential surfaces are always perpendicular to the electric field lines.

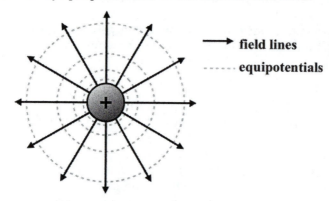

→ **field lines**

------ **equipotentials**

If a charge begins and ends its motion on the same equipotential, the work done to move that charge, *q*, between the points is $W = q\Delta V = 0$ regardless of the path taken.

The surface of a conductor is itself an equipotential surface.

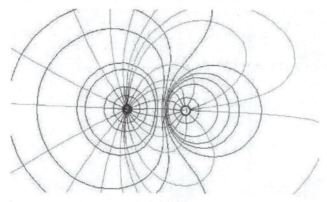

The lines of force and the equipotential lines associated with a point charges $+2q$ and $-q$ are shown above. The paths that terminate on either charge are electric field lines; the equipotentials from concentric shells around positive and negative charges.

## SAMPLE PROBLEM 15

An electric field of strength 10 N/C is shown below. Three points are shown in the field. Find the potential difference, $\Delta V$, between

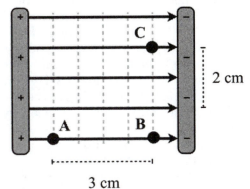

3 cm

(a) points A and B.
(b) points A and C.
(c) points B and C.

## SOLUTION TO PROBLEM 15

(a)  $\Delta V = -Ed$
$\Delta V = -10$ N/C(.03m)
$\Delta V = 0.30$ V

(b)  $\Delta V = 0.30$ V
The potential difference is based on the distance between the equipotential lines, not the distance between the two points. Therefore, the answer is the same as part (a).

(c)  $\Delta V = 0$ V
Point B and C are located on the same equipotential line, so there is the same potential at each point. There will be no potential difference.

# THE PARALLEL-PLATE CAPACITOR (PHYSICS 2 ONLY)

One of the most useful electrical devices in many aspects of electricity is the parallel-plate capacitor. The plates are metal and are separated by a plate gap, $d$. Connecting the capacitor to a battery charges the plates.

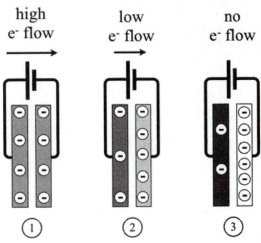

1. Initially, the capacitor is uncharged. There are an equal number of negative charges on both of the capacitor's plates along with an equal number of positive charges (not shown) to give a net charge of neutral. The battery can easily begin moving charge from one plate to the other. Here, electrons flow from the left plate, through the battery, and onto the right plate. Electrons flow very easily to or from the plates of an uncharged capacitor.

2. Some electrons have moved off the left plate onto the right plate. The battery cannot move electrons to the right plate quickly since so many extra negative charges are there already. The current slows down because the battery cannot move electrons as easily as before.

3. The capacitor is charged to the maximum amount given the battery that is connected. The battery cannot move more electrons to the right plate from the left plate. No more current is flowing.

A battery functions as an "electron pump." It removes electrons from what becomes the positive plate and places an equal number of electrons on what becomes the negative plate (charging a capacitor is done by electrostatic repulsion). It takes a brief period of time to charge the capacitor, and then the battery can be removed. The magnitude of the charges on the plates are equal, $|q_{plate\ 1}| = |q_{plate\ 2}|$. The amount of charge on each plate is proportional to the potential difference, $\Delta V$, across the plates

$$C = \frac{q}{V}$$

We call the constant of proportionality, $C$, the capacitance of the capacitor.

The SI unit of capacitance is the farad, F, and is defined as 1 F = 1 C/V. The farad is a very large unit. The usual capacitance of a capacitor is on the order of several microfarads, $\mu F$, to several picofarads pF. In short,

$1\ \mu F = 1 \times 10^{-6}\ F$

$1\ nF = 1 \times 10^{-9}\ F$

$1\ pF = 1 \times 10^{-12}\ F$

The upper plate in the diagram shown below has been charged positive. The field always points from positive charge to negative charge. Ignoring the edges of the plates, the electric field between the plates is uniform.

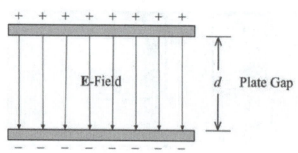

For a parallel-plate capacitor, the relationship between the electric field, $E$, between the plates, the potential difference across the plates and the plate gap is

$$V = Ed$$

## SAMPLE PROBLEM 16

A parallel plate capacitor has a potential difference of 100 V across its plates and a plate gap of 2.0 mm. What electric field, $E$, exists between the plates?

## SOLUTION TO PROBLEM 16

$$E = \frac{V}{d} = \frac{100 \text{ V}}{2 \times 10^{-3} \text{m}} = 5 \times 10^4 \text{ V/m}$$

The equipotential surfaces between the plates of a parallel capacitor are planes that are perpendicular to the field lines.

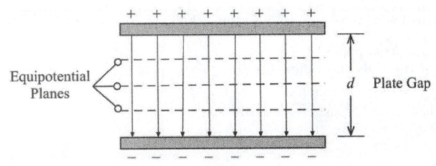

The capacitance of a parallel-plate capacitor can be calculated using

$$C = \varepsilon_o \frac{A}{d}$$

where $\varepsilon_o$ is the permittivity constant of free space and $A$ is the area of one of the plates. It gives us information on how well an $E$-field is set up in space. In SI units, $\varepsilon_o = 8.85 \times 10^{-12}$ C²/Nm². This unit is equivalent to farads per meter.

## SAMPLE PROBLEM 17

A set of capacitor plates each measures 10.0 cm by 12.0 cm, and they have a plate gap of 3.0 mm. Calculate the capacitance.

## SOLUTION TO PROBLEM 17

The area of one of the plates is

$$120 \text{ cm}^2 \times \frac{1 \text{ m}^2}{1 \times 10^4 \text{cm}^2} = 120 \times 10^{-4} \text{m}^2. \text{ Therefore,}$$

$$C = \varepsilon_o \frac{A}{d} = \frac{\left(8.85 \times 10^{-12} \text{ C}^2/\text{N} \cdot \text{m}^2\right)\left(120 \times 10^{-4} \text{m}^2\right)}{3 \times 10^{-3} \text{m}} = 35 \text{ pF}$$

## DIELECTRICS (PHYSICS 2 ONLY)

When an insulating material called a dielectric is inserted between the plates of a capacitor, the capacitance increases. The dielectric is represented by a dimensionless quantity called the dielectric constant, $\kappa$, which is different for different materials. With a dielectric inserted between the plates, the capacitance becomes

$$C = \kappa \varepsilon_o \frac{A}{d}$$

### SAMPLE PROBLEM 18

A parallel-plate capacitor of plate gap $d$ = 2.0 mm has a capacitance $C_0$ = 3.0 $\mu$F. A battery is used to charge the plates with a potential difference $\Delta V_0$ = 600 V. After the charging process, the battery is removed.
(a) What is the potential difference across the capacitor when a 2.0 mm thick slab of a dielectric of $\kappa$ = 7.5 is sandwiched between the plates?
(b) What is the new capacitance?
(c) What is the permittivity of the dielectric?

### SOLUTION TO PROBLEM 18

Proportionally, the dielectric constant is $\kappa = V_0/V$ and
(a) $V = V_0/\kappa$ = 600 V/7.5 = **80 V**

(b) $C = \kappa C_0$ = 7.5(3.0 × 10$^{-6}$ F) = **22.5 $\mu$F**

(c) $\varepsilon = \kappa \varepsilon_0$ = 7.5(8.85 × 10$^{-12}$ C$^2$/Nm$^2$) = **6.64 × 10$^{-11}$ C$^2$/Nm$^2$**

## ELECTRICAL ENERGY STORED IN A CAPACITOR (PHYSICS 2 ONLY)

Parallel-plate capacitors not only produce uniform electric fields between their plates and store charge on the plates but also store electrical energy. The energy stored in the field between the plates may be calculated by

$$U = \frac{1}{2}CV^2$$

$$\text{or} \quad U = \frac{1}{2}qV$$

$$\text{or} \quad U = \frac{1}{2}\frac{q^2}{C}$$

Only the first two are given on the formula sheet.

### SAMPLE PROBLEM 19

A parallel-plate capacitor with a plate gap of 0.6 mm has a capacitance of 6.0 $\mu$F. A battery charges the plates with a potential difference of 500 V and is then disconnected.
(a) Calculate the energy stored in the capacitor.
(b) What charge exists on the plates?

## SOLUTION TO PROBLEM 19

(a) Because the capacitance and potential difference are given, we will use $U = \frac{1}{2}CV^2$ to find the energy stored. Therefore,

(b) $U = \frac{1}{2}CV^2 = \frac{1}{2}(6.0 \times 10^{-6} \text{F})(500 \text{ V})^2 = \textbf{0.75 J}$.

(c) The charge is then $q = CV = (6.0 \times 10^{-6} \text{ F})(500 \text{ V}) = \textbf{3} \times \textbf{10}^{-3} \textbf{C}$

## COMBINATIONS OF CAPACITORS (PHYSICS 2 ONLY)

Electrical circuits frequently contain two or more capacitors grouped together to serve a particular function. In considering the effect of such groupings, it is convenient to use a circuit diagram. In such diagrams, electrical components are represented by symbols. The symbol for a battery is a set of unequal parallel lines. The high-potential terminal (+ terminal) of a battery is represented by the longer line. The capacitor is shown as a set of equal-length parallel lines. Wires are shown as lines that connect components.

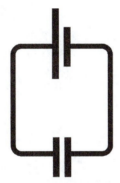

This diagram shows a battery connected to a capacitor.

When capacitors are arranged in parallel, their combined capacitance is the arithmetic sum of the individual capacitors.

$C_{eq} = C_1 + C_2 + C_3 + \ldots$

## SAMPLE PROBLEM 20

Three capacitors, $C_1 = 0.5 \ \mu\text{F}$, $C_2 = 0.3 \ \mu\text{F}$, and $C_3 = 0.2 \ \mu\text{F}$, are arranged in parallel. Calculate the equivalent capacitance.

## SOLUTION TO PROBLEM 20

We start with a circuit diagram and then substitute values.

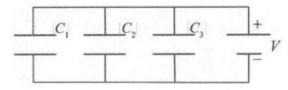

$C_{eq} = C_1 + C_2 + C_3 = (0.5 + 0.3 + 0.2) \ \mu\text{F} = \textbf{1.0} \ \mu\textbf{F}$

When capacitors are arranged in series, the reciprocal of the combined capacitance is the sum of the reciprocals of the capacitors.

$$\frac{1}{C_{eq}} = \frac{1}{C_1} + \frac{1}{C_2} + \frac{1}{C_3} + \dots$$

### SAMPLE PROBLEM 21

Three capacitors, $C_1 = 0.5\ \mu F$, $C_2 = 0.3\ \mu F$, and $C_3 = 0.2\ \mu F$, are arranged in series. Calculate the equivalent capacitance, $C_{eq}$.

### SOLUTION TO PROBLEM 21

Again, we start with a circuit diagram and substitute.

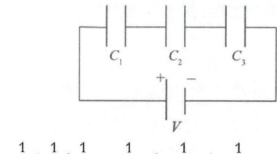

$$\frac{1}{C_{eq}} = \frac{1}{C_1} + \frac{1}{C_2} + \frac{1}{C_3} = \frac{1}{0.5\ \mu F} + \frac{1}{0.3\ \mu F} + \frac{1}{0.2\ \mu F}$$

$$C_{eq} = 0.097\ \mu F$$

### SAMPLE PROBLEM 22

(a) Calculate the total capacitance of the circuit shown below.
(b) Find the charge on each capacitor in the circuit.
(c) Calculate the potential difference across capacitor $C_2$.

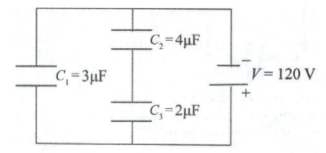

### SOLUTION TO PROBLEM 22

(a) Capacitors $C_2$ and $C_3$ are in series with each other, and their equivalent capacitance is

$$\frac{1}{C_{eq}} = \frac{1}{C_1} + \frac{1}{C_2} = \frac{1}{4\ \mu F} + \frac{1}{2\ \mu F}$$

$$C_{eq} = 1.33\ \mu F$$

The circuit is now reduced as shown below.

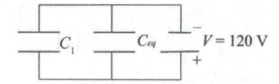

The two remaining capacitances are arranged in parallel, and
$C = C_1 + C_{eq} = 3 \ \mu F + 1.33 \ \mu F = \mathbf{4.33 \ \mu F}$.

(b) The total charge in the capacitors is $q = CV = (4.33 \ \mu F)(120 \ V) =$ 520 $\mu C$. The charge, $q_1$, on $C_1$ is $q_1 = QV = (3 \ \mu F)(120 \ V) = \mathbf{360 \ \mu C}$, and the remaining charge is $q - q_1 = 520 \ \mu C - 360 \ \mu C = 160 \ \mu C$. The 160 $\mu C$ must be deposited on capacitors $C_2$ and $C_3$. So $q_2 = q_3 = \mathbf{160 \ \mu C}$.

(c) The voltage drop is $V = q_2/C_2 = \mathbf{40 \ V}$.

# ELECTROSTATICS: STUDENT OBJECTIVES FOR THE AP® EXAM

- You should know that there are two fundamental charges and how charges interact.
- You should understand Coulomb's law for the fundamental electro-static force between point charges and know that the electric fields produced by these charges are vector quantities with magnitude and direction.
- You should know how to use the equations, identify the charges, and sketch the directions for the forces and field in solving problems that involve these quantities.
- You should know that potential difference and electrical potential, closely related concepts, are scalar quantities.
- You should know how to use the equations, identify the charges, and solve problems involving potential difference and electrical potential.
- You should know that the equations of potential difference and electrical potential can be used to solve conservation of energy and work-energy theorem problems.
- You should know the relationship between electric field lines and equipotential lines and be able to sketch them.
- You should understand how a capacitor works to store energy.

# MULTIPLE-CHOICE QUESTIONS

1. Two uncharged objects, a glass rod and a silk cloth, are rubbed together. The glass rod acquires a positive charge, while the silk cloth
   (A) remains uncharged.
   (B) becomes negatively charged.
   (C) is also positively charged.
   (D) has a negative charge of twice the magnitude of the positive charge transferred to the glass rod.

2. A positively charged glass rod is brought near but does not touch a neutral isolated conducting sphere.

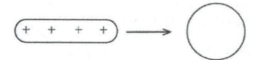

Which of the following illustrations is the correct representation of the charge distribution on the sphere?

| I | II | III | IV |
|---|---|---|---|

(A) I
(B) II
(C) III
(D) IV

3. A positively charged glass rod is brought near a neutral isolated conducting sphere as shown in the diagram below.

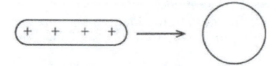

The positively charged glass rod is removed. The isolated conducting sphere will have
(A) a net positive charge.
(B) a net negative charge.
(C) charge separated: negative on one side of the sphere and positive on the other side.
(D) zero net charge.

4. Electrical charge that is transferred in the process of conduction when a charging wand touches an isolated body can produce charges on the body equal to

(A) $\pm\frac{1}{2}e$

(B) $\pm\frac{2}{3}e$

(C) $\pm\frac{3}{2}e$

(D) $\pm e$

5. A metal sphere, A, carries a charge of $-3Q$, and an identical sphere, B, has a charge of $+Q$. What is the magnitude of the force, $F$, that A exerts on B compared with the force that B exerts on A?

   (A) $\frac{1}{3}F$

   (B) $F$
   (C) $3F$
   (D) $9F$

6. A charge of $+Q$ is located in the center of a neutral ring that has an inner radius of $r_1$ and an outer radius of $r_2$ as shown below.

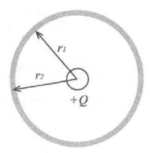

   The charge induced on the
   (A) inner surface of the ring is $-Q$.

   (B) inner surface of the ring is $-\frac{1}{2}Q$.

   (C) outer surface of the ring is $+\frac{1}{2}Q$.

   (D) outer surface of the ring is $-Q$.

7. An electron is placed in a uniform $E$-field between two charged parallel plates as shown in the diagram below.

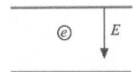

   What is the magnitude and the direction of the force on the electron given that the strength of the field is 2000 N/C?
   (A) $3.2 \times 10^{-16}$ N downward
   (B) $3.2 \times 10^{-16}$ N upward
   (C) $1.8 \times 10^{-27}$ N downward
   (D) $1.8 \times 10^{-27}$ N upward

8. The magnitude and the direction of the acceleration of the electron in Problem 7 is given correctly in which of the following choices?
   (A) $1.9 \times 10^{11}$ $m/s^2$ toward the upper plate

   (B) $1.9 \times 10^{11}$ $m/s^2$ toward the lower plate

   (C) $3.5 \times 10^{14}$ $m/s^2$ toward the upper plate

   (D) $3.5 \times 10^{14}$ $m/s^2$ toward the lower plate

9. Equipotential lines representing areas of equal electrical potential in a region are shown in the diagram below.

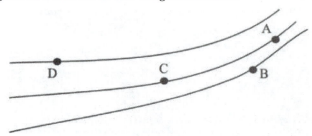

Rank the magnitude of the strength of the electric field from largest to smallest.
(A) D > C > B > A
(B) D = C = B = A
(C) D > B > C > A
(D) A > B > C > D

10. Three point charges, $+q_1$, $+q_2$, and $-q_3$, are spaced equally along the straight line shown below.

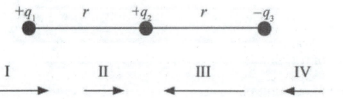

Which of the above vectors best represents the net force acting on the point charge $+q_2$?
(A) I
(B) II
(C) III
(D) IV

11. Two isolated identical spheres kept at rest at a separation of 0.10 m from center to center have charges as shown in the diagram below.

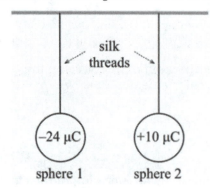

A copper wire is used to connect the two spheres and is then removed. The charge on each sphere after the copper wire was removed is
(A) $-14$ $\mu$C on sphere 1 and 0 on sphere 2.
(B) $-7$ $\mu$C on sphere 1 and $+7$ $\mu$C on sphere 2.
(C) $-7$ $\mu$C on sphere 1 and $-7$ $\mu$C on sphere 2.
(D) zero on each since the excess charge remains in the wire when it is removed.

12. During a physics laboratory experiment, an electron is placed in the electric field between the parallel plates shown in the diagram below.

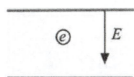

When the electron is released from rest, it will experience an acceleration due to the field that is
(A) less than the acceleration due to gravity and in the same direction as the gravitational acceleration.
(B) less than the acceleration due to gravity and in the opposite direction as the gravitational acceleration.
(C) much greater than the acceleration due to gravity and in the same direction as the gravitational acceleration.
(D) much greater than the acceleration due to gravity and in the opposite direction as the gravitational acceleration.

**Questions 13 to 15**
**Directions:** For each of the questions or incomplete statements below, two of the suggested answers will be correct. For each of these questions, you must select both correct choices to earn credit. No partial credit will be earned if only one correct choice is selected. Select the two that are best in each case.

13. Which of the following statements is true concerning the work done on an electron by the electric field when the electron is moved at constant speed between two points? Select two answers.
(A) The electron moved in the direction of the electric field; therefore, there was no change in its kinetic energy and thus no work done.
(B) The electron moved in the direction opposite the electric field; therefore, there was no change in its kinetic energy and thus no work done.
(C) The electron moved perpendicular to the electric field; therefore, there was no change in its kinetic energy and thus no work done.
(D) The electron moved along an equipotential line connecting the two points.

14. A charging wand is brought near four conducting spheres as shown in the illustration below. Which illustration indicates the correct charge on the sphere? Select two answers.

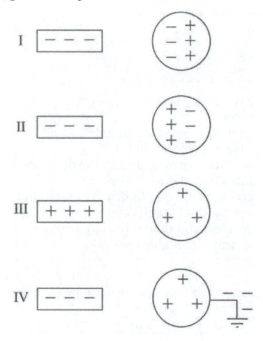

- (A) I
- (B) II
- (C) III
- (D) IV

15. In comparing the gravitational force with the Coulomb force, we find that (select two answers)
    - (A) the gravitational force acts at a distance while the Coulomb force is a contact force.
    - (B) they are inverse square laws.
    - (C) they are both forces of attraction.
    - (D) the Coulomb force acting between an electron and a proton is stronger than the gravitational force acting between the two.

## FREE-RESPONSE PROBLEMS

1. A tiny sphere of mass $8 \times 10^{-13}$ kg and charge $q = -4.8 \times 10^{-18}$ C is placed between the plates of a parallel-plate capacitor with plate gap $d = 20.0$ mm. The sphere remains suspended in the electric field of the capacitor as shown on the next page.

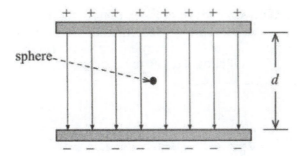

(a) On the diagram, show the forces acting on the sphere.
(b) How many electrons are in excess on the sphere?
(c) What is the strength of the electric field?
(d) Determine the potential difference across the plates.
(e) The charge is removed from the field. Both upper and lower plates are symmetrical, and each measures 25.0 cm by 25.0 cm. Find the capacitance of the capacitor.
(f) What charge exists on the lower plate of the capacitor?

2. Four identical charges $q = +6.0\ \mu C$ are arranged in a square that measures $R = 0.04$ m on a side as shown. $L$ is the distance from the center of the square to each of the charges.

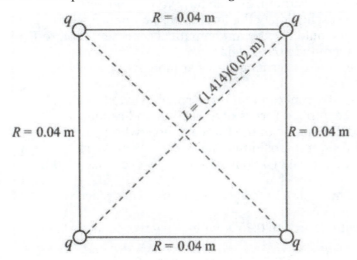

(a) Find the electric field at the center of the square.
(b) Determine the electric potential at the center of the square.

3. Two small identical Styrofoam spheres of mass $m$ connected by silk threads of length $L$ to a common point receive equal charges $q$ from a charging wand. After charging, the angle each silk thread makes with the vertical is 10° as illustrated in the diagram below.

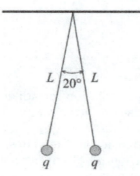

(a) i. Make a free-body diagram of the forces on either Styrofoam sphere.
   ii. If you drew a free-body diagram of the forces acting on the second sphere, how would they compare?
   iii. Does your free-body diagram depend on whether $q$ is + or –? Explain your answer using the correct scientific terms.
(b) Using your free-body diagram, explain how you would write a mathematical representation of the forces exerted on the sphere.
(c) Explain how you would use the mathematical representation to determine the magnitude of the charge.
(d) Explain how you would determine the separation, $r$, of the spheres.

4. In an experiment, students noted that three small isolated identical spheres A, B, and C exhibited forces of attraction and repulsion between various combinations of the three spheres when pairs of the spheres were brought to positions that were 0.05 m apart.
(a) A repelled B with a force of 1.00 N when they were placed 0.05 m apart.
(b) A attracted C with a force of 0.50 N when they were placed 0.05 m apart.
(c) B attracted C with a force of 0.25 N when they were placed 0.05 m apart.
(d) If the charge on A is $+2Q$, what are the charges on B and C? Explain each step of your reasoning.

# Answers

## MULTIPLE-CHOICE QUESTIONS

1. **B** Charge is conserved. If the glass rod acquired a positive charge during the process of the two materials being rubbed together, the silk cloth acquires the same magnitude of negative charge.

   (L.O. 1.B.1.1)

2. **A** The sphere is polarized due to the charging wand. The only charge that can move in a sphere is the negative charge. The electrons will

move to the side of the sphere closest to the positive charging wand. The side opposite the wand will have a positive charge.

(L.O. 1.B.1.2, 1.B.2.2)

3. **D** The realignment of charge is called polarization. When the positive charging wand in Problem 2 is removed, the negative charge will realign and the sphere will still have zero net charge.

(L.O. 1.B.2.3)

4. **D** The charge on an object is a whole number multiple of $e$.

(L.O. 1.B.3.1)

5. **B** Coulomb's law relates the magnitude of the force acting between two point charges, $F = k_e \dfrac{q_1 q_2}{r^2}$. The forces are action-reaction pair forces. $F_{AB} = -F_{BA}$. The force is attractive since the charges are $-3Q$ and $+Q$.

(L.O. 3.C.2.1)

6. **A** Charge is conserved. Since a charge of $+Q$ is located in the center of the conducting ring, the charge on the inner surface is $-Q$ and the charge on the outer surface is $+Q$.

(L.O. 5.A.2.1)

7. **B** The direction of the field is the direction that a positive charge would move when placed in the field. The electron will experience an upward force. The magnitude of the force is given by $F = Eq =$ (2000 N/C)(1.6 × 10⁻¹⁹ C) = 3.2 × 10⁻¹⁶ N.

(L.O. 2.C.1.1, 2.C.1.2)

8. **C** The direction of the field is the direction of the force on a positive charge that will move when placed in the field. The electron will move to the upper plate, against the field. The magnitude of the acceleration is given by

$$a = \frac{Eq}{m} = \frac{(2000\ \text{N/C})(1.6 \times 10^{-19}\ \text{C})}{9.11 \times 10^{-31}\ \text{kg}} = 3.5 \times 10^{14}\ \text{m/s}^2$$

(L.O. 3.A.1.1, 3.A.3.1)

9. **D** The points where the isolines are closer together indicate a stronger electric field in the region.

(L.O. 2.E.3.1)

10. **A** The forces between the charges, $q_1q_2$ and $q_2q_3$, have the same magnitude since the charges are of the same magnitude and the separation between $q_1q_2$ and $q_2q_3$ is the same. The force between $q_1q_2$ is repulsive, and the force between $q_2q_3$ is attractive, making vector I the proper choice.

(L.O. 3.A.3.4)

11. **C** Charge is conserved within an isolated system. When the wire connects the two spheres, negative charge moves from sphere 1 to sphere 2, cancelling the positive charge. The remaining charge redistributes due to repulsive forces between the charges. Since the spheres are identical, the charge on each is the same.

(L.O. 5.C.2.1)

12. **D** The electric field exerts a force on the electron that is much greater than the force exerted on it by the gravitational field. (Refer to the answer given for the numerical calculation in Problem 10 for the value of its acceleration.) The direction of the field determines whether the acceleration is in the same direction as the gravitational field or opposite that of the gravitational field. In the diagram, the field points downward and the electron will experience a force due to the electric field upward, against the field.

(L.O. 3.G.2.1)

13. **C and D** No work is done by a given force on a body when the angle between the force and the displacement is 90°. No work is done by the electric field in moving a charge along an equipotential in an electric field.

(L.O. 3.A.1.1)

14. **B and D** In a metal conductor or insulator, the only charge that can move is the negative charge. The spheres are isolated and are not in contact in examples I, II, and III; thus, charge cannot leave or move onto the spheres. In II, the sphere is polarized due to repulsion of the negative charge to the negatively charged wand.

The side closest to the negative wand will be the positive side. The sphere in IV is connected to ground so that charge can move to ground, leaving it positively charged.

(L.O. 4.E.3.1, 4.E.3.2)

15. **B and D** The two forces $F = k_e\dfrac{q_1q}{r^2}$ and $F = G\dfrac{m_1m_2}{r^2}$ are field forces that are inverse square laws. Gravitational forces are always forces of attraction where the Coulomb force can be either attractive (opposite charges) or repulsive (like charges). Electrical forces

between a proton and an electron are much stronger than gravitational forces between the two.

(L.O. 3.C.2.2)

# FREE-RESPONSE PROBLEMS

1.  (a)

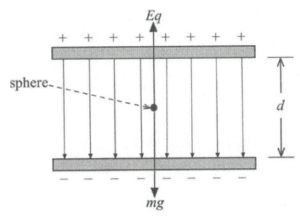

(b) Using the conversion factor for the charge on the electrons gives

$$4.8 \times 10^{-18} \text{ C} \times \frac{1\,e}{1.60 \times 10^{-19} \text{ C}} = \textbf{30 } e.$$

(c) In static equilibrium, the upward electrostatic force acting on the negative sphere equals the weight of the sphere. So $qE = mg$ and

$$E = \frac{mg}{q} = \frac{\left(8 \times 10^{-13} \text{ kg}\right)\left(9.8 \text{ m/s}\right)}{4.8 \times 10^{-18} \text{ C}} = \textbf{1.63} \times \textbf{10}^6 \textbf{ V/m}$$

(d) The potential difference can be found by
$V = Ed = (1.63 \times 10^6 \text{ V/m})(20 \times 10^{-3} \text{ m}) = \textbf{33 kV}.$

(e) The capacitance is

$$C = \varepsilon_0 \frac{A}{d} = \frac{\left(8.85 \times 10^{-12} \text{ F}/_{\text{m}}\right)(0.25 \text{ m})^2}{\left(20 \times 10^{-3} \text{ m}\right)} = \textbf{0.28 pF}$$

(f) The charge on the lower plate is
$q = CV = (0.28 \times 10^{-12} \text{ F})(1.63 \times 10^6 \text{ V}) = \textbf{4.56} \times \textbf{10}^{-5} \textbf{ C}.$

(L.O. 2.C.1.1, 2.C.1.2)

2.  (a) The distance, $L$, from the center of the square to each point charge is half the length of the diagonal of the square. So

$$L = \sqrt{2}\,\frac{R}{2} = (1.414)\left(\frac{0.04 \text{ m}}{2}\right) = 0.028 \text{ m}$$

The charges $q$ are identical, as are the distances $L$. Logically, the magnitude of the electric field vectors, $E$, for each charge will all be equal. The directions will differ and will be directed along the diagonals. Therefore, **by symmetry, the net $E$-field will be zero.**

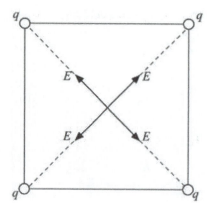

To calculate the resultant electric field at the center of the square, first, determine the x-component of the resultant E-vector:

$$E_x = E\left(\cos 45°\right) + E\left(\cos 135°\right) + E\left(\cos 225°\right) + E\left(\cos 315°\right)$$

$$E_x = E(0.707) + E(-0.707) + E(-0.707) + E(0.707) = 0$$

Next, find the y-component of E:

$$E_y = E\left(\sin 45°\right) + E\left(\sin 135°\right) + E\left(\sin 225°\right) + E\left(\sin 315°\right)$$

$$E_y = E(0.707) + E(0.707) + E(-0.707) + E(-0.707) = 0$$

Both the x-component and y-component of the resultant E are zero, which means that the resultant electric field intensity at the center of the square is itself zero and **E = 0**.

(b) Even though the field intensity at the center of the square is zero when identical charges are at each vertex, it does not necessarily mean that the electric potential is likewise zero. The electrical potential due to $i$ number of charges at a given point is found by $V = k\Sigma(q_i/r_i)$. Here, $i = 4$ charges, and the electric potential is calculated by

$$V = \frac{k}{L}(q_1 + q_2 + q_3 + q_4)$$

$$V = \left(\frac{9 \times 10^9 \, \text{N} \cdot \text{m}^2/\text{C}^2}{0.028 \, \text{m}}\right)(6.0 + 6.0 + 6.0 + 6.0) \times 10^{-6} \, \text{C}$$

$$= 7.71 \times 10^6 \, \text{V}$$

There is electric potential at the center of the square.

(L.O. 3.A.4.3)

3.  (a)  i.  The free-body diagram for the sphere on the right in the system is

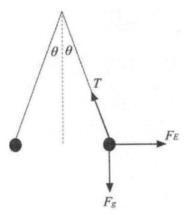

   ii.  The electrostatic forces between the charged Styrofoam spheres are action-reaction pairs. They are equal in magnitude and opposite in direction. $F_{12} = -F_{21}$. The free-body diagrams would be mirror images.

   iii.  Since the charges are equal in magnitude and repulsive, it makes no difference whether they are positive or negative.

   (b)  The forces acting on the Styrofoam are $F_g$, $F_E$, and the tension $T$ in the silk thread.

   These forces form a right triangle, and the $\tan\theta = \dfrac{F_E}{F_g}$.

   (c)  The sum of the force $x$-components and $y$-components are written as $\Sigma F_x = 0$ and $\Sigma F_y = 0$. The $x$-component becomes

   $\dfrac{kq^2}{r^2} = T\sin\theta$, and the $y$-component is $\dfrac{mg}{\cos\theta} = T$. Combining

   both equations by eliminating $T$ gives $k_e \dfrac{q^2}{r^2} = (mg)\tan\theta$.

   (d)  Since the silk threads have length $L$, the relationship between the length of the threads and the $\sin 10°$ will give the distance that each sphere moves from equilibrium position. The distance of separation of the two spheres $r$, is equal to $2L\sin 10°$.

   (L.O. 3.A.4.3, 3.C.2.3)

4.  A repels B and attracts C.
    B attracts C.
    Since A is positive, B is positive and C is negative.

    Set up ratios for the forces. Note that $\dfrac{k_e}{r^2}$ is constant and will not be used in the ratios.

    $\left|\dfrac{F_{AB}}{F_{AC}}\right| = \dfrac{Q_A Q_B}{Q_A Q_C}$

    Substituting the forces gives $\dfrac{1.00\ \text{N}}{0.50\ \text{N}} = \dfrac{Q_B}{Q_C}$, yielding $2Q_C = Q_B$.

A second ratio gives $Q_B$ in terms of $Q_A$.

$$\left|\frac{F_{AB}}{F_{BC}}\right| = \frac{Q_A Q_B}{Q_B \frac{Q_B}{2}}$$

Again, substitution of the forces gives $\dfrac{1.00\ \text{N}}{0.25\ \text{N}} = \dfrac{Q_A}{0.5 Q_B}$,

yielding $2Q_B = Q_A$.

Since A $= +2Q$, B $= +Q$ and C $= -\dfrac{1}{2}Q$.

(L.O. 3.C.2.1)

# 13

# DC Circuits

## Physics 1 and Physics 2

## Electrical Current

When there is a difference in potential, $\Delta V$, between two points along a conductor, electrical charge will move from an area of high potential to low potential. Movement of charge is called current, $I$. Current is a rate of charge per unit of time (i.e., how much charge is passing a particular point at a particular time).

$$I = \frac{q}{t}$$

The SI unit of current is the ampere, A. We define the ampere as 1 A = 1 C/s.

### AP® Tip

The direction of conventional current is always the same as the direction that positive charges move, even though current actually consists of a flow of electrons.

When current through a conductor is always in the same direction, it is called direct current, or DC.

## Sample Problem 1

A charge of 22.0 C flows through a cross section of wire every minute.
(a)   What is the current?
(b)   How many electrons go through the cross section each minute?

86570356666

SOLUTION TO PROBLEM 1

(a) $I = \dfrac{q}{t} = \dfrac{22\ C}{60\ s} = 0.37\ A$

(b) $22.0\ C \times \dfrac{1\ e}{1.60 \times 10^{-19}\ C} = 1.38 \times 10^{20}\ e$

# EMF

EMF, $\varepsilon$, was once known as electromotive force; however, it is not a force at all. It is defined as the work per unit charge done by a battery or generator on the charges in moving them around the circuit. You can think of a source of emf as a pump that brings charge to a higher potential energy.

The SI unit of emf is the volt, and it is defined as 1 V = 1 J/C. The most familiar source of emf is the battery. A battery should be thought of as a *charge pump* that sends conventional current into a closed circuit. The source of emf increases the electric energy of the charges flowing through the battery.

# RESISTANCE

Every material offers some resistance, $R$, to electric current. Good conductors like the metals copper, silver, and aluminum offer very little resistance to current. Nonconductive materials like rubber, plastic, and glass offer very high resistance to current.

The SI unit of resistance is the ohm ($\Omega$), which is defined as 1 $\Omega$ = 1 volt/ampere = 1 V/A.

# RESISTIVITY

Several factors determine the resistance of any section of wire: (1) the length, $L$; (2) the cross-sectional area, $A$; and (3) the resistivity, $\rho$, a property of the material of which the wire is composed. The resistivity indirectly gives a measure of how well a current will be conducted through a piece of wire. Resistivity is related to resistance by the relationship

$$R = \rho \frac{L}{A}$$

The resistivity of a wire is also affected by the temperature of the wire. The resistivity constant, $\rho$, changes with temperature. Temperature and resistance are proportional. If the temperature is raised, the resistance in the material will increase.

## Sample Problem 2

A 120 m long sector of circular wire has a diameter of 1.2 mm. The wire has a resistivity $\rho = 3.6 \times 10^{-8}\,\Omega \cdot m$.

(a) What is the resistance of this length of wire?

## Solution to Problem 2

The radius of the wire is $r = 0.6 \times 10^{-3}$ m.

(a) Then

$$R = \rho \frac{L}{A} = \frac{\left(3.6 \times 10^{-8}\,\Omega \cdot m\right)\left(120\ m\right)}{\pi \left(0.6 \times 10^{-3}\ m\right)^2} = 3.8\,\Omega$$

## Sample Problem 3

Describe how each of the following changes would affect the resistance of a conducting wire.

(a) Doubling the diameter of the wire
(b) Cutting the wire in half
(c) Increasing the temperature of the wire

## Solution to Sample Problem 3

(a) Doubling the diameter would also double the radius. This would quadruple the area. When the area is quadrupled, the resistance is now ¼ $R$.
(b) Cutting the wire in half will halve the resistance, ½ $R$.
(c) Increasing the temperature of the wire will increase the resistance in the wire.

# Ohm's Law

Ohm's law is the fundamental law in electricity that makes it possible to determine the potential difference, $\Delta V$, across the ends of a resistor when the current, $I$, through it and its resistance, $R$, are known. Many times the potential difference $\Delta V$ is represented as just $V$. Ohm's law for a resistor is

$V = IR$

## AP® Tip

Circuit diagrams greatly help in understanding a problem. Draw diagrams when working and solving problems.

Ammeters are electrical instruments designed to measure electrical current. An ammeter has low resistance and is connected in series so that all of the current passes through it. An ammeter must have low resistance, so it does not appreciably reduce the current. Voltmeters are designed to compare the potential at two different points in a circuit. Voltmeters have a high resistance and are connected in parallel with what they measure. Voltmeters have high resistance to ensure that only a miniscule amount of current can travel through the voltmeter. If a voltmeter's resistance were too low, the majority of the current would flow through the voltmeter instead of the actual circuit.

## SAMPLE PROBLEM 4

A battery is directly connected to a small lightbulb. The battery maintains a potential difference of 6.0 V across the bulb. If the current in the circuit is 0.5 A, what is the resistance of the lightbulb?

## SOLUTION TO PROBLEM 4

First, make a circuit diagram. The lightbulb is a resistor, and we show it in the circuit diagram as such.

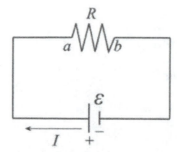

The voltage drop across $R$ is $V_{ab} = 6.0$ V. So $R = V/I = 6.0$ V/0.5 A = 12 Ω.

Although the resistance calculated in Problem 2 is assumed to be the resistance of the lightbulb, it really includes the resistance of the connecting wires and the resistance of the battery. In practice, the wires in a circuit have very low resistance, so this factor can be neglected in most calculations. In addition, when sources of emf have very low resistance, we can neglect the resistance in our calculations. Yet, there are times when we do account for the resistance of the emf source itself, the *internal resistance*. As charge moves through the battery, it experiences a resistance that causes a drop in potential. That is, it takes some amount of energy to move charge through a real battery.

$$\Delta V = \varepsilon - IR$$

$\Delta V$ is called the terminal voltage, the actual measured voltage of the battery when the circuit is in use and a current $I$ is flowing. $\varepsilon$ is the potential of the emf source, the voltage we measure when no current is running through the battery. $IR$ is the voltage cost to pump charge through the battery's internal resistance. When current equals zero,

$$\varepsilon = \Delta V.$$

Most materials we encounter in AP® Physics are treated as ohmic materials, meaning that they follow Ohm's law, $V = IR$. They have constant resistance over a wide range of voltages.

The current through an ohmic resistor is linear with the voltage across the resistor. Some materials do not obey Ohm's law and are considered *non-ohmic*. The current through a non-ohmic resistor increases with voltage, but the relationship is not linear. The resistance of a non-ohmic material changes as current changes. Sample graphs of ohmic and non-ohmic materials are shown on the next page.

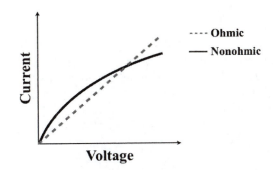

# POWER

Electrical power is the rate at which energy is produced or used. Sources of emf do work on charge and give it energy. The power output of an emf source, measured in watts, is given by

$$P_0 = I\mathcal{E}$$

Resistors use electrical energy, and the rate at which heat is dissipated in a circuit is called the power loss. The power loss in a resistor is given by

$$P = IV = I^2 R = \frac{V^2}{R}$$

The power output of the source of emf equals the power loss in the external circuit.

## SAMPLE PROBLEM 5

A resistor has a current of 12.5 A when connected across a 120 V emf source.
(a) Calculate the resistance of this electrical component.
(b) Calculate the power loss through the resistor.

## SOLUTION TO PROBLEM 5

(a) $R = V/I = 120$ V/12.5 A $= 9.6\ \Omega$
(b) $P = IV = (12.5$ A$)(120$ V$) = 1.5$ kW

## SAMPLE PROBLEM 6

A 420 W resistor with a resistance of 60 Ω is used in a circuit.
(a) Determine the potential difference across the ends of the resistor.
(b) What current passes through the resistor?

## SOLUTION TO PROBLEM 6

(a) $P = V^2 / R$ and $V = \sqrt{PR} = \sqrt{(420\ \text{W})(60\ \Omega)} = 159$ V
(b) $I = V/R = 159$ V/60 Ω $= 2.65$ A

## SAMPLE PROBLEM 7

A battery has an emf $\mathcal{E} = 24.0$ V. When connected in a circuit, it delivers a current of 3.4 A. The battery has an internal resistance of 1.5 Ω.
(a) What is the terminal voltage?
(b) There is an external resistor in the circuit. What is the potential difference across this resistor?
(c) What is the power loss in this external resistor?

## SOLUTION TO PROBLEM 7

(a)  $V_T = \mathcal{E} - Ir = 24.0 \text{ V} - (3.4 \text{ A})(1.5 \text{ }\Omega) = 18.9 \text{ V}$

(b)  The external resistor must use whatever battery voltage the battery puts in the external circuit. The external resistor has a potential difference of 18.9 V.

(c)  $P = IV = (3.4 \text{ A})(18.9 \text{ V}) = 64.3 \text{ W}$

## RESISTORS IN SERIES

When two or more resistances are connected in series in an electrical circuit, the current through all parts of the series combination is the same.

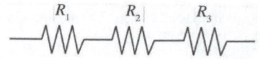

The sum of the voltage drops across the resistors in series is the sum of the voltage drop across each resistor. The equivalent resistance, $R_{eq}$, of the resistors in series is the sum of their resistances.

$$R_s = R_1 + R_2 + R_3 + \dots$$

## AP® Tip

The current through resistors in series is the same in each resistor, and the potential difference across them is additive.

## SAMPLE PROBLEM 8

The three resistors $R_1 = 12 \text{ }\Omega$, $R_2 = 6 \text{ }\Omega$, and $R_3 = 8 \text{ }\Omega$ are arranged in series.

(a)  Find the equivalent resistance of the combination.

(b)  The resistors are connected to a source of emf. The current through the resistors is 2.4 A. What is the $IR$ drop across junctions $a$ and $b$?

## SOLUTION TO PROBLEM 8

(a)  First, make a diagram.

For a set of resistors combined in a series, the equivalent resistance is the sum of the resistors as shown, and $R_s = R_{ab} = R_1 + R_2 + R_3 = 12 \text{ }\Omega + 6 \text{ }\Omega + 8 \text{ }\Omega = 26 \text{ }\Omega$. The three resistors behave like a single 26 Ω resistor.

(b)

26 Ω

$a$ ⎯⎯/\/\/\⎯ $b$

The *IR* drop between junctions *a* and *b* is $V_{ab} = IR_s = (2.4 \text{ A})(26 \text{ }\Omega)$
= 62.4 V.

## RESISTORS IN PARALLEL

When resistors are arranged in a parallel combination, the current splits. The sum of the currents in each branch is equal to the current that enters the combination. The voltage drop across the combination is equal to the voltage drop across each resistor in parallel.

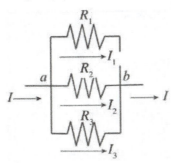

The reciprocal of the equivalent resistance, $R_{eq}$, of the resistors in parallel is the sum of the reciprocals of the individual resistances.

$$\frac{1}{R_{eq}} = \frac{1}{R_1} + \frac{1}{R_2} + \frac{1}{R_3} + \dots$$

### SAMPLE PROBLEM 9
The three resistors $R_1 = 12 \text{ }\Omega$, $R_2 = 6 \text{ }\Omega$, and $R_3 = 8 \text{ }\Omega$ are arranged in parallel.
(a) Calculate their equivalent resistance.
(b) Junctions *a* and *b* are connected across an emf. If a current of 2.4 A enters junction *a*, what is the IR drop across junctions *a* and *b*?
(c) What is the current through each resistor in the parallel bank?

### SOLUTION TO PROBLEM 9
(a) First, make a diagram of the resistor arrangement.

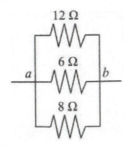

For a set of resistors combined in parallel, the reciprocal equivalent resistance is the sum of the reciprocals of the resistors.

$$\frac{1}{R_{eq}} = \frac{1}{R_{ab}} = \frac{1}{R_1} + \frac{1}{R_2} + \frac{1}{R_3} = \frac{1}{12 \text{ }\Omega} + \frac{1}{6 \text{ }\Omega} + \frac{1}{8 \text{ }\Omega} = 0.375 \frac{1}{\Omega}$$

Then

$$R_{eq} = R_{ab} = \frac{1}{0.375} \ \Omega = 2.67 \ \Omega$$

The three resistors in parallel function as a single 2.67 Ω resistor are shown.

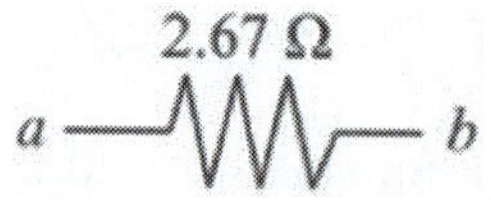

2.67 Ω

$a$ —/\/\/\— $b$

(b)  Current $I = 2.4$ A enters the resistor at point a. The $IR$ drop is $V_{ab}$ = $IR_s$ = (2.4 A)(2.67 Ω) = 6.4 V.

(c)  Each resistor in parallel undergoes the same voltage or $IR$ drop, which is 6.4 V. So the current in $R_1$ is

$$I_1 = \frac{V_{ab}}{R_1} = \frac{6.4 \ V}{12 \ \Omega} = 0.53 \ A$$

The current in $R_2$ is

$$I_2 = \frac{V_{ab}}{R_2} = \frac{6.4 \ V}{6 \ \Omega} = 1.07 \ A$$

The current in $R_3$ is

$$I_3 = \frac{V_{ab}}{R_3} = \frac{6.4 \ V}{8 \ \Omega} = 0.80 \ A$$

The sum of the currents going through the resistors in a parallel bank must equal the current entering the bank. As a check, add the currents of the resistors in parallel:

$$I = I_1 + I_2 + I_3 = 0.53 \ A + 1.07 \ A + 0.80 \ A = 2.4 \ A$$

They check.

## AP® Tip

The largest resistor in parallel carries the smallest current, and the smallest resistor in parallel carries the largest current. Current tends to take the path of least resistance.

## COMBINATION CIRCUITS

In most electrical circuits, resistors are wired partly in series and partly in parallel. In such circuits, the rules for determining the equivalent resistance are applied to each part of the circuit for circuit analysis.

### SAMPLE PROBLEM 10

The electrical circuit diagram shown on the next page contains a source of emf with zero internal resistance, three resistors, an ammeter, and a voltmeter.

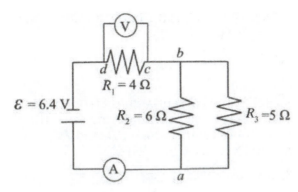

Perform the following analysis on the circuit.
(a)   Find the total resistance of the circuit.
(b)   What current will the ammeter A read?
(c)   Determine the current in each resistor.
(d)   What will the voltmeter V across resistor $R_1$ read?
(e)   Calculate the power output of the battery.
(f)   Find the power drop in each resistor.

## SOLUTION TO PROBLEM 10

Current $I$ passes from the positive terminal of the battery to the negative terminal. At junction $a$, the current splits and current $I_1$ passes through $R_2$ and $I_2$ passes through $R_3$. $I_1$ and $I_2$ merge at junction $b$. The full current passes through the ammeter. See the figure below.

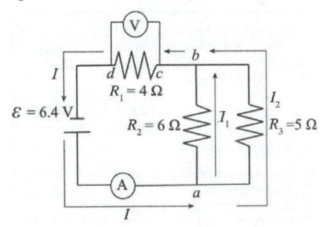

(a)   Resistors $R_1$ and $R_2$ are in parallel, and their equivalent resistance is

$$\frac{1}{R_{12}} = \frac{1}{R_1} + \frac{1}{R_2}$$

Using another approach, we can find that the equivalent resistance is the reciprocal of the sum of the reciprocals, or

$$R_{12} = \frac{1}{1/R_1 + 1/R_2} = \frac{1}{1/6\,\Omega + 1/5\,\Omega} = \frac{1}{11/30\,\Omega} = \frac{30\,\Omega}{11} = 2.73\,\Omega$$

Resistor $R_3$ is in series with the 2.73 $\Omega$, and the equivalent resistance, $R_{eq}$, of the circuit is

$$R_{eq} = R_3 + 2.73\,\Omega = 4\,\Omega + 2.73\,\Omega = 6.73\,\Omega$$

(b) The current through the battery, which is the current read by the ammeter, is found using Ohm's law

$$I = \frac{\mathcal{E}}{R_{eq}} = \frac{6.4 \text{ V}}{6.73 \text{ } \Omega} = 0.95 \text{ A}$$

(c) At junction $a$, the current splits and current $I_1$ goes through $R_1$ and current $I_2$ goes through $R_2$. All resistors suffer the same IR drop in parallel. The equivalent resistance, $R_{12}$, of $R_1$ and $R_2$ was found in part (a) to be $R_{12} = 2.73 \text{ } \Omega$. By Ohm's law,

$$V_{12} = IR_{12} = (0.95 \text{ A})(2.73 \text{ } \Omega) = 2.59 \text{ V}$$

$$I_1 = \frac{V_{12}}{R_1} = \frac{2.59 \text{ V}}{6 \text{ } \Omega} = 0.43 \text{ A}$$

$$I_2 = \frac{V_{12}}{R_2} = \frac{2.59 \text{ V}}{5 \text{ } \Omega} = 0.52 \text{ A}$$

(d) The voltmeter will read the IR drop across $R_1$. The 0.95 A passes through $R_1$, and by Ohm's law, $V_1 = IR_1 = (0.95 \text{ A})(4 \text{ } \Omega) = 3.80 \text{ V}$.

(e) The battery power output is the product of the current through the circuit and the emf: $P_{output} = I\mathcal{E} = (0.95 \text{ A})(6.4 \text{ V}) = 6.08 \text{ W}$.

(f) Resistors $R_1$ and $R_2$ are in parallel, and the power dissipated in each can be found by taking the product of the current squared and the resistance.

So for resistor 1: $P_1 = I_1^2 R_1 = (0.43 \text{ A})^2 (6 \text{ } \Omega) = 1.11 \text{ W}$.

And for resistor 2: $P_2 = I_2^2 R_2 = (0.52 \text{ A})^2 (5 \text{ } \Omega) = 1.35 \text{ W}$.

Resistor $R_3$ is in series, and the power output can be calculated taking the product of the IR drop and the current through the resistor. So $P_3 = IV_3 = (0.95 \text{ A})(3.80 \text{ V}) = 3.61 \text{ W}$.

# KIRCHHOFF'S RULES

In a complex circuit consisting of several resistors and more than one possible path through which current can travel, Ohm's law is not enough to describe the behavior of the circuit completely. Gustav Kirchhoff, in the nineteenth century, developed two more tools used to analyze the flow of current and the transfer of energy in a circuit. These are called *Kirchhoff's rules*.

### The Junction Rule

A junction is any point in the circuit at which three or more conducting wires connect. A junction could be where a wire branches into multiple paths or where paths meet and become one. The sum of the currents entering a junction must equal the sum of the currents exiting the junction. This rule is a statement of the law of conservation of electric charge.

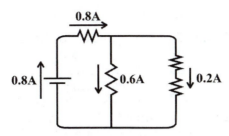

In the image above, the total current into the junction equals the total current through both branches. The current flowing through the battery and the topmost resistor is the same because there are no breaks or branches in the conducting wires connecting them. For the same reason, the current is 0.2A through both resistors on the right branch.

### The Loop Rule

The sum of all of the *IR* drops around a loop must equal the sum of all of the *emf's* around that loop. In the circuit shown below, the total energy gained by a charge as the charge travels through the battery is dissipated in the resistors. The sum of the *IR* drops in the resistors is 12 V no matter which path an electron follows.

This rule is a statement of the law of conservation of energy.

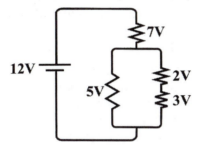

### SAMPLE PROBLEM 11

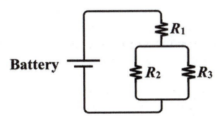

Complete the following table. Assume the battery has no internal resistance.

|  | **V** | **I** | **R** |
|---|---|---|---|
| Battery | 12 V | A |  |
| $R_1$ | 4 V | A | Ω |
| $R_2$ | V | 2 A | Ω |
| $R_3$ | V | 1 A | Ω |

## SOLUTION TO PROBLEM 11

First, we use the junction rule to find the current flowing through the battery and $R_1$. A current of 4 A flows through $R_2$, and a current of 2 A flows through $R_3$. The sum of those currents, 3 A, flows out of the junction and through the battery. The battery is in series with $R_1$, so 3 A of current also flows through $R_1$.

Next, we solve for the voltage across $R_2$ and $R_3$. Using the loop rule, we know that the total voltage drop in any loop will be 12 V. Through any path, there is a drop of 4 V across $R_1$, so there must be an 8 V drop across $R_2$ and $R_3$. This makes sense because $R_2$ and $R_3$ are in parallel.

Finally, we use Ohm's law to solve for the resistance of each resistor and the equivalent resistance of the circuit with $R = V/I$.

|         | *V*   | *I*  | *R*     |
|---------|-------|------|---------|
| Battery | 12 V  | 3 A  |         |
| $R_1$   | 4 V   | 3 A  | 1.33 Ω  |
| $R_2$   | 8 V   | 2 A  | 4 Ω     |
| $R_3$   | 8 V   | 1 A  | 8 Ω     |

# RC CIRCUITS (PHYSICS 2 ONLY)

RC circuits contain resistors and capacitors. The current through a capacitor changes over time as the capacitor charges or discharges, which affects the behavior of the rest of the circuit. The specific effect of a charging capacitor on the circuit depends on whether the capacitor is in series or in parallel with the circuit elements.

Electrons flow easily to or from the plates of an uncharged capacitor. In terms of current, an uncharged capacitor acts like a wire with zero resistance. As more and more charge is deposited on one plate and removed from the other, the current is reduced. Finally, when an emf has charged a capacitor as much as possible, there is no more current to the capacitor.

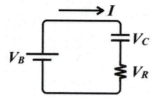

When a resistor, $R$, is in series with an initially uncharged capacitor and a voltage source, $V_B$, the amount of current throughout the capacitor is $I = V/R$. The IR drop across the resistor is the same as the emf of the voltage source. Initially, $V_C = 0$ and $V_R = V_B$. As the capacitor becomes charged, the voltage of the source will equal the voltage across the resistor plus the voltage across the capacitor, $V_B = V_R + V_C$. When the capacitor is fully charged, then there will be no more current through the circuit; $V_C = V_B$ and $V_R = 0$.

Once the capacitor is charged, it can be disconnected from the battery, and the energy it has stored can be used. As the capacitor discharges and releases its stored energy, the voltage across its plates and the amount of charge it pushes will decrease over time.

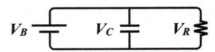

When an initially uncharged capacitor is connected in parallel with a resistor and a voltage source, none of the current in the circuit goes through the resistor. The current through the circuit branch with the capacitor is the same as the current through the battery. As more and more charge is deposited on one plate and removed from the other, the current in the capacitor's branch is reduced and current begins to flow through the resistor. When the capacitor is fully charged, current no longer runs in the capacitor's branch. All the current through the battery is also through the resistor.

# DC CIRCUITS: STUDENT OBJECTIVES FOR THE AP® EXAM

- You should know the definition for electric current in a conductor and be able to apply it.
- You should know that the direction of the current in a conductor is, by convention, the direction in which positive charge carriers flow.
- You should know and be able to apply Ohm's law and the power equations to all conductors as well as the entire DC circuit.
- You should know how to calculate the resistance of a conductor from its physical parameters.
- You should know how to draw and read schematic diagrams of the components in a direct-current circuit.
- You should know how to solve for equivalent resistance in series circuits, parallel circuits, and combination circuits.
- You should know how to solve for current, voltage, and power for an entire circuit and for each component of the circuit.
- You should be able to describe qualitatively and quantitatively how the flow of current through a circuit with a capacitor changes over time.
- You should know how to solve for terminal voltage in a DC circuit.
- You should be able to apply Kirchhoff's rules to more complex circuits.

## MULTIPLE-CHOICE QUESTIONS

1.  A simple circuit, shown in the diagram below, consists of a 20 Ω resistor in series with two resistors of 15 Ω and 30 Ω in parallel. The resistors are connected to a 12 V power supply. The current through the resistors is correctly identified in which of the following?

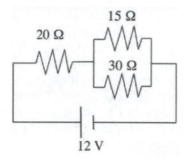

(A) $I_{15\,\Omega} = I_{20\,\Omega} = I_{30\,\Omega}$
(B) $I_{20\,\Omega} + I_{15\,\Omega} = I_{30\,\Omega}$
(C) $I_{20\,\Omega} = I_{15\,\Omega} + I_{30\,\Omega}$
(D) $I_{20\,\Omega} + I_{30\,\Omega} = I_{15\,\Omega}$

2.  In reference to the circuit in Problem 1, a voltmeter is placed across each of the resistors in turn and the electrical potential difference for each is indicated below. Which of the following gives the correct readings for the electrical potential difference across each of the resistors?

|       | **20 Ω** | **15 Ω** | **30 Ω** |
|-------|--------|--------|--------|
| (A)   | 6 V    | 6 V    | 6 V    |
| (B)   | 8 V    | 8 V    | 4 V    |
| (C)   | 8 V    | 4 V    | 8 V    |
| (D)   | 8 V    | 4 V    | 4 V    |

3.  A student wants to determine the resistivity of a conductor experimentally. The student needs to collect data to do this. The data taken should be the
    (A) potential difference across the ends of the conductor and the current through it.
    (B) type of material, the length of the conductor, and its cross-sectional area.
    (C) potential difference across ends of the conductor and its length.
    (D) potential difference across the ends of the conductor, the current through it, and the length and cross section of the conductor.

4. A circuit contains three resistors in different arrangements as shown in the schematic diagrams below. Rank from highest to lowest the current flowing through the 10 Ω resistor in the various arrangements.

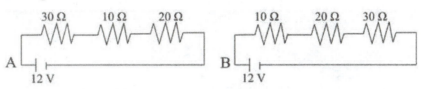

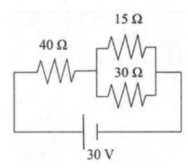

(A) C = D > A = B
(B) A = B > C = D
(C) A > B > D > C
(D) D > C > A > B

5. In the circuit shown below, what is the current in the 15 Ω resistor?

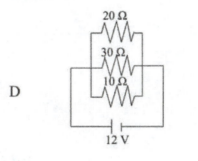

(A) 0.6 A
(B) 0.4 A
(C) 0.3 A
(D) 0.2 A

6. A wire whose resistivity is $\rho$ has dimensions of length $L$ and a radius of $r$, producing a resistance of $R_0$. The wire is replaced in the circuit with another wire whose resistivity $\rho$ is the same, but it has a length of 2.0 $L$ and a radius 0.5 $r$. What is the new resistance?
(A) $0.5R_0$
(B) $2R_0$
(C) $4R_0$
(D) $8R_0$

7. In the diagram shown below, rank the brightness of the bulbs from brightest to dimmest. All bulbs are identical.

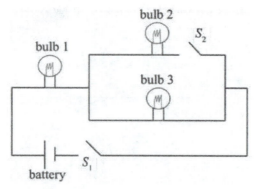

(A) Bulb 1 > Bulb 2 = Bulb 3
(B) Bulb 1 = Bulb 2 = Bulb 3
(C) Bulb 1 > Bulb 2 > Bulb 3
(D) Bulb 3 > Bulb 2 > Bulb 1

8. (Physics 2 Only) In the circuit shown below, what is the correct relationship for the currents through the resistors?

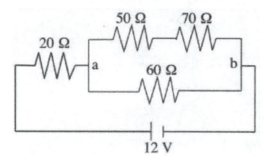

(A) $I_{20\,\Omega} = I_{50\,\Omega} + I_{60\,\Omega} + I_{70\,\Omega}$
(B) $I_{20\,\Omega} = I_{50\,\Omega} > I_{60\,\Omega} > I_{70\,\Omega}$
(C) $I_{20\,\Omega} = I_{60\,\Omega} + (I_{50\,\Omega} = I_{70\,\Omega})$
(D) $I_{20\,\Omega} = (I_{50\,\Omega} = I_{60\,\Omega}) + I_{70\,\Omega}$

9. (Physics 2 Only) An ammeter in the circuit shown below reads 0.10 A when connected in series with a 30.0 Ω resistor. What is the current in the 20.0 Ω resistor?

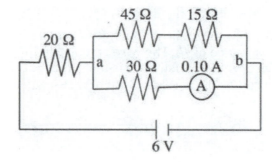

(A) 0.05 A
(B) 0.10 A
(C) 0.15 A
(D) 0.2 A

10. A group of physics students doing an experiment to determine whether Kirchhoff's loop rule applies to a simple circuit sets up a circuit containing a 12.0 V battery with an internal resistance of 0.5 Ω that is connected to a 56.0 Ω resistor as shown below. A voltmeter placed across the ends of the resistor as shown will indicate a potential difference equal to

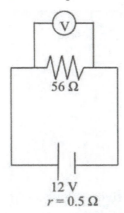

(A) the emf of the battery.
(B) a potential difference greater than the emf of the battery.
(C) a potential difference less than the emf of the battery.
(D) zero because no current will flow from the battery because of the internal resistance.

11. A variable resistor and a battery are arranged in series with each other. A graph of the power dissipated by the resistor versus the current flowing through the battery is shown. What is the terminal voltage of the battery?

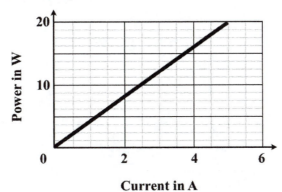

(A) 0.25 V
(B) 4 V
(C) 5 V
(D) 20 V

12. (Physics 2 Only) Three capacitors are shown below. A potential difference is applied across points A and B. The arrangement is allowed to reach its steady state. Rank the potential differences across the capacitors.

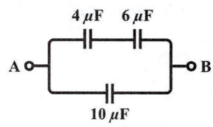

(A) $\Delta V_{10} = \Delta V_4 = \Delta V_6$
(B) $\Delta V_{10} > \Delta V_4 = \Delta V_6$
(C) $\Delta V_{10} > \Delta V_4 > \Delta V_6$
(D) $\Delta V_{10} > \Delta V_6 > \Delta V_4$

13. A wire with resistance, $R$, has a resistivity of $\rho$, a length $L$, and a cross-sectional area $A$. Which of the following would halve the resistance of the wire, $R/2$?
   (A) Doubling the length of the wire
   (B) Halving the temperature of the wire
   (C) Halving the area of the wire
   (D) Doubling the area of the wire

14. Several students conduct an experiment using a variable power supply, a voltmeter, an ammeter, a set of connecting wires, and a resistor to test Ohm's law. Their data produce the following graph. The graph indicates that

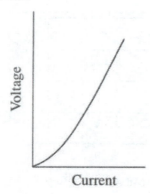

   (A) the variable power supply had an internal resistance the students could not measure.
   (B) Ohm's law was valid for the experiment.
   (C) the resistance was non-ohmic since the graph did not indicate a linear relationship.
   (D) the resistance was independent of both potential difference and current.

**Question 15**
**Directions:** For the question below, two of the suggested answers will be correct. You must select both correct choices to earn credit. No partial credit will be earned if only one correct choice is selected. Select the two that are best.

15. A circuit has three identical lightbulbs connected as shown in the schematic diagram below.

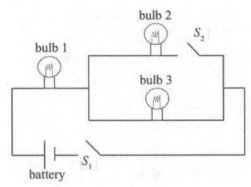

The two switches are initially closed in the circuit. When switch $S_2$ is opened, (select two answers)
(A) all three lightbulbs go out.
(B) only lightbulb 2 goes out.
(C) the brightness of lightbulbs 1 and 3 increases.
(D) the brightness of lightbulbs 1 and 2 decreases.

## FREE-RESPONSE PROBLEMS

1. (Physics 2 Only) Answer the following questions regarding the RC circuit below, where $R_2$ is double the resistance of $R_1$.

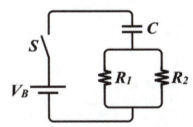

   (a) On the axes on the next page, sketch a graph to show how the currents through $R_1$ and $R_2$ change over time. Your time axis should begin when the switch, $S$, is closed and continue until the circuit reaches its steady state. Label each line as $R_1$ or $R_2$.

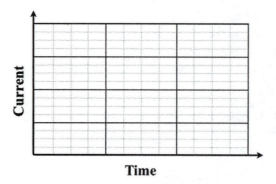

Time

(b) The capacitor is discharged, and the switch is opened. The capacitor and resistor 1 are exchanged as shown in the diagram below.

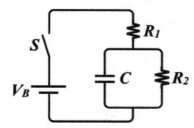

Draw how current changes over time in both resistors, $R_1$ and $R_2$, once the switch is closed. Label each line as $R_1$ or $R_2$.

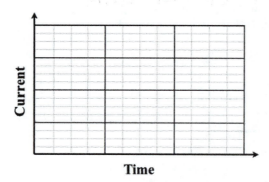

Time

2. (Physics 2 Only) The circuit diagram shown on the next page has a battery with an emf, $\varepsilon$. The ammeter in the circuit reads 2.4 A. Resistors $R_1 = R_2 = R_3 = 18\ \Omega$, $R_4 = 8.0\ \Omega$, and $R_5 = 4.0\ \Omega$. The internal resistance of the battery is negligible.
   (a) Calculate the emf of the battery.
   (b) Calculate the electrical potential difference between junctions $a$ and $d$.
   (c) Calculate the power loss between junctions $c$ and $d$.

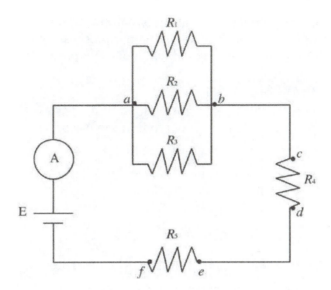

3. (a) In enough detail that another student could duplicate your experiment and obtain the same results, design an experiment to demonstrate that Kirchhoff's loop rule and junction rule are valid in a circuit using a battery, three resistors of the same known resistance, and connecting wires and meters.
   (b) Draw a schematic of your circuit, properly labeling the elements and placing the meters in their correct positions.
   (c) What measurements are you going to make and how are you going to use them in evaluating Kirchhoff's loop rule and junction rule for your circuit?
   (d) What are two possible sources of error in your experiment, and how might they affect the results you obtained?

4. You are given an aging 9.0 V battery, a known resistor, several segments of copper wire, an ammeter, and a voltmeter.
   (a) Explain how you can determine the internal resistance of the battery.
   (b) Draw the circuit diagram of how you plan to use the equipment, and indicate what readings you will take to find the internal resistance.
   (c) Write the mathematical solution that will permit you to solve for the internal resistance.

# Answers

## MULTIPLE-CHOICE QUESTIONS

1. **C** Charge is conserved in a circuit. The current $I = \dfrac{Q}{t}$ flowing through the 20 Ω resistor enters junction $a$ where it divides, flowing into the two parallel branches $I_{20\,\Omega} = I_{15\,\Omega} + I_{30\,\Omega}$, and recombines.

   (L.O. 1.B.1.2)

2.  **D** There can be only one potential difference between the same two points regardless of the number of paths. The parallel resistors will have the same voltmeter reading. The resistance in the parallel branch is found from $\dfrac{1}{15\ \Omega} + \dfrac{1}{30\ \Omega} = \dfrac{1}{R_{eq}}$. The 10 $\Omega$ equivalent resistance is in series with the 20 $\Omega$ resistor, giving a total resistance of 30 $\Omega$ for the circuit. Thus, the current in the circuit is $\dfrac{12\ V}{30\ \Omega} = 0.4\ A$. The voltmeter will read $V_{20\ \Omega} = (20\ \Omega)(0.4\ A) = 8\ V$; therefore, the voltage across the parallel resistors will read 4 V since Kirchhoff's loop rule applies.

    (L.O. 5.B.9.3)

3.  **D** The resistance of a conductor is given by $R = \rho\dfrac{L}{A}$. Data need to be taken concerning the length and the cross-sectional area. The resistance of the conductor can be calculated using a voltmeter to determine the potential difference across its ends and an ammeter that will record the current through it and using $R = \dfrac{V}{I}$.

    Substituting this resistance into the first equation will give the resistivity of the conductor.

    (L.O. 1.E.2.1)

4.  **A** In the schematics C and D, the resistors are in parallel connected to the 12 V battery, and the 10 $\Omega$ resistor will have the most current regardless of its position in the parallel network. In the series arrangement, the 10 $\Omega$ resistor has the same current regardless of its position in the series path. The parallel arrangement has a net resistance less than 10 $\Omega$ and therefore the greatest current from the source. Thus, the correct ranking is C = D > A = B.

    (L.O. 5.C.3.1)

5.  **B** The resistance for the parallel combination is found from $\dfrac{1}{15\ \Omega} + \dfrac{1}{30\ \Omega} = \dfrac{1}{R_{eq}}$. The equivalent resistance of 10 $\Omega$ is in series with a 40 $\Omega$ resistor, producing a total of 50 $\Omega$. The total current in the arrangement is $I = \dfrac{V}{R} = \dfrac{30\ V}{50\ \Omega} = 0.6\ A$. The potential difference across the 10 $\Omega$ equivalent is $V_{eq} = (10\ \Omega)(0.6\ A) = 6\ V$. The current in the 15 $\Omega$ resistor is thus $\dfrac{6\ V}{15\ \Omega} = 0.4\ A$.

    (L.O. 5.C.3.3)

6. **D** The original resistance is $R_0 = \rho \dfrac{L}{A} = \rho \dfrac{L}{\pi r^2}$. Replacing the wire with a new wire with the given dimensions produces a resistance that is $R = \rho \dfrac{2L}{\pi (0.5r)^2} = \rho \dfrac{8L}{\pi r^2}$; thus, $R = 8R_0$.

(L.O. 4.E.4.1)

7. **A** When both switches are closed, the entire current flows through the first lamp. It also has the largest potential difference across its ends. The potential on the parallel branch is less, and the current will divide between the two lightbulbs, the one with the lowest resistance getting a higher current. In this case, inspection will tell you that the smaller resistance gets $\frac{2}{3} I_T$ and the other bulb gets $\frac{1}{3} I_T$. Since the brightness of a bulb is related to power, the bulb with the most current is the brightest.

(L.O. 4.E.5.2)

8. **C** Kirchhoff's junction rule applies. The 20 Ω resistor is in series with a parallel arrangement that has two resistors in series in the upper branch; these two resistors have the same current. The current flow in and out of junction $a$ is thus $I_{20\,\Omega} = I_{60\,\Omega} + (I_{50\,\Omega} = I_{70\,\Omega})$. Current must be conserved in a circuit.

(L.O. 5.C.3.4)

9. **C** Kirchhoff's junction rule applies. The 30.0 Ω resistance has a current of 0.10 A as indicated from the ammeter. The potential difference across the ends of the resistor is 3.0 V. Since it is in parallel with an upper branch consisting of a 45.0 Ω resistor and a 15.0 Ω resistor giving an equivalency of 60 Ω, this combination must have the same potential difference, 3.0 V, as the lower branch. A current of 0.05 A exists in the upper branch. The total current flowing into the parallel arrangement is 0.15 A; thus, the current in the 20.0 Ω resistor is 0.15 A.

(L.O. 5.C.3.5)

10. **C** The total resistance in the circuit is $R + r$. This reduces the current in the circuit. Because the internal resistance in the battery dissipates some of the energy as internal energy (heat), the battery will deliver less energy to the circuit. The voltmeter will show a reading less than the emf of the source.

(L.O. 5.B.9.7)

11. **B** The slope of the graph will equal the emf (potential difference) of the battery because $P = IV$.

(L.O. 5.B.9.8)

12. **C** The 10 $\mu F$ capacitor is the only item on its parallel branch, so it has the greatest potential difference between its plates. The 4 $\mu F$ and the 6 $\mu F$ capacitors are in series with each other, so they have the same charge. Using $Q = CV$ and solving for $V$, we see that $V = Q/C$. The 4 $\mu F$ has the higher voltage because it has the lower capacitance.

(L.O. 5.C.3.7)

13. **D** The formula for resistance in a wire is $R = \rho L/A$. So doubling the area would halve the resistance.

(L.O. 4.E.4.1)

14. **C** A linear relationship will indicate that the resistor obeys Ohm's law. An internal resistance in the power supply would still give a linear relationship $V = I(R + r)$. The graph indicates a non-ohmic resistance.

(L.O. 5.B.9.7)

15. **B and D** When both switches are closed, the circuit consists of two bulbs in parallel connected in series to another bulb. When switch 2 is opened, bulb 2 goes out and the brightness of the two remaining bulbs decreases because the equivalent resistance in the circuit increased and thus the current in the circuit decreased.

(L.O. 4.E.5.1)

# FREE-RESPONSE PROBLEMS

1.  (a)

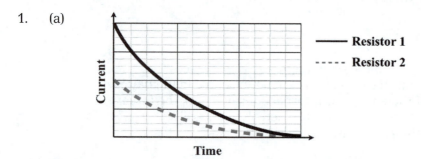

When the switch is closed, current begins flowing. As the capacitor charges, the current is gradually reduced until no more current flows. Because resistor 2 has double the resistance of resistor 1, only half as much current will flow through resistor 2.

(b)

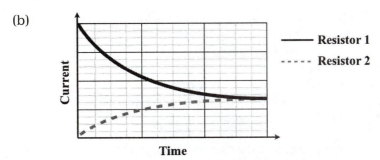

When the switch is closed, all of the current flowing through the battery flows through resistor 1 to the capacitor and no current flows through resistor 2. Effectively, the only resistance in the circuit is the resistance of resistor 1. As the capacitor charges, the current flowing to the capacitor decreases and the current flowing through resistor 2 increases. When no more current is flowing through the capacitor, resistor 2 is effectively in series with resistor 1. This triples the effective resistance of the circuit relative to its initial value, and the current decreases to one-third its initial value.

(L.O. 4.E.5.1, 4.E.5.2)

2.  (a) First, determine the equivalent resistance, $R_{eq}$, of the entire circuit. Starting with the parallel section,

$$R_{ab} = \frac{1}{1/R_1 + 1/R_2 + 1/R_3} = \frac{1}{1/18\ \Omega + 1/18\ \Omega + 1/18\ \Omega}$$

$$= \frac{1}{3/18\ \Omega} = \frac{18\ \Omega}{3} = 6\ \Omega$$

The remainder of the circuit is in series, and

$R_{eq} = R_{ab} + R_4 + R_5 = 6\ \Omega + 8\ \Omega + 4\ \Omega = 18\ \Omega$. So from Ohm's law,

$\varepsilon = IR_{eq} = (2.4\ \text{A})(18\ \Omega) = 43.2\ \text{V}$.

(b) The $IR$ drop across $ab$ is found by

$V_{ab} = IR_{ab} = (2.4\ \text{A})(6\ \Omega) = 14.4\ \text{V}$, and across $cd$, it is found by

$V_{cd} = IR_4 = (2.4\ \text{A})(8\ \Omega) = 19.2\ \text{V}$.

Therefore, between junctions $a$ and $d$, it is

$V_{ad} = V_{ab} + V_{cd} = 14.4\ \text{V} + 19.2\ \text{V} = 33.6\ \text{V}$.

(c) The power loss across $R_4 = R_{cd}$ is $P_{cd} = I^2 R_{cd} = (2.4\ \text{A})^2 (8\ \Omega)$ $= 46.1\ \text{W}$.

(L.O. 5.B.9.3, 5.B.9.5, 5.B.9.6, 5.C.3.3, 5.C.3.5)

3.  (a)  • Place the ammeter in series with the battery to measure the total current in the circuit. Record the reading.
        • Move the ammeter to other positions in the circuit, measuring the current in each resistor. Record the ammeter reading for each position of the meter.
        • Place the voltmeter in the circuit so that it is in parallel with each element in the circuit and record the reading across each element.

    (b) Connect the circuit according to the schematic diagrams shown below. With three resistors, there are four possible arrangements of the elements in the circuits.

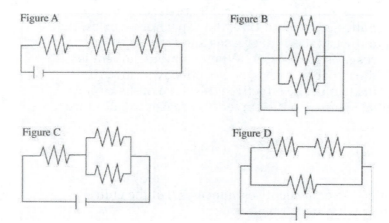

Figure A

Figure B

Figure C

Figure D

    (c) Measurements are indicated for the four possible arrangements.

    Figure A.  Three resistors in series.

    The voltmeter placed across each element, the battery, and the resistors should indicate that $\varepsilon = V + V_2 + V_3$.

    The ammeter placed in the circuit for the current leaving the battery and entering each resistor in turn should indicate $I_{Battery} = I_{R_1} = I_{R_2} = I_{R_3}$.

    Figure B.  Three resistors in parallel.

    The voltmeter placed across each element, the battery, and the resistors should indicate that $\varepsilon = V_1 = V_2 = V_3$.

    The ammeter placed in the circuit will read the current leaving the battery and entering each resistor in turn.

    Figure C.  One resistor in series with a parallel combination of resistors.

    The voltmeter placed across each element, the battery, and the resistors should indicate that $\varepsilon = V_1 + (V_2 = V_3)$.

    The ammeter placed in the circuit for the current leaving the battery and entering each resistor in turn should indicate $I_{Battery} = I_{R_1} = (I_{R_2} + I_{R_3})$.

    Figure D.  Two resistors in series connected in parallel with one additional resistor.

The ammeter will indicate that the resistors in the upper branch will each have the same current. The current from the battery into the junction will be $I_{total} = I_{top\ branch} + I_{bottom\ branch}$. The voltmeter will show that $\mathcal{E} = V_{bottom\ branch} + \Sigma V_{top\ branch}$.

(d) Possible Answers
- If the circuit operates for too long when readings are taken, the elements in the circuit could heat and the resistances in the circuit might increase. Thus, the meters will give readings that are not consistent in the experiment.
- The meters themselves will affect the readings for the experiment. A voltmeter should have an extremely high resistance with no current flow through the meter, but it does. Likewise an ammeter should have no resistance, but it does, and this will give readings that are slightly off.
- The connecting wires have resistance and will affect the readings.
- Resistors have a tolerance, and are only close to their labeled resistance.

(L.O. 5.C.3.1, 5.C.3.2, 5.C.3.4)

3. (a) Voltmeter readings will indicate either the potential difference across the known resistor or the terminal potential difference across the battery. The ammeter will indicate the current in the circuit. These can be used to determine the internal resistance.

(b) Two solutions are possible.

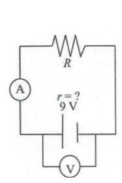 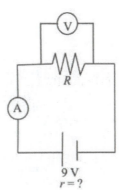

(c) Placing the voltmeter across the ends of the battery will give $V_{AB}$.

$V_{ab} = IR$, then $\mathcal{E} - V_{AB} = Ir$.

Placing the voltmeter across the ends of the resistor will give $V_R = IR$.

$\mathcal{E} = IR + Ir$

(L.O. 5.B.9.2, 5.B.9.6, 5.B.9.7)

# 14

# THE MAGNETIC FIELD

PHYSICS 2

## MAGNETS AND MAGNETISM

The electrical charges that make up matter are always in motion, which causes individual particles to be magnetic on a very small scale. This property, called *magnetic dipole moment*, is an intrinsic property of all electrons. In most materials, the dipole moments of individual electrons are randomly distributed, so the magnetic properties of nearby particles cancel each other out. However, ferromagnetic substances such as iron, nickel, and cobalt do have large-scale magnetic properties. Any sample of ferromagnetic material at room temperature is made up of microscopic *domains* (on the order of a few thousandths of a centimeter) in which the atoms are completely lined up. These domains are small magnets. If an object is magnetized spontaneously or is magnetized by an external magnetic field, the domains become closely aligned, and their individual magnetic effects add up.

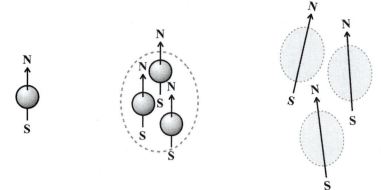

Left: An individual atom has a north and south pole because of the motion of its electrons.
Center: In a magnetic domain, nearby atoms line up. The domain itself acts as a magnet.
Right: In ferromagnetic materials, domains line up, causing whole objects to become magnetic.

If a magnet is broken into pieces, each piece still contains aligned magnetic domains that retain their magnetic properties. Magnets cannot be broken into a single N-pole and into a single S-pole.

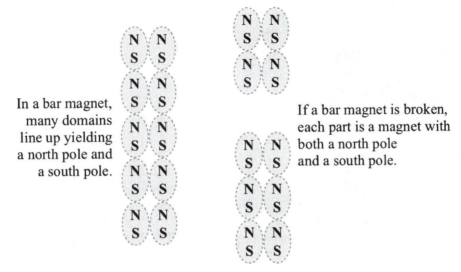

In a bar magnet, many domains line up yielding a north pole and a south pole.

If a bar magnet is broken, each part is a magnet with both a north pole and a south pole.

The north pole of a magnet is attracted to the south pole of another magnet. North poles repel north poles; south poles repel south poles. Unlike electrical charge, which is strictly conserved, magnets can be created and destroyed. Magnetism is a property that can be acquired or lost without regard to conservation laws.

Magnets affect the space around them because they create their own magnetic field, a **B**-field. Magnetic fields have vector properties; they have an intensity, $B$, and a direction in space. Just as electric field lines are useful in describing electric fields, magnetic field lines are useful for visualizing the magnetic field. Magnetic field lines leave the N-pole of a magnet and enter the S-pole. Unlike electric field lines, magnetic field lines do not have origins and end points—they form continuous loops.

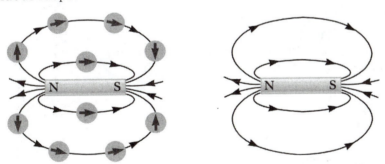

Left: Tracing the magnetic field of a bar magnet with compasses
Right: Several magnetic field lines of a bar magnet

Earth has its own magnetic field. The geographic north pole of Earth (where polar bears live) is actually a magnetic south pole. The geographic south pole (where penguins live) is a magnetic north pole. The north end of a compass points to the geographic north pole because it is attracted to the magnetic south pole.

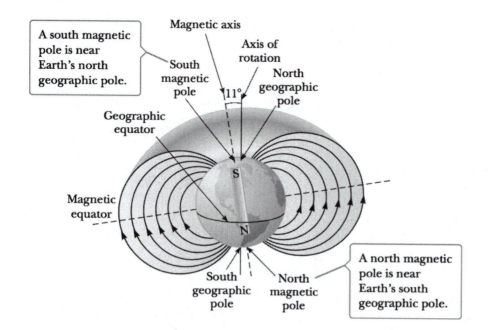

## FORCE ON A MOVING CHARGE

All moving charged particles generate a magnetic field. When a charged particle moves through a magnetic field, the fields interact and the particle experiences a force. This force is proportional to the amount of charge that is moving, the charge's velocity, the strength of the magnetic field, and the sine of the angle between the velocity and the field.

$$F_B = qvB \sin\theta$$

The SI unit of magnetic field, $B$, is the tesla, T, and it is defined as

$$1\ T = 1\ N/(C \cdot m/s) = 1\ N/A \cdot m$$

The tesla is not a fundamental unit. It is related to the newton, the ampere, and the meter. The newton and the ampere are in turn related to the fundamental units of length, mass, time, and charge.

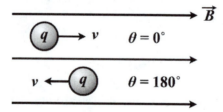

For both these moving charges, $F$ is zero.

The velocity $v$ and the magnetic field $B$ are both vectors; the force $F_B$ also is a vector perpendicular to both $v$ and $B$. When the velocity is parallel with the field, $\theta = 0°$ or $180°$ and $\sin 0° = 0$. The magnetic force on the charge moving parallel with a magnetic field is zero.

Motion in any other direction always produces a force. When $\theta = 90°$, $\sin 90° = 1$, making $v$ and $B$ perpendicular, and the magnetic force is a maximum at

$$F_B = qvB$$

Because our diagrams are only two-dimensional, we use a convention for drawing vectors in a third dimension. For vectors pointing into the page, we use ×; for vectors pointing out of the page, we use •.

The direction of magnetic force on a positive charge that is moving perpendicular to the plane of the velocity vector $v$ and the magnetic field vector $B$ is found by the _right-hand rule_: _Point the fingers of your right hand in the direction of the velocity and curl them toward the direction of the magnetic field. The direction of your thumb tells the direction of the resultant magnetic force._ For a negative charge, the direction of the force vector is reversed.

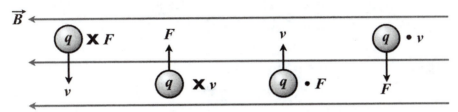

The right-hand rule shows which way magnetic force pushes a positive particle moving in a magnetic field.

Magnetic force always acts at a right angle to the velocity and the field vectors. Thus, magnetic force only changes a particle's direction. Magnetic force cannot cause a moving particle to speed up or slow down. Because of this, magnetic forces cannot change a particle's kinetic energy and do no work on moving particles. Acceleration caused by a magnetic field is centripetal acceleration and causes charged particles to move in circular or helical paths. The magnetic field exerts centripetal force on each moving charged particle of mass m.

$$F_C = \frac{mv^2}{R} = qvB$$

## AP® Tip

Since the magnetic force $\vec{F}_B$ is perpendicular to the velocity vector $\vec{v}$, it is a purely deflecting force. It changes the direction of $\vec{v}$, not the speed. Because there cannot be a component of magnetic force along the motion, there is no tangential acceleration. _No work can be done on the moving charge_ q _by the_ $\vec{B}$-_field, and no change in energy can take place._

Use your left hand to determine the force on a negative charge—an "e_left_ron."

## SAMPLE PROBLEM 1

A proton of charge $q = +e$, mass $1.67 \times 10^{-27}$ kg, and speed $v = 2.9 \times 10^6$ m/s orbits in a magnetic field of $B = 1.5$ T directed into the page.

(a) Calculate the radius of the orbit.

(b) Calculate the force the magnetic field exerts on the particle.

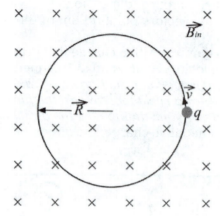

## SOLUTION TO PROBLEM 1

(a) The magnetic force provides the centripetal force to the orbiting particle.

$$F_B = \frac{mv^2}{R} = qvB$$

Solving for $R$,

$$R = \frac{mv}{qB} = \frac{\left(1.67 \times 10^{-27} \text{ kg}\right)\left(2.9 \times 10^6 \text{ m}\middle/\text{s}\right)}{\left(1.5 \text{ T}\right)\left(1.60 \times 10^{-19} \text{ C}\right)} = 0.02 \text{ m}$$

(b) The magnetic force is

$$F_B = qvB = \left(1.60 \times 10^{-19} \text{ C}\right)\left(2.9 \times 10^6 \text{ m}\middle/\text{s}\right)\left(1.5 \text{ T}\right) = 7.0 \times 10^{-13} \text{ N}$$

## SAMPLE PROBLEM 2

A *velocity selector* is a device that is operated in a vacuum and utilizes crossed $E$- and $B$-fields between a set of parallel-plate capacitors to select positively charged ions $+q$ of only one velocity. The ions are projected into the perpendicular fields at varying speeds. Particles with velocities sufficient to make the magnetic force equal and opposite to the electric force pass through the slit undeflected.

(a) Show that the speed of these particles is found from $v = \dfrac{E}{B}$.

(b) It is desired to have a beam of protons with a speed of $7.0 \times 10^3$ m/s. If $B = 0.25$ T, what should the strength of the electric field be?

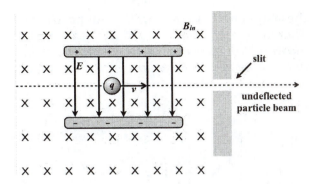

## SOLUTION TO PROBLEM 2

(a) The magnetic force on $+q$ is $qvB$. By the right-hand rule, we know it is upward. The electrostatic force on $+q$ is $Eq$. This force is downward along the field lines. Equating the upward and downward forces,

$$qvB = qE \quad \text{and} \quad v = \frac{E}{B}$$

(b) $E = vB = \left(7.0 \times 10^3 \text{ m}\!/\!_\text{s}\right)(0.25 \text{ T}) = 1.8 \times 10^3 \text{ N}\!/\!_\text{C}$

## SAMPLE PROBLEM 3

Ions of charge $q = +e$ are produced in the ion source of a mass spectrograph operated with a primary $B$-field of 0.40 T. The ion beam is directed into the velocity selector of Problem 2 and emerges with a velocity of $7.0 \times 10^3$ m/s. The ions enter the mass spectrograph where they travel a semicircular path of radius $7.2 \times 10^{-3}$ m. Calculate the mass of one of these ions.

## SOLUTION TO PROBLEM 3

The charge of an ion is $q = +e = +1.6010^{-19}$ C, and the mass is found by

$$m = \frac{BqR}{v} = \frac{(0.40 \text{ T})\left(1.60 \times 10^{-19} \text{ C}\right)\left(7.2 \times 10^{-3} \text{ m}\right)}{7.0 \times 10^3 \text{ m}\!/\!_\text{s}} = 6.6 \times 10^{-26} \text{ kg}$$

## FORCE ON A CURRENT-CARRYING WIRE

When a wire conducts a current through a magnetic field, the magnetic forces experienced by the individual charges in the current are exerted on the wire. A current-carrying wire in a magnetic field experiences a force

$$F_B = ILB$$

In the event $L$ makes an angle $\theta$ with the field, the relationship becomes

$$F_B = ILB \sin\theta$$

The direction of the quantities in the equation, as shown in the figure on the next page, can be found using the same right-hand rule as was used with a moving charged particle. The direction of magnetic force on a current-carrying wire that is placed perpendicular to the plane of the velocity vector *v* and the magnetic field vector *B* is found by the *right-hand rule*: Point the fingers of your right hand in the direction of the current and curl them toward the direction of the magnetic field. The direction of your thumb tells you the direction of the resultant magnetic force.

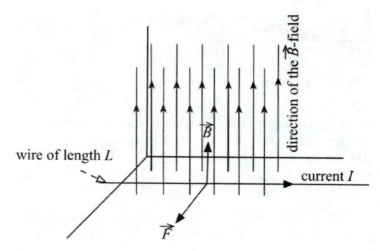

## SAMPLE PROBLEM 4

A wire connected across a potential difference carries a current $I = 2.4$ A. The wire is inserted into a magnetic field at a right angle to the field of $B = 0.5$ T that is directed into the page. If a length, $L = 0.5$ m, is exposed to the field, as illustrated below, what force acts on the wire?

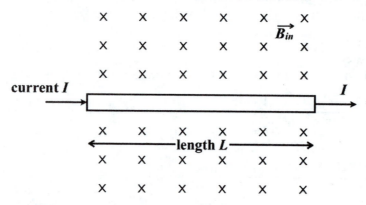

## SOLUTION TO PROBLEM 4

By the right-hand rule, the *B*-field exerts an upward force on the wire segment. The magnitude of the force is

$$F_B = ILB = (2.4 \text{ A})(0.5 \text{ m})(0.5 \text{ T}) = \textbf{0.6 N upward}$$

# MAGNETIC FIELD OF A LONG, STRAIGHT CURRENT-CARRYING WIRE

The magnetic field lines near a current-carrying wire are in the form of concentric circles. The relationship between the direction of the magnetic field and the direction of the current in the wire is given by the *right-hand current rule: when a wire is grasped by the right hand in such a way that the thumb points in the direction of the current, the fingers curl around the wire in the same direction as the magnetic field.*

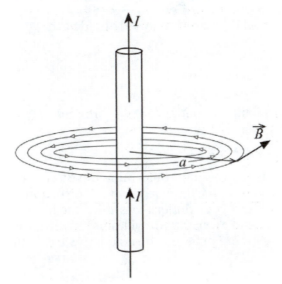

We can determine the magnitude of the magnetic field at some point near the wire if we know the current, $I$, in the wire and how far, $r$, the point is from the wire. This relationship can be shown by

$$B = \frac{\mu_0 I}{2\pi r}, \text{ where } \mu_o = 4\pi \times 10^{-7} \text{ T} \cdot \text{m/A}.$$

The symbol $\mu$ represents magnetic permeability, a measure of a material's ability to support a magnetic field. Specifically, $\mu_o$ represents a fundamental constant, the magnetic permeability of free space. Materials other than empty space have other $\mu$ values; for example, the permeability of ferromagnetic materials is much higher.

### SAMPLE PROBLEM 5

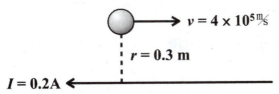

A proton is moving $4.0 \times 10^5$ m/s to the right of the page. At a location 0.3 m directly below the proton is a long current-carrying wire that carries 0.2 A toward the left side of the page.
(a) Draw the magnetic field above and below the wire.
(b) What is the magnitude and direction of the force exerted on the proton?

### SOLUTION TO PROBLEM 5

(a)    X  X  X  X  X  X  X  X
        X  X  X  X  X  X  X  X
        X  X  X  X  X  X  X  X
       ←———————————
        •  •  •  •  •  •  •  •
        •  •  •  •  •  •  •  •
        •  •  •  •  •  •  •  •

With our right thumb pointing in the direction of the current, our fingers curl into the page above the wire and out of the page below the wire.

(b) First, we find the strength of the wire's magnetic field at the location of the proton.

$$B = \frac{\mu I}{2\pi r} = \frac{4\pi \times 10^{-7}(0.2 \text{ A})}{2\pi(0.3 \text{ m})} = 1.33 \times 10^{-7} \text{ T}$$

Then using the strength of the magnetic field, we can find the magnitude of the force on the proton.

$$F = qvB = (1.6 \times 10^{-19} \text{ C})(4 \times 10^5 \text{ m/s})(1.33 \times 10^{-7} \text{ T}) = 8.5 \times 10^{-21} \text{ N}$$

The right-hand rule is then used to determine the direction of the force. The fingers point toward the right in the direction of the velocity of the proton. Curl your fingers into the page in the direction of the magnetic field. Your thumb will point toward the top of the page. The force on the proton is directed upward to the top of the page.

## MAGNETIC FIELD AT THE CENTER OF A COIL

The magnetic field produced by a current-carrying wire is greatly increased if the wire is wound into a circular coil with many turns. Following the right-hand current rule, we see that if our thumb follows the current around the loop, our fingers point through the middle of the loop. In this way, the magnetic field caused by each small section of current adds up. The magnetic field created by a flat coil carrying current $I$ is complex, as show below.

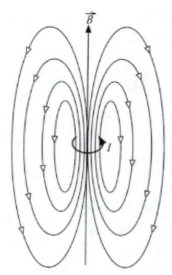

For a flat circular coil of $n$ turns and radius $a$, we can calculate the $B$-field at the very center of the coil, and it is

$$B = \frac{\mu_o n I}{2a}$$

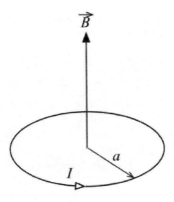

## SAMPLE PROBLEM 6

A flat circular coil of 10 turns of wire has a radius, *a*, of 4.0 cm. Determine the *B*-field at the center of the coil when a current of 4.0 A is circulated through the coil.

## SOLUTION TO PROBLEM 6

$$B = \frac{\mu_o nI}{2a} = \frac{\left(4\pi \times 10^{-7}\ \text{T} \cdot \text{m/A}\right)(10)(4.0\ \text{A})}{2(0.04\ \text{m})} = 6.3 \times 10^{-4}\ \text{T}$$

# FARADAY'S LAW OF INDUCTION

Electromagnetism provides the means by which electrical energy can be converted into mechanical work; magnetic induction provides the means by which mechanical work can be converted into electrical energy. All of our electric power comes from generators that operate on the principle of magnetic induction.

The magnetic flux $\Phi_B$ through a loop of wire is defined as the product of the magnetic field that goes through a loop of a given area, or

$$\Phi = BA \cos\theta$$

The magnetic flux through a loop of wire is shown in the diagram below.

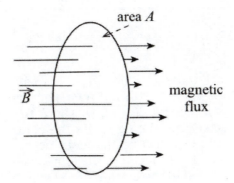

The SI unit of magnetic flux $\Phi_B$ is the weber, *Wb*, and it is defined as 1 Wb = 1 T · m².

Changing the flux through a loop of wire induces an emf in the loop. Faraday's law of magnetic induction states that when magnetic flux through a loop changes by an amount $\Delta\Phi_B$ in a time period $\Delta t$, the emf induced in the loop during time period $\Delta t$ is

$$\varepsilon = -\frac{\Delta\phi_M}{\Delta t}$$

The more rapidly the flux through a loop changes, the greater emf is induced.

There are three ways to change the magnetic flux through a loop of wire.
1.   Change $B$: Make the magnetic field strength weaker or stronger.
2.   Change $A$: Alter the wire to change its area.
3.   Change $\Phi$: Spin the loop of wire within the magnetic field.

When a conductor moves through a magnetic field and experiences a change in magnetic flux, the induced emf is called *motional emf*.

$$\varepsilon = BLv$$

where $\varepsilon$ is the induced potential, $B$ is the magnetic field strength, $L$ is the length of wire cutting through the field, and $v$ is the velocity of the wire.

The direction of the induced emf and the direction of any resulting induced current is determined by Lenz's law: **The direction of the induced emf is such as to oppose the change in magnetic flux that causes the induced emf.** The words change and flux are very important. Consider the word change. Lenz's law says that when the flux is increasing, the induced emf tries to make it decrease; when flux is decreasing the induced emf tries to make it increase. Note that Lenz's law does not say the induced emf opposes the magnetic field, but rather it says it opposes the change in magnetic flux.

## AP® Tip

If the magnitude of the flux through a circuit is *increasing*, the induced *I*-field is in the *opposite* direction of the primary field. If the magnitude of the flux through a circuit is *decreasing*, the induced field is in the *same* direction as the primary field.

Lenz's law is required by the law of conservation of energy.

## SAMPLE PROBLEM 7

Imagine moving a conductor of length $L = 0.2$ m sliding along a stationary conducting loop with a velocity $v = 2.0$ m/s. The loop, as shown on the next page, contains a resistor $R = 4.0\ \Omega$. The $B$-field has an intensity $B = 0.5$ T and is directed into the page. As the conductor slides to the right, the magnetic flux penetrating the loop increases as the area of the loop increases. An emf is induced in the loop as a result of this motion, and a current is induced in the circuit.
(a)   Calculate the magnitude of the motional emf.

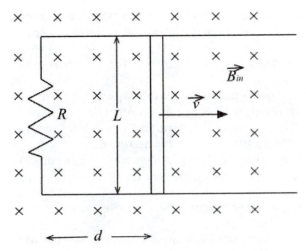

(b)   What current is induced in the loop?
(c)   What is the direction of this current?

## SOLUTION TO PROBLEM 7

(a)   The magnitude of the induced motional emf is

$$|\varepsilon| = \frac{\Delta \phi_B}{\Delta t} = \frac{\Delta(BA)}{\Delta t}$$

$B$ is constant and

$$|\varepsilon| = \frac{\Delta(BA)}{\Delta t} = \frac{B\Delta A}{\Delta t} = \frac{B\Delta(Ld)}{\Delta t}$$

The length of the conductor $L$ is constant, and

$$|\varepsilon| = BL\frac{\Delta d}{\Delta t} = BLv = (0.5\text{ T})(0.2\text{ m})(2.0\text{ m/s}) = \mathbf{0.2\ V}$$

(b)   The current is found using Ohm's law

$$I = \frac{\varepsilon}{R} = \frac{0.2\text{ V}}{4.0\ \Omega} = \mathbf{5 \times 10^{-2}\ A}$$

(c)   The direction of the induced current is counterclockwise. The magnetic flux is increasing into the page as the area increases. According to Lenz's law the loop will oppose this change by directing its magnetic field out of the page. Using the right-hand rule, point thumb in the direction of the induced magnetic field (out of the page) and curl fingers around the loop (counterclockwise).

# THE MAGNETIC FIELD: STUDENT OBJECTIVES FOR THE AP® EXAM

- You should know and be able to apply the expression used for the magnitude of the magnetic field.
- You should know that force, velocity, and magnetic field intensity are all vector quantities that are mutually perpendicular to one another.

- You should be able to apply the right-hand rule for a positive charge to determine the direction of the unknown parameter when the other two are known.
- You should be able to state and apply the equation for the magnetic force on a moving conductor.
- You should be able to relate the magnetic force on a moving charge in a magnetic field that causes the particle to alter its direction of travel to the centripetal force and to evaluate the equations to solve for any unknown parameter either mathematically or by diagram.
- You should know and be able to apply the second right-hand rule to the direction of the magnetic field produced by a long, straight wire carrying a current.
- You should be able to calculate the magnitude and direction of the B-field produced in a closed loop.
- You should know and be able to apply Faraday's and Lenz's laws for the magnitude and direction of the induced emf or current in closed loops.
- You should be able to draw or explain graphs involving changes in magnetic flux or induced emf as a function of time.

## MULTIPLE-CHOICE QUESTIONS

1. A proton traveling at $5.0 \times 10^5$ m/s crosses the page to the right.
   The proton enters a B-field of 0.1 T that is directed toward the top of the page. The magnitude and the direction of the force acting on the proton is
   (A) $8.0 \times 10^{-13}$ N into the page.
   (B) $8.0 \times 10^{-13}$ N out of the page.
   (C) $3.2 \times 10^{-13}$ N down the page.
   (D) $3.2 \times 10^{-13}$ N to the left on the page.

2. Two wires carrying currents in opposite directions produce a magnetic field between the wires that is best illustrated by which of the following diagrams?

   (A) I
   (B) II
   (C) III
   (D) IV

3. A pair of bar magnets is used to show the direction of the magnetic field relating to the north and south poles of the magnets. Iron filings sprinkled around the magnets will indicate that the field produces
   (A) concentric circles of iron filings around the magnet.
   (B) repulsion field when placed in the region between a north and south pole.
   (C) attraction when placed in the region between a north and south pole.
   (D) no interaction with either magnet.

4. Rank the domains in the ferromagnetic materials illustrated below that show the strength of the bar magnet from the weakest magnetic field to the strongest magnetic field.

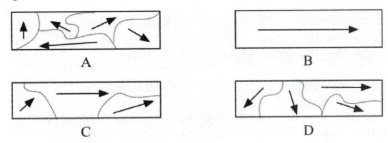

A    B

C    D

(A) B < C < A < D
(B) B < C < A < D
(C) A < D < C < B
(D) B < C < A = D

5. An electron moving with a constant velocity to the right along the positive $x$-direction enters a uniform magnetic field. The magnetic field is directed out of the page. The direction of the force acting on the electron as it moves through the field is shown by which of the force vectors shown below?

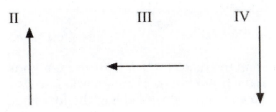

II    III    IV

(A) I
(B) II
(C) III
(D) IV

6. A bar magnet suspended from its midpoint in a horizontal plane will align itself with Earth's magnetic field with its north end pointing north. Which of the following statements is true?
(A) The force that Earth exerts on the bar magnet is greater than the force the bar magnet exerts on Earth because Earth's magnetic field is much larger.
(B) The force the bar magnetic exerts on Earth is larger because the bar magnet is suspended by a string and is free to rotate.
(C) The magnitude of the forces between the bar magnet and Earth is the same since they are action-reaction pairs.
(D) Only a gravitational force is acting between the magnet and Earth.

7. Iron filings sprinkled around a bar magnet are good materials to use to indicate the strength and direction of the magnetic field associated with a magnetic pole or a pair of magnetic poles because
   (A) the domains in the iron filings can be affected by the field of a strong magnet so that they orient themselves with the field of the magnet(s), showing the field direction around the magnet(s).
   (B) they are weakly magnetic and do not interfere with magnetic field of the bar magnet(s).
   (C) since they are weakly repelled by the external magnetic field of the bar magnet(s), they do not crowd together along the magnet field lines associated with the bar magnet(s).
   (D) they are nonmagnetic materials and will not affect the magnetic field of the bar magnet(s).

8. A circular coil is placed on a horizontal surface. Initially, a bar magnet is held above it with the N-pole pointing downward to the coil. When the magnet is moved toward the coil, it induces
   (A) a current in the coil that is counterclockwise as viewed from above.
   (B) a current in the coil that is clockwise as viewed from above.
   (C) no current in the coil since it is not connected to a power supply.
   (D) an alternating current in the coil.

9. Two loops of wire as shown in the diagram below have a common vertical axis. The current in the bottom loop is counterclockwise as shown and is increasing at a constant rate. Which of the following best describes the induced current in the upper loop?

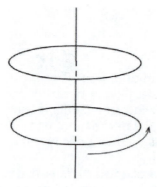

   (A) There is a clockwise current in the upper loop, but it will change due to the increasing current in the lower loop.
   (B) There is a counterclockwise current in the upper loop, but it will change due to the increasing current in the lower loop.
   (C) There is a constant clockwise current in the upper loop.
   (D) There is a constant counterclockwise current in the upper loop.

10. A proton, $+q$, moves to the right with a velocity of $1.0 \times 10^4$ m/s into the region between two parallel plates where both a uniform electric field of 1000 N/C and a magnetic field, $B = 0.1$ T, exist between the plates as shown in the diagram below.

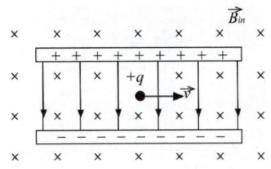

The electric and magnetic forces acting on the proton are
(A) in the same direction, causing the proton to accelerate upward.
(B) in opposite directions, causing the proton to accelerate downward.
(C) balanced, and the proton will be stationary in the region between the plates.
(D) balanced, and the proton will move to the right horizontally with a constant velocity.

11. A proton with a velocity, $v$, enters a region in which a magnetic field is directed into the page. The proton follows the path shown below. The magnetic field will

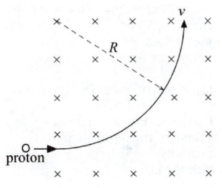

(A) produce an increase in the kinetic energy of the proton since the magnetic field works on moving the charge.
(B) decrease the kinetic energy of the proton since the magnetic field does not act parallel to the direction of motion of the moving charge.
(C) increase the angular momentum of the proton since the field produces a torque on the proton.
(D) only change the direction of motion of the proton since the velocity of the proton is perpendicular to the magnetic force.

12. Two horizontal wires carry a current *I* in opposite directions. Which diagram best represents the force exerted on each wire by the other?

(A)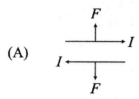

(B)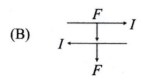

(C)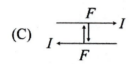

(D)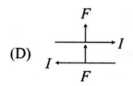

13. A region of space surrounding a moving electron contains
    (A) an electric field, a gravitational field, and a magnetic field.
    (B) an electric field and a gravitational field.
    (C) an electric field and a magnetic field.
    (D) a gravitational field and a magnetic field.

14. A bar magnet is pushed through a coil of wire. The induced emf is greatest when the
    (A) north end of the magnet is pushed through the wire.
    (B) south end of the magnet is pushed through the wire.
    (C) magnet is pushed through rapidly.
    (D) magnet is in equilibrium with the coil.

15. The phenomenon of magnetism is best understood in terms of
    (A) the force exerted on a charge in an electric field.
    (B) the magnetic field associated with moving charges.
    (C) force fields between electron and protons.
    (D) the existence of magnetic poles in materials.

## FREE-RESPONSE PROBLEMS

1.  Each electron in a beam of electrons travels with the same velocity of $1.50 \times 10^7$ m/s. The beam is bent 90° through an arc with a radius of 3.71 mm by a uniform magnetic field perpendicular to the beam path.

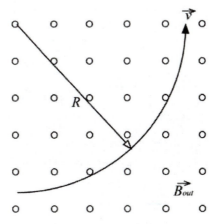

    (a) What is the magnetic field strength?
    (b) How much work is done by the magnetic field on a single electron in this beam? Justify your answer.
    (c) Through what potential difference were the electrons accelerated to acquire their velocity?

2.  A square loop of wire 20.0 cm on a side enters a region in space where a magnetic field of 1.6 T is directed into the page and confined in a square of dimensions 70.0 cm by 70.0 cm. The loop of wire has a constant velocity $v = 10.0$ cm/s perpendicular to the B-field as shown.

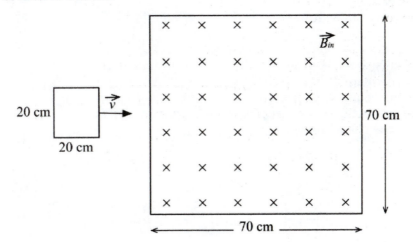

    (a) i.   What is the direction of the current in the loop as it enters the field?
        ii.  What is the direction of the current in the loop as it leaves the field?
        iii. Justify your answers.

(b) What is the magnitude of the induced emf as it enters and leaves the *B*-field?

(c) Sketch the emf as a function of time as the loop completely enters and completely emerges from the magnetic field. Justify your sketch.

3. A wire carrying a conventional current is placed in a magnetic field as illustrated below.

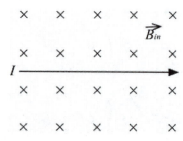

(a) Indicate the direction of the force acting on the wire shown below. Justify your answer.

⟶

(b) If the magnetic field is reversed, indicate the direction of the force on the wire shown below. Justify your answer.

⟶

(c) If the current in the wire is reversed, indicate the direction of the force on the wire shown below. Justify your answer.

⟵

(d) i. If we double the current in the wire, what happens to the force on the wire in part (a)?

___ Increases
___ Decreases
___ Remains constant

ii. Justify your answer.

4. (a) A bar magnet is shown below. Indicate the magnitude and direction of the magnetic field surrounding the bar magnet at the point identified in the diagram. Draw vectors to scale to represent the direction and strength of the magnetic field produced by the bar magnet.

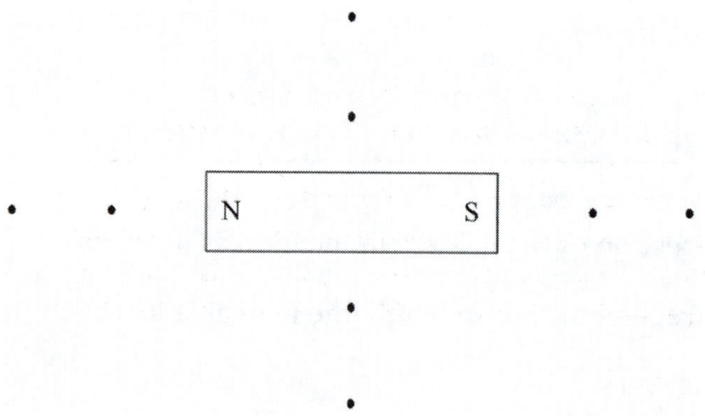

•

•

•  •  | N            S |  •  •

•

•

(b) Compare the magnetic field of the bar magnet above to the electric field due to an electric dipole. List at least two ways that they are
   i.  similar.
   ii. different.

# Answers

## MULTIPLE-CHOICE QUESTIONS

1.  **B** The magnitude of the force is given by $F = qvB$. The force is

    $$\left(1.6\times10^{-19}\ \text{C}\right)\left(5.0\times10^{5}\ \text{m/s}\right)(0.1\ \text{T}) = 8.0\times10^{-15}\ \text{N}$$

    The direction of the force is given by the *right-hand charge rule: The right hand is open, and the thumb points in the direction of the magnetic force. The fingers point in the direction of the velocity, and the palm of the hand points in the direction of the magnetic field.* The direction is out of the page.

    (L.O. 2.D.1.1, 3.C.3.1)

2.  **A** current-carrying wire produces a magnetic field that circles the wire. Use your right hand to determine the direction by placing your thumb in the direction of conventional (positive) current in the wire. The curl of your fingers will give the direction of the field associated with the current, × indicates into the paper, and o is used to show a direction out of the paper. Superposition applies to magnetic fields too. The wires carrying current in opposite directions will attract each other in the same manner as N- and S-poles of magnets.

    (L.O. 2.D.2.1)

3.  **C** The iron filings become temporary magnets when placed in the region around the bar magnets. The direction of the filings shows that an attractive field exists in the region between the north and south poles of the magnet.

    (L.O. 2.D.3.1)

4.  **C** Magnetic domains are regions in the bar magnet where spinning electrons create current loops that are oriented in the same direction. These domains can grow at the expense of other regions in the presence of an external magnetic field. When the domains are oriented in the same direction, they produce a strong magnetic field. The most disorganized ferromagnetic material is represented in A; the strongest, in B. The correct order is A < D < C < B.

    (L.O. 2.D.4.1)

5. **B** The direction of the force is given by the right-hand rule. The palm of your right hand will give the direction of the force on the charge. Since the charge is an electron, the direction of the force must be reversed.

   (L.O. 3.A.2.1, 3.C.3.1)

6. **C** The magnetic force between the bar magnet and Earth is equal in magnitude and opposite in direction. Earth and the bar magnet produce forces that are action-reaction pairs.

   (L.O. 3.A.3.3)

7. **A** Ferromagnetic materials such as iron filings are good materials to use to indicate the presence of an external magnetic field. They have permanent magnetic moments that can be temporarily aligned with the magnetic field of the magnet(s), showing the magnetic field lines of the bar magnet(s).

   (L.O. 4.E.1.1)

8. **A** Both Faraday's and Lenz's laws apply. As the north end of magnet is moved toward the coil, the magnetic field and flux increases downward in the coil. The induced current in the coil is counterclockwise in the coil, producing a $B$-field whose direction is upward to counter the increasing downward flux. Use the right-hand rule. If your fingers curl in the direction of the induced current, your thumb will point in the direction of the magnetic field produced.

   (L.O. 4.E.2.1)

9. **A** An increasing counterclockwise current in the bottom loop will produce a $B$-field that is increasing. Use the right-hand rule. If your fingers curl in the direction of the current, your thumb will point in the direction of the magnetic field. This behaves as if the north pole of a bar magnet is being moved toward the upper loop. Accordingly, the induced current in the upper loop will circle clockwise, producing a magnetic field that points downward to the lower loop. Because the current in the lower loop is increasing, the induced current in the upper loop will not be constant because the induced current depends on the rate change of flux.

   (L.O. 4.E.2.1)

10. **D** The vector sum of the forces acting on the proton are $\Sigma F = FE + FB$. $F_E = Eq$ acts downward, and $F_B = qvB$ acts upward (right-hand rule).

    $$-1000 \ {}^{N}\!/_{C}(q) + q\left(1 \times 10^4 \ {}^{m}\!/_{s}\right)(0.1 \ \text{T}) = \Sigma F$$

The net force is zero on the proton, and it will continue to move to the right at constant velocity.

(L.O. 2.C.1.1, 2.D.1.1, 3.A.2.1)

11. **D** Since the magnetic field produces a force that is perpendicular to the velocity and itself, the field does no work on the proton. A force that is perpendicular to the velocity can only change the direction of the particle. The direction of the force is given by the right-hand rule.

(L.O. 3.C.3.1)

12. **A** Each wire produces a magnetic field that circles the wire. First, we find the direction of the field by putting our thumbs in the direction of the current and curling our fingers in the direction of the field. In both cases, the magnetic field caused by one wire points into the page at the location of the other wire. Then we use the right hand that relates current, magnetic field, and force. The force on the top wire is up while the force on the bottom wire is down.

(L.O. 2.D.2.1)

13. **A** The electron has mass and creates a gravitational field. Since it has a charge, it also creates an electric field. It will have a magnetic field associated with it only when it is moving.

(L.O. 2.D.2.1)

14. **C** The induced emf given by Faraday's law relates the strength of the induced emf to the rate of change in the flux.

(L.O. 4.E.2.1)

15. **A** We define the properties of a magnetic field at a point in space in terms of the magnetic force $F_B$ exerted on a moving charge as $F_B = qvB\sin\theta$.

(L.O. 3.C.3.1)

## FREE-RESPONSE PROBLEMS

1.

(a) First, we write that magnetic force provides the centripetal force.

$$F_B = F_c$$

$$qvB = \frac{mv^2}{r}$$

$$B = \frac{mv}{er} = \frac{9.11 \times 10^{-31} \text{ kg} \left(1.5 \times 10^7 \text{ m/s}\right)}{(1.6 \times 10^{-19} \text{ C})(2.71 \times 10^{-3} \text{ m})} = 0.0315 \text{ T}$$

(b) The magnetic field does no work on the electron.
Work measures a change in energy, and the electron's kinetic energy does not change.

(c) An electric field's potential energy caused the electron's kinetic energy.

$$qV = \frac{1}{2}mv^2$$

$$V = \frac{mv^2}{2q} = \frac{\left(9.11 \times 10^{-31} \text{ kg}\right)\left(1.5 \times 10^7 \text{ m/s}\right)^2}{2\left(1.6 \times 10^{-19} \text{ C}\right)} = 640 \text{ V}$$

(L.O. 2.D.1.1, 3.A.2.1, 3.C.3.1)

2.

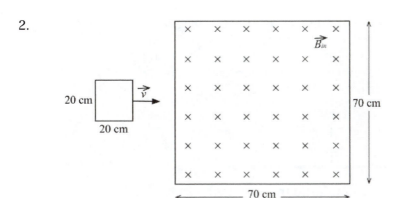

(a) i.  Counterclockwise
   ii.  Clockwise
   iii. Before the square loop enters the field, the flux through the loop is zero. As the loop enters the field, the flux increases until the left edge of the loop is just inside the field. The flux decreases to zero as the loop leaves the field.

   As the right side of the square loop enters the magnetic field, the magnetic flux directed into the page increases. According to Lenz's law, the induced current is counterclockwise because it must produce a magnetic field directed out of the page. The motional emf, *BLv*, arises from the magnetic force experienced by the charge carriers in the right side of the loop.

   Once the loop is entirely in the field, the change in the magnetic flux is zero. This happens because once the left side of the loop enters the field, the motional emf induced in it cancels the motional emf present in the right-hand side of the loop.

   As the right-hand side of the loop leaves the field, the flux inward begins to decrease, a clockwise current is induced, and the induced emf is −*BLv*. As soon as the left-hand side leaves the field, the emf decreases to zero.

(b) $\mathcal{E} = BLv = (1.6 \text{ T})(0.2 \text{ m})(0.1 \text{ m/s}) = \mathbf{3.2 \times 10^{-2} \text{ V}}$

(c) There is only a change in flux as the loop enters and leaves the magnetic field. A change in flux is required for an induced potential. During that time period when the loop is entering the field, an emf of 0.032 V is induced. While the loop is completely inside the magnetic field, there is no change in flux and no motional emf. When the loop leaves the magnetic field, an emf of −0.032 V is induced. This induced potential is in the opposite direction in accordance to Lenz's law. The graph should look like the one shown below.

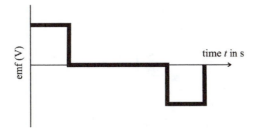

(L.O. 4.E.2.1)

3.  (a) The direction of the force on the wire is upward as shown on the diagram. The total force on the wire is given by the force on each charge times the number of charge carriers in the wire. $\Sigma F = (q v_{drift} B)(nAL)$ since the current in the wire is $I = nq v_{drift} A$. The force on the wire is $F = BIL$.

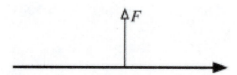

The right-hand rule applies to the wire in the same manner that applies to a charge moving in a magnetic field that is at right angles to the velocity.

    (b) Reversing the magnetic field reverses the direction of the force on the wire as shown in the diagram below. The right-hand rule applies.

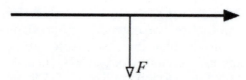

    (c) Reversing the direction of the current produces a force on the wire as shown in the diagram below. Again, the right-hand rule applies.

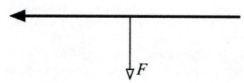

    (d) i.   The force increases.
        ii.  Increasing the current increases the force on the wire as is indicated in part (a) of this question.

    (L.O. 3.A.2.1, 3.C.3.1)

4.  (a) The diagram for the magnitude and direction of the field at the indicated points is shown.

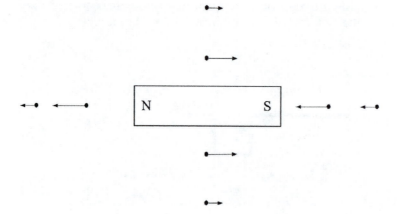

(b) i. Similarities between the magnetic field associated with the bar magnet and an electric dipole include the following:
- They are both inverse square laws; however, we usually do not solve for this when dealing with magnets.
- The field direction is always out from the surface of a positive charge and into the negative charge. For a magnet, the field is defined as being away from the N-pole of the magnet into the S-pole of the magnet.
- Magnets exert attractive and repulsive forces on other magnets. Electric dipoles exert attractive and repulsive forces on other charges.

ii. Differences between the magnetic field associated with the bar magnet and an electric dipole include the following:
- An electric dipole can be separated into two point charges that are positive and negative. Breaking a magnet into two pieces results in two smaller magnets with N- and S-poles. Magnetic monopoles do not exist in nature.
- Heating an electric dipole does nothing to the dipole, but heating a magnet reduces the magnetic properties since it puts random energy into the atoms of the magnetic material, reducing the alignment of the domains.
- Conservation of charge applies to electric dipoles, but the conservation laws do not apply to magnets.
- An electric dipole creates an electric field that acts on any charge placed it in. The magnetic field of a magnet will only affect a moving charge placed in the field at some angle with a component perpendicular to the field. It will have no effect on a stationary charge or a charge moving parallel to the field.

(L.O. 1.B.1.1, 1.B.3.1, 2.C.4.1, 2.D.1.1, 2.D.3.1)

# 15

# WAVES AND SOUND

## PHYSICS 1

## WAVES

Dropping a stone into a pool of still water creates ripples that spread outward. As one particle of water is disturbed, it bumps into another, causing that particle to move and so on. After the ripple passes, the particles return to their original positions, so they have zero displacement. This is an example of a *mechanical* wave. The water is the medium, and the falling stone is the disturbance.

The disturbance causes energy to be moved through the water. Waves transfer energy without transferring matter. The medium as a whole does not progress in the direction of motion of the wave. If there is no medium, mechanical waves cannot transfer energy.

There are several kinds of waves, which are classified by what disturbance causes the wave and how the energy moves through the medium. The most common types are *transverse* waves and *longitudinal* waves.

## TRANSVERSE WAVES

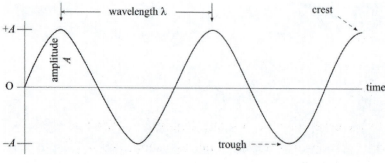

Sine Wave

For a mechanical wave to occur, it needs a *source* that produces a disturbance in a medium through which the disturbance can be transmitted. In a transverse wave, the disturbance is perpendicular to the movement of the energy. When a string under tension is moved rapidly up and back down, a *pulse* travels along the string. The displacement is up and down, while the motion of the pulse is along the length of the string. The pulse is transferred down the length of the string, yet the string returns to its original position.

The transverse pulse below travels to the right. Its velocity is a function of the material of the string and the tension the string is under. Note that the pulse moves along the string while the string stays in place.

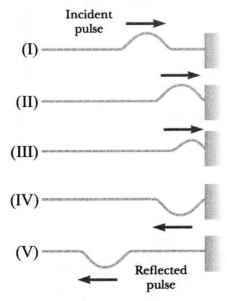

In II through IV, the pulse collides with a fixed barrier and the reflected pulse is inverted.

The string below is attached loosely to a pole. A transverse pulse traveling to the right is reflected in II and III and travels to the left without inversion.

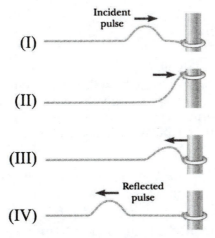

The speed of a transverse wave on a string depends on the linear density, $\mu$, of the string. $\mu$ measures the mass per unit length or $\mu = \dfrac{m}{L}$.

The tension in the string is $T$. The wave speed is

$$V = \sqrt{\frac{T}{\mu}} = \sqrt{\frac{TL}{m}}$$

The speed of a wave on a string increases with increasing tension. The speed decreases with increasing density.

## SAMPLE PROBLEM 1

A string 4.0 m in length has a mass of 3.0 g. The left end of the string is rigidly attached to a vertical wall. The other end hangs over a frictionless pulley with a 2.0 kg mass attached.
(a)   What is the speed of a transverse wave in the string?
(b)   What change could be made to the hanging mass to reduce the width of the wave pulse?

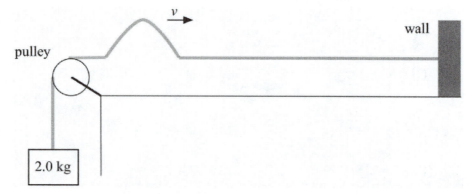

## SOLUTION TO PROBLEM 1

(a)   The tension in the string is $T = mg = (2.0 \text{ kg})(9.8 \text{ m}/_{s^2}) = 19.6 \text{ N}$

The wave speed is $v = \sqrt{\frac{TL}{m}} = \sqrt{\frac{(19.6 \text{ N})(4.0 \text{ m})}{(0.0030 \text{ kg})}} = 160 \text{ m}/_{s}$

(b)   If the wave traveled slower, it would not cover so much distance during a pulse. To make the wave slower, you reduce the tension by reducing the hanging mass.

## LONGITUDINAL WAVES

Another type of wave is shown below as it travels along a helical coil. Such a wave is called a *longitudinal wave* in which the displacements of the particles of the medium are parallel to the direction of energy transfer of the wave. A longitudinal wave is caused by oscillating a medium parallel to the direction the pulse will travel. Consider the stretched spring in diagram (a).

(a)

If the left end of the spring was suddenly pushed right for a moment and then jerked back into place, a pulse would travel down the length of the spring. In the pulse, individual pieces of the spring would mimic the disturbance that started the pulse, moving rapidly left and right. The spring returns to its original position, yet the wave passes along it.

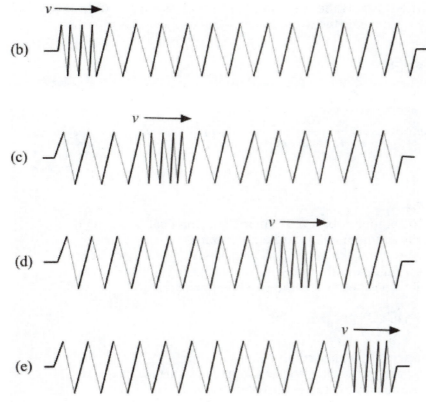

The region where the coils are close together is called a *compression*. The region where the coils are further apart is called a *rarefaction*.

## PROPERTIES OF WAVES

The *frequency, f,* of a wave describes how many times the wave repeats every second and is characteristic of how much energy the wave transfers. The *wavelength, λ,* is the distance between two identical, adjacent points on the wave. The velocity, *v,* of a wave describes the rate at which the pulse moves through the wave's medium. Velocity of the wave and its wavelength are directly related; therefore, the faster waves produce wider pulses or longer wavelengths.

When a wave passes from one medium into a second medium, the speed changes. In the process, the frequency remains the same, but the wavelength changes in proportion to the wave speed; if *v* increases, *λ* also increases.

The time it takes a single wave to pass a point is called the *period, T,* of the wave. Remember that $f = 1/T$. The *amplitude, A,* of the wave is the maximum displacement from the equilibrium position.

The relationship between wave speed, frequency, and wavelength are given by

$$v = f\lambda$$

### SAMPLE PROBLEM 2

A 250 Hz longitudinal wave is set up in a steel rod and passes from the rod into the air. The speed of the wave in the steel rod is $4.88 \times 10^3$ ms and 335.3 m/s in air. Calculate the wavelength of the sound wave in both mediums.

### SOLUTION TO PROBLEM 2

In steel, the wavelength is $\lambda = \dfrac{v}{f} = \dfrac{4.88 \times 10^3 \text{ m/s}}{250 \text{ s}^{-1}} = \textbf{19.5 m}$

In air, the wavelength is $\lambda = \dfrac{v}{f} = \dfrac{335.3 \text{ m/s}}{250 \text{ s}^{-1}} = \textbf{1.3 m}$

## SOUND

One of the most commonly observed types of mechanical waves is a sound wave. Usually the medium that transmits sound waves to our ears is the air that surrounds us. Air, as well as all gases, transmits sound as longitudinal waves. As a sound travels it creates pressure and density variations in the air. The speed of sound can be approximated by

$$v = 331 \text{ m/s} + \left( 0.6 \frac{\text{m/s}}{^\circ\text{C}} \right) T$$

where $T$ is the temperature measured in °C.

When a sound wave strikes a large, smooth, rigid surface, it will be reflected. The reflected wave is called an *echo.*

## SUPERPOSITION

When two wave pulses meet, their amplitudes add up in the region in which they overlap. The effect of two waves overlapping in space is called *interference.* If the waves are *in phase* and positive amplitude meets positive amplitude, the waves constructively interfere and their peaks add together during the moment they interact. If the waves are 180° out of phase with each other and positive amplitude meets negative amplitude, they destructively interfere. Constructive interference results in increased amplitude; destructive interference results in decreased amplitude. These types of interference are illustrated below. After the waves pass through each other and interfere, they continue unchanged along their original path.

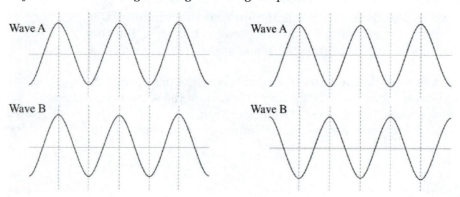

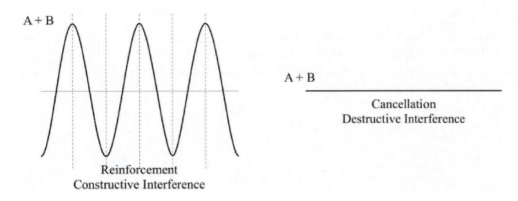

A + B

Reinforcement
Constructive Interference

A + B _____

Cancellation
Destructive Interference

## STANDING WAVES

Consider the waves set up by a vibrating string whose endpoints are fixed, as in the following diagram. The fixed end points represent *boundary conditions* that restrict the possible wavelengths that produce standing waves. In other words, the size of the region to which a wave is confined, not frequency of the source, determines the possible wavelengths in that region. Waves moving along the string and reflecting off the boundaries superimpose to produce a large amplitude *standing wave*, so called because it appears to be composed of pulses that don't move. Points on a standing wave where there is no displacement are called *nodes*. Destructive interference causes nodes. Halfway between two nodes are points called *antinodes*, at which displacement is a maximum. Antinodes are places of constructive interference.

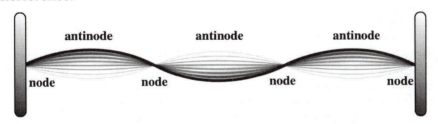

antinode        antinode        antinode

node        node        node        node

## VIBRATING STRINGS

A stretched, vibrating string can support many standing waves at more than one wavelength depending on the string's length. The frequencies at which standing waves are produced are called *resonance* or *natural frequencies*. The lowest frequency at which a standing wave is produced is called the *fundamental frequency*. It is also called the *first harmonic*. Other frequencies are called *overtones*.

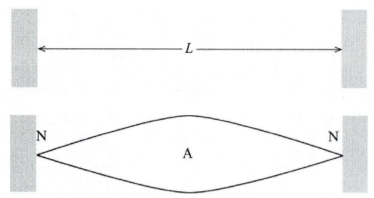

Fundamental mode or first harmonic.

Higher modes of oscillation occur for shorter wavelengths. From the following diagrams, it is noted that the allowable wavelengths are given by

$$\lambda_n = \frac{2L}{n} \qquad n = 1,\ 2,\ 3,\ \cdots$$

The frequency is determined from

$$f_n = n\frac{v}{2L} \qquad n = 1,\ 2,\ 3,\ \cdots$$

The entire series consisting of the fundamental and its overtones is known as a *harmonic series*.

First overtone or second harmonic.

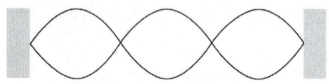

Second overtone or third harmonic.

Third overtone or fourth harmonic.

## SAMPLE PROBLEM 3

A 0.60 m long piece of steel piano wire has a mass of 10 g and is under a tension of 500 N. What is the frequency of its fundamental mode of vibration?

## SOLUTION TO PROBLEM 3

The frequency of the fundamental standing wave is $f_n = n\dfrac{v}{2L}$. However, we do not know the wave speed, $v$. Earlier, we saw that the wave speed in a wire under tension is $v = \sqrt{\dfrac{TL}{m}}$. Combining equations gives

$$f_1 = \frac{n}{2L}\sqrt{\frac{TL}{m}} = \frac{1}{2(0.60\text{ m})}\sqrt{\frac{(500\text{ N})(0.60\text{ m})}{(0.010\text{ kg})}} = \textbf{144 Hz}$$

## VIBRATING AIR COLUMNS

Standing waves can also be produced by the longitudinal vibrations of a column of air in an open or closed pipe. As in a vibrating string, the possible modes of vibration are determined by the boundary conditions. When a compression wave is set up in a closed pipe, the displacement of the air molecules at the closed end must be zero, a node.

### AP® Tip

The closed end of a pipe must be a displacement node.

Standing waves in air are longitudinal in character, and they are difficult to represent with any static drawing or diagram. For convenience only, it is customary to indicate the positions of nodes and antinodes as if they were transverse standing waves. Nodes and antinodes are labeled with N and A, respectively.

The various modes in which air columns may vibrate in open or closed pipes are shown in the following diagrams. With closed pipes, the lowest frequency is the fundamental, $f_1$, and the ones that follow are in odd multiples, $3f_1$, $5f_1$, $7f_1$, etc. No even-number harmonics can be sounded in a closed pipe. The wavelengths of possible waves depend on the length of the closed pipe and are given by $\lambda_n = \dfrac{4L}{n}$; therefore, their frequencies are $f_n = \dfrac{nv}{4L}$.

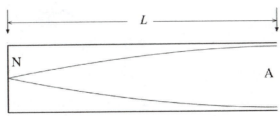

Fundamental mode or first harmonic.

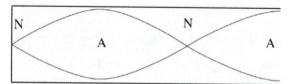

First overtone or third harmonic.

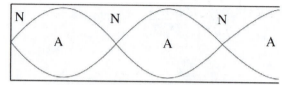

Second overtone or fifth harmonic.

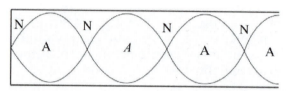

Third overtone or seventh harmonic.

## SAMPLE PROBLEM 4

Determine the frequencies of the fundamental and the first three overtones for a 14 cm long closed pipe on a day when the speed of sound is 342 m /s.

## SOLUTION TO PROBLEM 4

First, we need to find the fundamental frequency, and we do so by using $f_n = \dfrac{nv}{4L}$.

For the fundamental, $n = 1$ and $f_n = \dfrac{nv}{4L} = \dfrac{1(342 \text{ m/s})}{4(0.14 \text{ m})} = \textbf{611 Hz}$.

The first, second, and third overtones are the third, fifth, and seventh harmonics, respectively.

First overtone = $nf_1$ = 3(611 Hz) = **1833 Hz**

Second overtone = $nf_1$ = 5(611 Hz) = **3055 Hz**

Third overtone $nf_1$ = 7(611 Hz) = **4277 Hz**

## AP® Tip

Only the odd harmonics are allowed for a closed pipe.

A pipe open at both ends has different boundary conditions and different possible wavelengths than open only at one end. The air at the open end of a pipe has the greatest freedom of motion, so the displacement is a maximum at the open end. The wavelengths of possible waves depend on the length of the open pipe and are given by $\lambda_n = \dfrac{2L}{n}$, and the frequencies are $f_n = \dfrac{nv}{2L}$.

## AP® Tip

The open end of a pipe must be a displacement antinode.

With an open pipe, the lowest possible vibration frequency is called the fundamental frequency; the others, with whole-numbered multiples of the fundamental frequency, 2 $f_1$, 3 $f_1$, 4 $f_1$, etc., are possible harmonics.

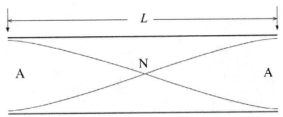

Fundamental mode or first harmonic.

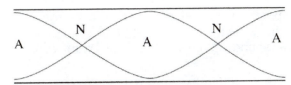

First overtone or second harmonic.

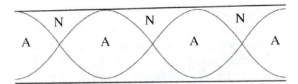

Second overtone or third harmonic.

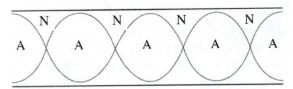

Third overtone or fourth harmonic.

### SAMPLE PROBLEM 5

What is the speed of sound on a day when a 30 cm long open pipe has a frequency of 1200 Hz as its first overtone?

### SOLUTION TO PROBLEM 5

The first overtone in an open pipe is the second harmonic, so $n = 2$. For the second harmonic $f_n = \dfrac{nv}{2L}$, and solving for $v$,

$v = Lf_2 = (0.30 \text{ m})(1200 \text{ Hz}) = \textbf{360 m/s}$.

## BEATS

When two sound waves with only slight differences in frequency interfere, they produce a sound that fluctuates in intensity, alternating between silence and a loud tone. The regular pulsations produced are called *beats*.

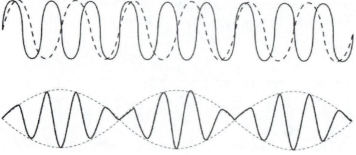

The superposition of the two sound waves, as shown in the diagram above, is the source of beats. The loud tones occur when the waves interfere constructively. The quiet tones are due to destructive interference. The number of beats, *N*, produced per second is determined by the difference in the frequencies. For example, when struck simultaneously, tuning forks of frequencies 340 Hz and 343 Hz emit sound that pulses three times per second.

## THE DOPPLER EFFECT

When a fire truck comes toward you, you hear a high-pitched sound. As the truck passes you and moves away from you, you hear the sound of the siren change to a lower pitch. This change in sound is due to the Doppler effect.

This phenomenon does not just happen if the thing making the sound is moving. If the source of sound is stationary, an observer moving toward the source will hear a similar rise in pitch. A moving observer leaving the sound source will hear a lower-pitched sound. The change in frequency due to relative motion between a source and an observer is called the *Doppler effect*.

The Doppler effect refers to the apparent change in frequency of a source when there is relative motion of the source and the observer. The diagram below shows how sound waves behave when a sound source and observer are stationary.

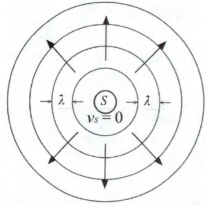

Representation of sound waves being emitted from a stationary sound source *S*.

Consider the sound source moving to the right toward a stationary observer at location A shown in the diagram below. As the source emits sound waves, it tends to overtake the bunched up waves moving in the same direction. Each successive sound wave is emitted at a point that is closer to the observer than its predecessor. The result is that the wavelength, $\lambda$, keeps shrinking as the sound waves bunch. A smaller wavelength means a higher frequency. Observer A hears a higher pitch compared to observer B who hears a lower pitch.

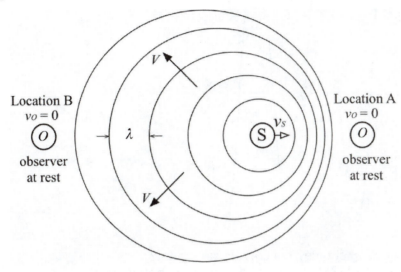

The waves show a Doppler shift. The waves in front of a moving sound source are closer together than the waves behind the source.

# WAVES AND SOUND: STUDENT OBJECTIVES FOR THE AP® EXAM

- You should be able to explain why the velocity of transverse waves differs from the velocity of longitudinal waves in a given medium.
- You should be able to explain why longitudinal waves can propagate in a gas but transverse waves cannot.
- You should be able to explain why waves of different frequencies may have different velocities in a given medium.
- You should be able to explain why increasing the tension in a string increases the velocity of the transverse waves.
- You should be able to discuss how interference could result from the overlapping of two waves with somewhat different amplitudes.
- You should be able to explain the phenomenon of beat frequency.
- You should be able to determine frequencies in open and closed pipes.
- You should be able to explain the Doppler effect.

## Multiple-Choice Questions

1.  Two transverse waves (1) and (2) travel with the same speed.

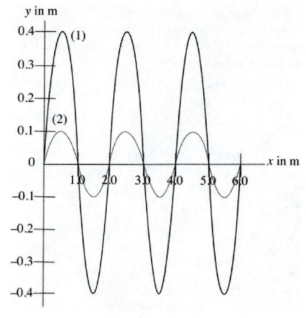

Which of the following choices is a correct statement?
(A) Both waves (1) and (2) have equal amplitudes; however, the wavelength of (1) is four times larger than the wavelength of (2).
(B) The waves have the same amplitude, but the wavelength of (1) is 0.4 m and the wavelength of (2) is 0.1 m.
(C) Both (1) and (2) have equal wavelength; however, the frequency of (1) is four times greater than the frequency of (2).
(D) Both (1) and (2) have equal wavelengths; however, the amplitude of wave (2) is one-fourth that of wave (1).

For questions 2 and 3, refer to the diagram below depicting the displacement of a vibrating string versus the position along the string at a particular instant. The wave on the string has a speed of 0.10 m/s.

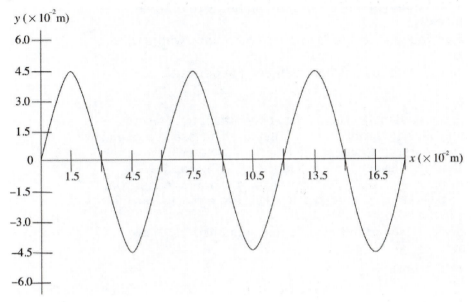

2.  What is the correct wavelength of the wave?
    (A) $3 \times 10^{-2}$ m
    (B) $6 \times 10^{-2}$ m
    (C) $9 \times 10^{-2}$ m
    (D) $12 \times 10^{-2}$ m

3.  If the frequency of the wave were doubled,
    (A) the wavelength would decrease to half its value.
    (B) the wavelength would also double.
    (C) the wavelength would be unchanged.
    (D) the wavelength would periodically shift between its current value and a new value.

4.  A transverse pulse is formed when a rope that is firmly attached at its ends between two barriers is pulled downward near one end and released. The pulse travels to the other end, and the reflected pulse's amplitude is
    (A) inverted and none of the energy of the pulse is transferred to the barrier.
    (B) inverted and some of the pulse's energy is transferred to the barrier.
    (C) not inverted and none of the energy of the pulse is transferred to the barrier.
    (D) not inverted and some of the energy of the pulse is transferred to the barrier.

5. Standing waves are produced in a string by the interference of two waves that have the same
   (A) frequency and amplitude moving in the same direction.
   (B) amplitude moving in the same direction but with different frequencies.
   (C) frequency and are moving in opposite directions with different amplitudes.
   (D) amplitude and frequency but are moving in opposite directions.

6. Several physics students performed an experiment to determine the speed of sound in the laboratory. They used a closed resonance tube that uses water to establish the length of the tube. By adjusting the water level in the resonance tube, they had two consecutive resonance points that were 18.0 cm and 54.0 cm from the open end when they sounded a 460.0 Hz tuning fork over the air column in the tube.
   The velocity of sound they should have obtained for their experiment was
   (A) 318 m/s.
   (B) 325 m/s.
   (C) 331 m/s.
   (D) 348 m/s.

7. A pulse in a rope approaches a fixed boundary as shown in the illustration below.

Which of the four choices shows the correct motion for the reflected wave?

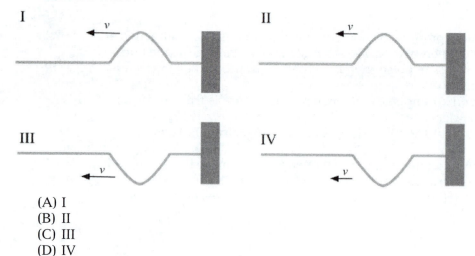

   (A) I
   (B) II
   (C) III
   (D) IV

8. Four waves are illustrated below. Rank the waves in order of energy from greatest to least.

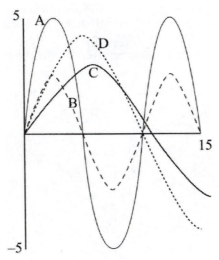

(A) B > C > D > A
(B) A > D > C > B
(C) D > C > A = B
(D) D > C > A > B

9. A 5.0 m cord is attached to a fixed support at one end, and the other end is under tension from a weight hanging over a pulley. A standing wave with 5 nodes is set up in the cord. If the velocity of the transverse wave is 15.0 m/s, what is the frequency of the wave?
(A) 3 Hz
(B) 6 Hz
(C) 9 Hz
(D) 12 Hz

10. If you increase the displacement amplitude for a given frequency of sound, the energy and the momentum of the wave carried by the cord will
(A) increase, causing the wave to travel faster in the medium.
(B) increase but the wave will slow as it moves through the medium.
(C) only increase the speed of the wave in the medium but not change the energy or the momentum carried by the wave.
(D) increase the momentum and energy carried by the wave, which will continue to travel at the same speed in the medium.

11. Sounds from a speaker radiate outward from the speaker in spherical waves. Two speakers producing sounds at the same frequency will produce spherical wave fronts that interfere with each other much like the interference pattern produced in water by two point sources oscillating with the same frequency. The diagram below shows such an interference pattern with three nodal lines identified.

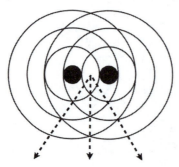

Two closely-spaced sources are emitting waves of the same wavelength. The dark circles represent crests. The dotted lines represent
(A) regions of constructive interference of the waves.
(B) regions of destructive interference of the waves.
(C) the amplitude of the wave.
(D) the frequency of the wave.

12. Students in a physics laboratory were experimenting with a resonance column over water and a tuning fork with a frequency of 520 Hz on a day when the speed of sound in the room was 343 m/s. The students move the column up and down until they find the shortest length at which they hear the tuning fork's note resonate loudly. What is the length of the column of air that resonates that note?
(A) 16.5 cm
(B) 33.0 cm
(C) 66.0 cm
(D) 82.5 cm

13. You are stopped in your car at an intersection when a fire truck, sounding its siren, passes you. The wave front produced by the moving fire truck that is passing you is illustrated in the diagram below.

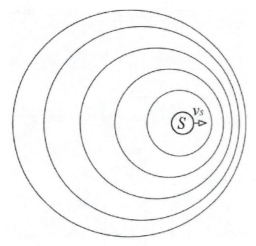

The frequency of the siren will
(A) decrease as the fire truck approaches and increase as it moves past you.
(B) increase as the fire truck approaches you and decrease as it moves past you.
(C) remain constant.
(D) produce a resonance pattern as it passes you.

14. Two sound waveforms are illustrated below.

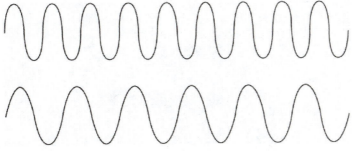

The combined waveform produces a beat pattern as shown below.

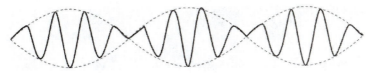

How many beats per second will an observer hear?
(A) 3
(B) 5
(C) 8
(D) 12

15. Two rectangular waves A and B approach each other as shown in the illustration shown below.

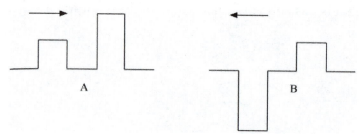

Which of the resulting waveforms best represents the interference of the two?

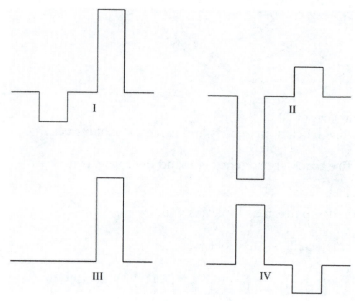

(A) I
(B) II
(C) III
(D) IV

# FREE-RESPONSE PROBLEMS

1.  (a) Using the waveforms shown in Figure (1), draw the superimposed waveforms as they pass through each other on the grid given below.

Figure (1)

(b) Using the waveforms shown in Figure (2), draw the superimposed waveforms as they pass through each other on the grid given below.

Figure (2)

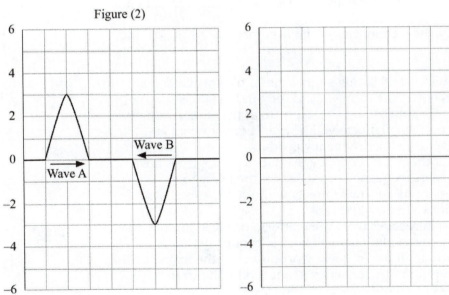

(c) Using the waveforms shown in Figure (3), draw the superimposed waveforms as they pass through each other on the grid given below.

Figure (3)

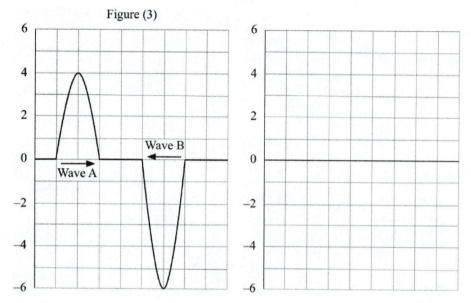

2. (a) On a day when the speed of sound is 343.0 m/s, what is the fundamental frequency in a pipe that is 2.50 m long when one end of the pipe is closed?
   (b) What are the frequencies of the first three overtones of this pipe?
   (c) If both ends of the pipe are opened, what is the fundamental frequency?
   (d) What are the frequencies of the first three overtones of the open pipe?

3. A train traveling at 50 km/hr approaches a platform at a station, sounding its horn. The frequency of the horn is 900 Hz.
   (a) As the train approaches the platform at the station, a person standing on the platform will hear a frequency that is
   ___ higher than the frequency produced by the train horn.
   ___ lower than the frequency produced by the train horn.
   ___ the same frequency produced by the train horn.
   Justify your answer qualitatively without equations or calculations.
   (b) As the train recedes from the platform of the station, a person standing on the platform will hear a frequency that is
   ___ higher than the frequency produced by the train horn.
   ___ lower than the frequency produced by the train horn.
   ___ the same frequency produced by the train horn.
   Justify your answer qualitatively without equations or calculations.
   (c) A passenger on the train will hear a frequency that is
   ___ higher than the frequency produced by the train horn.
   ___ lower than the frequency produced by the train horn.
   ___ the same frequency produced by the train horn.

(d) Sketch a wave front diagram that illustrates how the apparent wavelength is related to the relative motion of the source and the stationary observer. Label the wavelengths correctly for approach and recession.

4. Design a laboratory experiment to determine the relationship between the wavelength, frequency, and velocity of a standing wave in a string. You have available a string, a pulley and clamps, a mass hanger, masses, a meterstick, and an electrical string vibrator of constant frequency (120 Hz) as well as a platform balance.
   (a) Describe your experimental procedure in enough detail so that another student could perform your experiment and obtain results to determine the relationship between the wavelength and frequency. Include any measurement you will take and how you will use these measurements.
   (b) In enough detail that another student could duplicate your method of determining the frequency, describe how you will use the measurements to determine the frequency of the electrical vibrator used to produce the standing wave.
   (c) What assumption did you make about the design of the experiment that might affect your results in determining the frequency? How might this assumption alter your results?
   (d) If you graph your results, what would you graph? What would be the shape of the line?

# Answers

## MULTIPLE-CHOICE QUESTIONS

1. **D** Amplitude is the maximum displacement from the equilibrium position. From the scale, wave 1 has amplitude of about 0.4 m and wave 2 has amplitude of about 0.1 m. Their wavelengths are the same, and since they travel with the same speed, their frequencies are the same.

   (L.O. 6.A.3.1)

2. **B** The wavelength is the distance between two points that have the same amplitude and are in phase with each other (moving in the same direction). From the graph, this is $6 \times 10^{-2}$ m.

   (L.O. 6.B.2.1)

3. **D** Frequency is determined using $v = \lambda f$. For the same medium, doubling the driving frequency creates waves that are half as wide.

   (L.O. 6.B.1.1)

4.  **B** When the incident pulse strikes a fixed barrier, the wave is inverted (180°). Energy is always transferred to the medium of the barrier.

    (L.O. 6.B.1.1)

5.  **D** Standing waves are produced by waves that have the same frequency and amplitude but move in opposite directions, producing points where the displacement is always zero (nodes) and antinodes where the displacement is a maximum.

    (L.O. 6.D.3.4)

6.  **C** The open end of the tube is an antinode, and the closed end is a node. The fundamental wavelength is $\lambda = 4L$. If the first resonance is heard at 18.0 cm, the wavelength is 0.72 m.

    The speed of sound is then $v = \lambda f = (0.72 \text{ m})(460.0 \text{ Hz}) = 331 \text{ m}/\text{s}$

    The second resonance pattern occurred at $L = \dfrac{3}{4}\lambda$. The wavelength is

    $\lambda = \dfrac{4}{3}(0.54 \text{ m}) = 0.72 \text{ m}$ and produced the same velocity.

    (L.O. 6.D.3.1)

7.  **C** Since the pulse strikes a fixed barrier, it is reflected 180° out of phase. It returns with the same speed since speed is determined by the medium in which the pulse travels. When a pulse transfers energy, its amplitude decreases.

    (L.O. 6.D.1.1)

8.  **B** The energy carried by a wave is related to its amplitude. In the illustration, the amplitude of wave A is ± 5 units on the vertical axis, followed by D, C, and B.

    (L.O. 6.A.4.1)

9.  **B** Since the cord is under tension and supports a standing wave that has five nodes, the wavelength is 2.5 m (the ends of the cords must be nodes). Since the velocity of the wave is 15.0 m/s, the frequency is found from $v = \lambda f$ and $f = \dfrac{15.0 \text{ m}/\text{s}}{2.5 \text{ m}} = 6 \text{ Hz}$.

    (L.O. 6.D.4.2)

10. **D** Increasing the amplitude of vibration of a sound wave for a given frequency will increase the energy and the momentum of the particle that is displaced from the equilibrium. As the wave moves

out from the disturbance, the energy is carried with it, but not the medium. The speed of the wave will not change since the speed is determined by the medium.

(L.O. 6.A.2.1)

11. **A** The lines indicate regions where crests meet and maximum displacement from the equilibrium occurs.

(L.O. 6.D.2.1)

12. **D** The tube is closed at one end; therefore, only the odd harmonics are heard. The fundamental wavelength will be used to determine the length of the tube $520 \text{ Hz} = \dfrac{343 \text{ m/s}}{4L}$, and $L = 0.165$ m and then $\lambda = 4\,L = 0.66$ m.

The fifth harmonic corresponds to the third resonance position. The length of the air column for this resonance is $L = \dfrac{5}{4}\lambda$. This length is 0.825 m = 82.5 cm.

(L.O. 6.D.4.2)

13. **B** The Doppler effect relates the apparent frequency to the actual frequency due to relative motion between the source and the observer. When the source moves toward a stationary observer, the wave fronts crowd together. As a result, the stationary observer measures a smaller wavelength and thus a higher frequency. As the source moves away from the stationary observer, the observer measures a longer wavelength and therefore a lower frequency.

(L.O. 6.B.1.1)

14. **A** Two waveforms played at the same time will interfere with each other. Superposition tells us that variations in amplitudes will occur as the resulting wave oscillates in time. This variation produces beats. The beat frequency is the difference in the frequencies of the waveforms. $f_{beats} = |f_2 - f_1|$

(L.O. 6.D.5.1)

15. **A** The superposition principle applies to the combined waveform of A + B. Amplitudes must be added. When the leading edge of B combines with A, the amplitudes are +1 for A and –2 for B as estimated from the diagram. The amplitude of the combination is –1. The trailing edge's addition is +2 from A and +1 from B for a total of +3.

(L.O. 6.D.1.1, 6.D.2.1)

## FREE-RESPONSE PROBLEMS

1.  (a)  The two waves will superimpose with constructive interference. The combined amplitude when they are vertically aligned will be $y = +4$ units.

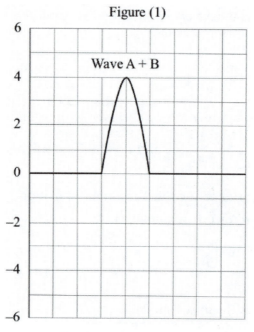

Figure (1)

(b)  The two waves will superimpose with totally destructive interference. The combined amplitude when they are vertically aligned will be $y = 0$ units.

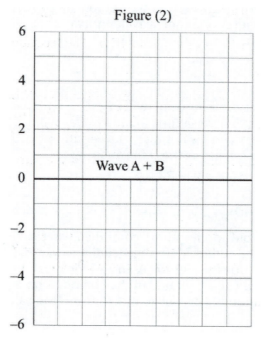

Figure (2)

(c) The two waves will superimpose for partial interference. The combined amplitude when they are vertically aligned will be $y = -2$ units.

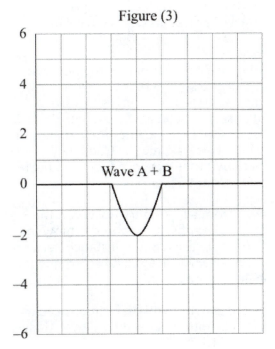

Figure (3)

(L.O. 6.A.3.1, 6.B.2.1, 6.D.1.1, 6.D.2.1)

2.  (a) If the pipe is closed at one end, the fundamental wavelength is $\lambda = 4L$.

    The wavelength is $\lambda = 4(2.50 \text{ m}) = 10.0 \text{ m}$. Since $v = \lambda f$, the fundamental frequency is $f = \dfrac{343.0 \text{ m/s}}{10.0 \text{ m}} = \textbf{34 Hz}$. The frequency can be calculated in one step from $f_n = n\dfrac{v}{4L}$.

    (b) If the pipe is closed at one end, only the odd harmonics will be heard. The frequencies of the first three overtones are $f_3 = 3f_1$, $f_5 = 5f_1$, and $f_7 = 7f_1$. The frequencies are $f_3 = \textbf{102 Hz}$, $f_5 = \textbf{170 Hz}$, and $f_7 = \textbf{238 Hz}$.

    (c) If the pipe is open at both ends, the fundamental wavelength is $\lambda = 2L$. The wavelength $\lambda = 2(2.50 \text{ m}) = 5.00 \text{ m}$, and the fundamental frequency is $f = \dfrac{343.0 \text{ m/s}}{5.00 \text{ m}} = \textbf{68.6 Hz}$

    (d) The pipe closed at both ends produces all of the harmonics. The frequencies for the first three overtones are $f_2 = 2f_1$, $f_3 = 3f_1$, and $f_4 = 4f_1$. The frequencies are $f_2 = \textbf{137 Hz}$, $f_3 = \textbf{206 Hz}$, and $f_4 = 273.6 \text{ Hz} = \textbf{274 Hz}$.

(L.O. 6.D.3.2, 6.D.4.2)

3. The Doppler effect relates the higher frequency heard by a stationary observer as a source approaches and the lower frequency heard when the source moves away from the observer. This relationship holds for a stationary source and a moving observer as well as both observer and source moving relative to each other.

(a) higher than the frequency

If both the source and observer are stationary, the observer hears the same frequency as the source because there is no change in the wavelength. As the source moves toward a stationary observer, the observer detects wave fronts that are closer together because the source is moving in the direction of the prior fronts. The observer hears a higher frequency because the wave fronts crowd together, producing a shorter wavelength than the sound from the source.

(b) lower than the frequency

As the source moves away from a stationary observer, the observer detects longer wavelengths. The source is moving away from prior wave fronts, increasing the distance between the fronts of the wave. The observer hears a lower frequency because the wave fronts have spread apart.

(c) the same frequency

The passenger in the train is moving with the same speed as the train. There is no relative motion between the two. Hence, the passenger hears a 900 Hz frequency.

(d) The diagram is shown below.

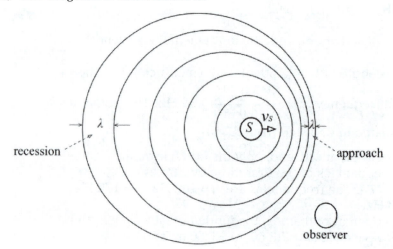

(L.O. 6.B.5.1)

4. A standing wave set up in the string is dependent on the wave speed and the wavelength. Knowing these two values, the frequency of the waveform can be obtained by substitution into $v = \lambda f$.

(a) 1. Measure the length of the entire string that will be used for the experiment. Determine the mass of the entire string and record both values.

2. Determine the linear mass density of the string. The linear mass density is its mass per unit length. Record the value.

3. Attach one end of the string to the string vibrator and pass the other end over a pulley you have clamped 1.00 m from the string vibrator. Attach a mass hanger to the string and add masses until you have the fundamental form of vibration for the standing wave in the string. Record the mass of the mass hanger and the masses on the hanger.

4. Measure the distance between the nodes. Since this is the fundamental wavelength, there are two nodes, one at each end of the string that is under tension. Record this value.

5. The fundamental wavelength is twice the length of the string between the nodes. Record this value.

6. Repeat for the experiment by changing the tension on the string, setting up standing waves with 3, 4, and 5 nodes. This requires removing mass from the hanger.

(b) 1. The fundamental wavelength is twice the length of the string between the nodes.

2. The velocity of the wave is determined by taking the square root of the tension in the string divided by the linear mass density of the string. (The tension on the string is the weight of the mass hanger and the masses on the hanger.)

3. Determine the frequency by dividing the velocity of the wave in the string by the wavelength.

(c) If the distance for the fundamental wavelength was measured incorrectly, the wavelength is wrong. The string must be measured from the point where it is attached to the vibrator to the point where the string passes over the pulley (at its axle). If it is measured to the end of the pulley, the length is too long. Then the wavelength is too large and the frequency will be smaller than it should be.

If the amplitude of vibration of the standing wave was not as large as possible, then the tension on the string was not accurate and the velocity of the wave is not accurate.

If the string stretched during the experiment, the linear mass density would decrease and the calculated velocity would increase. Thus, the experimental frequency would decrease.

(d) A graph of wavelength as a function of the square root of the tension in the string should be plotted. The graph should be linear.

(L.O. 6.D.1.2, 6.D.3.1 6.D.3.3, 6.D.4.2)

# 16

# GEOMETRICAL OPTICS

PHYSICS 2

## REFLECTION OF LIGHT

When a light ray shines on a reflective material, the angle of the incoming ray, the *angle of incidence*, $\theta_i$, equals the angle of the outgoing ray, the *angle of reflection*, $\theta_r$. Known as the law of reflection, this behavior demonstrates that one-way light acts as a particle. Incident light bounces off a surface like a ball in a collision. The convention for measuring the angles is to use a line that is perpendicular to the surface boundary that is called a *normal*, N.

The light reflected at the surface obeys the *laws of reflection*:
1. The angle of incidence equals the angle of reflection.
2. The incident ray, the reflected ray, and the normal to the surface are all in the same plane.

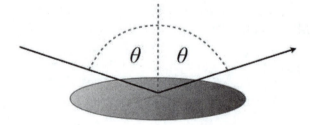

For visible light, the wavelength of light determines its color. If light of a single wavelength can be isolated, it is said to be *monochromatic*.

Light in a medium exhibits the following behaviors:
1. The velocity of light in free space is the same for all colors.
2. The velocity of light in any medium is always less than its velocity in a vacuum.

3.    The velocity of blue light in a given medium is less than the velocity of red light in that medium.

---

## AP® Tip

When working with reflection, angles are measured relative to the normal, not the surface.

---

## PLANE MIRRORS

When you look at yourself in a plane mirror, your image appears to be directly in front of you and on the other side of the mirror. Everything about your image is the same as you except for a left-to-right reversal. No light originates from the other side of the plane mirror; for this reason, your image is called *virtual*. A *virtual image* seems to be formed by light coming from the image, but no rays of light actually pass through it, whereas a *real image* is formed by actual rays of light that pass through it. Real images can be projected on a screen. Virtual images cannot be projected on a screen. A virtual image is your brain's interpretation of what might have caused light to strike your eyes based on light traveling only in straight lines.

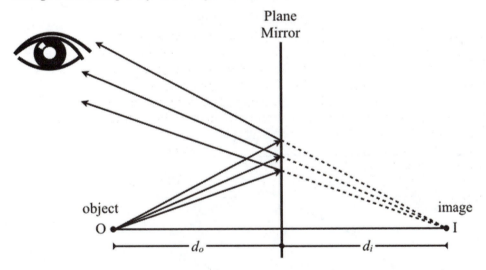

Light rays leave the object and are reflected by the mirror.
The eye perceives the light rays as
having traveled in straight lines from the image.

### SUMMARY OF THE PROPERTIES OF IMAGES IN PLANE MIRRORS
1.    The images are upright and virtual.
2.    The height of the image, $h_i$, is equal to the object's height $h_o$.
3.    The image distance, $s_i$, is equal to the object distance, $s_o$.

## SPHERICAL MIRRORS

Spherical mirrors are sections of a sphere. If the inside surface, the concave surface, is polished, the mirror is called a *concave mirror*. If the outside surface, the convex surface, is polished, the mirror is called a *convex mirror*. The *center of curvature*, C, is the center of the sphere of radius, *R*, of which the mirror is a section. The spherical mirrors we commonly see are so small compared to their radii that they appear nearly flat. The straight line connecting the center of curvature and the midpoint of the mirror is called the *principal axis*.

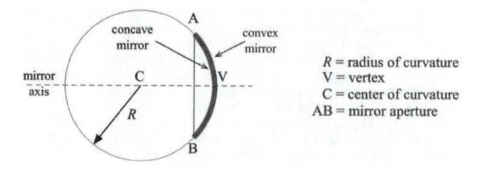

R = radius of curvature
V = vertex
C = center of curvature
AB = mirror aperture

Definition of Terms in Spherical Mirrors

The distance of the focal point, F, from the surface of the mirror is called the *focal length, f.* For a concave mirror, the focal length is positive and is

$$f = \frac{R}{2}$$

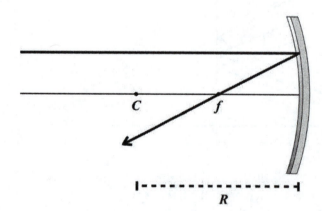

## CONCAVE MIRRORS

The reflections from a concave mirror can form real or virtual images depending on the relative position of the object and the mirror's focal point. To calculate where our eyes will see an image in a mirror, we use a set of sign conventions.

## SIGN CONVENTIONS FOR MIRRORS

1. Draw or sketch diagrams with light passing from left to right, placing the object to the left of the mirror.
2. The object distance, $s_o$, is positive.
3. The image distance, $s_i$, is positive when the image lies to the left (in front) of the mirror and negative when it lies to the right of (or behind) the mirror.
4. The object size, $h_o$, is positive.
5. The image size, $h_i$, is positive if the image is upright and negative if it is inverted.
6. Positive magnification means that the image is upright; negative magnification means that the image is inverted.
7. Magnification between 1 and –1 means that the image is smaller than the object, or *reduced*. Magnification greater than 1 or less than –1 means that the image is larger than the object, or *enlarged*.

## RAY TRACING FOR SPHERICAL MIRRORS

Ray tracing is a method for drawing and understanding where our eyes perceive images when looking at spherical mirrors. A ray diagram traces the path of several of the infinite light rays that reflect off an object and shows us how our eyes will interpret those rays. We use an arrow to represent the object and a vertical line to represent the mirror. If you prefer to draw the mirror on your diagram, include the vertical line too. The point where the reflected rays intersect determines the location of the image. Starting at the head of the object, we work with three rays whose paths are easily traced.

- *Concave Mirror Ray 1:* A ray that comes from the object and travels parallel to the principal axis reflects through the focal point of the concave mirror.

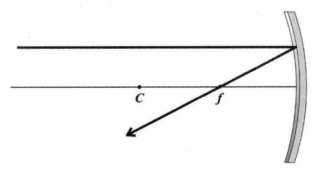

- *Concave Mirror Ray 2:* A ray that comes from the object and passes through the focal point of a concave mirror is reflected parallel to the principal axis.

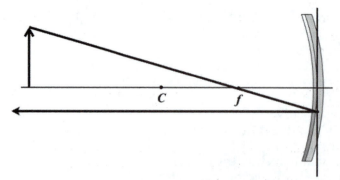

- *Concave Mirror Ray 3:* A ray that proceeds along the radius of the mirror is reflected back along its original path.

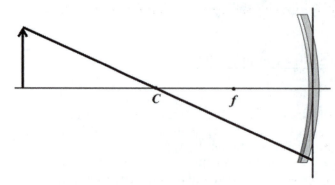

Tracing back the path of the reflected rays shows us the location and orientation of the image.

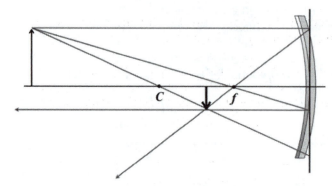

If we looked at the reflection of the object in this mirror, our eyes would see that its real image appears to be inverted and reduced.

In any particular analysis or problem, only two of the rays are necessary to locate the image of the object. In the diagrams, virtual rays and virtual images are represented by dotted lines.

Consider the following cases where images are formed by concave (converging) mirrors.

*Case 1* illustrates the image that is formed by an object, O, that is located beyond the center of curvature, C, of the mirror. The image is *real, inverted,* and *reduced* in size.

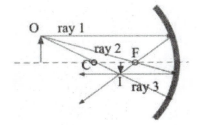

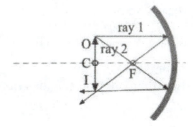

*Case 1:* Object is beyond C, the image is real, inverted, diminished in size, and is between F and C.

*Case 2:* Object is at C, the image is real, inverted, the same size, and is at C.

*Case 2* shows that the object is located at the center of curvature. The concave mirror forms an image at the center of curvature that is *real, inverted,* and the *same size* as the object.

*Case 3* positions the object between the center of curvature and the focal point. Ray tracing shows that the image is located beyond the center of curvature. The image is *real, inverted,* and *larger* than the object.

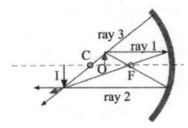

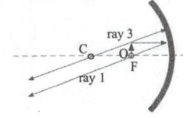

*Case 3:* Object is between C and F, the image is real, inverted, enlarged, and is beyond C.

*Case 4:* Object is at F, no image appears.

*Case 4* places the object at the focal point, F. All reflected rays are parallel and never intersect. No image is formed, or the image distance is infinite.

*Case 5* locates the object inside the focal point, F. If your eyes were to see the three rays reflecting off the mirror, the image would appear to be located behind the mirror. The image is *virtual, enlarged,* and *upright.*

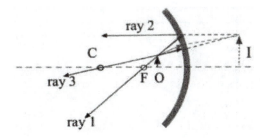

*Case 5:* Object is between F and mirror surface, the image is virtual, upright, and is enlarged.

## CONVEX MIRRORS

For a convex mirror, the focal length is negative. The negative sign means that the focal point, F, lies behind the mirror.

$$f = -\frac{R}{2}$$

*Convex Mirror Ray 1:* A ray that comes from the object and travels parallel to the principal axis that reflects through it came from the focal point of a convex mirror.

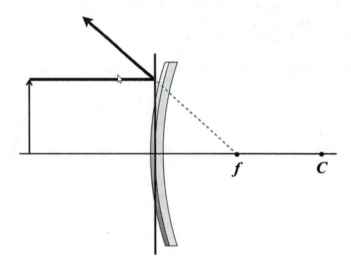

*Convex Mirror Ray 2:* A ray that is directed toward the focal point of a convex mirror gets reflected parallel to the principal axis.

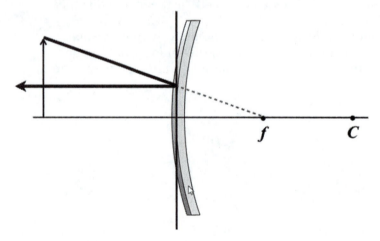

*Convex Mirror Ray 3:* A ray that is directed toward the center is reflected back along its original path.

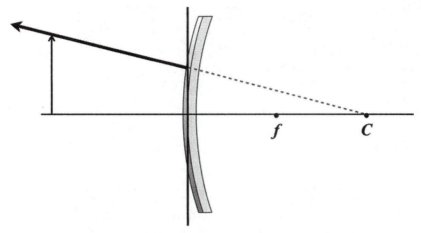

Tracing back the path of the three reflected rays shows us the location and orientation of the image.

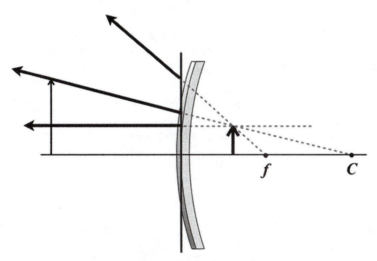

If we looked at the reflection of the object in this mirror,
our eyes would see that its virtual image appears to be upright and reduced.

The following are a few things to remember about images and curved mirrors:

- An image created by a concave mirror is real if $s_o > f$ and virtual if $s_o < f$. The focal length of a concave mirror is positive.
- An image created by a convex mirror is always upright, virtual, and reduced. The focal length of a convex mirror is negative.
- Virtual images are made of diverging light rays.
- Virtual images appear to come from behind a mirror, so they have negative image distance $s_i$.
- Virtual images are always upright.
- Real images are made of converging light rays and have a positive image distance $s_i$.
- Real images are always inverted.
- Upright images are always virtual; $h_i$ is positive for an upright image.
- Inverted images are always real; $h_i$ is negative for an inverted image.

## AP® Tip

An image from a convex mirror cannot be projected on a screen. Virtual images cannot be projected because they are not formed by actual/real light rays.

## THE MIRROR EQUATION

Ray diagrams are useful for getting a feel for the formation of images. However, a set of analytical methods exist that will allow us to calculate the features of these images. The first of these is the *mirror equation*.

$$\frac{1}{s_o} + \frac{1}{s_i} = \frac{1}{f}$$

Images formed by spherical mirrors may be equal in size, larger, or smaller compared to the object itself. We define *magnification, M,* as the ratio of the image size, $h_i$, to the object size, $h_o$, or

$$M = \frac{h_i}{h_o}$$

Another useful relationship is using object distance $s_o$ and image distance $s_i$.

$$M = -\frac{s_i}{s_o}$$

## AP® Tip

Inverted images have negative magnification.
Upright images have positive magnification.

### SAMPLE PROBLEM 1

A 6 cm tall golf tee is 60 cm from a concave mirror whose focal length is 20 cm. Calculate the position and size of the image.

Is the image upright or inverted?

Is the image real or virtual?

### SOLUTION TO PROBLEM 1

First, find the image distance, $s_i$.

Write the mirror equation $\frac{1}{s_i} + \frac{1}{s_o} = \frac{1}{f}$ and solve for $s_i$.

$$s_i = \left(\frac{1}{f} - \frac{1}{s_o}\right)^{-1} = \left(\frac{1}{20 \text{ cm}} - \frac{1}{60 \text{ cm}}\right)^{-1} = \textbf{30 cm}$$

Since the image distance, $s_i$, is positive, the image is **real**.

To find the magnification, write $M = -\dfrac{s_i}{s_o}$

Solving for the image's height, $h_i = -\dfrac{s_i h}{s_o} = -\dfrac{(30\text{ cm})(6\text{ cm})}{60\text{ cm}} = -3\text{ cm}$

The negative sign implies that the image is **inverted**. The magnification is **−0.5**.

## SAMPLE PROBLEM 2

If a convex spherical mirror has a radius of curvature of 40 cm, where must an AA battery be positioned to form an image one-half the length of the battery?

## SOLUTION TO PROBLEM 2

Find the focal length of the mirror: $f = \dfrac{R}{2} = \dfrac{-40\text{ cm}}{2} = \textbf{−20 cm}$

The negative sign occurs because of the convex (diverging) mirror. Such a mirror always forms an upright image, reduced in size, and the magnification in this case is $+\dfrac{1}{2}$.

Then $M = -\dfrac{s_i}{s_o} = +\dfrac{1}{2}$ and $s_i = -\dfrac{s_o}{2}$

Rearranging the mirror equation, $s_i = \dfrac{s_o f}{s_o - f}$. Combining equations,

$$\dfrac{s_o f}{s_o - f} = -\dfrac{s_o}{2}$$

Dividing by $s_o$ gives $\dfrac{f}{s_o - f} = -\dfrac{1}{2}$ and $2f = -s_o + f$.

$s_o = -f = -(-20\text{ cm}) = 20\text{ cm}$

When an object is held at a distance equal to the focal length from a convex mirror, the image size is one-half the object size.

# THE REFRACTION OF LIGHT

When light strikes the boundary between two media such as air and water shown in the diagram on the next page, some of the incident light on the glass surface is reflected and some of it will pass into the glass. The light entering the glass will be partially transmitted and partially absorbed. If the light enters the glass at an angle, the transmitted light undergoes a change in direction, called *refraction*.

The velocity of light in a vacuum is $3.0 \times 10^8$ m/s. However, when light passes through transparent materials, its velocity is reduced because of the electrical characteristics of the atoms and molecules and the chemical bonds connecting them. Recall that light is an electromagnetic wave. Since various media have different bonding properties, the velocity of light varies from one medium to another. The variation also depends to some extent on the wavelength of the light.

The ratio of the velocity of light, *c*, in a vacuum to its velocity, *v*, in a given medium is defined as the *index of refraction*, *n*, of that medium.

Or

$$n = \frac{c}{v}$$

The index of refraction of plate glass is 1.52. This means that light travels 1.52 times as fast in vacuum as it does in plate glass.

## AP® Tip

The index of refraction of air is 1.0003, indicating that for our purposes, we can consider the velocity of light in air to be the same as in a vacuum.

### SAMPLE PROBLEM 3

What is the velocity of yellow light in water if the index of refraction of water for yellow light is 1.33?

### SOLUTION TO PROBLEM 3

Write $n = \dfrac{c}{v}$ and solve for the velocity in water.

$$v = \frac{c}{n} = \frac{3.0 \times 10^8 \text{ m}/\text{s}}{1.33} = 2.25 \times 10^8 \text{ m}/\text{s}$$

In a given uniform medium, light travels in straight lines at a constant speed. When the medium changes, the speed of the light also changes and the light travels in a straight line along a new path. The bending of a light ray as it passes obliquely from one medium to another medium is known as *refraction*. The principle of refraction is illustrated in the following figure for a light ray entering water from the air. The angle of incidence $\theta_i$ the ray makes with the normal, $N$, to the surface is called the *angle of incidence*. The angle $\theta_r$ between the refracted ray and the normal is referred to as the angle of refraction. In passing into the denser medium, the angle between the incident ray and the normal is decreased. The limiting value for the angle between a ray and a normal is 90°.

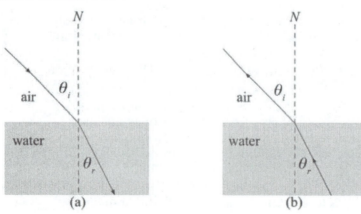

(a)                    (b)

The Laws of Refraction: (a) The incident ray, the refracted ray, and the normal to the surface all lie in the same plan. (b) The path of a ray refracted at the surface between two media is exactly reversible.

The relationship between the angles of incidence, refraction, and the velocities in the two media in question can be stated in the form of *Snell's law*, which states that the ratio of the sine of the angle of incidence to the sine of the angle of refraction is equal to the ratio of the velocities of light in the incident medium to that of the refracted medium, or

$$\frac{\sin\theta_1}{\sin\theta_2} = \frac{v_1}{v_2}$$

From above, $v_1 = \dfrac{c}{n_1}$ and $v_2 = \dfrac{c}{n_2}$; then substituting gives

$$n_1\sin\theta_1 = n_2\sin\theta_2$$

Regardless of medium, the frequency of a light ray is constant. Frequency is a property of the light itself. The wavelength, $\lambda$, of a wave, however, depends on the medium through which the wave travels.

$$n_1\lambda_1 = n_2\lambda_2$$

## AP® Tip

Since the sine of an angle increases as the angle increases, an increase in the index of refraction results in a decrease in the angle and vice versa.

## SAMPLE PROBLEM 4

A light ray passes from air into an organic liquid whose index of refraction is 1.39. Find the speed of light in the liquid and the angle of refraction if the light ray enters the liquid with an angle of incidence of 55°.

## SOLUTION TO PROBLEM 4

$$v = \frac{c}{n} = \frac{3.0\times10^8 \text{ m/s}}{1.39} = 2.16\times10^8 \text{ m/s}$$

The angle of refraction is found from Snell's law: $n_1\sin\theta_1 = n_2\sin\theta_2$

$$\sin\theta_2 = \frac{n_1\sin\theta_1}{n_2} = \frac{(1)(\sin 55°)}{1.39} \text{ and } \theta_2 = \sin^{-1}(0.5893) = 36.1°$$

## TOTAL INTERNAL REFLECTION

*Total internal reflection* occurs when light encounters a boundary between a medium with a higher index of refraction, $n_1$, and one with a lower index of refraction, $n_2$. On the next page, a light ray travels in water and meets the boundary between the water and air, where $n_w$ is greater than $n_a$. Refracted rays bend away from the normal because $n_w$ is greater than $n_a$. At some particular angle of incidence, $\theta_c$, called the *critical angle,* the refracted light ray moves parallel to the boundary so that $\theta_a = 90°$. For angles greater than $\theta_c$, the light ray is entirely reflected at the boundary, as is light ray 3 in the diagram below.

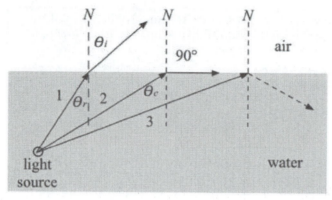

Critical angle of incidence

Total internal reflection is possible when $n_1 > n_2$; the critical angle is calculated using

$$\sin \theta_c = \frac{n_2}{n_1}$$

## AP® Tip

Total internal reflection occurs only when light is incident on the boundary of a medium having a lower index of refraction than the medium in which it is traveling.

### SAMPLE PROBLEM 5

Determine the critical angle for light passing from glass, $n_1 = 1.54$, to water, $n_2 = 1.33$.

### SOLUTION TO PROBLEM 5

By definition and substituting, $\sin \theta_c = \frac{n_2}{n_1} = \frac{1.33}{1.54} = 0.863$.

Solving for $\theta_c$: $\theta_c = \sin^{-1}(0.863) = $ **59.7°**

## CONVEX LENSES

A convex lens or *converging lens* refracts and converges parallel light rays to a focal point beyond the lens.

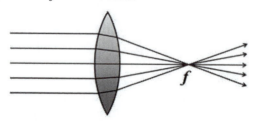

## SUGGESTED SIGN CONVENTIONS FOR LENSES

1. Always make diagrams with light passing from left to right. Always place the object to the left of the surface being studied.
2. Object and image distances are taken as positive for real objects and images and negative for virtual objects and images.
3. The image distance, $s_i$, is positive when the image lies to the right of the lens and negative when it lies to the left of the lens.
4. Consider the focal length, $f$, to be positive for a convex (converging) lens and negative for a concave (diverging) lens.
5. Indicate the height of the object by $h_o$ and consider it positive when it is upright. Let the image size be represented by $h_i$, which also is positive if it is upright and negative if it is inverted.
6. Note that the magnification $M = 1$ means that the image has the same size as the object. $M > 1$ means that the image is enlarged. Reduced, $M < 1$ does not mean negative magnification, but fractional magnification. This means the image is smaller than the object. Negative magnification merely means that the image is inverted with respect to the object.

## RAY TRACING FOR LENSES

Converging lenses have two focal points. We define the first focal point, $F_1$, as the one that is located on the same side of the lens as the source of incident light. The second focal point, $F_2$, is located on the opposite side or the far side of the lens. There are three principal rays that we use to trace through a lens.

*Ray Tracing:* As with mirrors, we start at the head of the object and work with three rays whose paths are easily traced.

- *Ray 1:* A ray parallel to the principal axis of the lens passes through the second focal point, $F_2$, of the converging lens or seems to come from the first focal point, $F_1$, of a diverging lens.
- *Ray 2:* A ray that passes through the first focal point, $F_1$, of a converging lens or proceeds toward the second focal point, $F_2$, of a diverging lens is refracted parallel to the principal axis of the lens.
- *Ray 3:* A ray that proceeds through the geometric center of a lens will not be deviated.

*Case 1:* The object, O, is located beyond twice the focal point, $2F_1$. The image, I, forms on the other side of the lens between $F_2$ and $2F_2$. It is *real, inverted,* and *diminished* in size.

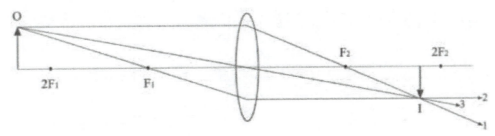

*Case 2:* The object, O, is located at twice the focal point, $2F_1$. The image, I, forms on the other side of the lens, and it is *real, inverted,* and the *same size* as O.

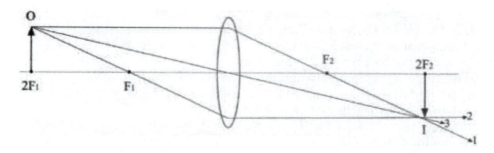

*Case 3:* The object, O, is located between $2F_1$ and $F_1$. The image, I, forms beyond $2F_2$ on the other side of the lens, and it is *real, inverted,* and *larger* than the object.

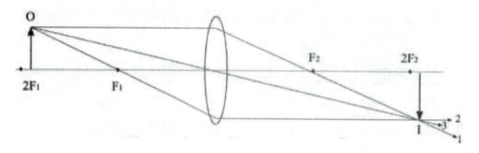

*Case 4:* The object, O, is located at the focal point, $F_1$. All reflected rays are parallel and never intersect. No image is ever formed.

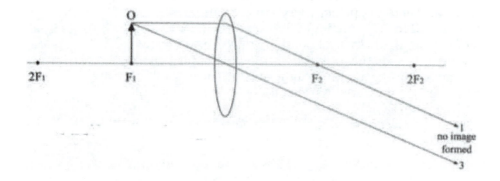

*Case 5:* The object, O, is inside the focal point, F₁. The image, I, appears to be on the same side of the lens as O, and is *virtual, enlarged,* and *upright.*

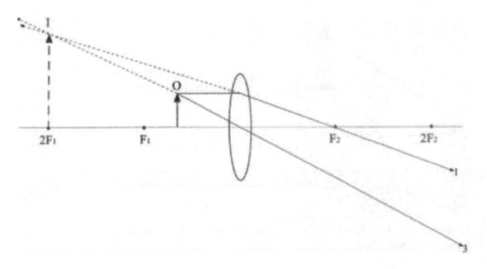

Images formed by *convex (converging) lenses* are similar to the images formed by *concave mirrors.* Both converge light rays.

## CONCAVE (DIVERGING) LENSES

A *diverging lens* refracts and diverges parallel light rays from a point located in front of the lens.

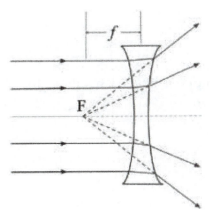

Diverging lenses refract light so that it appears to come from a point in front of the lens.

Images of real objects that are formed by diverging lenses are *always* virtual, upright, and diminished in size. Diverging lenses are frequently used to reduce the effects of converging lenses.

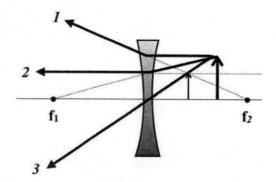

A viewer on the left looking through the diverging lens would perceive an upright, reduced, virtual image that is closer to the lens than the object.

## AP® Tip

To avoid confusion, you should identify both mirrors and lenses as either *converging* or *diverging*.

## THE LENS EQUATION

The nature of the images formed by lenses can be determined by the lens equation

$$\frac{1}{s_i} + \frac{1}{s_o} = \frac{1}{f}$$

where $f$ = focal length of the lens
  $s_o$ = object distance
  $s_i$ = image distance

When a lens produces an image of an object, the image is rarely the same size as the object. The ratio of the image's height to the object's height is called the *magnification*.

$$\text{Magnification} = M = \frac{\text{image size}}{\text{object size}}$$

Or

$$M = \frac{h_i}{h_o}$$

We can also relate this to the object distance $s_o$ and image distance $s_i$.

$$M = -\frac{s_i}{s_o}$$

## AP® Tip

A positive magnification indicates that the image is upright, whereas a negative magnification only occurs when the image is inverted.

### SAMPLE PROBLEM 6

A thin converging lens has a focal length of 20 cm. An object that is 4 cm tall is placed 10 cm from the lens. What are the nature, size, and position of the image?

### SOLUTION TO PROBLEM 6

The units are not changed into meters since they will divide out. The image distance is found using

$$\frac{1}{s_i} + \frac{1}{s_o} = \frac{1}{f}$$

$$\frac{1}{s_i} + \frac{1}{10} = \frac{1}{20}$$

$s_i = -20$ cm

The negative sign means that the image is virtual. Magnification is

$$M = \frac{h_i}{h_o} = -\frac{s_i}{s_o}$$

$$\frac{h_i}{4\ cm} = -\frac{-20\ cm}{10}$$

$h_i = +8$ cm

The positive sign means that the image is upright.

### SAMPLE PROBLEM 7

A diverging lens has a focal length of –16 cm. This lens is held 10 cm from an object. In cm, where is the image located? What is the magnification of this lens?

### SOLUTION TO PROBLEM 7

Direct substitution into the lens equation gives

$$\frac{1}{s_i} + \frac{1}{s_o} = \frac{1}{f}$$

$$\frac{1}{s_i} + \frac{1}{10\ cm} = \frac{1}{-16\ cm}$$

$s_i = -6.15$ cm

The negative sign indicates that the image is virtual. Magnification is

$$M = -\frac{s_i}{s_o} = -\frac{-6.15\ cm}{10\ cm} = +0.62$$

Positive magnification means that the image is upright.

## IMAGES PRODUCED BY TWO LENSES

Systems of two or more lenses are generally used for magnification purposes, as in the microscope and the telescope. The magnification in these cases is, in general

$$M = -\frac{s_i}{s_o} = \frac{-(-6.15\,\text{cm})}{10\,\text{cm}} = +0.62$$

To find the location of an image produced by a pair of lenses, solve for the location of the image produced by the first lens. Then use the image's location as the object distance for the second lens. That is, the image of the first lens is the object of the second.

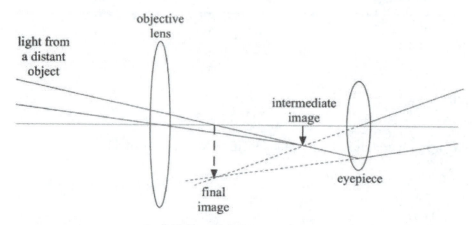

Basic Refracting Telescope Design

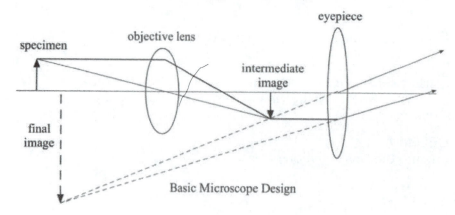

Basic Microscope Design

## SAMPLE PROBLEM 8

The objective lens of a microscope has a focal length of 8 mm. The eyepiece has a focal length of 40 mm. When the two lenses are 200 mm apart, the final image appears to be at a distance of 250 mm from the eyepiece.

(a)  In mm, how far is the object from the objective lens?

(b)  What is the magnification of the microscope?

## SOLUTION TO PROBLEM 8

(a) The units will be left in mm. Label the objective lens as 1 and the eyepiece as 2. We will find the position of the intermediate image first.

$$\frac{1}{s_{i2}} + \frac{1}{s_{o2}} = \frac{1}{f_2}$$

$$\frac{1}{s_i} + \frac{1}{-250\ mm} = \frac{1}{40\ mm}$$

$s_o = 34.5$ mm

The negative sign was used for the image distance $s_{i2}$ because it was measured to a virtual image. Since $s_{o2}$ is known, we can find the image distance $s_{i1}$ for the first image.

$s_{i1} = 200$ mm – 34.5 mm = 165.5 mm

The object distance $s_{o1}$ is then

$$\frac{1}{s_{i1}} + \frac{1}{s_{o1}} = \frac{1}{f_1}$$

$$\frac{1}{165.5\ mm} + \frac{1}{s_{o2}} = \frac{1}{8\ mm}$$

$s_o = 8.41$ mm

(b) The magnification is the product of the individual magnifications

$$M = M_1 \times M_2 = \frac{s_{i1}s_{i2}}{s_{o1}s_{o2}} = \frac{(165.5\ mm)(-250\ mm)}{(8.41\ mm)(34.5\ mm)} = -143$$

The negative magnification implies that the final image is inverted.

# GEOMETRICAL OPTICS: STUDENT OBJECTIVES FOR THE AP® EXAM

- You should be able to determine how the speed and wavelength of light change when light passes from one medium into another.
- You should be able to show on a diagram the directions of reflected and refracted rays.
- You should be able to use Snell's law to relate the directions of the incident ray and the refracted ray and the indices of refraction of the media.
- You should be able to identify conditions under which total internal reflection will occur.
- You should understand image formation by plane or spherical mirrors.
- You should be able to relate the focal point of a spherical mirror to its center of curvature ($R = 2f$).
- You should be able, given a diagram of a curved mirror with the focal point shown, to locate by ray tracing the image of a real object and

determine whether the image is real or virtual, upright or inverted, enlarged or reduced in size.

▪ You should understand image formation by concave or convex lenses.
▪ You should be able to determine whether the focal length of a lens is increased or decreased as a result of a change in curvature of its surface or in the index of refraction of the material of which the lens is made or the medium in which it is immersed.
▪ You should be able to determine by ray tracing the location of the image of a real object located inside or outside the focal point of the lens and state whether the resulting image is upright or inverted, real or virtual.
▪ You should be able to use the thin lens equation to relate the object distance, image distance, and focal length for a lens and to determine the image size in terms of the object size (i.e., magnification).
▪ You should be able to analyze simple situations in which the image formed by one lens serves as the object for another lens.

## MULTIPLE-CHOICE QUESTIONS

1. A student uses a plane mirror to study image formation. Using pins and ray tracing the student is able to locate the images of the pins in the mirror for each position of the pin in front of the mirror. The student wants to make a graph of pin distance from the front of the mirror versus distance of the image of the pin to the mirror. Which of the following graphs best describe the relationship between object distance and image distance?

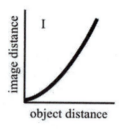

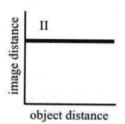

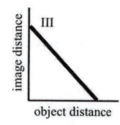

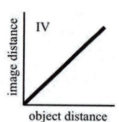

(A) I
(B) II
(C) III
(D) IV

2. When light is incident on a surface, all of the electromagnetic radiation is
   (A) either reflected from the surface or transmitted.
   (B) either reflected from the surface or absorbed.
   (C) reflected, transmitted, and absorbed at a surface.
   (D) either reflected from the surface or absorbed at the surface.

3. Light incident on the interface between air ($n = 1$) and water ($n = 1.33$) along the normal to the interface
   (A) slows and is bent toward the normal.
   (B) slows and is bent away from the normal.
   (C) slows but is not bent since it travels along the normal.
   (D) continues into the water at constant speed but is bent toward the normal.

4. Several students designed an experiment to determine the radius of curvature of an unknown mirror. They used an optical bench, a mirror, a film arrow, and a screen. Part of their data is shown below.

| $s_o$ (cm) | $s_i$ (cm) | $h_{obj}$ (cm) | $h_{image}$ (cm) |
|---|---|---|---|
| 15.0 | 30.0 | 1.00 | 1.99 |
| 20.0 | 20.0 | 1.00 | 1.00 |
| 25.0 | 16.7 | 1.00 | 0.67 |
| 40.0 | 13.3 | 1.00 | 0.33 |

   Using the data, the mirror is
   (A) concave with a radius of curvature of 10 cm.
   (B) concave with a radius of curvature of 20 cm.
   (C) convex with a radius of curvature of 10 cm.
   (D) convex with a radius of curvature of 20 cm.

5. A ray of light in air strikes the interface between air and water at some angle with the normal. In water, the ray has a
   (A) larger wavelength.
   (B) larger frequency.
   (C) smaller wavelength.
   (D) smaller frequency.

6.  Students obtained the following data during an experiment using an optical bench, a convex lens, a film arrow, and a screen.

| $s_o$ (cm) | $s_i$ (cm) |
|---|---|
| 15.0 | 30.0 |
| 20.0 | 20.0 |
| 25.0 | 16.7 |
| 30.0 | 15.0 |
| 40.0 | 13.3 |
| 50.0 | 12.5 |

Which of the graphs shown below gives the correct relationship between $s_o$ and $s_i$?

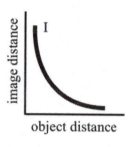

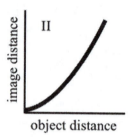

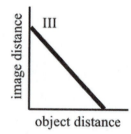

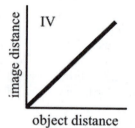

(A) I
(B) II
(C) III
(D) IV

7.  A layer of water ($n = 1.33$) covers a block of glass with an index of refraction of $n = 1.52$. Total internal reflection at the interface between the two media
    (A) occurs whenever the ray of light goes from the glass to the water because the speed of light increases in the water.
    (B) occurs whenever the ray of light goes from the water to the glass because the speed of light increases in the glass.
    (C) may occur when the ray of light goes from the glass to the water because the speed of light increases in the water.
    (D) may occur when the ray of light goes from the water to the glass because the speed of light increases in the glass.

8. A real image twice as large as an object is produced when the object is placed 20.00 cm away from a converging lens. What is the focal length of the lens?
   (A) 13.3 cm
   (B) 15.5 cm
   (C) 25.0 cm
   (D) 40.0 cm

9. A 3.0 cm tall object is placed 25.0 cm from a diverging lens whose focal length is –10.0 cm. The image formed by this lens is
   (A) real and inverted with a magnification $|M| = 0.72$.
   (B) virtual and inverted with a magnification of $|M| = 0.72$.
   (C) real and upright with a magnification $|M| = 0.29$.
   (D) virtual and upright with a magnification $|M| = 0.29$.

10. Which of the following materials will produce a convex lens that has the longest focal length?
    (A) Crown glass with $n = 1.52$
    (B) Flint glass $n = 1.66$
    (C) Fused quartz $n = 1.458$
    (D) Zircon $n = 1.923$

11. An object is placed in front of four different optical devices: two lenses and two mirrors. Which of the following will produce a real image of the object?

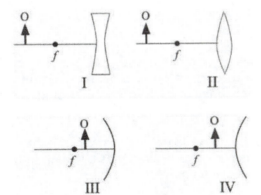

    (A) I
    (B) II
    (C) III
    (D) IV

12. A 580 nm beam of light traveling in air is incident on a cube of quartz ($n = 1.46$). In the quartz,
    (A) the speed of light is constant and the wavelength decreases because the frequency increases.
    (B) the speed of light decreases and the wavelength also must decrease because the frequency remains constant.
    (C) the speed of light decreases and the wavelength must increase because the frequency also decreases.
    (D) there is no change in the wavelength since both the speed of light and the frequency of the beam decrease.

13. On physics day at an amusement park, two physics students stand in front of a concave mirror with a radius of curvature of 4.00 m. Their images are upright and appear to be 2.50 times taller than their actual height. How far are they standing from the mirror?
    (A) 1.20 m
    (B) 1.50 m
    (C) 2.20 m
    (D) 4.00 m

14. A ray of light originates in the lower medium water ($n = 1.33$), striking the boundary between the water and air. Which of the four possible rays gives a correct possible direction for the ray in the air?

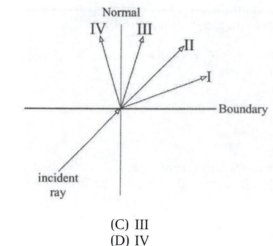

    (A) I          (C) III
    (B) II         (D) IV

**Directions:** For the question below, two of the suggested answers will be correct. You must select both correct choices to earn credit. No partial credit will be earned if only one correct choice is selected. Select the two that are best.

15. The diagram below shows the possible paths for a ray of light striking the interfaces between air, water, and glass. Which of the labeled rays is possible? Select two answers.

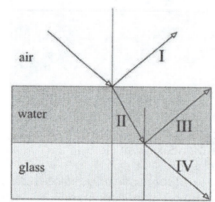

    (A) I
    (B) II
    (C) III
    (D) IV

# FREE-RESPONSE PROBLEMS

1. A student designs an experiment to determine how the path of a ray of light from air incident on the upper surface of a liquid contained in a plastic box changes as it enters the liquid. The student uses a monochromatic laser $\lambda = 740$ nm and a protractor in the experiment. The student obtains the following data.

| Angle of incidence $\theta_1$ | Angle of refraction $\theta_2$ |
|---|---|
| 10° | 7° |
| 20° | 13° |
| 30° | 20° |
| 40° | 26° |
| 50° | 31° |
| 60° | 35° |

(a) Graphically, determine the index of refraction in the glass by plotting on the same graph

   i. $\theta_1/\theta_2$ as a function of $\theta_1$.

   ii. $\dfrac{\sin\theta_1}{\sin\theta_2}$ as a function of $\theta_1$.

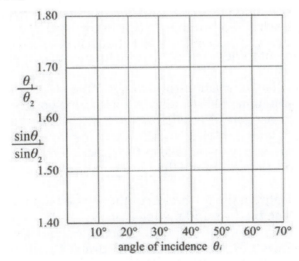

(b) Explain which relationship, (i) or (ii), if either, is more constant.
(c) Justify, without equations, why it is possible to neglect the refraction of the box containing the liquid.
(d) Clearly explain, without equations, how the angles of incidence and refraction of the beam of light at the entrance boundary are related to the angles of incidence and refraction of the beam of light at the exit boundary.
(e) How would you determine, without equations, what the wavelength of the beam of light was in the liquid?

2.  (a)  You are walking toward a full-length flat mirror at 1/40 m/s.
        Your image appears to be
        ___ approaching.
        ___ receding.
        ___ stationary.
        Explain your reasoning.
    (b)  What is the relative velocity between you and your image?
        Explain.

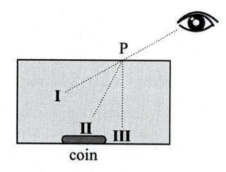

3.  (a)  A glass paperweight ($n = 1.52$) has a decorative coin inside of
        it. Someone looking at point P will see the coin at
        approximately which position?
        _____ position I    _____ position II _____ position III
        Justify your answer.
    (b)  If the coin is placed in flint glass ($n = 1.61$), the coin will appear to
        ___ be higher in the flint glass paperweight.
        ___ be lower in the flint glass paperweight.
        ___ have no change in its position in the flint glass
        paperweight.
        Justify your answer.

4.  Design an experiment to determine the focal length of a convex
    lens.
    (a)  Make a diagram of your laboratory setup, correctly labeling
        each piece of equipment in the diagram. Describe your
        experimental procedure in enough detail that another student
        could duplicate your experiment and obtain the same result.
    (b)  What measurements will you take in the experiment, and how
        will you use them to determine the focal length of the convex
        lens?
    (c)  Show by ray diagram the formation of the image when the
        object is between the lens and the focal point.
    (d)  Another student performed a similar experiment and obtained
        the following graph. How can you use the graph to determine
        the focal length of the lens?

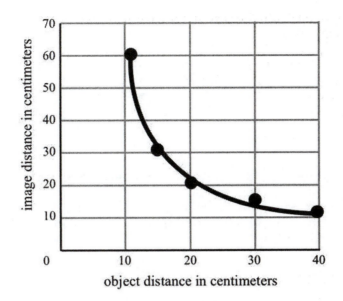

# Answers

## MULTIPLE-CHOICE QUESTIONS

1. **D** The image in a plane mirror is virtual, upright, and the same distance behind the mirror as the object in front of it ($s_o = s_i$). The image also has the same height. The graph will show a linear relationship between the two. The graph is a straight diagonal line slanting upward to the right.

    (L.O. 6.E.2.1)

2. **C** When light from one medium strikes the surface of the second medium, some or all of the light may be turned back into the first medium—this is called reflection. If the surface is transparent or translucent, some of the light will be transmitted, undergoing a change in speed at the interface. Some of the light is absorbed by the surface as internal energy.

    (L.O. 6.E.1.1)

3. **C** A ray of light from air into water traveling along the normal between the two materials slows but is not bent as it enters the water.

    (L.O. 6.E.3.1)

4. **B** The image distance is positive; therefore, the mirror is concave. The image distance for a convex mirror is always negative.

Substitution into the mirror equation $\frac{1}{s_i} + \frac{1}{s_o} = \frac{1}{f}$ from any of the data values listed will give a focal length of +10.0 cm. The focal length is related to the radius of curvature by $f = \frac{R}{2}$. The radius of curvature is 20.0 cm.

(L.O. 6.E.4.1)

5. **C** The frequency of light inside the medium is the same as it is outside the medium because the number of waves entering and leaving the interface is the same. Since the index of refraction is greater in the water than in the air and the ray slows down, the wavelength must reduce as well.

(L.O. 6.E.3.1)

6. **A** The equation for the relationship between $f$, $s_i$, and $s_o$ is given by $\frac{1}{s_i} + \frac{1}{s_o} = \frac{1}{f}$. The lens remains the same in the experiment; thus,

$$\frac{1}{s_{i1}} + \frac{1}{s_{o1}} = \frac{1}{s_{i2}} + \frac{1}{s_{o2}}$$

The product of $s_o s_i$ is a constant. The graph is hyperbolic—graph A.

(L.O. 6.E.5.2)

7. **C** Total internal reflection can occur only when light passes from a medium of greater optical density to one of lower optical density at an angle greater than the critical angle. When light travels from a medium where its speed is lower to one in which its speed is higher, total internal reflection occurs with light incident on the glass-water interface at the critical angle and the ray is bent along the interface. At angles less than the critical angle, the light leaves the interface, bending away from the normal.

(L.O. 6.E.3.3)

8. **A** The magnification of a thin lens produces an image that is twice as large as the object when the object is placed 20.0 cm from the convex lens, $M = -\frac{s_i}{s_o}$ gives $2s_o = s_i$.

Substitution into $\frac{1}{s_i} + \frac{1}{s_o} = \frac{1}{f}$ is $\frac{1}{40.0\ \text{cm}} + \frac{1}{20.0\ \text{cm}} = \frac{1}{f}$

The focal length of the convex lens is 13.3 cm.

(L.O. 6.E.5.1)

9. **D** All images produced by an object placed in front of a diverging lens are upright and virtual with a magnification $|M| < 1$.

Substitution into $\dfrac{1}{s_i} + \dfrac{1}{s_o} = \dfrac{1}{f}$ is $\dfrac{1}{s_i} + \dfrac{1}{25.0\,\text{cm}} = \dfrac{1}{-10.0\,\text{cm}}$

which gives $s_i = -7.14$ cm. The magnification $M = -\dfrac{s_i}{s_o}$ produces a magnification of 0.2857, which gives the correct answer of 0.29.

(L.O. 6.E.5.1)

10. **C** The longest focal length will be produced by the material that has the largest index of refraction when light from air enters the substance—bent the least because the speed of light is higher in the substance compared to the other materials listed. This is the material that has the smallest index of refraction. For the materials listed, fused quartz has the smallest index of refraction.

(L.O. 6.E.3.1)

11. **B** Only concave mirrors or convex lenses can produce a real image of an object placed in front of it. The object is placed between the focal point and the concave mirror in figure III and produces a virtual image. The only diagram that can produce a real image of the object is diagram II.

(L.O. 6.E.4.2, 6.E.5.1)

12. **B** The frequency of light inside the quartz is the same as it is outside the quartz because the number of waves entering and leaving the interface is the same. Since the index of refraction is greater than 1, the beam slows as it enters the interface; thus, the wavelength of light must decrease.

(L.O. 6.E.3.1)

13. **A** The radius of curvature is related to the focal length by $f = \dfrac{R}{2}$. Thus, the focal length of the mirror is 2.00 m. The magnification $M = -\dfrac{s_i}{s_o} = 250$. The magnification gives $s_i$ a value of $s_i = -2.50 s_o$ since the image is upright and therefore behind the mirror.

Substitution into $\dfrac{1}{s_i} + \dfrac{1}{s_o} = \dfrac{1}{f}$ is $\dfrac{1}{-2.5 s_o} + \dfrac{1}{s_o} = \dfrac{1}{2.00\,\text{m}}$; thus, $s_o = 1.2$ m.

(L.O. 6.E.4.2)

14. **A** The ray travels from water to air where the speed of the ray will increase. The ray will bend away from the normal as it crosses the boundary between water and air.

(L.O. 6.E.4.2)

15. **A and B** The angle of incidence between the air and water interface shows a reflected angle $\theta_i = \theta_r$ so that that ray is possible. The angle of refraction between the air and water interface shows the ray bending toward the normal. This ray is possible. The reflected ray at the interface between the water and the glass shows a ray parallel to the reflected ray at the upper surface. This solution is not possible. The last ray shows light bending away from the normal as the ray travels from water to glass, where the index of refractions would indicate that the ray slows down in glass and should bend toward the normal. This is not possible either.

(L.O. 6.E.4.2)

## FREE-RESPONSE PROBLEMS

1. (a)

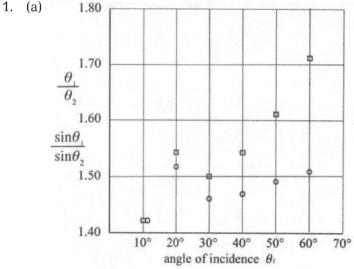

(b) The ratio of the $\sin\theta_1 / \sin\theta_2$ as a function of $\theta_1$ is more constant than the ratio of $\theta_1 / \theta_2$.

The average value for $\sin\theta_1 / \sin\theta_2$ is 1.48. Snell's law applies.

The index of refraction for air is 1. Thus, the index of refraction for the liquid is the ratio of $\sin\theta_1 / \sin\theta_2$.

(c) From the top, the light refracts from air into plastic and then into liquid. Then the light refracts from liquid into plastic into air out of the bottom. The refractions at the upper boundaries are opposite the refraction at the lower boundaries. Regardless of the box's thickness, the speed changes will be the same.

(d) The box has parallel sides; thus, the ray emerges from the lower surface parallel to the ray striking the upper surface. The explanation may be done with a diagram showing the rays at the interfaces.

(e) The index of refraction is the ratio of the speed of light in air divided by the speed of light in the liquid. The frequency in air is the same as the frequency in the liquid since the number of waves crossing the boundary will not change. Knowing that the product of the wavelength times the frequency is equal to the speed of light in the specific media, the wavelength in the liquid is equal to the wavelength in air divided by the index of refraction.

(L.O. 6.E.3.1, 6.E.3.2, 6.E.3.3)

2. (a) You are walking toward a full-length flat mirror at 1.40 m/s. Your image appears to be ___√___ approaching. The image seen in a flat mirror is virtual, erect, and as far behind the mirror as the object in front of it. If you are approaching the mirror, your image is approaching you.

(b) The image will be approaching the mirror at the same rate you are approaching it. The relative velocity between you and your image will be twice as great as your velocity.

(L.O. 6.E.2.1)

3. Position I
The ray of light from the coin refracts away from the normal as it leaves the glass and enters the air due to the change in speed at the interface between glass and air. The eye traces the ray of light from the coin back from the apparent source in a straight line. We perceive the coin to be located at position I.

(a) The correct line checked is ___√___ higher in the flint glass paperweight.

The index of refraction in the flint glass is higher than the index of refraction of the glass. The light passing from the flint glass into the air will be bent farther away from the normal because there is greater change in the speed. Tracing the apparent position of the coin in a straight line would make the image of the coin appear to be closer to the surface in the flint glass paperweight.

(L.O. 6.E.2.1)

4. (a) Diagram of the experimental apparatus.

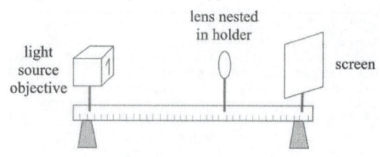

lens nested
in holder

light
source
objective

screen

1. Set up the equipment as shown in the diagram above. Make sure the pieces of equipment are centered in their respective holders.
2. Place the lens between the object (arrow) and the screen and move either the screen or the lens until a sharp real image is formed on the screen.
3. Measure the distance between the object and the lens (object distance $s_i$) and record this value in the data table. (Make sure the distances are measured from the center of the holders.)
4. Measure the distance between the image on the screen and the lens (image distance $s_o$) and record this value in the data table. (Make sure the distances are measured from the center of the holders.)
5. Repeat for other values.

(b) Using the values for the object distance $s_o$ and the image distance $s_i$ and substituting into $\dfrac{1}{s_i} + \dfrac{1}{s_o} = \dfrac{1}{f}$ will give an experimental value for the focal length of the lens.

Averaging the values for the several trials will be considered the experimental focal length of the lens.

(c) Image formation for an object placed between the focal point and the lens:

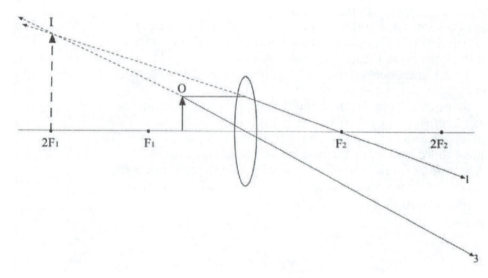

(d) Using data sets from the graph, the focal length can be determined from $\dfrac{1}{s_i} + \dfrac{1}{s_o} = \dfrac{1}{f}$.

Looking at a few data sets, indicate, for example,

$$f = \frac{450 \text{ cm}^2}{45 \text{ cm}} = 10 \text{ cm} \quad \text{or} \quad f = \frac{720 \text{ cm}^2}{72 \text{ cm}} = 10 \text{ cm}.$$

(L.O. 6.E.5.1, 6.E.5.2)

# WAVE OPTICS

PHYSICS 2

## DISPERSION

When a thin beam of sunlight passes through a glass prism, the initial beam separates into beams of various colors. Observing this, Newton concluded that *white light* is actually a mixture of light of all colors. The colorful array of refracted beams that emerges from the glass prism is a *spectrum*. Light of all wavelengths is reduced in speed in glass, but the degree of refraction depends on the light's wavelength. The shortest wavelength visible light, violet light, travels relatively slowly and refracts the most. The longest wavelength visible light, red light, travels relatively quickly and refracts the least. This effect is called *dispersion*. The diagram shows the result of directing a narrow beam of white light at one face of a glass prism.

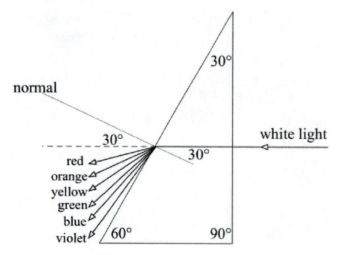

# ELECTROMAGNETIC WAVES

Light waves are *electromagnetic waves* composed of two perpendicular transverse waves that come from an electric field and a magnetic field. Like all transverse waves, the movement of an electromagnetic wave is perpendicular to the movement of the energy. Electromagnetic waves do not need a medium and can travel in a vacuum as well as in air, water, or glass.

All electromagnetic waves share the basic properties of frequency $f$ and wavelength $\lambda$. All electromagnetic waves travel at the same speed in vacuum, the speed of light $c$. The speed of light is related to these properties by $c = f\lambda$.

Because of light's small wavelength, we use the *nanometer* (nm) or the *picometer* (pm) to measure wavelength. One nanometer (1 nm) is one billionth of a meter, 1 nm = $10^{-9}$ m, and 1 picometer (1 pm) is one trillionth of a meter, 1 pm = $10^{-12}$ m.

The names given to various waves in the electromagnetic spectrum are convenient for describing the region of the spectrum where they are found. There is no sharp dividing point between one type of electromagnetic radiation and the next, and in many cases, there is considerable overlap between classifications.

| Electromagnetic Radiation | Sources | Wavelength Range |
|---|---|---|
| *Radio waves* | Radio waves are the result of accelerating electrons through a transmitting antenna by an oscillator. Radio waves are the basis of TV and radio communication. | From more than $10^4$ m to about 0.1 m. |
| *Microwaves* | Electronic devices produce microwaves. Radar systems generate microwaves, and microwave ovens, which produce a wavelength of $\lambda = 12.2$ cm, are a common household device. | From approximately 0.3 m to $10^{-4}$ m. |
| *Infrared waves* (IR) | Molecules and room-temperature objects produce infrared waves. Some IR is produced by outer electron transitions in excited atoms. | From $10^{-3}$ m to 700 nm. |
| *Visible light* | Visible light can be detected by the human eye. It is produced by electron transitions in the outer electron shell of excited atoms and molecules. The sensitivity of the human eye is at a maximum of 550 nm at yellow green. | From 700 nm for red light to 400 nm for violet light. |
| *Ultraviolet waves* (UV) | UV waves are produced by electron transitions in the outer shells of excited atoms. The surface of the sun produces great quantities of UV. | From about 400 nm to 0.6 nm. |
| *X-rays* | The most common source of X-rays is the deceleration of high-energy electrons bombarding a heavy metal target in a vacuum tube. X-rays are also produced by the electron transitions that follow the removal of electrons from the inner electron shells of heavy metal atoms. | From approximately $10^{-8}$ m to $10^{-12}$ m. |
| *Gamma rays* | Gamma rays are the result of the rearrangement of neutrons and electrically charged protons in the nucleus after alpha or beta radioactive decay. Gamma rays are a component of high-energy cosmic radiation that enters Earth's atmosphere from deep space. Matter-antimatter interactions also produce gamma rays. | From about $10^{-10}$ m to less than $10^{-14}$ m. |

## SAMPLE PROBLEM 1

(a)   Calculate the frequency of yellow light, $\lambda = 600$ nm.

(b)   Find the wavelength of radio waves whose frequency is $f = 1.00$ MHz.

## SOLUTION TO PROBLEM 1

(a)   The frequency is found using

$$f = \frac{c}{\lambda} = \frac{3.00 \times 10^8 \ \text{m/s}}{600 \times 10^{-9} \ \text{m}} = 5.00 \times 10^{14} \ \text{Hz}$$

(b)   Solving for wavelength,

$$\lambda = \frac{c}{f} = \frac{3.00 \times 10^8 \ \text{m/s}}{1.00 \times 10^6 \ \text{s}^{-1}} = 300 \ \text{m}$$

## POLARIZATION

An ordinary beam of light is composed of many waves, each of which came from an atom oscillating in its own orientation. The superposition of these transverse waves creates an *unpolarized* wave in which the electric field oscillates in all directions. Removing all waves from a beam except for those oscillating in one particular direction creates a *polarized* wave.

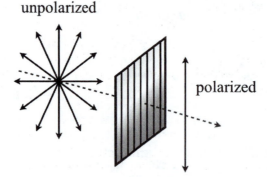

Unpolarized light oscillates in all orientations.
After a filter allows only one orientation of wave to pass through,
the light is polarized.

## INTERFERENCE

We have seen that mechanical waves can combine constructively or destructively as the amplitudes of separate waves add up at the same point in space. Interference between light waves is an interaction of the electromagnetic fields of the individual waves. This ability to interfere is a property of all waves, including electromagnetic waves. Only waves exhibit interference behavior, so the interference of light waves is evidence that light behaves like a wave.

## DIFFRACTION

As waves travel through a uniform medium, they move in straight lines. They can only change direction when the medium changes; then they undergo refraction and reflection. Another wave phenomenon occurs when a wave passes by an edge and bends around the edge into the shadow region. The spreading of waves into the region behind an obstacle is called *diffraction*. As light shines through an open door into a dark room, the light does not take the shape of the doorway. Instead the light bends around the corners of the door into the room. Diffraction occurs in accordance with Huygens' principle; this interference pheno-menon can occur in any wave, including both light and sound. Without diffraction, we could not hear around a corner without seeing the speaker because the sound waves would only travel in a straight line.

Any obstacle placed in a light beam coming from a point source will cause diffraction effects under proper conditions. A slit, a thin wire, and a razor edge are examples of such objects. A razor's edge illuminated by a beam of monochromatic light from a point source casts a shadow that is not geometrically sharp. A small amount of light bends around the edge into the geometric shadow, and a series of alternating light and dark bands are found to border the shadow.

If a slit in a piece of cardboard is illuminated by monochromatic light, the image formed on a screen some distance away is not a single well-defined slit of light but rather a series of light and dark spots. The middle bright spot will have the highest intensity with a dark spot on either side of it. Further from the middle bright spot, the intensity of subsequent bright spots is much less. The pattern produced through the single slit is a result of diffraction of light waves. As the waves are bent around the slit's edges, the waves can interfere constructively or destructively, with other waves creating the diffraction pattern.

## YOUNG'S DOUBLE-SLIT EXPERIMENT

In 1801, Thomas Young conducted an experiment investigating the wave properties of light. He directed a thin beam of monochromatic light through narrow slits and onto a screen. If light were a stream of particles, the particles would have to go through one slit or the other, but the pattern of light on the screen made it clear that light was simultaneously passing through both slits. Only a wave can pass through two separate openings simultaneously.

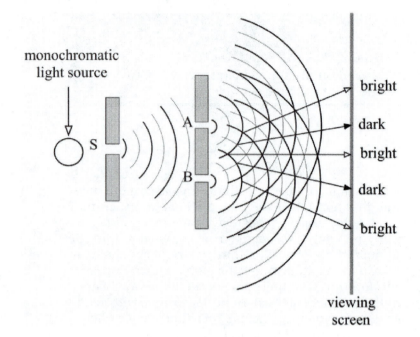

The screen in the diagram above shows a pattern of bright and dark bands called fringes. The light fringes are areas of constructive interference where two crests or two troughs coincide. The dark fringes are areas of destructive interference where crests meet troughs.

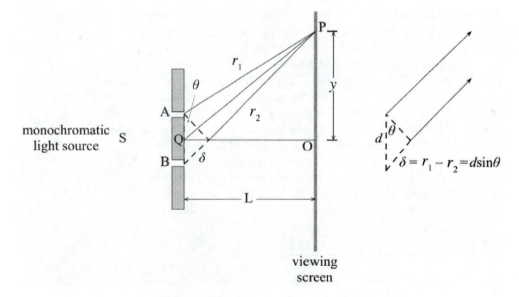

In the diagram above, we see monochromatic light shining from the left through two closely-spaced slits. Slits A and B both act like emitters of waves. The waves emitted by slits A and B travel along different paths and arrive at a screen a distance L away. At points where the difference between $r_1$ and $r_2$ is an integer multiple of a wavelength, there is constructive interference.

The condition for constructive interference and bright fringes is

$$d \sin \theta_{bright} = m\lambda \qquad \text{where } m = 0, \pm 1, \pm 2, \cdots$$

The number $m$ is the *order number*. The central bright fringe at $\theta_{bright} = 0$ is the zeroth-order maximum. The maximum on each side of the zeroth-order maximum where $m = \pm 1$ is called the first-order maximum. Other integer values of $m$ find other bright fringes.

When the path difference is an odd multiple of $\lambda/2$, the two waves arriving at point P (shown in the diagram above) are completely out of phase. There, crests meet troughs and cause destructive interference. Dark fringes occur between bright fringes; to locate them, use $m = \pm 0.5$, $\pm 1.5$, $\pm 2.5$, and so on.

constructive interference

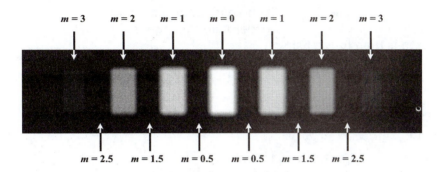

destructive interference

The spacing in the diagram above is somewhat exaggerated. The spacing between the fringes is not actually uniform and increases as $\theta\lambda$ increases. For small angles of less than $5°$, the bright spots are approximately evenly spaced.

For small angles of only a few degrees, we can use

$$y_{bright} = \frac{\lambda L}{d} m \qquad \text{where } m = 0, \ \pm 1, \ \pm 2, \ \cdots$$

The results of Young's double-split experiment are significant not only because they provide compelling proof for the wave theory of light but also because they give us a practical way to make very precise measurements on the nano scale. Diffraction patterns tell us about many systems with repeating structures on the very small scale, such as the surface of a Blu-ray disc.

If we know the wavelength of the incident light, then measuring the spacing of bright fringes tells us the spacing of the slits or other repeating structures. The spacing between the slits and the spacing between the fringes are inversely related. The closer together the slits are, the further apart the bright fringes will be.

If we know the spacing of the slits, we can use this relationship to determine the wavelength of the incident light. The greater the wavelength of light is, the further apart the bright fringes will be.

## AP® Tip

Constructive interference occurs when the waves arrive in phase with one another. Destructive interference occurs when the phase difference is a half-integral number of wavelengths or out of phase.

## SAMPLE PROBLEM 2

In a Young's double-slit experiment, a blue laser shines on two slits that are 0.02 mm apart. The screen is located 1.80 m from the slits, and the third-order bright fringe is located 12.50 cm from the central fringe.

(a) Determine the wavelength of the blue laser.
(b) Where will the first dark fringe appear?
(c) How would the spacing of the bright fringes change if red light was used?

## SOLUTION TO PROBLEM 2

(a) Since $\theta$ is small, we can use the small angle identity, $\sin\theta \cong \tan\theta$, and substitute $y/L$ for $\theta$. For a third-order bright fringe, $m = 3$. The wavelength is then found using

$$\lambda = \frac{y_{bright}d}{mL} = \frac{(0.125 \text{ m})(2.00 \times 10^{-5} \text{ m})}{3(1.80 \text{ m})} = 4.63 \times 10^{-7} \text{ m} = 463 \text{ nm}$$

(b) For the first dark fringe, $m = 0$. The position of the first dark fringe is found from

$$y_{dark} = \frac{\lambda L}{d}\left(m + \frac{1}{2}\right) = \frac{(4.63 \times 10^{-7} \text{ m})(1.80 \text{ m})}{2(2.00 \times 10^{-5} \text{ m})} = 2.08 \text{ cm}$$

(c) Red light has a longer wavelength than blue light, so the distance between bright fringes would be greater.

## THIN FILMS AND OPTICAL COATINGS

Interference effects are responsible for the swirling rainbows in thin layers of oil on water and the thin surface of a soap bubble. These thin materials, with an index of refraction greater than the surrounding air, are called *films*. The varied colors result from the interference of light waves reflected from the upper and lower surfaces of the film.

When light shines on a thin film of thickness $t$, as illustrated on the next page, the incident ray is partially reflected at the top of the film, surface A, as it travels from a material of index of refraction $n_1$ to the film that has an index of refraction of $n_2$. For clarity, the reflected ray, ray 1, is displaced slightly to the left.

The remainder of the incident ray is transmitted through the film and reaches its bottom, surface B, where it encounters another medium with an index of refraction $n_3$ and is partially reflected and partially transmitted into the new material of index of refraction $n_3$.

The reflected ray, ray 2, then travels back to the top surface of the film where it is transmitted into the material of original index $n_1$. Thus, ray 1 and ray 2 travel together away from the upper surface of the film. These two waves add up by superposition and interfere with each other.

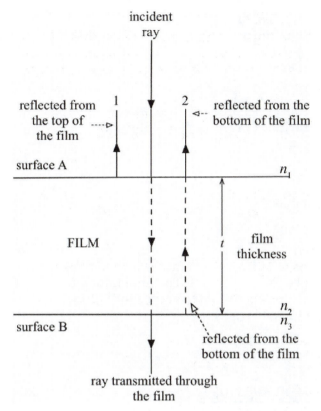

If a thin film has a higher index of refraction than the materials it touches, like a soap bubble with air on either side, it is referred to simply as a film. If the film has an index of refraction between the two materials it touches, like a thin layer of oil between air and water, it is considered a *coating*. Whether the reflected rays interfere constructively or destructively depends on the indices of refraction of the various materials and the thickness of the material compared to the wavelength of the light in the material. When light traveling through a material with index of refraction $n_1$ strikes a boundary with a material with an index of $n_2$, its reflection may undergo a *phase shift* if $n_2$ is greater than $n_1$. When a phase shift occurs, the wave pattern is inverted.

To find the wavelength of light in the material, we use the relationship $n_1\lambda_1 = n_2\lambda_2$. Then we compare $\lambda_2$ to the thickness of the film or coating.

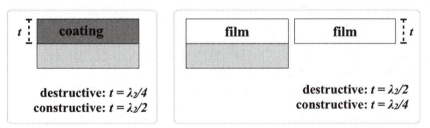

Light undergoes a phase shift when it reflects at the lower surface of a film but not at the lower surface of a coating.

## SAMPLE PROBLEM 3

A thin anti-reflective oil ($n = 1.75$) covers a glass lens ($n = 1.6$). The reflection of sunshine in the glasses has a strong green tint, with a wavelength of 510 nm reflecting the most.

   (a)   Does this oil function as a film or a coating?

   (b)   What is the wavelength of the green light when it is inside the oil?

   (c)   What is the minimum thickness of the oil?

## SOLUTION TO PROBLEM 3

   (a)   The oil functions as a film. The index of refraction on either side of the oil is less than the index of refraction of the oil.

   (b)   $n_1 \lambda_1 = n_2 \lambda_2$, so

$$\lambda_2 = \frac{n_1 \lambda_1}{n_2}$$

$$\lambda_2 = \frac{(1.00)\,(510\text{ nm})}{(1.75)}$$

$$\lambda_2 = 291 \text{ nm}$$

   (c)   The question references a "strong green tint" which implies constructive interference for green light. The minimum thickness of oil that would cause constructive interference for light with a 291 nm wavelength in the oil is given by

$$t = \frac{\lambda}{4} = \frac{291\ nm}{4} = 73\ nm$$

# WAVE OPTICS: STUDENT OBJECTIVES FOR THE AP® EXAM

- You should know that all electromagnetic waves travel with the same speed, *c*, in a vacuum and be able to state and relate the relationship between speed, wavelength, and frequency of these waves.
- You should know the difference between parts of the electromagnetic spectrum and their sources.
- You should be able to explain why dispersion occurs for the visible part of the electromagnetic spectrum.
- You should know and be able to state what is necessary for two or more light waves to produce interference phenomena.
- You should be able to explain why diffraction, dispersion, and polarization are evidence for a wave model.
- You should be able to explain how wavelength, distance, and slit separation affect the spacing of a diffraction pattern in *Young's double-slit experiment*.
- You should be able to discuss constructive and destructive interference in terms of waves.

## MULTIPLE-CHOICE QUESTIONS

A double-slit interference pattern, as illustrated in the diagram below, is to be used to answer questions 1 and 2.

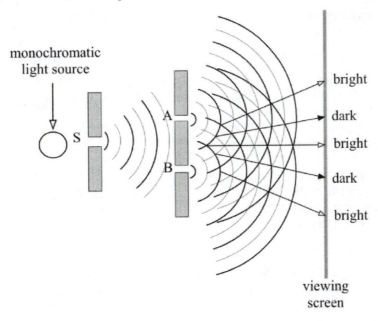

1. The slits A and B in the above diagram can be moved relative to each other. Based on this, which of the following is true?
   (A) Decreasing the slit separation increases the width of the fringes.
   (B) Increasing the slit separation increases the width of the fringes.
   (C) Increasing the slit separation increases the intensity of the bright fringes.
   (D) Changing the slit separation does not alter the width of the fringes.

2. In the double-slit interference pattern shown above, when the path difference between two waves is
   (A) even multiples of $\lambda/2$, destructive interference occurs, producing a dark fringe.
   (B) odd multiples of $\lambda/2$, destructive interference occurs, producing a dark fringe.
   (C) even multiples of $\lambda$, destructive interference occurs, producing a bright fringe.
   (D) odd multiples of $\lambda$, destructive interference occurs, producing a bright fringe.

3.  In a double-slit interference pattern, using different wavelengths of light to produce fringes, rank the wavelength of light illuminating the slits that will produce fringe patterns on the screen whose spacing is closer together to further apart.
    (A) red, orange, green, blue
    (B) green, blue, orange, red
    (C) blue, red, green, orange
    (D) blue, green, orange, red

4.  In a single-slit experiment, increasing the width of the slit results in
    (A) widening the diffraction pattern, moving the secondary bands farther from a wider central band.
    (B) narrowing the diffraction pattern, moving the secondary bands closer to a narrower central band.
    (C) no change in the central band while moving wider secondary bands farther away from the central band.
    (D) increasing the width of the central band while decreasing the width of the secondary bands.

5.  White light falling on a prism as shown in the diagram below refracts the light and produces a series of colors known as a visible spectrum. This is due to the fact that the index of refraction, $n$, which relates the speed of light in air to the speed of light in the prism,

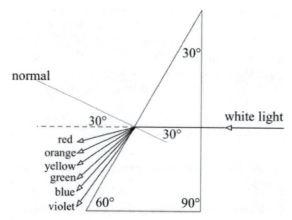

    (A) is independent of the wavelength of the light in the prism.
    (B) is dependent on the angle of incidence.
    (C) is dependent on the wavelength of the light in the prism with shorter wavelengths refracted the least.
    (D) is dependent on the wavelength of the light in the prism with shorter wavelengths refracted the most.

6.  Rank the frequency of the light in the prism shown in the previous question from highest to lowest.
    (A) violet > blue > green > red
    (B) violet = blue = green = red
    (C) red > green > blue > violet
    (D) red > green > blue = violet

7.  Colors are observed in soap bubbles suspended in air. These are caused by interference between the light reflected from the upper and lower surfaces of the soap bubble and the thickness of the soap bubble. (Assume near normal angles for the surfaces of the film and uniform thickness of the film.) Which of the following statements is correct?

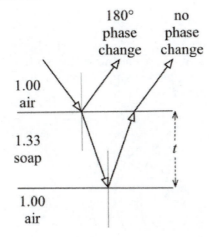

    (A) For constructive interference to occur for a given color, the path length in the film should be $t = \lambda/4$ of the wavelength for that color.
    (B) For constructive interference to occur for a given color, the path length in the film should be $t = \lambda/2$ of the wavelength for that color.
    (C) For destructive interference to occur for a given color, the path length in the film should be $t = 3\lambda/8$ of the wavelength for that color.
    (D) For destructive interference to occur for a given color, the path length in the film is $t = 3\lambda/16$ of the wavelength for that color.

8. A 765 nm beam of light passes through several narrow slits, producing the diffraction patterns shown below. Rank the pattern from the widest slit to the narrowest.

A           B

C           D

(A) A > B > C > D
(B) C > B > A > D
(C) A > D > B > C
(D) D > A > B > C

9. Two sources that are in phase are used to create a disturbance in a ripple tank, producing circular waves spreading outward from the sources. The high points of the ripples are shown expanding in circles from their respective sources.

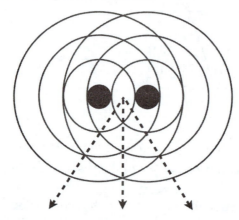

The pattern produced is similar to that observed in Young's double-slit experiment. The dotted lines represent
(A) constructive interference of the water waves.
(B) destructive interference of the water waves.
(C) polarization of the light passing through the water.
(D) dispersion of the light passing through the water.

10. Light coming from a monochromatic light source passes through an opening between two obstacles whose spacing is comparable in size to the wavelength of light illuminating the opening spreads out carrying energy with the wave. What best explains the pattern of bright and dark areas seen on a screen to the right of the openings?

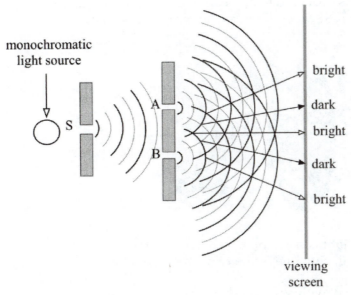

(A) A series of bright and dark fringes of the superposition of Huygens' wavelets from each part of the slit have produced constructive and destructive interference some distance to the top and bottom of the slit, carrying energy outward from the central bright as indicated by the bright fringes.

(B) Most of the energy of the wave is carried in a very narrow region where a single band of light appears; very little is transferred around the obstacles.

(C) Light refracts as it passes through the opening, producing a series of bright and dark fringes as shown in the diagram, carrying energy with it as it refracts.

(D) As the light passes through the opening, polarization of the wave occurs causing the electric field vector $\vec{E}$ and the magnetic field vector $\vec{B}$ to separate, carrying energy to the right and left of the opening between the obstacles, reaching distances far away from the opening.

11. Two electromagnetic waves are superimposed on each other with their electric fields oscillating along the axes shown. Although both waves consist of oscillating electric and magnetic fields, only their electric fields, $E_1$ and $E_2$, are shown. The combined waves travel as a beam with intensity $I_0$ and approach a pair of identical filters. Both filters are in the same orientation. The beam coming from the first filter is observed to have an intensity of ½ $I_0$. The beam coming from the second filter is also observed to have an intensity of ½ $I_0$.

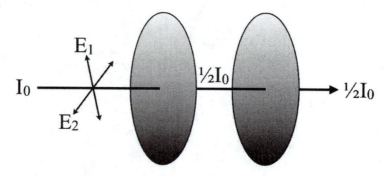

When the beam of light is between the two filters, it is
(A) linearly polarized, with one of the two waves completely blocked.
(B) partially polarized, with a reduction in the electric and magnetic field vectors of both of the incoming waves.
(C) unpolarized but the filter absorbed half of the energy, converting it to internal energy in the filter.
(D) unpolarized since the intensity dropped to 0.5 $I_0$. If it were linearly polarized, I would be zero.

12. Which of the following phenomena involving light cannot be explained by the wave model of light?
(A) When light passes through a small opening approximately as wide as the light's wavelength, the light spreads and makes a pattern of light and dark on a screen.
(B) When sunlight reflects off anti-glare sunglasses, the reflection appears to be mostly one color.
(C) When high-energy light shines on a small particle of matter, the light collides with the matter and transfers some of its momentum.
(D) Light from the sun travels to Earth through the vacuum of space.

13. An electromagnetic wave showing an oscillating electric field vector $\vec{E}$ and magnetic field vector $\vec{B}$ is indicated in the diagram below.

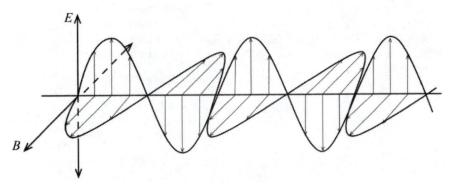

Which of the following diagrams gives a correct representation of the electromagnetic wave as it passes through a polarizer?

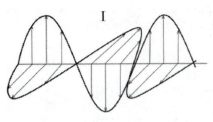 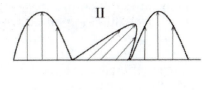

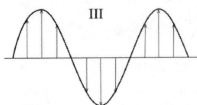 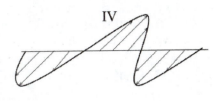

(A) I
(B) II
(C) III
(D) IV

14. In a single-slit experiment, increasing the width of the slit, $d$, from a fraction of a wavelength to a wavelength results in
(A) increasing the diffraction pattern.
(B) narrowing the diffraction pattern.
(C) no change in the diffraction pattern.
(D) reversing the diffraction pattern.

**Question 15**
**Directions:** For the question below, two of the suggested answers will be correct. You must select both correct choices to earn credit. No partial credit will be earned if only one correct choice is selected. Select the two that are best.

15. Which of the following statements are true of both waveforms shown below? Select two answers.

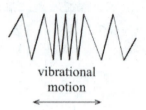

vibrational
motion

They both
(A) transfer energy and momentum from one part of the wave to another without transferring matter as the disturbance moves through the medium.
(B) have properties associated with their motion, such as wavelength, frequency, wave speed, and amplitude, that remain constant as they travel through a medium of uniform density.
(C) can travel in solids, liquids, gases, and vacuum with constant speed depending on the density of the medium.
(D) can show reflection and refraction, but superposition only applies to transverse wave forms.

## FREE-RESPONSE PROBLEMS

1. A student conducts an interference experiment using a laser source of monochromatic light of wavelength = 635 nm.
   (a) The light is incident on a double slit having a separation of $d = 0.04$ mm. At what angle from the central maximum will the second-order
      i.   maximum occur?
      ii.  minimum occur?
   (b) The light is incident on a single slit of separation $a = 0.04$ mm. At what angle will the second-order
      i.   maximum occur?
      ii.  minimum occur?
   (c) If the screen is 1.50 m from the double slits, what is the linear separation between the first-order maximum and the fourth-order maximum?
   (d) The student places the double-slit film in water ($n = 1.33$) and illuminates it with monochromatic light of $\lambda_{air} = 635$ nm. What is the angle from the central maximum of the second-order maximum?

2. Two laser beams of the same wavelength pass through a plastic sheet that has many repeating slits in two dimensions. The beams produce the diffraction patterns on a nearby wall shown below.

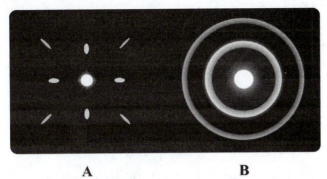

        A                          B

(a) Which of these diffraction patterns is produced by polarized laser light?
(b) Justify your answer to part (a).
(c) Which diffraction pattern provides stronger evidence of the wave nature of light?
(d) Justify your answer to part (c).

3. In enough detail that another student could perform your experiment and duplicate your results, design an experiment to determine the wavelength of light from a coherent source such as a laser pointer falling on a diffraction grating.
(a) Make a diagram of your experimental setup, labeling all of the equipment you will use in your experiment.
(b) What measurements will you take and how will you use them to determine the wavelength of the source of coherent light used in your experiment?
(c) What assumptions did you make, and how might these assumptions affect your results?
(d) What determines the maximum number of orders that are seen in the experiment?

4. (a) Why is it possible for radio waves to diffract around buildings when visible light waves that are also electromagnetic radiation traveling with the same velocity do not?
(b) While another class is in session, you leave the cafeteria early to ask your teacher about an assignment. How is it possible that you can hear the conversation in the room while you are standing in the hall away from the open door in the room?

# Answers

## MULTIPLE-CHOICE QUESTIONS

1. **A** In a double-slit interference pattern, when the separation, $d$, between the slits is decreased, the width of the fringes increases.

$$y_{bright} = \frac{\lambda L}{d}$$

(L.O. 6.C.2.1, 6.C.3.1)

2. **B** For destructive interference to occur between two waves in a Young's double-slit experiment producing a dark fringe, the path length must be an odd multiple of $\lambda/2$.

If the order number $m = 0$, then the path difference $\delta = \lambda/2$; if the order number is $m = 1$, the path difference is $\delta = 3\lambda/2$, etc.

(L.O. 6.C.2.1, 6.C.3.1)

3. **D** The smaller the wavelength, the closer the fringe patterns are together, $y_{bright} = \frac{m\lambda L}{d}$.

The correct ranking is blue, green, orange, red.

(L.O. 6.C.2.1, 6.3.C.1, 6.E.1.1)

4. **B** As the width of the slit in a single-slit diffraction pattern is increased, the pattern narrows. The central band becomes narrower, and the secondary bands move closer to the central band.

(L.O. 6.C.2.1)

5. **D** Light of all wavelengths is reduced in speed in glass, but violet light with a wavelength of $\lambda \cong 400$ nm, which is refracted the greatest amount, travels more slowly than red light, $\lambda \cong 650$ nm, which is refracted least. This effect is called *dispersion*.

(L.O. 6.E.3.3)

6. **A** The frequency of light in the prism is determined from $v = \lambda f$. Since red is refracted the least, its speed is the greatest. Since its speed in the prism is the largest, its frequency is the smallest.

(L.O. 6.E.3.3)

7. **A** The ray of light in the diagram traveling from air to the soap bubble undergoes reflection and refraction due to a speed change at the upper boundary. The ray is reflected 180° out of phase at the upper surface of the soap bubble (acts as a fixed boundary since $n$ in the soap bubble is greater than $n$ in air), producing a path difference of $\lambda/2$ between the reflected and refracted rays at the upper surface. The ray striking the lower surface and reflected back to the upper surface is reflected in phase since the ray interacting at the lower surface travels from the soap bubble to air and there is no phase change at the lower boundary (free boundary). To cause constructive interference of the two rays from the upper and lower boundaries, the ray in the soap bubble must travel a distance $t$ equal to $\lambda/4$ down and $\lambda/4$ back to the upper surface to be in phase with the ray at the upper surface. Thus, the minimum thickness of the film must be $\lambda/4$.

(L.O. 6.E.3.1, 6.E.3.3)

8. **D** The narrower the slit the light passes through, the more the diffracted light is spread out. The central maximum is widest in C, so this is the narrowest slit. The correct order from widest to narrowest is D > A > B > C.

(L.O. 6.E.3.1)

9. **B** The nodal lines indicate regions where no displacement from the equilibrium occurs. Therefore, it is destructive interference, corresponding to a dark fringe in a Young's double-slit experiment.

(L.O. 6.C.1.1)

10. **A** When the width of the opening between obstacles is comparable to the wavelength of light illuminating the opening, the wave spreads out around the obstacles, carrying energy with it. This is very much like water waves in a pond bending around an opening in a barrier, carrying energy to the other end of the pond or being able to hear sound coming from the other side of an open door.

(L.O. 6.C.4.1)

11. **A** The first filter polarizes the beam so that only one of the component waves passes through. The second polarizing filter is in the same orientation as the first, so it does not affect the beam.

(L.O. 6.A.1.3)

12. **C** Only particles carry momentum and have collisions. Answers (A) and (B) describe behaviors that are unique to waves, not particles. Answer (D) applies to both waves and particles.

(L.O. 6.B.3.1)

13. **A** Electromagnetic waves with an oscillating electric field vector $\vec{E}$ and a magnetic field vector $\vec{B}$ have random planes of vibration before they pass through a polarizer. The polarizer will transmit the EM waves that vibrate in the same plane—linearly or plane polarized.

(L.O. 6.B.3.1)

14. **B** As the slit opening $d$ increases, the pattern narrows. The central band becomes narrower, and the secondary bands move closer to the central band.

(L.O. 6.E.3.1)

15. **A and B** Both waveforms transfer energy as they move through the medium. The energy causes each particle in the wave to oscillate about the equilibrium position as the wave front advances through the medium. In a medium of uniform density, $v = \lambda f$ will give the relationship among the three. Amplitude is related to energy; the more energy there is, the greater the particle will move in its oscillation from the equilibrium position. Transverse waves such as electromagnetic waves, unlike longitudinal waves, can travel in vacuum. Superposition applies to all waveforms.

(L.O. 6.A.1.2)

## FREE-RESPONSE PROBLEMS

1. (a) i. The angle for the second-order maximum is given by

$$d\sin\theta_{bright} = (m\lambda) \qquad \text{where } m = 0, \pm 1, \pm 2, \dots$$

Solving for $\theta$,

$$\theta_{bright} = \sin^{-1}\left(\frac{m\lambda}{d}\right) = \sin^{-1}\left[\frac{2(6.35\times10^{-7}\text{m})}{4.0\times10^{-5}\text{m}}\right] = \mathbf{1.82°}$$

   ii. The angle for the second-order minimum is given by

$$d\sin\theta_{dark} = \left(m + \frac{1}{2}\right)\lambda \qquad \text{where } m = 0, \pm 1, \pm 2, \dots$$

The second order minimum occurs at

$$\theta_{dark} = \sin^{-1}\left(\frac{1.5\lambda}{d}\right) = \sin^{-1}\left[\frac{1.5\left(6.35\times10^{-7}\text{m}\right)}{4.0\times10^{-5}\text{m}}\right] = \mathbf{1.36°}$$

(b) The equations for part (a) i. $\sin\theta_{bright} = \dfrac{m\lambda}{d}$ and for part (b) i.

$\sin\theta_{dark} = \dfrac{m\lambda}{a}$ have the same form, *but* they do not mean the same thing. In part (a) the equation describes the *bright* regions in a two-slit interference pattern, where $d$ is the separation of the slits A and B.

In part (b), the equation describes the *dark* regions in a single-slit interference pattern where $a$ is the single-slit width.

i. $a\sin\theta_{bright} = \left(m+\dfrac{1}{2}\right)\lambda$ where $m = 0, \pm1, \pm2, ...$

Solving for $\theta$,

$$\theta_{bright} = \sin^{-1}\left(\frac{1.5\lambda}{a}\right) = \sin^{-1}\left[\frac{1.5\left(6.35\times10^{-7}\text{m}\right)}{4.0\times10^{-5}\text{m}}\right] = \mathbf{1.36°}$$

ii. $$\theta_{dark} = \sin^{-1}\left(\frac{m\lambda}{a}\right) = \sin^{-1}\left[\frac{2\left(6.35\times10^{-7}\text{m}\right)}{4.0\times10^{-5}\text{m}}\right] = \mathbf{1.82°}$$

(c) The linear separation is given by

$$y_{bright} = \frac{\lambda L m}{d} \text{ and } \Delta y = (y_4 - y_1) = \frac{6.35\times10^{-7}\text{m}\times1.5\text{m}\times(4-1)}{4.0\times10^{-5}\text{m}} =$$

**0.071 m**

(d) The angle for the second-order maximum is

$$\lambda_n = \frac{\lambda}{n} = \frac{635 \text{ nm}}{1.33} = 477 \text{ nm}$$

$$\theta = \sin^{-1}\left(\frac{m\lambda}{d}\right) = \sin^{-1}\left[\frac{2\left(4.77\times10^{-7}\text{m}\right)}{4.0\times10^{-5}\text{m}}\right] = \mathbf{1.37°}$$

(L.O. 6.E.3.1)

2. (a) Diffraction pattern A is made by polarized light.
   (b) Image A shows diffraction in only two dimensions from the two-dimensional slits. Image B shows diffraction in many orientations because unpolarized light is a superposition of many orientations.
   (c) Image A provides better evidence for the wave nature of light.
   (d) While both A and B show diffraction, image A goes further, demonstrating polarization.

   Diffraction and polarization are unique to waves.

(L.O. 6.B.3.1)

3. (a) Design of the experimental procedure:
   - Record the ruling on the diffraction grating.
   - Determine the spacing, *d*, between the rulings of the diffraction grating in the experiment. Record *d* in the data chart.

   (Note to student: The ruling is provided on the diffraction grating or can be obtained from your teacher. For example, if the grating is marked 100 lines/mm, the spacing between the lines is the reciprocal of the ruling—in this case, the spacing is $10^{-5}$ m/line.)

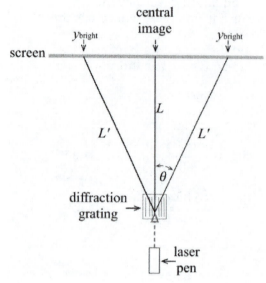

   - Measure the distance L from the diffraction grating to the screen. (White paper taped to a wall or whiteboard.) Record.
   - Project the laser pointer so that the light passing through the diffraction grating falls on the white paper, producing the central image and orders to the left and right of the central image.
   - Mark the position of the central image on the white paper.
   - Mark the positions of the first-order images right and left from the central image on the white paper.
   - Continue marking for the second- and third-order images right and left from the central image.
   - Mark other order images if you can obtain them using your diffraction grating.
   (b) The wavelength of the laser light is determined from $m\lambda = d \sin\theta_{bright}$ where $m = 0, \pm1, \pm2, ...$ If the distance from the grating is large enough, the $\tan\theta_{bright} = \dfrac{y_{bright}}{L}$ can be substituted for $\sin\theta_{bright}$ and you can eliminate the use of a protractor in the experiment.

   Measurement will include the distance from the grating to the screen locating the central image. Distance from the central image to each order image will be measured as $y_{bright}$ and recorded in the data chart.

   The distance to the right and left of the central bright image will be averaged for each order seen and marked.

Calculate the wavelength of the laser pointer's light by substitution into $m\lambda = d\sin\theta_{bright}$ or $m\lambda = \dfrac{d(y_{bright})}{L}$.

(c) The laser pointer does not produce one specific wavelength, but a narrow range. The range for a red laser pointer is 630 nm to 670 nm. For green, the range is 500 nm to 550 nm. Error should be calculated from the range, not a specific number as marked on some inexpensive pointers.

Substitution of $\tan\theta_{bright}$ for $\sin\theta_{bright}$ will introduce error if the angle is greater than 5°.

(d) The maximum angle of deviation is 90°. If the second order is seen at 76°, there will be no third-order image.

Alternatives to the experiment may involve using a mercury vapor lamp as the source of a discrete spectrum of light since the wavelengths in the mercury vapor spectrum are known to one decimal place.

Discharge tubes that produce discrete spectrums, such as hydrogen, helium, and mercury, can be used in the experiment.

A showcase bulb will produce a continuous spectrum, and it will be difficult to analyze a specific wavelength even if a colored cellophane filter is wrapped around the bulb.

(L.O. 6.C.2.1, 6.C.3.1)

4. (a) The requirement for diffraction of Huygens' wavelets is the opening through which the plane waves pass to be about the same size as the wavelength of the wave passing through the opening. If the opening is small, the wavelet cannot reproduce a straight wave front; some spreads out and diffracts through the opening.

In the case of radio waves, the wavelength $\lambda$ is about $10^3$ m to about $10^{-2}$ m, whereas visible light has a wavelength $\lambda$ ranging from 400 nm to 700 nm. For visible light, the opening around buildings is too large to cause diffraction.

(b) Compressional or longitudinal sound waves diffract in the same manner as transverse waveforms. When sound waves pass through an aperture such as a door or past the edge of an obstacle, they bend to some degree into a region not directly exposed to the sound or light source. Diffraction is the ability of waves to bend around edges and obstacles placed in their path. For waves to diffract around the edges of an open door, the size of the opening must be about the same size as the wave. For human speech, the wavelengths for sound waves are about 1 m, close to the size of the open door.

(L.O. 6.C.2.1, 6.E.1.1)

# 18

# Atomic and Quantum Physics

## Physics 2

## The Bohr Theory of the Hydrogen Atom

By the beginning of the twentieth century, the physics community had developed a model of the atom that explained most of the observations scientists had made in the past. However, two major anomalous observations remained for which the model could not account. First, the model predicted that excited atoms should produce a continuous spectrum of light of all colors. However, when the spectrum of the simplest element, hydrogen, was measured, it was missing specific frequencies. Although the missing frequencies seemed to fit a pattern, the existing atomic models couldn't explain why frequencies were missing or why there should even be a pattern. Second, scientists had also observed that a charged particle moving in a curved path should lose energy and spiral inward, yet electrons were clearly observed to orbit nuclei in stable orbits. The early twentieth-century model proved not to be a viable model for a stable atom.

It was evident to Niels Bohr that the atom had to consist of a heavy nucleus orbited at relatively great distance by electrons. Bohr proposed a remarkable set of postulates as the basis for a new atomic model.

Electrons can orbit a nucleus only at certain allowed radii that correspond to the limited subset of energy levels the electron can have. Bohr postulated that when a hydrogen atom is given outside energy by absorbing a photon, the electron absorbs that energy and moves to a higher orbit where it remains, circling the nucleus for a very short time. The electron then emits a photon and makes a *quantum jump* to a lower orbit.

**494**

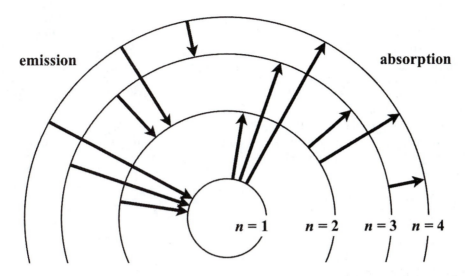

emission                                                    absorption

$n = 1$      $n = 2$      $n = 3$   $n = 4$

In Bohr's model of the hydrogen atom, as electrons emit energy as a photon, they fall to a lower energy level shown on the left. On the right, electrons absorb energy from a photon and move to a higher energy level. Notice that electrons are not located between the energy levels.

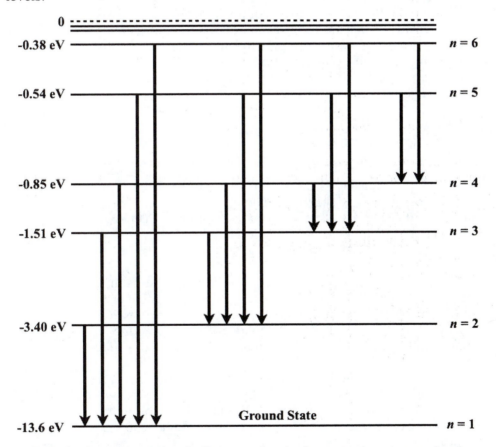

This is the energy level diagram for hydrogen. An energy level diagram is a useful way to illustrate energy transfers of electrons. Each energy level is given a value in eV. The difference in the energy levels can correspond to the frequency of the photon emitted through use of $E = hf$.

A convenient way to describe the transitions between orbits or stationary states is in terms of energy level diagrams. In these, the energy levels are plotted but not to scale, as illustrated on the previous page. The transitions are shown by arrows that run from the initial excited state of the atom to a final state.

## SAMPLE PROBLEM 1

Consider an excited hydrogen atom with an electron in the $n = 6$ energy level. The electron undergoes a quantum jump to the $n = 3$. Using the diagram on the previous page, determine the energy carried away by the photon created in the jump.

## SOLUTION TO PROBLEM 1

Find the energy difference

$$\Delta E = E_{final} - E_{initial}$$

$$\Delta E = E_{n=3} - E_{n=7}$$

$$\Delta E = (-1.51 \text{ eV}) - (-0.38 \text{ eV})$$

$$\Delta E = -1.13 \text{ eV}$$

The negative sign means that 1.23 eV of energy is carried away in the transition, the quantum jump.

It was soon found that Bohr's description of the hydrogen atom was only partially successful in calculating wavelengths of spectral lines emitted by other elements. The more complex the atom, the greater the failure of the Bohr model. The Bohr model of the atom, although now replaced with the more powerful *quantum mechanics*, remains an effective pictorial way to discuss the concept of energy levels.

## PLANCK'S QUANTUM THEORY

In 1903, Max Planck theoretically showed that light comes from light sources as tiny lumps of energy that he called *quanta* and that each quantum (quantum is the singular form of quanta) of light has an energy related to the frequency of the light. That relationship is

$$E = hf,$$

where $f$ is the frequency of the light and $h$ is called Planck's constant, which has the value $h = 6.63 \times 10^{-34}$ J/Hz. Such a constant measures energy per vibration, which is often written as J·s. For the previous 200 years, science had thought that light was made of electromagnetic waves, yet Planck's data suggested that light was made of particles that carried energy.

So far, we have calculated energy in units of joules, which are useful for human-scale interactions. To express the small energies we encounter in electron transitions, we use a smaller unit. Electron volts, eV, are convenient units to use at the atomic scale. 1 eV = $1.602 \times 10^{-19}$ J. Expressed in these terms, $h = 4.14 \times 10^{-15}$ eV·s.

## SAMPLE PROBLEM 2

Determine the energy of a single quantum of yellow light that has a wavelength of 590 nm. Express the answer in both J and eV.

## SOLUTION TO PROBLEM 2

First, we write the Planck equation $E = hf$. Recall that wavelength and frequency are related by $c = \lambda f$ or $f = c/\lambda$. Substitute for frequency in the Planck equation and

$$E = \frac{hc}{\lambda} = \frac{(6.63 \times 10^{-34} \text{ J} \cdot \text{s})(3 \times 10^8 \text{ m}/\text{s})}{590 \times 10^{-9} \text{ m}} = \mathbf{3.37 \times 10^{-19} \text{ J}}$$

Converting J to eV, we need a conversion factor

$$3.37 \times 10^{-19} \text{ J} \times \frac{1 \text{ eV}}{1.602 \times 10^{-19} \text{ J}} = \mathbf{2.10 \text{ eV}}$$

## SAMPLE PROBLEM 3

A quantum of infrared light carries an energy of 0.90 eV. Wha t is the wavelength of the associated electromagnetic wave? What is its frequency?

## SOLUTION TO PROBLEM 3

First, convert eV to J as $0.90 \text{ eV} \times \dfrac{1.602 \times 10^{-19} \text{ J}}{1 \text{ eV}} = 1.44 \times 10^{-19} \text{ J}$

Next, write the expanded form of Planck's equation $E = \dfrac{hc}{\lambda}$ and solve for wavelength.

$$\lambda = \frac{hc}{E} = \frac{(6.64 \times 10^{-34} \text{ J} \cdot \text{s})(3 \times 10^8 \text{ m}/\text{s})}{1.44 \times 10^{-19} \text{ J}} = \mathbf{1.38 \times 10^{-6} \text{ m}}$$

To find the frequency, use $c = f\lambda$.

$$f = \frac{c}{\lambda} = \frac{3 \times 10^8 \ m/s}{1.38 \times 10^{-6} \ m} = 4.14 \ Hz$$

## SAMPLE PROBLEM 4

Use the diagram showing energy level transitions to analyze a transition from the $n = 3$ orbit to the $n = 2$ orbit.
(a)  What energy does the electron lose in making this jump?
(b)  What wavelength of light is produced in this jump?
(c)  What color of light is produced?

## SOLUTION TO PROBLEM 4

(a)  Using the energy level diagram, the energy lost as the electron jumps from $n = 3$ to $n = 2$ is

$$\Delta E = E_2 - E_3 = (-1.51 \text{ eV}) - (-3.40 \text{ eV}) = \mathbf{1.89 \text{ eV}}$$

(b)  Depending on the energy units used,

$$h = 6.63 \times 10^{-34} \text{ J} \cdot \text{s} = 4.14 \times 10^{-15} \text{ eV} \cdot \text{s}$$

$$E = hf = \frac{hc}{\lambda} \text{ and}$$

$$\lambda = \frac{hc}{\Delta E} = \frac{(4.14 \times 10^{-15} \text{ eV} \cdot \text{s})(3 \times 10^8 \text{ m}/\text{s})}{1.89 \text{ eV}} = \mathbf{6.57 \times 10^{-7} \text{ m}}$$

(c)  A wavelength of 656 nm corresponds to **red light**.

## MODERN MODEL OF THE ATOM

According to our current model of the atom, electrons, neutrinos, photons, and quarks are considered *fundamental particles* that have no internal structure and cannot be subdivided. Neutrons and protons are composed of quarks, tiny charged particles that carry fractions of the elementary charge of the electron.

## ATOMIC SPECTRA OF HYDROGEN

When a sample of a low-density atomic gas or a simple molecular gas is heated or excited electrically and the spectra of these glowing gases are directed into a spectrometer, a series of discrete colored lines are observed. The line spectra are different for each element and provide a unique way of identifying the elements present in the sample of a material. No two elements have the same atomic spectrum. The spectrometer not only separates atomic spectral lines but also measures the wavelength of these lines.

When the full spectrum of light shines on an atom's electrons, only certain frequencies of photons can be absorbed by the electrons and cause an energy level transition. When these frequencies are absorbed, the rest of the spectrum can be seen with the absorbed frequencies missing. This is an *absorption spectrum*.

When an excited electron falls to a lower energy level, its energy is given off as a photon with a specific frequency that matches the energy transition. When these electron transitions are seen through a spectrometer, an *emission spectrum* is seen as a set of colored lines on a black background. The emission spectrum for an element is the opposite of the absorption spectrum, and both can be used to identify specific elements.

Line spectra served as the key to developing a model for atomic structure since the model had to explain line spectra and predict their wavelengths.

## THE DUAL NATURE OF MATTER

So far, we have treated all matter as particles. We used standard Newtonian mechanics to describe cars, the sun, and people. However, we cannot use traditional mechanics to describe fully the domain of small particles, like electrons, moving at great speeds. An electron behaves like a wave and a particle under varying circumstances. Throughout the early twentieth century, various experiments by Louis de Broglie, Erwin Schrödinger, Clinton Joseph Davisson, Lester Germer, and George Paget Thomson have supported both the particle and wave definitions of matter.

Davisson and Germer projected a stream of electrons through a crystal. The electrons diffracted and created the same type of diffraction pattern as X-rays diffracted in the crystal. The experiment showed that high-energy electrons have wavelengths. Thomson also confirmed the wave properties of electrons by showing they could be diffracted through a double slit. The ability to diffract is a property unique to waves.

## DE BROGLIE WAVES

In 1924, Louis de Broglie made the suggestion that this dual particle wave character of light should apply to moving objects as well. His suggestion was based on the general observation that nature often reveals a physical or mathematical symmetry. His reasoning was that the wavelength, $\lambda$, and the momentum, $p$, of a particle is related by the equation

$$\lambda = \frac{h}{p} = \frac{h}{mv}$$

The de Broglie equation states that as the momentum of a particle increases, the *de Broglie wavelength* decreases.

De Broglie's wave description of an electron fits with Bohr's description of electron orbits in hydrogen. If an electron can be modeled as a standing wave using de Broglie's relationship and if that electron is confined to a certain geometry by Bohr's model, then an electron's orbits can be modeled as the allowed resonant harmonics of a wave. In this way, the electron's quantum jumps were also changes in the wavelength of its standing wave.

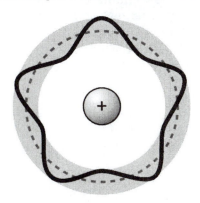 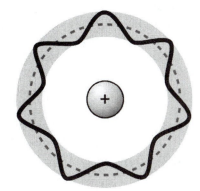

Left: Five de Broglie wavelengths fit in a Bohr orbit.
Right: Seven de Broglie wavelengths fit in a Bohr orbit.

### SAMPLE PROBLEM 5

Find the de Broglie wavelength of an electron that is traveling at $1.31 \times 10^6$ m/s.

### SOLUTION TO PROBLEM 5

$$\lambda = \frac{h}{p} = \frac{h}{mv} = \frac{6.62 \times 10^{-34} \, J \cdot s}{(9.11 \times 10^{-31} \, kg)(1.31 \times 10^6 \, m/s)} = 5.55 \times 10^{-10} \, m$$

### SAMPLE PROBLEM 6

A stream of electrons with the de Broglie wavelength found in Problem 5 is projected through a crystal, creating a diffraction pattern on a phosphorescent screen. How would the spacing between interference maxima of the diffraction pattern change if the speed of the electron beam was increased?

## Solution to Problem 6

As the speed of the electrons in the beam was increased, the momentum of the electrons would increase as $p = mv$. When the momentum increased, the wavelength would decrease due to $\lambda = \dfrac{h}{p}$. The spacing between the maxima would also decrease since $y = m\dfrac{\lambda L}{d}$.

## The Dual Nature of Light

Before the twentieth century, our models of light treated light as either a particle or a wave. Yet neither class of model was about to explain all of light's observed behaviors fully. It was believed that treating light as a particle or a wave was mutually exclusive. Today, we know that light has a dual nature, sometimes demonstrating its wavelike properties and sometimes acting more like a particle. The wave nature of light can be easily observed when the size of the interaction is comparable to or larger than the wavelength of light. A wave model of light explains phenomena such as diffraction and polarization, but these classical models cannot explain all aspects of photons.

When the energy transferred during an interaction is on the scale of the energy of a single photon, photons act as particles, countable objects that carry momentum. Collisions can occur only when action-reaction forces are shared by a pair of particles. The clearest evidence of photons' particle nature is that they can collide with other particles and transfer their momentum. In the case of a photon, momentum and energy can be related to wavelength and frequency.

We find a photon's momentum using de Broglie's equation, $p = \dfrac{h}{\lambda}$.

And we find a photon's energy using Planck's equation, $E = hf$.

## Sample Problem 7

Find the momentum of an X-ray photon with a wavelength of 5.0 nm.

## Solution to Problem 7

$$p = \frac{h}{\lambda} = \frac{6.63 \times 10^{-34}\,J \cdot s}{5.0 \times 10^{-9}\,m} = 1.33 \times 10^{-25}\,kg\frac{m}{s}$$

## The Photoelectric Effect

In his papers on special relativity, Einstein stated that light is emitted in the form of particles he called photons. As described by Planck's data, each photon carries energy $E = hf$. Scientists had observed that when certain types of light shone on metallic surfaces, the surfaces acquired positive electrical charge; however, with only the wave definition of light, the phenomena could not be explained. Einstein suggested a solution to a problem. Einstein stated that photons struck surface electrons and were absorbed by them. Surface electrons are bound by an energy we call the work function, $\phi$, which is specific to individual photoactive metals. If the incoming photons had enough energy, surface electrons not only would leave the surface of the metal, but also would leave with a kinetic energy. As the electrons were emitted, the metals would be left with a positive charge. Photons with less energy than the work function would not transfer enough energy to the electrons to cause them to be ejected. The amount of kinetic energy with which an electron will leave the surface of a metal is found using Einstein's photoelectric equation

$K = hf - \phi$

In the experimental apparatus, electrons are allowed to travel from the metal's surface to a collector. The voltage difference between the metal and the collector can be adjusted, and experimenters can observe the effect of the collector voltage on the rate at which charges arrive at the collector. The emitted electrons have a range of kinetic energies, but their maximum kinetic energy can be measured by looking at how much voltage is necessary to stop them. The potential (voltage) at which electrons with the maximum kinetic energy can be stopped is called the *stopping potential.*

This explanation only makes sense if we think of light as a particle. Individual photons are absorbed by individual electrons. If light were only a wave, a continuous beam of light would eventually give an electron enough energy to leave the metal's surface regardless of the light's energy. Also, if light were only a wave, adjusting the brightness of the light would have some effect on the kinetic energy of the ejected electrons. However, the brightness or intensity of the incident light, which can be considered as the rate at which photons collide with the surface, affects only the rate at which electrons are ejected rather than the electrons' energy.

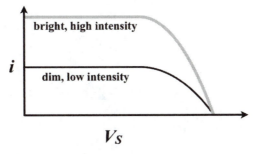

For light of the same frequency, increasing the intensity of the incident light causes more electrons to be ejected, but the ejected electrons have the same maximum kinetic energy.

Changing the intensity affects the current but does not affect the stopping potential.

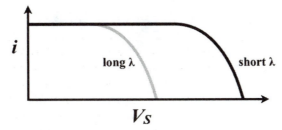

For light of the same intensity, increasing the wavelength of the incident light causes ejected electrons to have more energy, yet increasing the energy of the light doesn't dislodge an increased number of electrons.

Changing the wavelength affects the stopping potential but does not affect the current.

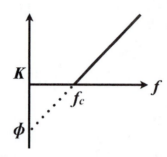

Increasing the frequency of the incident light increases the maximum kinetic energy of the ejected electrons.

Light below the threshold frequency or critical frequency *fc* jostles electrons around but cannot eject them.

The slope of this graph is Planck's constant. The *y*-intercept is equal to the work function.

## SAMPLE PROBLEM 8

What is the maximum kinetic energy of an electron released when a potassium wafer is irradiated with 250 nm of light. Potassium has a work function of $\phi = 2.21$ eV.

## SOLUTION TO PROBLEM 8

First, write the Einstein photoelectric equation

$$K = hf - \phi$$

$$K = \frac{hc}{\lambda} - \phi$$

$$K = \frac{\left(6.63 \times 10^{-34} \text{ J} \cdot \text{s}\right)\left(3.00 \times 10^{8} \frac{\text{m}}{\text{s}}\right)}{\left(250 \times 10^{-9} \text{ m}\right)} - 2.21 \text{ eV} \times \frac{1.602 \times 10^{-19} \text{ J}}{1 \text{ eV}}$$

$$K = 4.42 \times 10^{-19} \text{ J}$$

**Alternative Solution**

Express *h* in terms of eV · s.

$$h = 6.63 \times 10^{-34} \text{ J} \cdot \text{s} \times \frac{1.00 \text{ eV}}{1.602 \times 10^{-19} \text{ J}} = 4.14 \times 10^{-15} \text{ eV} \cdot \text{s}$$

$$K = hf - \phi$$

$$K = \frac{hc}{\lambda} - \phi$$

$$K = \frac{\left(4.14 \times 10^{-15} \text{ eV} \cdot \text{s}\right)\left(3 \times 10^{8} \frac{\text{m}}{\text{s}}\right)}{\left(250 \times 10^{-9} \text{ m}\right)} - 2.21 \text{ eV}$$

$$K = 2.76 \text{ eV}$$

As a check,

$$4.42 \times 10^{-19} \text{ J} \times \frac{1.00 \text{ eV}}{1.602 \times 10^{-19} \text{ J}} = 2.76 \text{ eV}$$

# THE COMPTON EFFECT

In 1923, Arthur H. Compton discovered that a high-energy photon can collide with an electron and bounce off with reduced energy in another direction. In this collision, just like any other collision, the incident photon loses momentum while the electron gains momentum. This is analogous to the collision between two billiard balls. His experiment demonstrated that photons had momentum and could cause collisions, proving that light behaves as both a wave and a stream of particles.

The momentum of a photon is given by de Broglie's relationship, $p = \frac{h}{\lambda}$.

## SAMPLE PROBLEM 9

A gamma ray with a wavelength of 6.48 pm collides with a stationary electron. The electron rebounds with a de Broglie wavelength of 26.3 pm.
(a) Calculate the momentum of the electron.
(b) Calculate the speed of the electron.
(c) Calculate the wavelength of the photon after the collision.

## SOLUTION TO PROBLEM 9

(a) We can find the electron's momentum using de Broglie's equation.

$$p = \frac{h}{\lambda} = \frac{6.62 \times 10^{-34} \text{ J} \cdot s}{26.3 \times 10^{-12} \text{ m}} = 2.52 \times 10^{-23} \text{ kg} \frac{m}{s}$$

(b) If we know a particle's momentum and mass, we can find its speed using

$$v = \frac{p}{m} = \frac{2.52 \times 10^{-23} \text{ kg} \frac{m}{s}}{9.11 \times 10^{-31} \text{ kg}} = 2.77 \times 10^{7} \frac{m}{s}$$

(c) The momentum gained by the electron was lost by the photon, so

$$P_{final} = P_{initial} - P_{electron}$$

$$P_{final} = \frac{h}{\lambda} - P_{electron} = \frac{6.62 \times 10^{-34} \text{ J} \cdot s}{6.48 \times 10^{-12} \text{ m}} - 2.52 \times 10^{-23} \text{ kg} \frac{m}{s}$$

$$P_{final} = 7.70 \times 10^{-23} \text{ kg} \frac{m}{s} \qquad \frac{h}{\lambda_f} = \frac{h}{\lambda_{initial}} - \frac{h}{\lambda_{electron}}$$

$$P_{final} = \frac{h}{\lambda} - P_{electron} = \frac{6.62 \times 10^{-34} \text{ J} \cdot s}{6.48 \times 10^{-12} \text{ m}} - 2.52 \times 10^{-23} \text{ kg} \frac{m}{s} = 7.70 \times 10^{-23} \text{ kg} \frac{m}{s}$$

Momentum is related to wavelength by de Broglie's equation.

$$\lambda = \frac{h}{p} = \frac{6.62 \times 10^{-34}\ \text{J}\cdot\text{s}}{7.70 \times 10^{-23}\ \text{kg}\frac{\text{m}}{\text{s}}} = 8.60\ \text{pm}$$

## QUANTUM MECHANICS

Classical physics, or Newtonian physics, is said to be *deterministic.* If we know the position, velocity, momentum, and kinetic and potential energies of an object at a particular time, then its future position, velocity, momentum, and kinetic and potential energies can be determined precisely if the forces acting on the object are known. Quantum mechanics, however, is *probabilistic.* In 1926, Erwin Schrödinger showed the *quantum mechanical model* of the atom and demonstrated that the electron orbits proposed by Bohr were incorrect. On its smallest scale matter—electrons in this case—is governed by the laws of probability. Electrons are located in regions of probability, also known as electron clouds, described by the wave function, $\psi$. The probability of finding an electron at a specific location is proportional to the square of the wave function. Schrödinger argued that we cannot say exactly where an electron is; we can only determine the probability that an electron may be at a particular region in space. For objects on the macroscopic scale, quantum mechanics predicts with a high probability that they behave according to Newtonian mechanics.

Quantum mechanics, which deals with the microscopic world of atoms, electrons, photons, and all quantum phenomena, is accepted as the fundamental theory underlying all physical processes.

Classical physics produces accurate results for the motion of ordinary bodies from grains of salt to Saturn. For objects moving at high speeds, we use special relativity.

## ATOMIC AND QUANTUM PHYSICS: STUDENT OBJECTIVES FOR THE AP® EXAM

- You should have an understanding of the basics of Planck's quantum theory.
- You should know and use Planck's quantum equation to solve problems.
- You should know and use the photoelectric effect equation and be able to apply it to graphical solutions.
- You should have an understanding of the Compton scattering effect.
- You should have an understanding of the origins of atomic spectra.
- You should have an understanding of the Bohr hydrogen atom.
- You should have an understanding of the de Broglie wave-particle duality concept.

## MULTIPLE-CHOICE QUESTIONS

1. An electron in a hydrogen atom drops from an energy state of
   –0.56 eV to an energy state of –3.40 eV, emitting a photon of
   electromagnetic radiation as it drops. The energy of this photon is
   (A) 3.96 eV because it cannot transition between allowed energy
       levels unless it has the total energy associated with these levels.
   (B) 3.40 eV because it can only reach this state if it radiates the
       energy at this level.
   (C) 2.86 eV because energy must be conserved in the drop between
       allowed states.
   (D) 0.56 eV because this is the maximum energy the electron
       received as it was boosted to a higher state.

2. The orbital radius of an electron in the lowest energy level of a
   hydrogen atom is $a_0$. The probability finding an electron in the
   ground state is greatest when
   (A) the de Broglie wavelength is $a_0$.
   (B) the de Broglie wavelength is $\pi a_0$.
   (C) the de Broglie wavelength is $2\pi a_0$.
   (D) the de Broglie wavelength is $4\pi a_0$.

3. Compton scattering produces a shift in the wavelength of the
   incident photon when it collides with an electron. The
   interaction is
   (A) elastic with a decrease in the wavelength of the scattered photon.
   (B) elastic with an increase in the wavelength of the scattered photon.
   (C) inelastic with a decrease in the wavelength of the scattered
       photon.
   (D) inelastic with an increase in the wavelength of the scattered
       photon.

4. According to the classical wave model of light, light travels as a
   continuous wave. In this model, the intensity of the wave depends
   on amplitude, not frequency. Experimental observations showed
   that when electromagnetic radiation struck the surface of certain
   metals, electrons were ejected from the surface. This can be
   explained by
   (A) the fact that electromagnetic radiation is both wave and
       particle in nature; the energy of the radiation is contained in
       packets called photons associated with frequency that can
       transfer both energy and momentum.
   (B) the fact that electromagnetic radiation is a wave in nature, the
       energy transfer taking some measurable time to transfer
       enough energy to free an electron.
   (C) the fact that electromagnetic radiation is both wave and
       particle in nature; the energy of the radiation is contained in
       packets called photons associated with amplitude of the
       intensity of the wave that can transfer both energy and
       momentum.
   (D) the fact that Newton's corpuscular theory was correct and that
       light consists of extremely small particles moving at high
       velocity.

5. The graph below shows the wave function, $\psi$, of a particle moving in the region $0 < x < 100$ pm. At which of the following positions is the probability of finding the particle greatest?

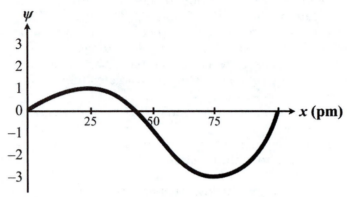

(A) 0 pm
(B) 25 pm
(C) 75 pm
(D) 100 pm

6. An energy level diagram is illustrated below for a certain atom. Rank the transitions for the wavelength $\lambda$ of the emitted electromagnetic radiation from longest to shortest wavelengths.

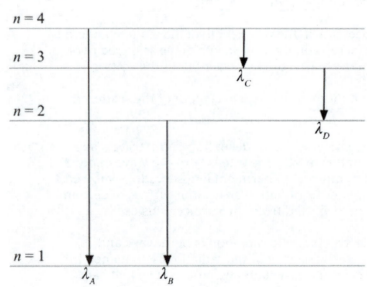

(A) $\lambda_C > \lambda_D > \lambda_B > \lambda_A$
(B) $\lambda_A > \lambda_B > \lambda_D > \lambda_C$
(C) $\lambda_A > \lambda_B > \lambda_C > \lambda_D$
(D) $\lambda_D > \lambda_C > \lambda_B > \lambda_A$

7. What is the de Broglie wavelength of an electron traveling with velocity of 0.300 c?
(A) $8.08 \times 10^{-12}$ m
(B) $2.42 \times 10^{-12}$ m
(C) $3.97 \times 10^{-14}$ m
(D) $1.32 \times 10^{-15}$ m

8. Bohr's theory of the hydrogen atom requires that the angular momentum of the electron be quantized. The de Broglie wave particle duality shows that the electron has a wave character given by $\lambda = \dfrac{h}{p}$; thus, the electron in allowed orbits forms a standing wave in the orbit. These standing waves are whole number multiples of the
   (A) frequency of the permitted radiation emitted or absorbed by the electron.
   (B) wavelength associated with electron's motion.
   (C) angular momentum of the electron.
   (D) circumference of the orbit.

9. The emission of a photoelectron from the surface of a certain metal as it is illuminated by light of a certain frequency and intensity is the indication that the electron
   (A) has absorbed the energy of several photons and after sufficient time will be able to leave the surface of the metal.
   (B) will leave the surface of the metal with a de Broglie wavelength equal to the wavelength of the light illuminating the surface.
   (C) will leave with excess kinetic energy if the light illuminating the surface has enough intensity to overcome the binding energy of the electron to the metal.
   (D) will leave with excess kinetic energy if the light illuminating the surface has enough energy associated with the photon to overcome the binding energy of the electron to the metal.

10. An electron and a proton are moving in the *x*-direction with a speed of either 0.10c or 0.30c as shown in the illustration below.

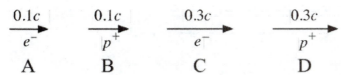

   Rank the de Broglie wavelengths associated with the particles from longest to shortest wavelength.
   (A) D > B > C > A
   (B) B > D > A > C
   (C) C > D > A > B
   (D) A > C > B > D

11. A graph of the maximum kinetic energy of the photoelectrons emitted as a function of the frequency of the incident light from the surface of a certain metal plate is shown for data obtained in a photoelectric effect experiment. The range of the electromagnetic radiation needed to produce the photoelectron from this metal was in what part of the electromagnetic spectrum?

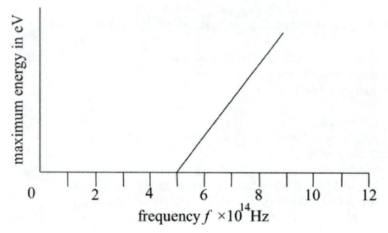

(A) infrared
(B) ultraviolet
(C) visible
(D) X-ray

12. An atom has the energy states pictured below. An electron in the ground state absorbs a photon with a wavelength of 248 nm. What will be electron's energy state after absorbing the photon?

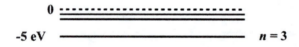

(A) $n = 1$
(B) $n = 2$
(C) $n = 3$
(D) The photon has enough energy to ionize the atom, so the electron will be liberated.

13. In the energy level diagram illustrated below for a certain atom, four transitions are shown. Rank the transitions for the frequency of the emitted electromagnetic radiation from largest to smallest.

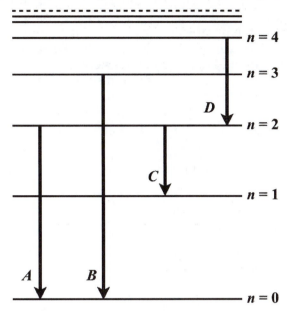

(A) $f_A > f_B > f_C > f_D$
(B) $f_C > f_D > f_A > f_B$
(C) $f_A = f_B = f_D > f_C$
(D) $f_B > f_A > f_D > f_C$

14. An X-ray photon collides with an electron at rest in a crystal, causing the electron to be ejected from the crystal with a gain in kinetic energy as the photon is deflected from its path. If the collision is considered to be elastic, then the photon must carry momentum away from the crystal moving with
(A) the same frequency and wavelength that it had prior to the collision.
(B) a smaller wavelength and a larger frequency that it had before the collision.
(C) the same wavelength but a reduction in speed since it transferred energy to the electron.
(D) a larger wavelength since momentum is carried away with the deflected photon.

15. Experimental proof that matter has wave properties is given by
(A) the Compton scattering of photons striking electrons in crystals.
(B) the photoelectric effect in which photons of electromagnetic radiation eject electrons from the surface of a material.
(C) diffraction of low-energy electrons striking crystals as they exhibit de Broglie waves associated with their velocity.
(D) transitions of electrons in the Bohr atom when they absorb or radiate electromagnetic radiation.

## Free-Response Problems

1. Physics students performed a photoelectric experiment in which they determined the relationship between the electrical stopping potential, $V_s$, needed to stop the photoelectron with the maximum kinetic energy and the frequency of light illuminating the surface. To make evaluation of their data easier, they converted each electrical stopping potential into energy units in joules by multiplying each $V_s$ by the charge on an electron, $1.6 \times 10^{-19}$ C.

| $K_{max}$ ($1.6 \times 10^{-19}$ J) | frequency ($f \times 10^{15}$ hz) |
|---|---|
| 0.00 | 0.50 |
| 3.31 | 1.00 |
| 5.51 | 1.25 |
| 6.62 | 1.50 |
| 9.94 | 2.00 |
| 12.6 | 2.25 |
| 13.3 | 2.50 |
| 16.6 | 3.00 |

(a) Plot the data points on the graph below.

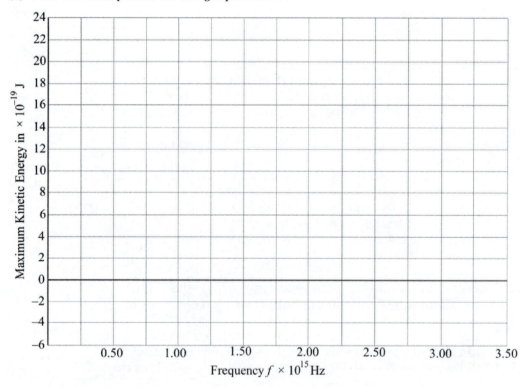

(b) What is the physical significance of the slope of the line?
(c) Determine the threshold frequency from the graph.
(d) What does the $y$-intercept of the graph represent?

2. A certain metal has a work function of 3.00 eV. When illuminated by light of intensity $I_0$ and frequency $f_0$, the maximum kinetic energy of the photoelectrons is 2.00 eV.
   (a) Doubling the intensity from $I_0$ to $2I_0$ results in
   ___ increasing the kinetic energy of the photoelectrons.
   ___ decreasing the kinetic energy of the photoelectrons.
   ___ no change in the kinetic energy of the photoelectrons.
   Justify your answer.
   (b) Doubling the frequency from $f_0$ to $2f_0$ results in
   ___ increasing the maximum kinetic energy of the photoelectrons.
   ___ decreasing the maximum kinetic energy of the photoelectrons.
   ___ no change in the maximum kinetic energy of the photoelectrons.
   Justify your answer.
   (c) Explain without mathematics how you can determine the threshold frequency for this metal.

3. The first four energy levels of a Bohr-like atom are given below.

   –0.85 eV ——————————— $n = 4$

   –1.51 eV ——————————— $n = 3$

   –3.40 eV ——————————— $n = 2$

   –13.6 eV ——————————— $n = 1$

   (a) What are the possible transitions that will produce unique spectral lines?
   (b) Which of these transitions will be in the visible part of the spectrum, and what are the energies of the emitted photons in eV?
   (c) What are the possible wavelengths in the visible part of the spectrum?

4. An electron is placed in the electric field shown on the next page and then released from rest at the surface of the negative plate. The potential difference across the plates is 50.0 V.

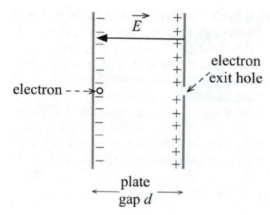

(a) What is its gain in kinetic energy as it accelerates across the plate gap $d$ and exits the plates through the hole in the positive plate?
(b) What is its increase in momentum upon exiting the field?
(c) Determine the de Broglie wavelength associated with the momentum of the electron.
(d) Would the electron cause diffraction if it interacted with a crystal whose spacing between layers is 0.10 nm?
___ would cause diffraction
___ would not cause diffraction
___ not enough information is given
Justify your answer.

# Answers

## MULTIPLE-CHOICE QUESTIONS

1. **C** The energy of the photon in transition is determined from $E = E_i - E_f$. The energy of the photon $E = -0.54$ eV $- (-3.40$ eV$)$ is 2.86 eV. Energy must be conserved in the transitions between allowed energy levels.

    (L.O. 5.B.8.1)

2. **C** The greatest probability of finding an electron in the hydrogen atom is in a region where there are an integer number of de Broglie wavelengths fitting into the circumference of the electron orbit. This probability would be greatest where the electron is closest to the nucleus and the circumference one wavelength.

    (L.O. 7.C.1.1, 7.C.2.1)

3. **B** The photon carries both momentum and energy that must be conserved in the collision. Thus, the collision is elastic. The photon transfers momentum and energy to the electron; therefore, the energy of the photon must decrease. Since the frequency decreases, the wavelength of the scattered photon must increase.

    (L.O. 5.D.1.6)

4.  **A** Electromagnetic radiation exhibits both a particle and wave nature. The photoelectric effect is an aspect of the particle nature in which the energy of the radiation is related to its frequency by $E = hf$. The very small discrete wave packets are called photons of light. The photons carry energy and momentum; interaction of the photon and matter will change these quantities. In the photoelectric effect, when the photon is absorbed, its momentum drops to zero, and if sufficient energy to overcome binding of the electron to the surface is absorbed, the photoelectron moves from the surface with some maximum kinetic energy determined by the frequency of the photon.

    (L.O. 6.F.3.1)

5.  **C** The probability of finding an electron is proportional to the square of the wave function.

    (L.O. 7.C.1.1)

6.  **A** The largest drop is the electromagnetic radiation that has the greatest frequency and thus the smallest wavelength. The energy is related to the wavelength by $\Delta E = hf = \dfrac{hc}{\lambda}$.

    (L.O. 5.B.8.1)

7.  **A** The velocity of the electron is $0.300c = 9.00 \times 10^7$ m/s. The de Broglie wavelength is

    $$\lambda = \frac{h}{mv} \quad \lambda = \frac{\left(6.63 \times 10^{-34} \text{ J} \cdot \text{s}\right)}{\left(9.11 \times 10^{-31} \text{ kg}\right)\left(9.00 \times 10^7 \text{ m/s}\right)} = 8.08 \times 10^{-12} \text{ m}$$

    (L.O. 1.D.1.1)

8.  **D** The Bohr theory requires that an integer multiple of the electron's wavelength equal the orbital circumference of the electron in that radius.

    (L.O. 7.C.2.1)

9.  **D** The emission of a photoelectron from the surface of a certain metal requires that the photon of light has sufficient energy to overcome the forces binding the electron to the surface. If the photon carries energy in excess of the binding energy, the photoelectron will leave the surface with excess kinetic energy. $K = hf - \phi$

    (L.O. 6.F.3.1)

10. **D** The de Broglie wavelength is determined from $\lambda = \dfrac{h}{mv}$. The slower the particle moves, the longer the de Broglie wavelength is. A more massive particle will have a shorter wavelength. Thus, compared to the protons, both electrons will have longer wavelengths. The proton traveling at 0.30c will have the shortest wavelength.

(L.O. 6.G.1.1)

11. **C** The threshold frequency obtained from the graph is $5 \times 10^{14}$ hz. Since $c = \lambda f$, the wavelength $\lambda = \dfrac{3.0 \times 10^8 \text{ m/s}}{5 \times 10^{14} \text{ hz}} = 600$ nm. This puts it in the range of visible light.

(L.O. 6.F.3.1)

12. **B** The incident photon has a wavelength of 248 nm so it has an energy of

$$E = \frac{hc}{\lambda} = \frac{1240 \text{ eV} \cdot \text{nm}}{248 \text{ nm}} = 5 \text{ eV}$$

After absorbing the photon, the electron's energy will increase by 5 eV up to −10 eV.

(L.O. 5.B.8.1)

13. **D** The greatest energy is for the longest transitions. Since $E = hf$, the highest energy has the highest frequency and shortest wavelength. The correct ranking is $f_B > f_A > f_D > f_C$.

(L.O. 7.C.4.1)

14. **D** Momentum must be conserved in a collision. The momentum of the X-ray photon is given by $\vec{p} = \dfrac{hf}{c} = \dfrac{h}{\lambda}$. Since momentum is transferred to the electron that gains kinetic energy in the collision, $\dfrac{h}{\lambda} = \dfrac{h}{\lambda'} + m_e v$. The X-ray photon is electromagnetic radiation moving with a velocity of $c$; thus, the wavelength must decrease.

(L.O. 5.D.1.6)

15. **C** X-rays passing through a crystal produce diffraction patterns. Low-energy electrons with a de Broglie wavelength in the order of magnitude of X-rays will produce a diffraction pattern similar to the one observed with X-rays. This is the experimental evidence that confirmed de Broglie's matter waves.

(L.O. 6.G.2.1)

## FREE-RESPONSE PROBLEMS

1.  (a)  The graph of the points is shown below.

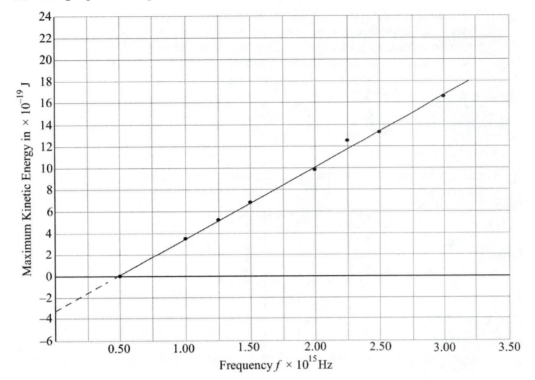

(b)  The slope of the line is Planck's constant.

$$h = \frac{\left(13.25 \times 10^{-19}\ \text{J} - 3.31 \times 10^{-19}\ \text{J}\right)}{\left(2.50 \times 10^{15}\ \text{Hz} - 1.00 \times 10^{15}\ \text{Hz}\right)} = 6.63 \times 10^{-34}\ \text{J} \cdot \text{s}$$

(c)  The $y$-intercept will give the work function for the material. The intercept gives $\phi \cong 3.5 \times 10^{-19}$ J.

Calculations from a substitution of data points into $K_{max} = hf - \phi$ produce a work function

$\phi = (6.624 \times 10^{-34}\ \text{J} \cdot \text{s})(2.00 \times 10^{15}\ \text{Hz}) - 9.94 \times 10^{-19}\ \text{J} = 3.31 \times 10^{-19}\ \text{J}$

(L.O. 6.F.3.1)

2.  (a)  The correct line checked is
___ √___ no change in the kinetic energy of the photoelectrons.
The maximum kinetic energy is independent of the intensity of light. Changing the intensity of the light just produces a greater number of photoelectrons per unit time, increasing the current so that a photoelectric blub will be brighter.

(b)  The correct line checked is
___ √___ increasing the kinetic energy of the photoelectrons.
Increasing the frequency means that more energy is added to the photoelectron since the energy of the incident photon is given by $E = hf$.

(c) The threshold frequency can be determined from setting the kinetic energy in the photoelectric equation to zero. The binding energy holding the electron to the metal has been overcome. Then since the kinetic energy is zero, the work function is equal to the product of Planck's constant times the threshold frequency. Converting the work function from electron volts into joules by multiplying by the change of the electron will result in consistent units for the division of the work function by Planck's constant. This will give the frequency in Hertz.

(L.O. 6.F.3.1)

3. (a) The six transitions that will produce unique spectral lines are shown below.

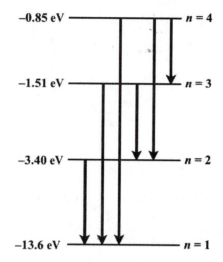

$n = 2$ to $n = 1$
$n = 3$ to $n = 1$
$n = 4$ to $n = 1$
$n = 3$ to $n = 2$
$n = 4$ to $n = 2$
$n = 4$ to $n = 3$

(b) In order to be in the visible spectrum, the emitted photon needs a wavelength between 700 nm and 400 nm. We can find the energy of visible photons by finding the energy of a 700 nm photon and a 400 nm photon. $E = \dfrac{hc}{\lambda}$

$$E_{700\,nm} = \frac{hc}{\lambda} = \frac{1240\ eV \cdot nm}{700\ nm} = 1.77\ eV$$

$$E_{400\,nm} = \frac{hc}{\lambda} = \frac{1240\ eV \cdot nm}{400\ nm} = 3.1\ eV$$

The only two transitions that emit photons in this energy range are $n = 4$ to $n = 2$ and $n = 3$ to $n = 2$. To find the energy of these photons, we compare the initial and final energies of the electron.

For $n = 4$ to $n = 2$, $\Delta E = (-3.40 \text{ eV}) - (-0.85 \text{ eV}) = -2.55 \text{ eV}$

For $n = 3$ to $n = 2$, $\Delta E = (-3.40 \text{ eV}) - (-1.51 \text{ eV}) = -1.89 \text{ eV}$

(c) $E = \dfrac{hc}{\lambda}$

$$\lambda = \dfrac{hc}{E} = \dfrac{1240 \text{ eV} \cdot \text{nm}}{2.55 \text{ eV}} = 486 \text{ nm}$$

$$\lambda = \dfrac{hc}{E} = \dfrac{1240 \text{ eV} \cdot \text{nm}}{1.89 \text{ eV}} = 656 \text{ nm}$$

(L.O. 5.B.8.1)

4. (a) The electron gains kinetic energy given by $W = \Delta K = q\Delta V$.

The gain in $\Delta K = e(50.0 \text{ V}) = \textbf{50.0 eV}$ or

$$\left(1.609 \times 10^{-19} \text{ C}\right)\left(50.0 \text{ J}/\text{C}\right) = \textbf{8.05} \times \textbf{10}^{-18} \textbf{ J}$$

(b) The momentum of the electron is $\vec{p} = m\vec{v}$ since it starts from rest.

Solve for the velocity from $\Delta K = \dfrac{1}{2}mv^2$. Thus, $v = \sqrt{\dfrac{2\Delta K}{m}}$

$$v = \sqrt{\dfrac{2\left(8.05 \times 10^{-18} \text{ J}\right)}{9.11 \times 10^{-31} \text{ kg}}} = 4.21 \times 10^6 \text{ m}/\text{s}$$

The momentum is

$$p = \left(9.11 \times 10^{-31} \text{ kg}\right)\left(4.21 \times 10^6 \text{ m}/\text{s}\right) = \textbf{3.83} \times \textbf{10}^{-24} \textbf{ kg} \cdot \textbf{m}/\textbf{s}$$

(c) The de Broglie wavelength is $\lambda = \dfrac{h}{mv} = \dfrac{h}{\vec{p}}$.

$$\lambda = \dfrac{6.624 \times 10^{-34} \text{ J} \cdot \text{s}}{3.83 \times 10^{-24} \text{ kg} \cdot \text{m}/\text{s}} = \textbf{1.73} \times \textbf{10}^{-10} \textbf{ m}$$

The de Broglie wavelength is **0.173 nm**.

(d) The correct line checked is

___ √ ___ would cause diffraction.

To cause diffraction of the crystal by the electron, the de Broglie wavelength must be approximately the size of the layer separation. Since the layer separation is given as 0.10 nm and the de Broglie wavelength for this problem is 0.244 nm, the electron would produce a diffraction pattern similar to one produced by X-ray diffraction of the crystal.

(L.O. 6.G.1.1, 6.G.2.1, 6.G.2.2)

# 19

# RELATIVITY AND NUCLEAR PHYSICS

PHYSICS 2

## SPECIAL RELATIVITY

In the spring of 1905, Albert Einstein's *special theory of relativity* gave the scientific community an entirely new way of seeing the universe. The theory introduced relativistic mass, time dilation, relativistic length, and the equivalency of mass and energy.

Einstein showed that no material object can travel as fast as light, which has a set speed limit of $3 \times 10^8$ m/s. As objects accelerate to near light speed, they exhibit behavior we don't observe at our everyday scale. Material objects increase in mass as they approach the speed of light and acquire *relativistic mass m*.

Relativistic mass is calculated from $m = \dfrac{m_0}{\sqrt{1 - \dfrac{v^2}{c^2}}}$

where $m_0$ is the *rest mass* of a body when it is at rest on the surface of Earth. The value $c$ is the speed of light, $3 \times 10^8$ m/s, and $v$ is the speed of the body relative to an observer at rest.

Applying this relationship, we see that a 2.0 kg object that was initially at rest traveling with relativistic speed $v$ of 86.6% $c$ will have a relativistic mass of 4.0 kg. The extra 2.0 kg of mass is actually the mass of the kinetic energy the body acquired as it was accelerated to $0.886c$. The relativistic mass equation has been verified many times since particle accelerators were built in the 1940s. It is very common at places like CERN and Fermilab to accelerate particles such as protons to $0.999c$ and slightly beyond.

**518**

Einstein also showed that time changes as clocks travel near the speed of light. Time intervals, $\Delta t$, undergo a relativistic effect. The faster a body travels, the slower its clock runs. The value $\Delta t_0$ is a time interval relative to the clock of an observer at rest. For example, a subatomic particle called a muon is quite unstable and decays in just 2.21 $\mu s$ when at rest in a laboratory on Earth. When muons are given speeds of 86.6% $c$ in particle accelerators, they are observed to "live" for 4.42 $\mu s$ as predicted by the relativistic time dilation equation

$$\Delta t = \frac{\Delta t_0}{\sqrt{1 - \frac{v^2}{c^2}}}$$

Relativistic time dilation has been verified many times.

Length also undergoes relativistic change as an object travels very close to the speed of light. This *relativistic contraction* occurs only in the direction of motion. Relativistic length, $L$, is determined from

$$L = L_0 \sqrt{1 - \frac{v^2}{c^2}}$$

In the equation, $L_0$ is the length of the body relative to the observer at rest. A 1.00 m rod traveling at $0.866c$ would have a length of 0.50 m.

The result of all investigations that have confirmed Einstein's interpretation of relativity is that observers in different reference frames moving at different velocities can disagree about time, mass, and distance.

## MASS-ENERGY EQUIVALENCE

Before special relativity, physicists and chemists had considered mass and energy as separate quantities that had to be conserved separately. Special relativity considers mass and energy to be the same thing but in two different forms; mass can be converted into energy, and energy can be converted into mass. Energy is measured in J; mass, in kg. Einstein found the conversion factor between them to be the speed of light squared.

$E = mc^2$

A very tiny mass can yield a very large quantity of energy. The usual SI unit of energy is the joule, J, but the electron volt is a more convenient way to express energy in quantum, atomic, and nuclear physics. Once again, we define the electron volt as

1 eV = $1.602 \times 10^{-19}$ J, or better yet, 1 MeV = $1.602 \times 10^{-13}$ J

### SAMPLE PROBLEM 1
Calculate the rest mass-energy of an electron in both joules and MeV.

### SOLUTION TO PROBLEM 1
Write the Einstein mass-energy relationship $E = mc^2$ and substitute.

$$E = \left(9.11 \times 10^{-31} \text{ kg}\right)\left(3.00 \times 10^8 \ \frac{\text{m}}{\text{s}}\right)^2 = 8.20 \times 10^{-14} \text{ J}$$

Converting to MeV,

$$8.20 \times 10^{-14} \text{ J} \times \frac{1 \text{ eV}}{1.602 \times 10^{-19} \text{ J}} \times \frac{1 \text{ Mev}}{1 \times 10^6 \text{ eV}} = \mathbf{0.51 \text{ MeV}}$$

The reverse is sometimes true as well; energy can be converted to matter under certain conditions. When a high-energy photon converts into particles, the mass of the particles comes from the energy of the photon. In an event like this, the *hf* of the photon becomes the *mc²* of the particles. Two particles can also combine to create photons.

### SAMPLE PROBLEM 2

What energy is released when a proton annihilates an anti-proton to form two identical gamma ray photons?

### SOLUTION TO PROBLEM 2

$p^+ + \overline{p}^- \rightarrow \gamma + \gamma$. The energy release is

$$E = mc^2 = 2m_p c^2$$

$$E = 2(1.67 \times 10^{-27} \text{kg})(3 \times 10^8 \text{ m}/\text{s})^2 = \mathbf{3.01 \times 10^{-10} \text{ J}}$$

## THE ATOMIC MASS UNIT

In the study of nuclear physics, the masses of nuclei and nucleons are expressed in terms of the *unified atomic mass unit*, u, which is $\frac{1}{12}$ the mass of the C-12 atom.

$$1.000\ 000 \text{ u} = 1.660540 \times 10^{-27} \text{ kg}$$

It is useful to express the unified atomic mass unit in terms of its rest mass energy.

> ## AP® Tip
>
> Calculations involving nuclear reactions involve working with precise masses expressed to six decimal places, yet the table of information lists masses to only three significant figures. If a particular question includes the masses of sub-atomic particles written to many significant figures, use them all and do not round.

### SAMPLE PROBLEM 3

What is the rest mass energy, in MeV, of the unified mass unit, u?

### SOLUTION TO PROBLEM 3

Write the Einstein mass-energy relationship

$$E = mc^2$$

$$E = (1.660540 \times 10^{-27} \text{ kg})\left(2.997924 \times 10^8 \ \frac{\text{m}}{\text{s}}\right)^2$$

$$E = 1.492419 \times 10^{-10} \text{ J} \times \frac{1.00 \text{ MeV}}{1.602218 \times 10^{-13} \text{ J}}$$

**1.00 u = 931.494 MeV**

---

## AP® Tip

Use the conversion factor $931.494 \frac{\text{MeV}}{\text{u}}$ to all three decimal places when determining the energy in nuclear reactions.

---

## NUCLEI AND NUCLEONS

The nucleus of an atom is composed of protons and neutrons, which are called nucleons due to their location. Nuclear reactions involve nucleons, not electrons. The atomic number, $Z$, of an atom tells us the number of protons within the nucleus. All atoms of the same element have the same number of protons. The number of neutrons can vary between atoms of the same element. We can use the mass number, $A$, to find how many neutrons are in the atom. The mass number, $A$, tells us the total number of nucleons in the atom, so $A - Z = N$, where $N$ is the number of neutrons.

Nuclear species are symbolically expressed as ${}^A_Z X$. The O-16 nuclear symbol is then ${}^{16}_8 O$. The O-16 nucleus contains 16 nucleons, of which 8 are protons and 8 are neutrons.

Atoms of the same element with differing number of neutrons are called isotopes. C-12, ${}^{12}_6 C$, and C-13, ${}^{13}_6 C$, are isotopes.

## RADIOACTIVITY

A stable nucleus will not spontaneously change. A nucleus is said to be radioactive if, after some time, the nucleus does spontaneously change. In the lightest nuclei, the number of protons and neutrons are about equal ($Z \approx N$). As the number of protons and neutrons increases in higher $Z$ nuclei, the number of neutrons becomes greater than the proton population ($N > Z$). Radioactive nuclei stabilize themselves by undergoing radioactive decay. There are several modes of radioactive decay.

In every nuclear decay event, nucleons change and rearrange, yet the total number of nucleons is the same before and after the event due to the *conservation of mass*. Likewise, the total amount of charge is the same before and after the event due to the *conservation of charge*.

Alpha decay is a spontaneous nuclear reaction that takes place in heavy, neutron-rich (too many neutrons) radioactive nuclei ($Z > 61$). A parent nucleus $X$ emits an alpha particle ${}^4_2 He$ or ${}^4_2 \alpha$ and forms a daughter nucleus $Y$. The nuclear reaction for this decay is ${}^A_Z X \rightarrow {}^4_2 He + {}^{A-4}_{Z-2} Y$.

### SAMPLE PROBLEM 4

Write the balanced alpha decay reactions for ${}^{239}_{94} Pu$, ${}^{238}_{92} U$, ${}^{230}_{90} Th$, ${}^{222}_{86} Rn$, and ${}^{214}_{84} Po$.

## SOLUTION TO PROBLEM 4

$$^{239}_{94}Pu \rightarrow {}^{4}_{2}\alpha + {}^{235}_{92}U$$

$$^{235}_{92}U \rightarrow {}^{4}_{2}\alpha + {}^{231}_{90}Th$$

$$^{230}_{90}Th \rightarrow {}^{4}_{2}\alpha + {}^{226}_{88}Ra$$

$$^{222}_{86}Rn \rightarrow {}^{4}_{2}\alpha + {}^{218}_{84}Po$$

$$^{218}_{84}Po \rightarrow {}^{4}_{2}\alpha + {}^{214}_{82}Pb$$

Notice that each alpha decay results in a new nucleus that is two atomic numbers lower on the periodic table.

### AP® Tip

The alpha particle is a helium nucleus and its nuclear notation is commonly expressed either as $^{4}_{2}He$ or $^{4}_{2}\alpha$. You should be familiar with both notations.

Negative beta decay takes place in neutron-rich (too many neutrons), radioactive nuclei when a neutron is converted into a proton while emitting a beta particle and antineutrino. A parent nucleus X decays by emitting a beta particle $^{0}_{-1}e$ or $^{0}_{-1}\beta$ and an antineutrino $^{0}_{0}\bar{v}$ and forms a daughter nucleus Y. The nuclear reaction for this decay process is $^{A}_{Z}X \rightarrow {}^{0}_{-1}e + {}^{0}_{0}\bar{v} + {}^{A}_{Z+1}Y$. The beta particle is an electron. Note that electrons (beta particles) do not exist in the nucleus; the beta particle and antineutrino are created in the decay process. Both leave the surface of the nucleus at near the speed of light.

### AP® Tip

The positive beta particle or positron is an anti-electron, and its nuclear notation is commonly expressed as either $^{0}_{-1}e$ or $^{0}_{-1}\beta$. You should be familiar with both notations.

## SAMPLE PROBLEM 5

Write balanced reactions for the beta decay of $^{239}_{92}U$, $^{239}_{93}Np$, $^{90}_{38}Sr$, $^{14}_{6}C$, and $^{3}_{1}H$.

## SOLUTION TO PROBLEM 5

$$^{230}_{92}U \rightarrow {}^{0}_{-1}\beta + {}^{0}_{0}\bar{v} + {}^{239}_{93}Np$$

$$^{239}_{93}Np \rightarrow {}^{0}_{-1}\beta + {}^{0}_{0}\bar{v} + {}^{239}_{94}Pu$$

$$^{90}_{38}Sr \rightarrow {}^{0}_{-1}\beta + {}^{0}_{0}\bar{v} + {}^{90}_{39}Y$$

$$^{14}_{6}C \rightarrow {}^{0}_{-1}\beta + {}^{0}_{0}\bar{v} + {}^{14}_{7}N$$

$$^{3}_{1}H \rightarrow {}^{0}_{-1}\beta + {}^{0}_{0}\bar{v} + {}^{3}_{2}He$$

Each beta decay results in a new nucleus that is one atomic number higher on the periodic table.

## AP® Tip

The beta particle is an electron and its nuclear notation is commonly expressed either as $_{-1}^{0}e$ or $_{-1}^{0}\beta$. You should be familiar with both notations.

Positron decay takes place in neutron-poor (too few neutrons) radioactive nuclei. These nuclei are not found in nature—they are artificially made in nuclear laboratories. Basically, a proton is converted into a neutron with the emission of the positron and neutrino. A parent nucleus X decays by emitting a positron particle $_{+1}^{0}e$ or $_{+1}^{0}\beta$ or a neutrino $_{0}^{0}\nu$ and forms a daughter nucleus Y. The nuclear reaction for this decay is $_{Z}^{A}X \rightarrow {}_{+1}^{0}e + {}_{0}^{0}\bar{\nu} + {}_{z-1}^{A}Y$.

The positron is an anti-electron. Note that positrons do not exist in the nucleus; the positron and neutrino are created in the decay process. Both leave the surface of the nucleus at near the speed of light.

### SAMPLE PROBLEM 6

Write the positron decay reactions for $_{26}^{53}Fe$, $_{17}^{33}Cl$, $_{15}^{28}P$, $_{8}^{14}O$ and $_{6}^{11}C$.

### SOLUTION TO PROBLEM 6

$$_{26}^{53}Fe \rightarrow {}_{+1}^{0}\beta + {}_{0}^{0}\bar{\nu} + {}_{25}^{53}Mn$$

$$_{17}^{33}Cl \rightarrow {}_{+1}^{0}\beta + {}_{0}^{0}\bar{\nu} + {}_{16}^{33}S$$

$$_{15}^{28}P \rightarrow {}_{+1}^{0}\beta + {}_{0}^{0}\bar{\nu} + {}_{14}^{28}Si$$

$$_{8}^{14}O \rightarrow {}_{+1}^{0}\beta + {}_{0}^{0}\bar{\nu} + {}_{7}^{14}N$$

$$_{6}^{11}C \rightarrow {}_{+1}^{0}\beta + {}_{0}^{0}\bar{\nu} + {}_{5}^{11}B$$

Each positron decay results in a new nucleus that is one atomic number lower on the periodic table.

The nucleons of a nucleus that has undergone alpha or some form of beta decay are left in an excited state. To stabilize itself, the excited nucleus discards the excitation energy as a gamma ray photon $\gamma$. The excited state is short-lived and emits a gamma ray in about $10^{-8}$ s. A star (*) indicates an excited state, and both the daughter nucleus and the parent nucleus have the same nucleon and proton populations. The nuclear reaction for this decay process is $_{Z}^{A}X^{*} \rightarrow {}_{Z}^{A}Y + \gamma$.

## RADIOACTIVE HALF-LIFE

Over time, a sample of radioactive isotopes undergoes change. The parent radioisotopes decay into other elements, although the quantity of material remains the same. The rate of decay of a particular sample of a radioisotope is measured by the *half-life*, $T_{1/2}$ defined as the time it takes half of the parent nuclei to decay. After 1 half-life, half of the parent nuclei has decayed into other elements and half of the parent nuclei remains. After 2 half-lives, only a quarter of the parent nuclei remains. After 3 half-lives, only an eighth of the parent nuclei remains.

## SAMPLE PROBLEM 7

Carbon-14 is used to date elements based on its nuclear decay. With a half-life of 5370 years, C-14 decays to its daughter isotope of N-14.
(a)   If a sample originally contains 10,000 molecules of U-235, how many parent molecules will remain after 2.5 half-lives?
(b)   How many molecules of N-14 will the sample contain after the same period of time?

## SOLUTION TO PROBLEM 7

(a)   After 1 half-life, 5000 molecules of U-235 will remain. One-half of the remaining molecules will remain after the second half-life has passed, which will leave 2500 molecules. One-fourth of the remaining molecules will decay after 0.5 half-life has passed, leaving 1875 remaining molecules.
(b)   The molecules that decayed turn into N-14, so 8125 molecules of N-14 should be present in the sample.

# NUCLEAR FISSION

When a nucleus of fissionable material such as $^{235}_{92}U$ or $^{239}_{94}Pu$ captures an additional neutron, the nucleus becomes more unstable and splits. When fission occurs, there is a large release of energy. The energy released is measured in electron volts, eV. When nuclear fission occurs, two fission fragments $A$ and $B$ are formed. We can write the fission reaction as

$$^{235}_{92}U + {}^{1}_{0}n \rightarrow A + B + x{}^{1}_{0}n + energy$$

Fissionable materials have a high density, and the fission fragments are quickly stopped, transforming their energy into thermal energy. Nuclear fission releases gigantic quantities of heat.

## SAMPLE PROBLEM 8

Write the nuclear reactions for the formation of Pu-239 from U-238 in the breeder reactor. In the breeder reactor, U-238 absorbs a neutron and forms the short-lived U-239 isotope. U-239 is a beta emitter and forms element 93 that is also a short-lived beta emitter. The end product is Pu-239.

## SOLUTION TO PROBLEM 8

Write the reaction for the absorption of the neutron by U-238.
$$^{238}_{92}U + {}^{1}_{0}n \rightarrow {}^{239}_{92}U$$

The beta decay is $^{239}_{92}U \rightarrow {}^{0}_{-1}e + {}^{0}_{0}\bar{\nu} + {}^{239}_{93}Np$

and the next beta decay $^{239}_{93}Np \rightarrow {}^{0}_{-1}e + {}^{0}_{0}\bar{\nu} + {}^{239}_{94}Pu$.

# THERMONUCLEAR FUSION

Nuclear fusion is a nuclear process in which light nuclei fuse or join to form a more massive nucleus with an extremely large energy release.
Thermonuclear fusion reactions drive stars and fuel the fireball of a thermonuclear device—a hydrogen bomb. Lithium-6 deuteride is a fuel used in thermonuclear devices. Below is an example of a fusion reaction.

$$^{6}_{3}Li + {}^{2}_{1}H \rightarrow {}^{4}_{2}He + {}^{4}_{2}He + energy$$

# NUCLEAR PHYSICS: STUDENT OBJECTIVES FOR THE AP® EXAM

- You should know and be able to use the mass-energy relationship.
- You should know and understand the symbols for atomic number $Z$, neutron number $N$, and mass number $A$ and be able to use them correctly in nuclear reactions.
- You should know the processes for (1) alpha, (2) beta, and (3) gamma decay and be able to balance nuclear reactions.
- You should understand the meaning of radioactive half-life.
- You should know the difference between nuclear fission and fusion and be able to balance their nuclear reactions.
- You should know that measurements of distance, speed, and time depend on the observer's speed.
- You should know that observers in different reference frames will not always agree on measurements of distance, speed, and time.

## MULTIPLE-CHOICE QUESTIONS

1. A nucleus of $^{30}_{14}\text{Si}$ contains

|       | protons | neutrons |
|-------|---------|----------|
| (A)   | 14      | 30       |
| (B)   | 14      | 16       |
| (C)   | 30      | 14       |
| (D)   | 30      | 16       |

2. In the nuclear reaction $^{224}_{88}\text{Ra} \rightarrow {}^{220}_{86}\text{Rn} + ?$, the particle(s) ejected
   (A) is an alpha particle, $^{4}_{2}\text{He}$.
   (B) is an alpha particle, $^{4}_{2}\text{He}$, and a conservation particle, the antineutrino, $\bar{v}$
   (C) is an alpha particle, $^{4}_{2}\text{He}$, and a conservation particle, the neutrino, $v$.
   (D) are two deuterons.

3. A certain radioactive isotope has a half-life of 2.40 days. If the initial mass of the sample is 6.00 $\mu$g, what is the mass of the remaining isotope after 5 half-lives?
   (A) 1.50 $\mu$g
   (B) 0.750 $\mu$g
   (C) 0.375 $\mu$g
   (D) 0.188 $\mu$g

4. In the nuclear decay $^{239}_{93}\text{Np} \rightarrow {}^{239}_{94}\text{Pu} + \bar{v} + ?$, what is the missing particle?
   (A) $^{4}_{2}\text{He}$
   (B) $^{1}_{0}\text{n}$
   (C) $e^{-}$
   (D) $e^{+}$

5. In the process of pair production, a high-energy photon becomes an electron and a positron.

$$\gamma \rightarrow e^+ + e^-$$

Why is this process possible?

(A) The initial kinetic energy of the incident photon equals the total kinetic energy of the electron and the positron.

(B) Total charge is conserved because photons carry no charge.

(C) Photons and electrons are essentially massless, so total mass is conserved in this process.

(D) This process is not possible because the total number of particles is not conserved.

6. What is the missing particle in the following nuclear reaction?

$$^{27}_{13}\text{Al} + {}^4_2\text{He} \rightarrow {}^{30}_{15}\text{P} + ?$$

(A) proton

(B) deuteron

(C) alpha particle

(D) neutron

7. Consider the following nuclear fission reaction:

$$^{235}_{92}\text{U} + {}^1_0\text{n} \rightarrow {}^{102}_{42}\text{Mo} + {}^{130}_{50}\text{Sn} + ?.$$

What is the missing quantity?

(A) $^4_2\text{He}$

(B) $2\left({}^1_1\text{H}\right)$

(C) $^4_0\text{n}$

(D) $4\left({}^1_0\text{n}\right)$

8. The conservation laws that govern all nuclear reactions include the conservation of

(A) charge and momentum.

(B) charge and mass-energy.

(C) mass energy and momentum.

(D) charge, mass energy, and momentum.

9. The half-life of a radioactive material is the time for the number of unstable nuclei of a given isotope of an element to decrease by a factor of 2. After 3 half-lives, the amount of radioactive isotope remaining is

(A) $\frac{1}{8}$ the initial amount.

(B) $\frac{1}{6}$ the initial amount.

(C) $\frac{1}{4}$ the initial amount.

(D) $\frac{1}{3}$ the initial amount.

10. Consider the following two-step nuclear reactions—first the $\beta^-$ decay of $^{137}_{55}Cs$,

$$^{137}_{55}Cs \rightarrow \, ^{137}_{56}Ba^* + \beta^- + \bar{v}$$

followed by the decay of

$$^{137}_{56}Ba^* \rightarrow \, ^{137}_{56}Ba + ?$$

What is the unknown particle?
(A) $\alpha$
(B) $\beta^+$
(C) $\gamma$
(D) $n$

11. An alpha decay occurring from the parent isotope of $^{238}_{92}U$ initially at rest forms the daughter isotope $^{234}_{90}Th$. In the process, the alpha particle, $^{4}_{2}He$, recoils from the thorium nucleus $^{234}_{90}Th$. The velocity of the center of mass of the system
(A) moves with the recoiling parent $^{238}_{92}U$ as the alpha particle moves away from the parent in the decay.
(B) moves in the direction of the thorium isotope since it is much larger.
(C) remains at rest since momentum is conserved in the process and the parent $^{238}_{92}U$ was at rest before the decay.
(D) moves in the direction of the alpha particle since it is smaller and carries the majority of the kinetic energy.

12. Scientists use carbon dating to date organic material since the ratio of $^{14}_{6}C$ to $^{12}_{6}C$ molecules in our atmosphere is constant. All living things have the same ratio. When they die, they no longer absorb $^{14}_{6}C$. The decay of the $^{14}_{6}C$ isotope produces which of the following particles?
(A) $\alpha$
(B) $\beta$–
(C) $\beta^+$
(D) $\gamma$

13. The fission of a heavy isotope into two smaller fission fragments is generally initiated using a thermal neutron since the neutron is
(A) unstable outside the nucleus with a half-life of about 10.4 minutes.
(B) electrically neutral and will not have be repelled by the nucleus as it is absorbed by the heavy isotope.
(C) about the same size as the fissionable nucleus so that it can cause the nucleus to fragment easily.
(D) basically massless and will not transfer much energy into the heavy isotope.

14. The number of neutrons in the atom of $^{11}_{5}B$ is
(A) 0.
(B) 5.
(C) 6.
(D) 11.

15. The mass of a stable nucleus is less than the sum of the constituent protons and neutrons that make up that nucleus. The difference is accounted for because
    (A) the protons and neutrons are arranged in the nucleus energy levels much like the electrons in their energy levels outside the nucleus.
    (B) the process of fission converts some mass into energy.
    (C) differences exist between the $Z$ and $A$ numbers.
    (D) the devices used to measure mass in the nuclear range are inaccurate.

## FREE-RESPONSE PROBLEMS

1.  A spacecraft rapidly orbiting Earth has a clock aboard; a video camera is aimed at the clock. A scientist on Earth watches the live video feed of the clock and compares it to an identical clock in a laboratory on Earth. Which of the following statements describes the observation the scientist would make about the clocks?
    (a) Both clocks record the same time.
    (b) The clock on the spacecraft shows an earlier time than the clock on Earth.
    (c) The clock on the spacecraft shows a later time than the clock on Earth.
    (d) The clock on the spacecraft alternately shows an earlier time or a later time than the clock on Earth.
    Explain your answer.

2.  Balance the following nuclear reactions. Justify your answers.
    (a) $^{14}_{7}\text{N} + {}^{4}_{2}\text{He} \rightarrow {}^{1}_{1}\text{H} + ?$
    (b) $^{2}_{1}\text{H} + {}^{12}_{6}\text{C} \rightarrow ? + {}^{10}_{5}\text{B}$
    (c) $^{1}_{1}\text{H} + {}^{1}_{0}\text{n} \rightarrow \gamma + ?$
    (d) $^{6}_{3}\text{Li} + {}^{1}_{1}\text{H} \rightarrow {}^{4}_{2}\text{He} + ?$
    (e) $^{2}_{1}\text{H} + {}^{3}_{1}\text{H} \rightarrow {}^{3}_{2}\text{He} + ? + ?$

3.  Write complete nuclear reactions for parts (a) through (e). Justify your answers as to the correct equation for each.
    (a) beta decay of $^{14}_{6}\text{C}$
    (b) gamma emission of $^{9}_{4}\text{Be}^{*}$
    (c) alpha decay of $^{240}_{94}\text{Pu}$
    (d) positron decay of $^{11}_{6}\text{C}$
    (e) nuclear fusion of two deuterons, $^{2}_{1}\text{H}$

4.  A canister holds 600 g of Ra-226, which has a half-life of 1620 years.
    (a) What mass of radium will be left in the canister after 3240 years if the canister is still around?
    (b) Compare your answer to part (a) with the mass of the original sample. If the two masses are different, explain whether this violates the law of conservation of mass.

# Answers

## MULTIPLE-CHOICE QUESTIONS

1. **B** An atom of an element denoted as $^A_Z X$, $Z$ is the proton number, and $A$ is the mass number. The atomic number gives the sum of the protons and neutrons in the isotope of the element. In this case, there are 14 protons and 30 – 14, or 16, neutrons.

   (L.O. 1.A.2.1)

2. **A** The decay is an alpha decay. Charge and mass must balance in the equation; thus, $^{224}_{88}\text{Ra} \rightarrow {}^{220}_{86}\text{Rn} + ?$. (No other particle is emitted with the alpha particle.)

   (L.O. 1.A.5.1)

3. **D** Half-life is the time it takes for half of the given number of radioactive nuclei to decay. $N = N_0 e^{-\lambda t}$ is the equation for the number of radioactive atoms undergoing decay. (Count rate $R$ in disintegrations /sec dps or activity A could be used as well.) Since the half-life is related to the decay constant $T_{1/2} = \dfrac{0.692}{\lambda}$, the equation can be written as $N = N_0 e^{-0.693t/T_{1/2}}$. The substitution into this equation gives $N = 6.00\ \mu g \left( e^{-0.693(12.0\ d)/2.4\ d} \right)$; then $N = 6.00\ \mu g (e^{-3.465})$ = 0.1876 $\mu$g. An alternative method of doing this without the equations above is as follows.

   | $T_{1/2}$ in days | Sample size in $\mu$g |
   |---|---|
   | 0 | 6.00 |
   | 2.40 | 3.00 |
   | 4.80 | 1.50 |
   | 7.20 | 0.750 |
   | 9.60 | 0.375 |
   | 12.00 | 0.1857 |

   which gives 0.188 $\mu$g, answer (D).

   (L.O. 7.C.3.1)

4. **C** In a nuclear reaction, the $Z$ number as well as the $A$ number must be conserved.

   In terms of the $Z$ number, $93 \rightarrow 94 + 0 - 1$; thus, the particle must have a charge ($Z$ number) of –1. (The antineutrino $\bar{\nu}$ is a conservation particle with zero charge and "zero" mass and is required for conservation of energy as well as spin.)

Conservation of mass number A gives $239 \rightarrow 239 + 0 + 0$. Thus, the particle missing in the equation must be a beta particle $e^-$. The beta particle can also be written as $\beta^-$ or $_{-1}^{0}\beta$.

(L.O. 7.C. 3.1)

5. **B** Photons are particles of light; they carry no charge. Electrons are negatively charged, and positrons are positively charged. The total charge before and after pair production is zero.

(L.O. 1.C.4.1, 5.G.1.1)

6. **D** In any nuclear reaction, both the $Z$ number and the $A$ number must be conserved.

$_{13}^{27}\text{Al} + {}_{2}^{4}\text{He} \rightarrow {}_{15}^{30}\text{P} + ?$ The $Z$ number is $13 + 2 = 15 + 0$. The $A$ number is $27 + 4 = 30 + 1$.

The particle has a mass $A = 1$ and $Z = 0$; thus, it is a neutron $_{0}^{1}\text{n}$.

(L.O. 5.C.1.1)

7. **D** In any nuclear reaction, both the $Z$ number and the $A$ number must be conserved.

$_{92}^{235}\text{U} + {}_{0}^{1}\text{n} \rightarrow {}_{42}^{102}\text{Mo} + {}_{50}^{130}\text{Sn} + ?$ The $Z$ number is $92 = 42 + 50 + 0$.

The particle must be a neutron. The $A$ number gives $235 + 1 = 102 + 130 + 4$.

Since the mass is 4 atomic mass units and the charge is zero, four neutrons are needed to complete the reaction.

(L.O. 5.G.1.1)

8. **D** Momentum is conserved in all interactions. Charge must be conserved in all nuclear reactions. Conservation of mass energy is also conserved in nuclear reactions. In nuclear reactions, $E = mc^2$ relates the conversions between mass and energy.

(L.O. 1.C.4.1, 5.G.1.1)

9. **A** After 1 half-life, the amount that remains is $\frac{1}{2}$. After 2 half-lives, the amount remaining is $\frac{1}{2} \times \frac{1}{2} = \frac{1}{4}$. This can be written as $\left(\frac{1}{2}\right)^2$. After 3 half-lives, the amount remaining is $\left(\frac{1}{2}\right)^3 = \frac{1}{8}$.

(L.O. 7.C.3.1)

10. **C** In all nuclear reactions, $Z$ and $A$ numbers must balance. The $Z$ numbers for barium remain as 56; the mass number A, as 137. This reaction indicates that $^{137}_{56}\text{Ba}^*$ is in an excited state and decayed to a more stable state by giving off energy in the form of a gamma ray, $\gamma$.

    (L.O. 5.G.1.1)

11. **C** The alpha decay of $^{238}_{92}\text{U} \rightarrow {}^{234}_{90}\text{Th} + {}^{4}_{2}\text{He}$ results in two new particles, $^{234}_{90}\text{Th}$ and $^{4}_{2}\text{He}$. Since the initial momentum of the system was zero, the total momentum after decay is also zero to conserve momentum. This means that the velocity of center of mass of the system will also be zero.

    (L.O. 5.G.1.1)

12. **C** The decay is $\beta^-$. There are six protons and eight neutrons in the nucleus of $^{14}_{6}\text{C}$. The neutron is transformed into a proton that remains in the nucleus. thus increasing the $Z$ number to 7 and producing nitrogen. The reaction is $^{14}_{6}\text{C} \rightarrow {}^{14}_{7}\text{N} + \beta^- + \overline{v}$.

    (L.O. 5.C.1.1)

13. **B** The neutron has no charge and will not be deflected by the orbital electrons or the positively charged nucleus—there is no Coulombic barrier for the neutron. When it is absorbed into the nucleus (easier to capture a slow-moving neutron), all of its initial energy will be used to disturb the nucleus. Thus, $^{1}_{0}\text{n} + {}^{235}_{92}\text{U} \rightarrow {}^{236}_{92}\text{U}^*$. The unstable uranium in the product then fissions and splits into smaller fragments (other elements) and several additional neutrons are also released.

    (L.O. 5.G.1.1)

14. **C** In the $^{11}_{5}\text{B}$ nucleus, there are $A - Z$ neutrons. $11 - 5 = 6$ neutrons.

    (L.O. 1.A.2.1)

15. **B** The total energy of a bound system is less than the energy of its separated nucleons. This difference is the $\Delta m$ of the system converted into binding energy produced in the formation of the nucleus. It is calculated from $E = mc^2$.

    (L.O. 4.C.4.1)

## FREE-RESPONSE PROBLEMS

1. The correct answer is (b): The clock on the spacecraft shows an earlier time than the clock on Earth.

$$\Delta t = \frac{\Delta t_0}{\sqrt{1 - \frac{v^2}{c^2}}}$$

   An object's speed affects the $\Delta t$ of the object. In the time dilation equation, increasing $v$ decreases the value of $\Delta t$, which is the time interval of the orbiting clock. The clock in orbit is moving much faster than the clock on Earth, so time will run slower on the clock in orbit.

   (L.O. 6.D.3.1)

2. Nuclear reactions:
   In any nuclear reaction, the $Z$ number as well as the $A$ number must be conserved.
   (a) $^{14}_{7}\text{N} + ^{4}_{2}\text{He} \rightarrow ^{1}_{1}\text{H} + ^{17}_{8}\text{O}$
      The $Z$ number is $7 + 2 = 1 + 8$; the $A$ number is $14 + 4 = 1 + 17$; thus, the missing nuclei is $^{17}_{8}\text{O}$.
   (b) $^{2}_{1}\text{H} + ^{12}_{6}\text{C} \rightarrow ^{4}_{2}\text{He} + ^{10}_{5}\text{B}$
      The $Z$ number is $1 + 6 = 2 + 5$; the $A$ number is $2 + 12 = 4 + 10$. The missing nucleus is 2He or $a$.
   (c) $^{1}_{1}\text{H} + ^{1}_{0}\text{n} \rightarrow \gamma + ^{2}_{1}\text{H}$
      The gamma ray, $\gamma$, has zero charge and zero mass. Conservation of $Z$ number gives $1 + 0 = 0 + 1$. Conservation of $A$ number is $1 + 1 = 0 + 2$. The unknown in the reaction is $^{2}_{1}\text{H}$. This is a deuteron and may be written as D in the reaction.
   (d) $^{6}_{3}\text{Li} + ^{1}_{1}\text{H} \rightarrow ^{4}_{2}\text{He} + ^{3}_{2}\text{He}$
      In this reaction, conservation of charge, $Z$ gives $3 + 1 = 2 + 2$. The $A$ number is determined as $6 + 1 = 4 + 3$. The unknown is an isotope of helium, $^{3}_{2}\text{He}$. It is *not* an alpha particle.
   (e) $^{2}_{1}\text{H} + ^{3}_{1}\text{H} \rightarrow ^{3}_{2}\text{He} + ^{1}_{0}\text{n} + ^{1}_{0}\text{n}$
      Conservation of $Z$ number is $1 + 1 = 2 + 0$. $A$ number is $2 + 3 = 3 + 2$. The only particle that could satisfy this equation is the neutron, $^{1}_{0}\text{n}$. Two neutrons are emitted. They are written separately or as $2\left(^{1}_{0}\text{n}\right)$. It is *not* written as $^{2}_{0}\text{n}$.

   (L.O. 5.C.1.1, 5.G.1.1)

3. Nuclear decays:
   (a) $^{14}_{6}\text{C} \rightarrow ^{0}_{-1}\beta + ^{0}_{0}\overline{\nu} + ^{14}_{7}\text{N}$   or   $^{14}_{6}\text{C} \rightarrow ^{0}_{-1}\text{e} + ^{0}_{0}\overline{\nu} + ^{14}_{7}\text{N}$
      For $Z$, the charge, $6 = -1 + 7$, and A requires $11 = 0 + 0 + 11$ since two particles are emitted in a beta negative decay: the beta particle and the conservation particle, the antineutrino.

(b) $^{9}_{4}\text{Be}^{*} \rightarrow ^{9}_{4}\text{Be} + \gamma$

This reaction indicates that $^{9}_{4}\text{Be}^{*}$ is in an excited state and decayed to a more stable state by giving off energy in the form of a gamma ray, $\gamma$, which has $Z = 0$ and $A = 0$.

(c) $^{240}_{94}\text{Pu} \rightarrow ^{236}_{92}\text{U} + ^{4}_{2}\alpha$ or $^{240}_{94}\text{Pu} \rightarrow ^{236}_{92}\text{U} + ^{4}_{2}\text{He}$

Substitution for $Z$ is $94 = 92 + 2$; $A$ produces $240 = 236 + 4$. No other particle is emitted with the alpha particle.

(d) $^{11}_{6}\text{C} \rightarrow ^{0}_{+1}\beta + ^{0}_{0}\nu + ^{11}_{5}\text{B}$ or $^{11}_{6}\text{C} \rightarrow ^{0}_{+1}\text{e} + ^{0}_{0}\nu + ^{11}_{5}\text{B}$

For $Z$, the charge, $6 = 1 + 5$, and $A$ requires $11 = 0 + 0 + 11$ since two particles are emitted in a beta positive or positron decay: the beta particle and the conservation particle, the neutrino.

(e) $^{2}_{1}\text{H} + ^{2}_{1}\text{H} \rightarrow ^{4}_{2}\text{He}$

In a fusion reaction, one heavier nuclide is produced. Thus, $A$ gives $2 + 2 = 4$, and $Z$ gives $1 + 1 = 2$, which is $^{4}_{2}\text{He}$ or $\alpha$.

(L.O. 5.C.1.1, 5.G.1.1, 7.C.3.1)


4. (a) 3240 years is equivalent to 2 half-lives of Ra-226. The first half-life will decrease the Ra-226 by half to 300 g. The next half-life will decrease the Ra-226 in the sample by another half to 150 g.

(b) The sample will still have 600 g. The Ra-226 will decay into Rd-222. Conservation of mass still holds.

(L.O. 7.C.3.1)

# AP® TABLES OF INFORMATION*

ADVANCED PLACEMENT PHYSICS 1 TABLE OF INFORMATION

| Constants and Conversion Factors | |
|---|---|
| Proton mass, $m_p = 1.67 \times 10^{-27}$ kg | Electron charge magnitude, $e = 1.60 \times 10^{-19}$ C |
| Neutron mass, $m_n = 1.67 \times 10^{-27}$ kg | Couloub's law constant, $k = \dfrac{1}{4}\pi\varepsilon_0 = 9.0 \times 10^9 \, \text{N} \cdot \text{m}^2 / \text{C}^2$ |
| Electron mass, $m_e = 9.11 \times 10^{-31}$ kg | Universal gravitational constant, $G = 6.67 \times 10^{-11}$ m³/kg · s² |
| Speed of light, $c = 3.00 \times 10^8$ m/s | Acceleration due to gravity at Earth's surface, $g = 9.8$ m/s² |

| Unit Symbols | meter | m | kelvin | K | joule | J | volt | V |
|---|---|---|---|---|---|---|---|---|
| | kilogram | kg | hertz | Hz | watt | W | ohm | Ω |
| | second | s | newton | N | coulomb | C | degree Celsius | °C |
| | ampere | A | | | | | | |

| Prefixes | | |
|---|---|---|
| Factor | Prefix | Symbol |
| $10^{12}$ | tera | T |
| $10^{9}$ | giga | G |
| $10^{6}$ | mega | M |
| $10^{3}$ | kilo | k |
| $10^{-2}$ | centi | c |
| $10^{-3}$ | milli | m |
| $10^{-6}$ | micro | $\mu$ |
| $10^{-9}$ | nano | n |
| $10^{-12}$ | pico | P |

| Values of Trigonometric Functions for Common Angles | | | | | | | |
|---|---|---|---|---|---|---|---|
| $\theta$ | 0° | 30° | 37° | 45° | 53° | 60° | 90° |
| $\sin\theta$ | 0 | 1/2 | 3/5 | $\sqrt{2}/2$ | 4/5 | $\sqrt{3}/2$ | 1 |
| $\cos\theta$ | 1 | $\sqrt{3}/2$ | 4/5 | $\sqrt{2}/2$ | 3/5 | 1/2 | 0 |
| $\tan\theta$ | 0 | $\sqrt{3}/3$ | 3/4 | 1 | 4/3 | $\sqrt{3}$ | ∞ |

The following conventions are used in this exam.

I.   The frame of reference of any problem is assumed to be inertial unless otherwise stated.

II.  Assume air resistance is negligible unless otherwise stated.

III. In all situations, positive work is defined as work done <u>on</u> a system.

IV.  The direction of current is conventional current: the direction in which positive charge would drift.

V.   Assume all batteries and meters are ideal unless otherwise stated.

---

AP® is a trademark registered by the College Board, which is not affiliated with, and does not endorse, this product.

## ADVANCED PLACEMENT PHYSICS 1 EQUATIONS

### Mechanics

$$v_x = v_{x0} + a_x t$$

$$x = x_0 + v_{x0}t + \tfrac{1}{2}a_x t^2$$

$$v_x^2 = v_{x0}^2 + 2a_x(x - x_0)$$

$$\vec{a} = \frac{\sum \vec{F}}{m} = \frac{\vec{F}_{net}}{m}$$

$$\left|\vec{F_f}\right| \le \mu \left|\vec{F_n}\right|$$

$$a_c = \frac{v^2}{r}$$

$$\vec{p} = m\vec{v}$$

$$\Delta \vec{p} = \vec{F}\Delta t$$

$$K = \frac{1}{2}mv^2$$

$$\Delta E = W = F_\parallel d$$

$$= Fd\cos\theta$$

$$P = \frac{\Delta E}{\Delta t}$$

$$\theta = \theta_0 + \omega_0 t + \tfrac{1}{2}\alpha t^2$$

$$\omega = \omega_0 + \alpha t$$

$$x = A\cos(2\pi ft)$$

$$\vec{\alpha} = \frac{\sum \vec{\tau}}{I} = \frac{\vec{\tau}_{net}}{I}$$

$$\tau = r_\perp F = rF\sin\theta$$

$$L = I\omega$$

$$\Delta L = \tau\Delta t$$

$$K = \frac{1}{2}I\omega^2$$

$$\left|\vec{F_s}\right| = k\left|\vec{x}\right|$$

$$U_s = \frac{1}{2}kx^2$$

$$\rho = \frac{m}{V}$$

$$\Delta U_g = mg\Delta y$$

$$T = \frac{2\pi}{\omega} = \frac{1}{f}$$

$$T_s = 2\pi\sqrt{\frac{m}{k}}$$

$$T_p = 2\pi\sqrt{\frac{\ell}{g}}$$

$$\left|\vec{F_g}\right| = G\frac{m_1 m_2}{r^2}$$

$$\vec{g} = \frac{\vec{F_g}}{m}$$

$$U_G = -\frac{Gm_1 m_2}{r}$$

$a$ = acceleration
$d$ = distance
$E$ = energy
$F$ = force
$f$ = frequency
$h$ = height
$I$ = rotational inertia
$K$ = kinetic energy
$k$ = spring constant
$L$ = angular momentum
$\ell$ = length
$m$ = mass
$P$ = power
$p$ = momentum
$r$ = radius or separation
$T$ = period
$t$ = time
$U$ = potential energy
$V$ = volume
$v$ = speed
$W$ = work done on a system
$x$ = position
$\alpha$ = angular acceleration
$\mu$ = coefficient of friction
$\theta$ = angle
$\tau$ = torque
$\omega$ = angular speed

### Electricity

$$\left|\vec{F_E}\right| = k\frac{|q_1 q_2|}{r^2}$$

$$I = \frac{\Delta q}{\Delta t}$$

$$R = \frac{\rho\ell}{A}$$

$$I = \frac{\Delta V}{R}$$

$$P = I\Delta V$$

$$R_s = \sum_i R_i$$

$$\frac{1}{R_p} = \sum_i \frac{1}{R_i}$$

$A$ = area
$F$ = force
$I$ = current
$\ell$ = length
$P$ = power
$q$ = charge
$R$ = resistance
$r$ = separation
$t$ = time
$V$ = electric potential
$\rho$ = resistivity

### Waves

$$\lambda = \frac{v}{f}$$

$f$ = frequency
$v$ = speed
$\lambda$ = wavelength

### Geometry and Trigonometry

Rectangle $A = bh$

Triangle $A = \frac{1}{2}bh$

Circle
$A = \pi r^2$
$C = 2\pi r$

Rectangular solid
$V = \ell w h$

Right triangle
$c^2 = a^2 + b^2$

$$\sin\theta = \frac{a}{c}$$

$$\cos\theta = \frac{b}{c}$$

$$\tan\theta = \frac{a}{b}$$

$A$ = area
$C$ = circumference
$V$ = volume
$S$ = surface area
$b$ = base
$h$ = height
$\ell$ = length
$w$ = width
$r$ = radius

Cylinder
$$V = \pi r^2 \ell$$
$$S = 2\pi r\ell + 2\pi r^2$$

Sphere
$$V = \frac{4}{3}\pi r^3$$
$$S = 4\pi r^2$$

## ADVANCED PLACEMENT PHYSICS 2 TABLE OF INFORMATION

### Constants and Conversion Factors

| | |
|---|---|
| Proton mass, $m_p = 1.67 \times 10^{-27}$ kg | Electron charge magnitude, $e = 1.60 \times 10^{-19}$ C |
| Neutron mass, $m_n = 1.67 \times 10^{-27}$ kg | 1 electron volt, $1\ eV = 1.60 \times 10^{-19}$ J |
| Electron mass, $m_e = 9.11 \times 10^{-31}$ kg | Speed of light, $c = 3.00 \times 10^8$ m/s |
| Avogadro's number, $N_0 = 6.02 \times 10^{23}$ mol$^{-1}$ | Universal gravitational constant, |
| Universal gas constant, $R = 8.31$ J / (mol · $K$) | $G = 6.67 \times 10^{-11}$ m³/kg · s² |
| Boltzmann's constant, $k_B = 1.38 \times 10^{-23}$ J/K | Acceleration due to gravity at Earth's surface, $g = 9.8$ m/s² |

| | |
|---|---|
| 1 unified atomic mass unit, $1\ u = 1.66 \times 10^{-27}$ kg $= 931$ MeV/$c^2$ | Coulomb's law constant, $k = 1/4\pi\varepsilon_0 = 9.0 \times 10^9$ N · m²/C² |
| Planck's constant, $h = 6.63 \times 10^{34}$ J · s $= 4.14 \times 10^{15}$ eV · s | Vacuum permeability, $\mu_0 = 4\pi \times 10^{-7}$ (T · m) / A |
| $hc = 1.99 \times 10^{-25}$ J · m $= 1.24 \times 10^3$ eV · nm | Magnetic constant, $k' = \mu_0 / 4\pi = 1 \times 10^{-7}$ (T · m) / A |
| Vacuum permittivity, $e_0 = 8.85 \times 10^{-12}$ C² /N · m² | 1 atmosphere pressure, $1$ atm $= 1.0 \times 10^5$ N/m² $= 1.0 \times 10^5$ Pa |

| Unit Symbols | | | | | | | | |
|---|---|---|---|---|---|---|---|---|
| | meter | m | mole | mol | watt | W | farad | F |
| | kilogram | kg | hertz | Hz | coulomb | C | tesla | T |
| | second | s | newton | N | volt | V | degree Celsius | °C |
| | ampere | A | pascal | Pa | ohm | Ω | electron volt | eV |
| | kelvin | K | joule | J | henry | H | | |

### Prefixes

| Factor | Prefix | Symbol |
|---|---|---|
| $10^{12}$ | tera | T |
| $10^9$ | giga | G |
| $10^6$ | mega | M |
| $10^3$ | kilo | k |
| $10^{-2}$ | centi | c |
| $10^{-3}$ | milli | m |
| $10^{-6}$ | micro | $\mu$ |
| $10^{-9}$ | nano | n |
| $10^{-12}$ | pico | P |

### Values of Trigonometric Functions for Common Angles

| $\theta$ | 0° | 30° | 37° | 45° | 53° | 60° | 90° |
|---|---|---|---|---|---|---|---|
| $\sin\theta$ | 0 | 1/2 | 3/5 | $\sqrt{2}/2$ | 4/5 | $\sqrt{3}/2$ | 1 |
| $\cos\theta$ | 1 | $\sqrt{3}/2$ | 4/5 | $\sqrt{2}/2$ | 3/5 | 1/2 | 0 |
| $\tan\theta$ | 0 | $\sqrt{3}/3$ | 3/4 | 1 | 4/3 | $\sqrt{3}$ | $\infty$ |

The following conventions are used in this exam.

I. The frame of reference of any problem is assumed to be inertial unless otherwise stated.

II. In all situations, positive work is defined as work done on a system.

III. The direction of current is conventional current: the direction in which positive charge would drift.

IV. Assume all batteries and meters are ideal unless otherwise stated.

V. Assume edge effects for the electric field of a parallel plate capacitor unless otherwise stated.

VI. For any isolated electrically charged object, the electric potential is defined as zero at infinite distance from the charged object.

## ADVANCED PLACEMENT PHYSICS 2 EQUATIONS

### Mechanics

$$v_x = v_{x0} + a_x t$$

$$x = x_0 + v_{x0}t + \tfrac{1}{2}a_x t^2$$

$$v_x^2 = v_{x0}^2 + 2a_x(x - x_0)$$

$$\vec{a} = \frac{\sum \vec{F}}{m} = \frac{\vec{F}_{net}}{m}$$

$$\left|\vec{F}_f\right| \le \mu \left|\vec{F}_n\right|$$

$$a_c = \frac{v^2}{r}$$

$$\vec{p} = m\vec{v}$$

$$\Delta \vec{p} = \vec{F}\Delta t$$

$$K = \frac{1}{2}mv^2$$

$$\Delta E = W = F_\parallel d$$

$$= Fd \cos \theta$$

$$P = \frac{\Delta E}{\Delta t}$$

$$\theta = \theta_0 + \omega_0 t + \frac{1}{2}\alpha t^2$$

$$\omega = \omega_0 + \alpha t$$

$$x = A \cos(\omega t)$$

$$= A \cos(2\pi f t)$$

$$x_{cm} = \frac{\sum m_i x_i}{\sum m_i}$$

$$\vec{\alpha} = \frac{\sum \vec{\tau}}{I} = \frac{\vec{\tau}_{net}}{I}$$

$$\tau = r_\perp F = rF \sin \theta$$

$$L = I\omega$$

$$\Delta L = \tau \Delta t$$

$$K = \frac{1}{2}I\omega^2$$

$$\left|\vec{F}_s\right| = k\left|\vec{x}\right|$$

$$U_s = \frac{1}{2}kx^2$$

$$\Delta U_g = mg\Delta y$$

$$T = \frac{2\pi}{\omega} = \frac{1}{f}$$

$$T_s = 2\pi\sqrt{\frac{m}{k}}$$

$$T_p = 2\pi\sqrt{\frac{\ell}{g}}$$

$$\left|\vec{F}_g\right| = G\frac{m_1 m_2}{r^2}$$

$$\vec{g} = \frac{\vec{F}_g}{m}$$

$$U_G = -\frac{Gm_1 m_2}{r}$$

$a$ = acceleration
$d$ = distance
$E$ = energy
$F$ = force
$f$ = frequency
$h$ = height
$I$ = rotational inertia
$K$ = kinetic energy
$k$ = spring constant

$L$ = angular momentum
$\ell$ = length
$m$ = mass
$P$ = power
$p$ = momentum
$r$ = radius or separation
$T$ = period
$t$ = time
$U$ = potential energy
$v$ = speed
$W$ = work done on a system
$x$ = position
$\alpha$ = angular acceleration
$\mu$ = coefficient of friction
$\theta$ = angle
$\tau$ = torque
$\omega$ = angular speed

### Electricity and Magnetism

$$\left|\vec{F}_E\right| = \frac{1}{4\pi\varepsilon_0}\frac{|q_1 q_2|}{r^2}$$

$$\vec{E} = \frac{\vec{F}_E}{q}$$

$$\left|\vec{E}\right| = \frac{1}{4\pi\varepsilon_0}\frac{|q|}{r^2}$$

$$\Delta U_E = q\Delta V$$

$$V = \frac{1}{4\pi\varepsilon_0}\frac{q}{r}$$

$$\left|\vec{E}\right| = \left|\frac{\Delta V}{\Delta r}\right|$$

$$\Delta V = \frac{Q}{C}$$

$$C = \kappa\varepsilon_0\frac{A}{d}$$

$$E = \frac{Q}{\varepsilon_0 A}$$

$$U_c = \frac{1}{2}Q\Delta V = \frac{1}{2}C(\Delta V)^2$$

$$I = \frac{\Delta Q}{\Delta t}$$

$$R = \frac{\rho \ell}{A}$$

$$P = I\Delta V$$

$$I = \frac{\Delta V}{R}$$

$$R_s = \sum_i R_i$$

$$\frac{1}{R_p} = \sum_i \frac{1}{R_i}$$

$$C_p = \sum_i C_i$$

$$\frac{1}{C_s} = \sum_i \frac{1}{C_i}$$

$$B = \frac{\mu_0}{2\pi}\frac{I}{r}$$

$$\vec{F}_M = q\vec{v}\times\vec{B}$$

$$\left|\vec{F}_M\right| = \left|q\vec{v}\right|\left|\sin\theta\right|\left|\vec{B}\right|$$

$$\vec{F}_M = I\vec{\ell}\times\vec{B}$$

$$\left|\vec{F}_M\right| = \left|I\vec{\ell}\right|\left|\sin\theta\right|\left|\vec{B}\right|$$

$$\Phi_B = \vec{B}\cdot\vec{A}$$

$$\Phi_B = \left|\vec{B}\right|\cos\theta\left|\vec{A}\right|$$

$$\varepsilon = -\frac{\Delta\Phi_B}{\Delta t}$$

$$\varepsilon = B\ell v$$

$A$ = area
$B$ = magnetic field
$C$ = capacitance
$d$ = distance

$E$ = electric field
$\varepsilon$ = emf
$F$ = force
$I$ = current
$\ell$ = length
$P$ = power
$Q$ = charge
$q$ = point charge
$R$ = resistance
$r$ = separation
$t$ = time
$U$ = potential (stored) energy
$V$ = electric potential
$v$ = speed
$\rho$ = resistivity
$\theta$ = angle
$\Phi$ = flux

## ADVANCED PLACEMENT PHYSICS 2 EQUATIONS

| Fluid Mechanics and Thermal Physics | | Waves and Optics | |
|---|---|---|---|
| $\rho = \dfrac{m}{V}$ | $A$ = area | $\lambda = \dfrac{v}{f}$ | $d \sin \theta = m\lambda$ |
| $P = \dfrac{F}{A}$ | $F$ = force | | $d$ = separation |
| | $h$ = depth | $n = \dfrac{c}{v}$ | $f$ = frequency or focal length |
| $P = P_0 + \rho g h$ | $k$ = thermal conductivity | | $h$ = height |
| $F_b = \rho V g$ | $K$ = kinetic energy | $n_1 \sin \theta_1 = n_2 \sin \theta_2$ | $L$ = distance |
| $A_1 v_1 = A_2 v_2$ | $L$ = thickness | | $M$ = magnification |
| | $m$ = mass | $\dfrac{1}{s_i} + \dfrac{1}{s_o} = \dfrac{1}{f}$ | $m$ = an integer |
| $P_1 + \rho g y_1 + \dfrac{1}{2}\rho v_1^2$ | $n$ = number of moles | | $n$ = index of refraction |
| | $N$ = number of molecules | $\left| M \right| = \left| \dfrac{h_i}{h_o} \right| = \left| \dfrac{s_i}{s_o} \right|$ | $s$ = distance |
| $\quad = P_2 + \rho g y_2 + \dfrac{1}{2}\rho v_2^2$ | $P$ = pressure | | $v$ = speed |
| | $Q$ = energy transferred to a system by heating | $\Delta L = m\lambda$ | $\lambda$ = wavelength |
| $\dfrac{Q}{\Delta t} = \dfrac{kA\Delta T}{L}$ | $T$ = temperature | | $\theta$ = angle |
| $PV = nRT = Nk_B T$ | $t$ = time | | |
| | $U$ = internal energy | | |
| $K = \dfrac{3}{2}k_B T$ | $V$ = volume | | |
| | $v$ = speed | | |
| $W = -P\Delta V$ | $W$ = work done on a system | | |
| $\Delta U = Q + W$ | $y$ = height | | |
| | $\rho$ = density | | |

| Modern Physics | Geometry and Trigonometry | |
|---|---|---|
| $E = hf$ | Rectangle $A = bh$ | $A$ = area |
| $K_{max} = hf - \phi$ | Triangle $A = \dfrac{1}{2}bh$ | $C$ = circumference |
| | | $V$ = volume |
| $\lambda = \dfrac{h}{p}$ | Circle | $S$ = surface area |
| | $A = \pi r^2$ | $b$ = base |
| $E = mc^2$ | $C = 2\pi r$ | $h$ = height |
| $E$ = energy | Rectangular solid $V = \ell w h$ | $\ell$ = length |
| $f$ = frequency | Right triangle | $w$ = width |
| $K$ = kinetic energy | $c^2 = a^2 + b^2$ | $r$ = radius |
| $m$ = mass | $\sin\theta = \dfrac{a}{c}$ | Cylinder |
| $p$ = momentum | $\cos\theta = \dfrac{b}{c}$ | $V = \pi r^2 \ell$ |
| $\lambda$ = wavelength | $\tan\theta = \dfrac{a}{b}$ | $S = 2\pi r \ell + 2\pi r^2$ |
| $\phi$ = work function | | Sphere |
| | | $V = \dfrac{4}{3}\pi r^3$ |
| | | $S = 4\pi r^2$ |

# Part III

## Practice Tests

# AP® PHYSICS 1
# PRACTICE EXAM A

AP® PHYSICS 1
Section I
50 Multiple-Choice Questions
Time—90 Minutes

**Note:** To simplify calculations, you may use $g = 10 \text{ m/s}^2$ in all problems.

**Directions:** Each of the questions or incomplete statements below is followed by four suggested answers or completions. Select the one that is best in each case.

1. Marks made on a timing tape every 0.02 s apart are shown below. Which diagram best illustrates uniform acceleration of the body attached to the timing tape?

   I   •  •   •   •   •   •   •   •

   II  •  • • •    •   • • •     •

   III •    •        •           •

   IV • • •      •       •       •

   (A) I
   (B) II
   (C) III
   (D) IV

2. An object has an initial velocity of 10.0 m/s. How long must it accelerate at a constant rate of 2.00 m/s² to have an average velocity equal to twice its initial velocity?
   (A) 10.0 s
   (B) 12.0 s
   (C) 16.0 s
   (D) 20.0 s

3. A graph of the velocity of four objects as a function of time is shown below.

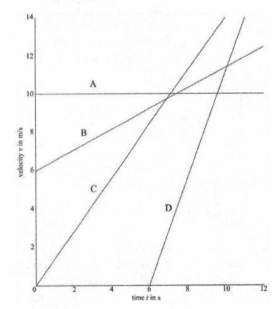

   Rank the graphs for the accelerations from greatest to least.
   (A) A > B > C > D
   (B) B > C > D > A
   (C) C > D > B > A
   (D) D > C > B > A

**GO ON TO NEXT PAGE**

4.  A 1.00 kg body and a 2.00 kg body are dropped simultaneously from the same height above the ground. In the absence of friction,
    (A) the 2.00 kg body reaches the ground first with the larger velocity since the gravitational force acting on it is larger.
    (B) the 1.00 kg body reaches the ground first with the larger velocity since it has a smaller inertia and is easier to accelerate.
    (C) they reach the ground at the same time with the same velocity since they are in the same gravitational field.
    (D) they reach the ground at the same time but the smaller body has the larger speed since it has the smaller inertia.

5.  An object is projected at some velocity $v_0$ at an angle $\theta$ above the horizontal. When the body is at the highest position in its flight, which of the following diagrams best represents the correct velocity and acceleration vectors for the body?

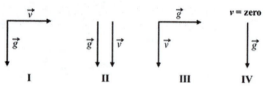

I    II    III    IV

    (A) I
    (B) II
    (C) III
    (D) IV

**Questions 6 and 7**
Base your answers to questions 6 and 7 on the following information.

6.  A 1.00 kg ball with a speed of 14.0 m/s is projected horizontally from the roof of a building. If the height of the building is 60.0 m and air resistance is negligible, the approximate time the ball is in the air is
    (A) 4.29 s.
    (B) 3.50 s.
    (C) 2.93 s.
    (D) 1.76 s.

7.  As the ball in Problem 6 falls freely under the influence of gravity to Earth's surface, it will gain an equal amount of
    (A) momentum for each meter through which it falls.
    (B) speed for each meter through which it falls.
    (C) momentum during each second it falls.
    (D) kinetic energy for each second it falls.

**Questions 8 and 9**
Base your answers to questions 8 and 9 on the following information.

8.  A hockey puck with a mass of 0.170 kg is sliding on ice at 10.0 m/s. Neglect friction on the surface of the ice. The hockey puck leaves the ice and moves onto a sidewalk around the ice rink. If the sidewalk has a coefficient of friction $\mu_k = 0.42$, which of the following choices best represents the forces acting on the hockey puck when it moves onto the sidewalk?

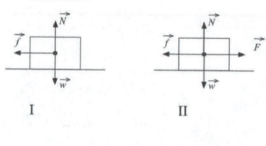

I    II

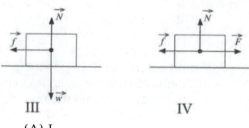

III    IV

    (A) I
    (B) II
    (C) III
    (D) IV

9. Using information from question 8, determine the distance the hockey puck will travel before coming to rest.
   (A) 12.1 m
   (B) 26.5 m
   (C) 71.5 m
   (D) 143 m

10. A 3.00 kg body is lifted 1.60 m above a floor and then moved a horizontal distance of 2.00 m. What is the work done by the gravitational force on the body?
    (A) –165 J
    (B) –118 J
    (C) –58.8 J
    (D) –47.0 J

11. A 2.00 kg body is dropped from a tall building. Ignoring friction,
    (A) its momentum remains constant since the only force acting on it is the force of gravity.
    (B) only its momentum increases because of the external force of gravity.
    (C) its momentum remains constant but its kinetic energy increases since the gravitational force does work on the body.
    (D) both its momentum and the kinetic energy increase because of the external force of gravity.

12. The momentum of a car rounding a curve at constant speed is
    (A) constant because its speed is constant.
    (B) constant because the acceleration of the car is constant.
    (C) not constant because an external force is acting on the car to change its direction.
    (D) not constant because work is done on the car by the external force.

13. An object moves under the action of a variable force as shown in the graph below.

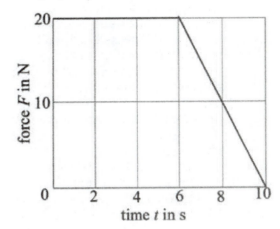

The change in the momentum of the body over the 10 s interval is closest to
(A) 100 N · s.
(B) 120 N · s.
(C) 160 N · s.
(D) 200 N · s.

14. An object initially moving to the right on a flat horizontal frictionless surface with a velocity of 1.20 m/s is subjected to several forces, as shown in the diagram, for a period of 10.0 s.

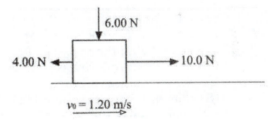

The momentum of the body will
(A) decrease because a force is applied to the body opposite its initial motion during the 10.0 s interval.
(B) increase because the net force is in the direction of the initial velocity.
(C) remain constant since the net force acting on the body is zero during the 10.0 s interval.
(D) be undetermined since the mass of the body is not given.

**GO ON TO NEXT PAGE**

15. A 1.80 kg box is held in place at the top of a 30° incline and then released. It slides down the incline, which is 1.20 m in length. If the coefficient of kinetic friction between the box and the incline is 0.20, what is the work done by the frictional force on the box as it slides completely down the incline?
    (A) −3.67 J
    (B) −2.12 J
    (C) 2.12 J
    (D) 3.67 J

16. A 0.200 kg block is attached to a horizontal spring $k = 150.0$ N/m on a frictionless surface. On the opposite side of the block, the spring is attached to a wall. The block-spring system is compressed 0.100 m and then released. What is the maximum velocity of the system, and where does it occur?
    (A) 7.50 m/s at $x = \pm A$
    (B) 7.50 m/s at $x = 0$
    (C) 2.74 m/s at $x = \pm A$
    (D) 2.74 m/s at $x = 0$

17. A 2.50 kg body slides across a frictionless horizontal tabletop that is 1.20 m above the level of the floor with a speed of 2.10 m/s. Relative to the tabletop, the mechanical energy associated with the body is
    (A) only kinetic energy as it slides on the frictionless surface.
    (B) only gravitational potential energy since the table is 1.20 m above the floor.
    (C) the sum of the kinetic energy and the gravitational potential energy since the surface is frictionless.
    (D) the difference between the kinetic energy and the gravitational potential energy.

18. A 1.50 kg body slides off a frictionless horizontal tabletop that is 1.00 m above the floor with a speed of 2.00 m/s. As it strikes the ground, its energy is
    (A) 11.7 J.
    (B) 14.7 J.
    (C) 16.2 J.
    (D) 17.7 J.

19. A small object is attached to a horizontal spring which is connected to a stationary wall. The spring is compressed to a position $x = -A$ and then released. In one complete cycle, the object moves a distance of
    (A) 4$A$.
    (B) 2$A$.
    (C) $A$/2.
    (D) $A$/4.

20. A block of ice sliding across a frozen pond at 3.00 m/s enters a region of rough ice where a frictional force of 1.20 N acts. As the block of ice travels over the rough patch, the block's kinetic energy will
    (A) increase because the friction force is applied in the direction the block's motion.
    (B) decrease because the friction force is applied opposite the direction of the block's motion.
    (C) remain constant since the block maintains its speed of 3.00 m/s.
    (D) initially decrease and then remain constant once the block reaches a speed less than 1.2m/s.

21. The period of a simple pendulum 1.20 m long at the surface of Earth is 2.20 s. What is its period if it is taken to a planet whose mass is $3M_E$ and radius is $2R_E$?
    (A) 1.90 s
    (B) 2.25 s
    (C) 2.54 s
    (D) 3.11 s

22. Three masses $m_1 = 10.0$ kg, $m_2 = 20.0$ kg, and $m_3 = 30.0$ kg are located on a horizontal line as shown below.

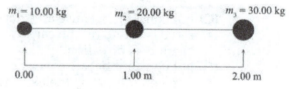

The gravitational force on $m_3$ due to the other two masses is closest to
    (A) $5.0 \times 10^{-9}$ N to the left.
    (B) $4.0 \times 10^{-8}$ N to the right.
    (C) $4.5 \times 10^{-8}$ N to the left.
    (D) $5.8 \times 10^{-8}$ N to the right.

23. A person in a car at an intersection is waiting for the light to turn green when an ambulance sounding its siren with a frequency $f_0 = 1000$ Hz approaches the car from the rear at 15 m/s. As the ambulance approaches the car, the person in the car perceives

(A) a lower frequency because the wavelength between successive waves is shorter as the ambulance approaches the stationary car with a velocity relative to the person in the car.

(B) a higher frequency because the wavelength between successive waves is shorter as the ambulance approaches the stationary car with a velocity relative to the person in the car.

(C) no change in frequency since the person is stationary with respect to the ambulance.

(D) a frequency that increases as the ambulance approaches because the distance between the successive waves decreases as the ambulance approaches the car.

24. A projectile shot with some velocity $\vec{v}_0$ at an angle $\theta$ with respect to the horizontal explodes when it is at a distance $h$ above the ground.

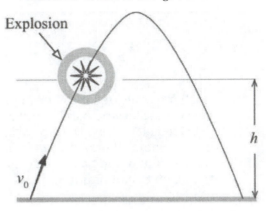

Which of the four illustrations best indicates the path of the fragments?

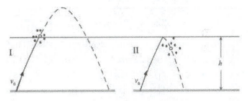

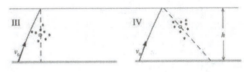

(A) I
(B) II
(C) III
(D) IV

25. A beam of gas molecules is incident on a barrier at 30° relative to the horizontal surface as shown.

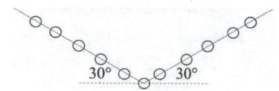

Each molecule of a mass $m$ and a speed $v$ strikes the barrier and is reflected elastically. During this event, what is the total change in momentum of the gas molecules in the beam containing $n$ particles?

(A) $nmv$
(B) $2nmv$
(C) $nmv(\sin 30°)$
(D) $nmv(\cos 30°)$

26. An object is lifted in the gravitational field associated with a planet of radius $2R_E$ and mass $3M_E$. Which of the graphs below best represents the relationship between the mass and the elevation it is lifted?

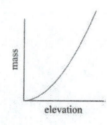

I

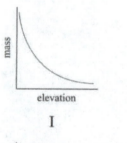

II

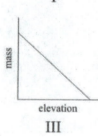

III

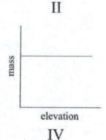

IV

(A) I
(B) II
(C) III
(D) IV

27. A 2.00 kg block sliding across a frictionless surface strikes a spring whose spring constant is $k = 15.0$ N/m, compressing it 0.12 m. The initial velocity of the block is closest to

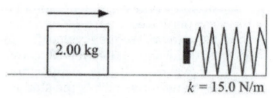

(A) 0.22 m/s.
(B) 0.33 m/s.
(C) 0.57m/s.
(D) 0.95 m/s.

28. A set of horizontal rods have their axis of rotation at the left end of the rod, and forces are applied to the rods as shown. Rank the forces from the greatest to least torque acting on the rods.

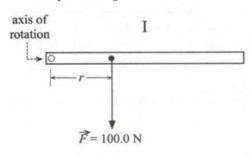

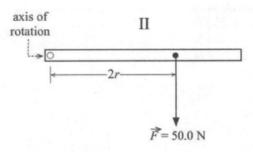

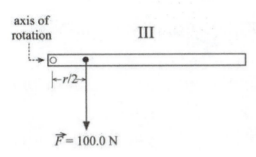

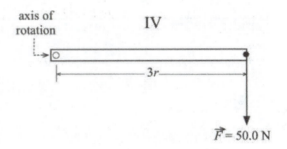

(A) IV > (I = II) > III
(B) IV > II > I > III
(C) I > III > IV > II
(D) I > II > III > IV

29. A figure skater with arms extended starts spinning on ice at $\omega_0 = 1.50$ rad/s. When the skater draws her arms in against her body,

(A) her angular momentum remains constant but her kinetic energy increases since work is done by the skater pulling her arms inward.

(B) her angular momentum increases but her kinetic energy decreases since she changed her moment of inertia.

(C) both her kinetic energy and the angular momentum increase since she did work in pulling her arms inward.

(D) neither her kinetic energy nor the angular momentum increased since no net external torque was applied to her.

30. A 1.50 kg block is attached to a spring of $k = 25.0$ N/m that oscillates on a frictionless horizontal surface with an amplitude of 0.08 m. Doubling the

(A) mass of the block and the amplitude without changing the spring constant will increase the period of vibration by a factor of 2.

(B) mass of the block and the amplitude without changing the spring constant will increase the period of vibration by a factor of $\sqrt{2}$.

(C) spring constant and the amplitude while holding the mass constant will increase the period of vibration by a factor of 2.

(D) spring constant and the amplitude while holding the mass constant will increase the period of vibration by a factor of $\sqrt{2}$.

**GO ON TO NEXT PAGE**

31. A 2.50 kg mass hung at rest on the end of a spring held vertically from a support rod stretches the spring 0.065 m. When the spring-mass system is set into vibration, the frequency of oscillation is
(A) 0.624 Hz.
(B) 1.96 Hz.
(C) 3.92 Hz.
(D) 12.3 Hz.

32. What is the wavelength of the wave shown below?

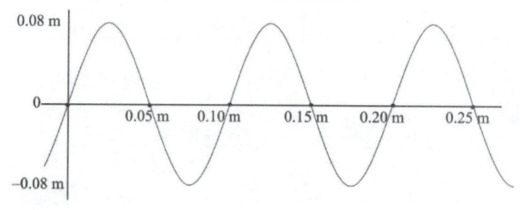

(A) 0.20 m
(B) 0.16 m
(C) 0.10 m
(D) 0.05 m

33. A wave moving through a medium is shown in the graph below.

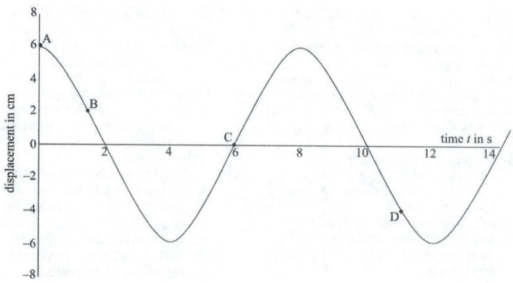

Rank the magnitude of the displacement of the labeled points in the waveform shown from greatest to least.
(A) D > C > B > A
(B) A > D > B > C
(C) C > A > B > D
(D) C > B > D > A

34. A standing wave is set up in a string as shown below.

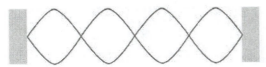

The waveform has
(A) five nodes and four antinodes.
(B) four nodes and five antinodes.
(C) four nodes and three antinodes.
(D) three nodes and four antinodes.

35. One end of a string is attached to a 60.0 Hz mechanical oscillator; the other end passes over a frictionless pulley to a known weight. The distance of the string between the tuning fork and the pulley is 1.00 m long. Observation of the string shows that the standing wave vibrates in four loops under the known weight. The speed of the wave in the string is
(A) 15.0 m/s.
(B) 30.0 m/s.
(C) 60.0 m/s.
(D) 120 m/s.

36. Two forces are applied to a wheel and axle as shown. $\vec{F}_1 = 25.0$ N is applied to the outer cylinder of radius $r_1 = 0.300$ m. $\vec{F}_2 = 40.0$ N is applied to the inner cylinder of radius $r_2 = 0.100$ m.

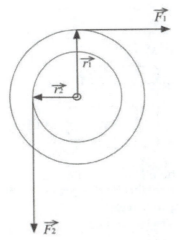

The net torque acting on the body is
(A) 3.50 N · m clockwise.
(B) 3.50 N · m counterclockwise.
(C) 11.5 N · m clockwise.
(D) 11.5 N · m counterclockwise.

37. Two charges $q_1 = 9.00 \times 10^{-9}$ C and $q_2 = -5.00 \times 10^{-9}$ C are located 0.01 m apart in air. The force acting between them is
(A) $4.05 \times 10^{-3}$ N and is attractive.
(B) $4.05 \times 10^{-3}$ N and is repulsive.
(C) $4.05 \times 10^{-5}$ N and is attractive.
(D) $4.05 \times 10^{-5}$ N and is repulsive.

38. Electrical charge that is transferred in the process of conduction when a charging wand touches an isolated body can produce charges on the body equal to multiples of
(A) $\pm 0.25e$.
(B) $\pm 0.5e$.
(C) $\pm e$.
(D) $\pm 1.5e$.

39. A small particle carrying a static charge is considered to be positive when it has
(A) dipole properties.
(B) excess protons.
(C) a deficiency of electrons.
(D) a deficiency of protons.

40. A flat disk with a moment of inertia $I_0$ is rotating with an angular velocity $w_0$ when a small piece of modeling clay drops vertically on the edge of the disk. The angular velocity of the system will
(A) increase because the modeling clay imparted a torque on the disk in the same direction of the initial angular velocity of the disk.
(B) decrease because the modeling clay imparted a frictional force on the disk that was opposite the initial motion of the disk.
(C) decrease because the moment of inertia of the system increased from $I_0$ to $I_0 + I_{clay}$.
(D) remain constant because no net torque was applied to the disk when the small piece of clay dropped vertically on the disk.

**GO ON TO NEXT PAGE**

41. In the circuit shown below, the current in the 30 Ω resistor is closest to

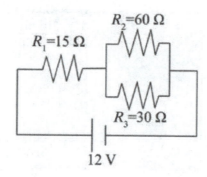

(A) 0.11 A
(B) 0.23 A
(C) 0.34 A
(D) 0.44 A

42. A negatively charged wand is brought near but does not touch an isolated neutral electroscope. Which of the drawings best indicates the correct arrangement of charges induced on the electroscope?

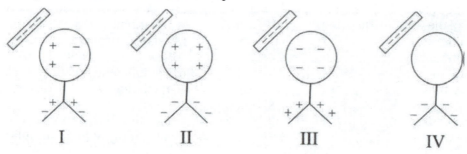

(A) I
(B) II
(C) III
(D) IV

43. Torques are applied to the cylinder as shown for 5.00 s. What is the change in the angular momentum of the cylinder?

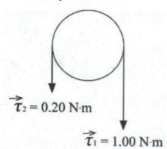

(A) 0.80 N · m · s clockwise
(B) 4.00 N · m · s clockwise
(C) 6.00 N · m · s clockwise
(D) 8.00 N · m · s clockwise

44. A circuit is shown below.

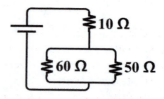

Rank the electrical potential difference across the resistors from highest to lowest.
(A) $V_{60\Omega} > V_{50\Omega} > V_{10\Omega}$
(B) $V_{60\Omega} > V_{50\Omega} > V_{70\Omega}$
(C) $V_{60\Omega} = V_{50\Omega} > V_{10\Omega}$
(D) $V_{10\Omega} = V_{60\Omega} > V_{50\Omega}$

45. A circuit consists of two resistors in a series connected to a 9.0 V battery.

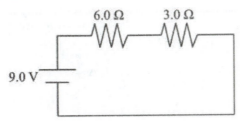

Which of the following graphs shows the correct energy changes within the circuit?

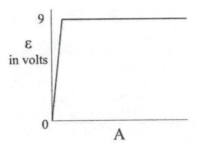

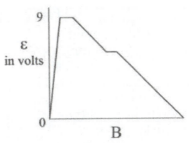

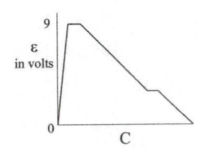

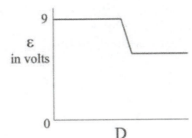

(A) A
(B) B
(C) C
(D) D

**Questions 46 to 50**
**Directions:** For each of the questions or incomplete statements below, two of the suggested answers will be correct. For each of these questions, you must select both correct choices to earn credit. No partial credit will be earned if only one correct choice is selected. Select the two that are best in each case.

46. A particle of mass $m$ moving in a straight line on a smooth frictionless surface experiences an increase in velocity. This increase in velocity indicates (select two answers)
    (A) the presence of a force with a component in the same direction as the direction of the velocity vector.
    (B) that no net force is acting on the mass.
    (C) that work was done on the body, increasing its kinetic energy.
    (D) that only the momentum of the body increases.

**GO ON TO NEXT PAGE**

47. A satellite travels in an orbit around the sun as shown below.

Which of the following two statements are true?
(A) The gravitational force acting on the satellite at point A is greater than the gravitational force acting on the satellite at point B because it is closer to the sun.
(B) The force acting on the body at point B is greater than at point A since the gravitational force acting between the sun and the satellite must be larger to move the satellite back toward the sun.
(C) The angular momentum of the satellite is larger at point A than at point B.
(D) The angular velocity of the satellite is larger at point A because the body is closer to the sun.

48. A graph of the force applied to an object versus its displacement is shown below.

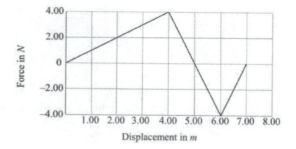

Which of the following two statements are true?
(A) The kinetic energy of the body over the displacement of 7.00 m will increase since the net amount of work on the body is positive.
(B) The kinetic energy of the body will decrease since negative forces acted on the body.
(C) The momentum of the body will increase since the total work done by the variable force was positive.
(D) The momentum of the body will not change because the force was variable. For the momentum to change, a constant force must be applied.

49. One end of a wire, 1.20 m long, is attached to a mechanical oscillator; the other end, to a known weight passing over a frictionless pulley. A standing wave is set up in the wire with a wave speed of 180 m/s. When the first harmonic is produced in the wire, (select two answers)
(A) there are two complete waves in the region between the end points of the wire.
(B) there is ½ of a complete wave in the region between the end points of the wire.
(C) the frequency of the wave is 75 Hz.
(D) the frequency of the wave is 300 Hz.

50. The mechanical waveforms shown in the diagrams below are to be used to answer this question concerning the vibration of the particles in the waves and the direction of the wave front advancing through the medium.

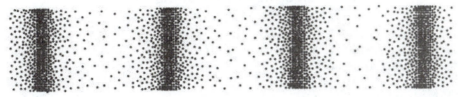

Figure 1

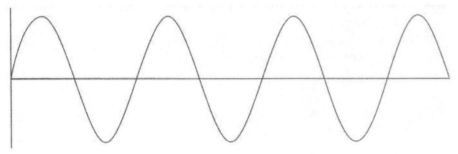

Figure 2

Which of the following two statements are true?

(A) In a transverse wave, the particles move back and forth parallel to the direction of propagation of the wave disturbance illustrated in Figure 1.

(B) In a longitudinal wave, the particles move back and forth parallel to the direction of propagation of the wave disturbance illustrated in Figure 1.

(C) In a transverse wave, the particles oscillate about the equilibrium position perpendicular to the direction of propagation of the wave disturbance as illustrated in Figure 2.

(D) In a longitudinal wave, the particles oscillate about the equilibrium position perpendicular to the direction of propagation of the wave disturbance as illustrated in Figure 2.

## STOP
## END OF SECTION I

*__IF YOU FINISH BEFORE TIME IS CALLED, YOU MAY CHECK YOUR WORK ON THIS SECTION. DO NOT GO ON TO SECTION II UNTIL YOU ARE TOLD TO DO SO.__*

## AP® PHYSICS 1
### Section II
### 5 Free-Response Problems
### Time—90 Minutes

**Directions:** Solve each of the following problems. Unless the directions indicate otherwise, respond to all parts of each question.

1.  **(12 Points)** Your teacher tells you to use the equipment shown in the diagram to design an experiment to determine the acceleration due to gravity. She tells you to use a total mass of 500 g to 1000 g on the mass hangers. You have access to other standard equipment in the room.

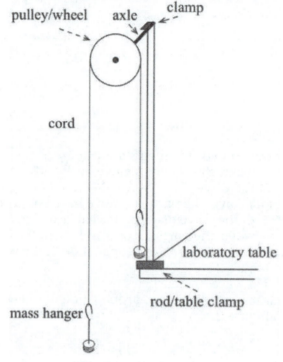

(a) Design a laboratory experiment in enough detail that another student could duplicate your experiment and obtain the same results.
(b) What measurements will you take and how will you use them to answer the question posed?
(c) If you plot a graph, what will you plot and how will you use the graph to determine an experiment value for the acceleration due to gravity?
(d) What assumption(s) did you make in the experiment, and how might they affect your results?

2. **(7 Points)** A uniform rod, whose moment of inertia is $I = \frac{1}{3}mL^2$ about its end, is supported at the left end as shown.

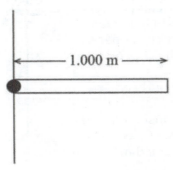

(a) What is the initial angular acceleration of the rod when it is released in terms of $g$ and $L$?

(b) When the rod is held and is placed on a pivot located at $\frac{L}{4}$ as shown, its new moment of inertia becomes $I = \frac{7}{48} mL^2$. The rod is then released.

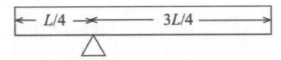

The initial angular acceleration of the rod when released, compared to the angular acceleration in part (a), will
___ increase.
___ decrease.
___ remain constant.
Justify your answer.

(c) A uniform meterstick whose weight is 0.980 N is placed on a pivot at the 0.250 m mark as shown. What are the direction and the point of application of a 1.50 N force that will place the meterstick in equilibrium?

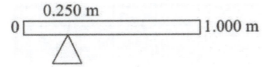

3. **(12 Points)** As shown in the diagram, a circuit consists of three lamps: $R_{50} = R_{60} = R_{70}$.
   (a) When the switch $S$ between $R_{50}$ and $R_{70}$ is closed, the current in the circuit is
      ___ greater than when the switch is open.
      ___ the same as when the switch is open.
      ___ less than when the switch is open.
      Explain your reasoning.
   (b) When the switch is closed, the power consumed in $R_{60}$ is
      ___ greater than the power consumed in $R_{50}$.
      ___ the same as the power consumed in $R_{50}$.
      ___ less than the power consumed in $R_{50}$.
      Justify your reasoning without calculations.
   (c) Assuming that the switch is closed, rank the brightness of the lamps from brightest to least bright.
      Explain your ranking without calculations.

4. **(7 Points)** A 1.53 kg block is attached to a fixed spring whose spring constant is $k = 15.0$ N/m and oscillates on a flat frictionless surface with an amplitude of vibration of 0.120 m.
   (a) What is the period of vibration?
   (b) What is the maximum velocity of the block?
   (c) What is the maximum acceleration of the block?
   (d) Complete the graphs of the following:
      i. displacement of the body versus time for at least two cycles of the subsequent simple harmonic motion

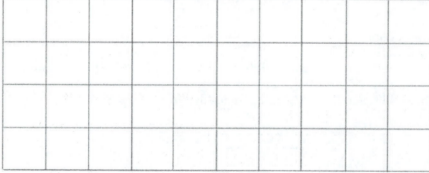

      ii. velocity of the body versus time for two cycles of its motion

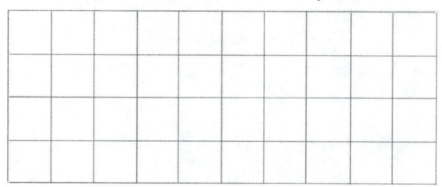

iii. acceleration of the body versus time for two cycles of its motion

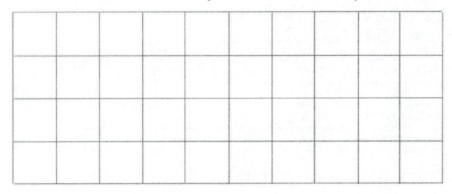

5. **(7 Points)** A 2.00 kg block is released from the top of a frictionless ramp as shown in the drawing below.

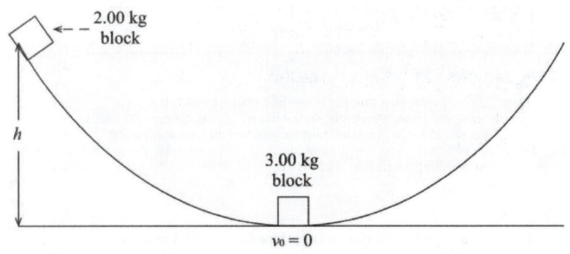

(a) If the height of the ramp is 2.0 m, what will be the velocity of the 2.0 kg block right before it hits the 3.0 kg block?

(b) Students predict that the collision will be perfectly inelastic before the 2.0 kg block is released. What should the velocity of the two-block system be immediately after the collision if the collision is perfectly inelastic?

(c) When the blocks collide, they do not stick together. In a short paragraph, describe what information is needed to determine the type of collision and how the collision type should be justified.

## *END OF EXAMINATION*

# Answers

### ANSWER KEY FOR MULTIPLE-CHOICE QUESTIONS

| | | | | |
|---|---|---|---|---|
| 1. C | 11. D | 21. C | 31. B | 41. B |
| 2. A | 12. C | 22. C | 32. C | 42. B |
| 3. D | 13. C | 23. B | 33. B | 43. B |
| 4. C | 14. B | 24. A | 34. A | 44. C |
| 5. A | 15. A | 25. A | 35. B | 45. C |
| 6. B | 16. D | 26. D | 36. A | 46. A & C |
| 7. C | 17. A | 27. B | 37. A | 47. A & D |
| 8. A | 18. D | 28. A | 38. C | 48. A & C |
| 9. A | 19. A | 29. A | 39. C | 49. B & C |
| 10. D | 20. B | 30. B | 40. C | 50. B & C |

### EXPLANATIONS FOR THE MULTIPLE-CHOICE ANSWERS

1.  **C** Tape III indicates that the distance between the dots is increasing as time increases. Tapes II and IV show increases and decreases in the spacing of the dots and would give a variable acceleration for each tape. Tape I indicates that speed is fairly constant and thus has no acceleration.

    (L.O. 3.A.1.1)

2.  **A** If the body is uniformly accelerated, the average velocity is $v_{avg} = \dfrac{v_f + v_0}{2}$. For this question, the average velocity is twice the initial velocity of 10.0 m/s, or 20.0 m/s. If the average is 20.0 m/s, the final must be 30.0 m/s. Then using $v_f = v_0 + at$, the time is $\dfrac{(30.0 \text{ m/s} - 10.0 \text{ m/s})}{2.00 \text{ m/s}^2} = 10.0$ s.

    (L.O. 3.A.1.1)

3.  **D** The acceleration is the slope of the line. Line A indicates a constant velocity and therefore zero acceleration. The slopes of the other three indicate that D has the greatest slope in the times indicated followed by C and B in that order.

    (L.O. 3.A.1.1)

4.  **C** In the absence of friction, they reach the ground at the same time with the same velocity since they are in the gravitational field of Earth. The gravitational acceleration is the same for both.

    (L.O. 3.A.1.1)

5.  **A** The body moves in the gravitational field of Earth, whose acceleration points downward. In the absence of friction, the horizontal velocity remains constant, but at maximum height, the vertical component of the initial velocity is zero. The two vectors are perpendicular to each other.

    (L.O. 3.B.2.1)

6.  **B** The body is projected with a horizontal velocity that in the absence of friction is constant. It starts from rest in the vertical direction, so its time of flight is given by $y = \frac{1}{2} at^2$. Substitution into the equation is $-60.0 \text{ m} = \frac{1}{2}(9.80 \text{ m/s}^2)t^2$. Solving gives the time as 3.50 s.

    (L.O. 3.A.1.1)

7.  **C** The ball's speed increases by 9.8 m/s every second, and its mass remains constant. Therefore, the ball's momentum, $mv$, increases uniformly every second.

    (L.O. 3.D.2.2)

8.  **A** The correct free-body diagram for the hockey puck when it moves onto the sidewalk is I. A frictional force acts between the hockey puck and the sidewalk, slowing the hockey puck. Since the frictional force is parallel to the surfaces in contact, the vectors that represent the weight and the normal are equal in magnitude, $\Sigma F_y = 0$, and opposite in direction.

    (L.O. 3.B.2.1)

9.  **A** When the hockey puck leaves the ice and moves onto the sidewalk, the net force acting on it is friction $f_k = ma$. Expanding the equation $-\mu_k mg = m\left(\frac{v_f^2 - v_i^2}{2x}\right)$, then $-(0.42)(-9.80 \text{ m/s}^2) = \frac{-(10.0 \text{ m/s})^2}{2x}$. The distance the hockey puck travels is 12.1 m.
    The distance traveled also may be determined from the reduction of the kinetic energy as the frictional force acts on the hockey puck. This is determined from the work-kinetic energy theorem, $W_f = \Delta K$. Since the frictional force is opposite the distance the hockey puck travels, we can write $-f \cdot x = \frac{1}{2} m(v_f^2 - v_0^2)$ and solve for the distance the hockey puck travels. Substitution follows as in prior solution shown above.

    (L.O. 3.B.1.3, 4.C.1.1)

10. **D** The work done by the gravitational force on the body elevated in the field is negative since the direction of the force is opposite the displacement. $W = F_g d(cos180°)$. The work done is $(9.80 \text{ m/s}^2)(1.60 \text{ m})cos180° = -47.0 \text{ J}$. The gravitational force does no work on the body as it is moved horizontally 2.00 m since the gravitational force is perpendicular to the displacement.

(L.O. 4.C.1.1)

11. **D** The downward velocity increases. Therefore, both the kinetic energy and the momentum increase because the gravitational force is in the same direction as its velocity during the time it falls.

(L.O. 3.D.2.2, 3.E.1.1)

12. **C** A frictional force acts on the car toward the center of the curve, producing an acceleration that will change the direction of the car while maintaining its speed. Since the impulse provided by the centripetal force acts on the car $F\Delta t = m\Delta v$, the speed but not the velocity is constant. The velocity changes direction; thus, the momentum (which is a vector pointing in the direction of the car's velocity) changes.

(L.O. 3.D.2.2)

13. **C** The area under the force vs. time graph is the change in the momentum of the body. The total area is the sum of the two areas under the curve that will give the change in momentum,
$A_1 = 20 \text{ N}(6 \text{ s}) = 120 \text{ N} \cdot \text{s}$ and $A_2 = \frac{1}{2}(20 \text{ N})(4 \text{ s}) = 40 \text{ N} \cdot \text{s}$.
The total area is 160 N · s.

(L.O. 4.B.2.1)

14. **B** The net force applied to the body in the direction of motion is 6.00 N. The normal to the surface will increase because a vertical force of 6.00 N is applied to the body, but it will not produce a change in the horizontal forces acting on the body since the surface is frictionless.

(L.O. 3.D.2.2)

15. **A** The work done by friction is negative $W_f = -f\Delta x$ because the angle between the frictional force vector and the displacement of the box as it moves is 180° and the cosine of 180° is –1. The friction acting on the box is found by solving $f_k = \mu_k N = \mu_k mg \cos30°$. Solving for the frictional force gives $f_k = 3.06$ N. The work done by friction is then $W_f = -3.06$ N $(1.20$ m$) = -3.67$ J.

(L.O. 5.B.3.3)

16. **D** The block has maximum velocity as it passes though the equilibrium position.

    Its velocity determined from conservation of energy $U_s + K$ is a constant since the surface is frictionless.

    $$\frac{1}{2}kx_i^2 + \frac{1}{2}mv_i^2 = \frac{1}{2}kx_f^2 + \frac{1}{2}mv_f^2.$$

    Substitution into the equation is

    $\frac{1}{2}(150.0 \text{ N/m})(0.100 \text{ m})^2 = \frac{1}{2}(0.200 \text{ kg})v^2$. Solving gives $v = 2.74$ m/s.

    (L.O. 3.B.3.1, 5.B.3.2, 5.B.3.3, 5.B.4.2)

17. **A** Since the reference position is the tabletop, the body has no gravitational potential energy. It possesses only kinetic energy.

    (L.O. 5.B.1.1)

18. **D** Conservation of energy applies: $K_i + U_i = K_f + U_f$

    $\frac{1}{2}(1.50 \text{ kg})(2.00 \text{ m/s})^2 + (1.50 \text{ kg})(9.80 \text{ m/s}^2)(1.00 \text{ m}) = 17.7$ J

    (L.O. 4.C.1.1)

19. **A** After one complete cycle, the body has returned to its initial position. It moves from $-A$ to $+A$ and back, a distance of $4A$.

    (L.O. 3.B.3.1)

20. **B** Negative work is done on the block of ice by the frictional force. Since no other force is applied to the block of ice in the direction of motion, the kinetic energy will decrease. $W_f = \Delta K$

    (L.O. 3.E.1.1)

21. **C** The gravitational acceleration on the planet is $\dfrac{g_E}{g_P} = \dfrac{M_E / R_E^2}{3M_E / (2R_E)^2}$.

    Solving gives the ratio $\dfrac{g_E}{g_P} = \dfrac{4}{3}$. Since the acceleration on the planet

    is $\dfrac{3g_E}{4} = g_P$, the period of the pendulum is

    $T_P = 2\pi \sqrt{\dfrac{1.20 \text{ m}}{7.35 \text{ m/s}^2}} = 2.54$ s.

    (L.O. 2.B.2.2, 3.B.3.1)

22. **C** The gravitational force between the masses is a force of attraction. The net force on $m_3$ is toward the left. Its magnitude is calculated from $|F_{13}| + |F_{23}| = |F_{net}|$.

$$|F_{13}| = 6.67 \times 10^{-11} \frac{\text{N} \cdot \text{m}^2}{\text{kg}^2} \left( \frac{10.00 \text{ kg} \cdot 30.00 \text{ kg}}{(2.00 \text{ m})^2} \right)$$
$$= 4.00 \times 10^{-8} \text{ to the left}$$

$$|F_{23}| = 6.67 \times 10^{-11} \frac{\text{N} \cdot \text{m}^2}{\text{kg}^2} \left( \frac{20.00 \text{ kg} \cdot 30.00 \text{ kg}}{(1.00 \text{ m})^2} \right)$$
$$= 5.00 \times 10^{-9} \text{ to the left}$$

then $|F_{net}| = 5.00 \times 10^{-9}\text{N} + 4.00 \times 10^{-8} \text{ N} = 4.5 \times 10^{-8}\text{N}$ to the left.

(L.O. 3.C.1.1)

23. **B** The perceived shift in frequency occurs due to relative motion between the stationary observer and the moving ambulance. The wave fronts crowd together as waves produced by the siren on the moving ambulance approach waves that were already traveling in the medium. As the apparent wavelength decreases, the perceived frequency increases.

(L.O. 6.B.1.1)

24. **A** The center of mass of the fragments of the projectile after the explosion will follow the trajectory of the projectile had it not exploded. The explosion was a chemical process inside the projectile, and no net force was applied to it.

(L.O. 5.D.3.1)

25. **A** The net change in momentum arises from the change in the horizontal and vertical components of the momentum when an impulse $F\Delta t = \Delta p$ is applied. Solving for the change of momentum in the $y$-direction,

$$\Delta p_y = n(mv \sin 30^o) - n(-mv \sin 30^o) = n(2mv \sin 30^o).$$

In the $x$-direction, $\Delta p_x = n(mv \cos 30^o) - n(mv \cos 30^o) = 0$.

Thus, the change in momentum is

$$\Delta p = \Delta p_x + \Delta p_y = 0 + n(2mv \sin 30^o) = nmv.$$

(L.O. 3.D.2.1)

26. **D** As the body is lifted in a gravitational field, the mass of the body will not change as it is lifted.

(L.O. 2.B.1.1, 2.B.2.1)

27. **B** Conservation of energy applies: $K_i + U_{s_i} = K_f + U_{s_f}$.

    The equation is written as $\frac{1}{2}mv_i^2 = \frac{1}{2}kx_f^2$. The initial velocity of the

    block is $v = \sqrt{\dfrac{(15.0 \text{ N/m})(0.12\text{m})^2}{2.00 \text{ kg}}} = 0.33$ m/s.

    (L.O. 4.C.1.1)

28. **A** Torque is defined as $\tau = rF\sin\theta$. (In all cases, the angle is 90°.) IV has the largest torque since it has the longest moment arm and its torque is $3r(50 \text{ N})$. Figures I and II have the same value $r(100 \text{ N})$, and III has the least with a value of $r(50 \text{ N})$.

    (L.O. 3.F.1.2)

29. **A** No outside torque is applied to the skater; thus, her angular momentum is constant. Reducing her moment of inertia by bringing her arms in toward her body increases her angular velocity. The skater does work as she pulls her arms inward, thereby increasing the kinetic energy.

    (L.O. 3.E.1.1, 5.E.1.1)

30. **B** The period of oscillation for a spring-mass system does not depend on the amplitude of vibration. It does depend on the mass of the body attached to the spring and the spring constant.

    $T = 2\pi\sqrt{\dfrac{m}{k}}$. Doubling the mass without changing the spring

    constant will change the period of vibration by a factor of the $\sqrt{2}$.

    (L.O. 3.B.3.1)

31. **B** The spring constant is determined from the restoring force, which is equal to the weight of the mass at rest on the end of the spring stretching it 0.065 m. $F_g = mg$. The restoring force is 24.5 N. The spring constant is then found by substituting into $F_s = -kx$. The

    spring constant is $k = \dfrac{24.5 \text{ N}}{0.065 \text{ m}} = 377$ N/m. Solving the equation, the

    frequency is then $f = \dfrac{1}{2\pi}\sqrt{\dfrac{377 \text{ N/m}}{2.50 \text{ kg}}} = 1.96$ Hz.

    (L.O. 3.B.3.4)

32. **C** The wavelength of the wave is the distance between two points on a wave that have the same amplitude and the same phase (particles on the wave form are moving in the same direction).

    (L.O. 6.A.4.1)

33. **B** The displacement of a point on a wave is the distance of the vibrating particle from the equilibrium position and is measured above and below the equilibrium. In the wave shown, point A is at maximum amplitude and has the largest displacement. Since the question asked for the magnitude of the displacement, point D is ranked above point B. Point C is at the equilibrium; thus, its displacement is zero.

    (L.O. 6.A.3.1)

34. **A** A standing wave is set up in the string due to interference between the incident waves and the reflected waves. This standing wave has nodes at the fixed boundaries. There are two complete waves in the diagram; thus, there are five nodes and four antinodes.

    (L.O. 6.D.1.1, 6.D.3.2)

35. **B** There are four loops in the standing wave, which are two complete waves in the region. The string is 1.00 m long, and the wavelength is $\lambda = (1.00 \text{ m})/2 = 0.500$ m. The velocity of the wave can be calculated from $v = \lambda f$. The velocity is 60.0 Hz (0.500 m) = 30.0 m/s.

    (L.O. 6.D.4.2)

36. **A** The net torque acting on the cylinder (where we define the clockwise direction to be negative) is $\Sigma \tau = -(25.0 \text{ N})(0.300 \text{ m}) + (40.0 \text{ N})(0.100 \text{ m}) = -3.5 \text{ N} \cdot \text{m}$. Since the net torque is negative, the rotation is clockwise.

    (L.O. 3.F.1.3)

37. **A** Coulomb's law gives the magnitude of the force acting between the charges $= k\dfrac{q_1 q_2}{r^2}$. The magnitude of the force is $9 \times 10^9$ $\dfrac{\text{N} \cdot \text{m}^2}{\text{C}^2} \dfrac{(9.00 \times 10^{-9}\text{C})(5.00 \times 10^{-9} \text{ C})}{(0.01 \text{ m})^2} = 4.05 \times 10^{-3}\text{N}$. Since the charges are opposite in sign, the force is attractive.

    (L.O. 3.C.2.1)

38. **C** The charge on the electron, *e*, has one value, *e*. Only whole number of charges ±*e*, ±2*e*, ±3*e*, etc., are transferred.

    (L.O. 1.B.3.1)

39. **C** A static charge on the surface of a small particle is positive when there is a deficiency of electrons on its surface.

    (L.O. 1.B.1.1)

40. **C** Conservation of angular momentum applies. The small piece of modeling clay is a point mass and is in the inelastic collision with the rotating disk $I_0$ the moment of inertia of the system increases to $I_0 + I_{clay}$. Thus, the angular velocity must decrease.

    (L.O. 5.E.1.1)

41. **B** The parallel branch consists of a 30 Ω resistor and a 60 Ω resistor. The equivalent resistance for this part of the circuit is $\frac{1}{R} = \frac{1}{30\ \Omega} + \frac{1}{60\ \Omega}$, or 20 Ω. The equivalent of 20 Ω is in series with the 15 Ω resistor for a total of 35 Ω for the circuit. Solving $V = IR$ will give the total current for the circuit. $\frac{12\ V}{35\ \Omega} = 0.34$ A. Kirchhoff's junction rule applies. The current 0.34 A will divide in the parallel branch with the 30 Ω resistor carrying $\frac{2}{3}I_{total}$, or 0.23 A. (The equivalent resistance is 20 Ω for the parallel branch: the division of current is $\frac{20\ \Omega}{30\ \Omega}I_{total} = \frac{2}{3}I_{total}$ for the 30 Ω resistor and $\frac{20\ \Omega}{60\ \Omega} = \frac{1}{3}I_{total}$ for the 60 Ω resistor.)

    (L.O. 5.C.3.3)

42. **B** The only charge that can move on the surface of the isolated conductor is the negative charge. Due to the electrostatic repulsion between the negative charge on the wand and the electrons on the electroscope, the electrons will move to the leaves of the electroscope and the charge on the surface will be positive. The leaves of the electroscope will have identical excess negative charge.

    (L.O. 1.B.1.2)

43. **B** The change in the angular momentum can be determined from $\Sigma\tau = \frac{\Delta L}{\Delta t}$.

    The two torques applied to the cylinder give a net torque of −1.00 N · m + 0.200 N · m = −0.800 N · m. The change in the angular momentum is then $-4.00 \text{ N} \cdot \text{m} \cdot \text{s}$. Rotation in a clockwise manner is defined as negative.

(L.O. 4.D.3.1)

44. **C** In the circuit, the parallel branch consists of two resistors of 60 Ω and 50 Ω. The resistance for the parallel part of the circuit is found from $\dfrac{1}{R} = \dfrac{1}{50 \text{ Ω}} + \dfrac{1}{60 \text{ Ω}}$. The parallel branch has a total equivalent resistance of 27.3 Ω. This 27.3 Ω resistance is in series with a 10 Ω resistance for a total resistance in the circuit of 37.3 Ω. The total current in the circuit—and thus in the 10 Ω resistor—is found from $V = IR$. The total current in the circuit is $\dfrac{20 \text{ V}}{27.3 \text{ Ω}} = 0.53$.

Kirchhoff's loop rule applies. The electrical potential difference is greatest over the parallel part of the circuit since it has the largest resistance, $V = IR$. The electrical potential difference across the 60 Ω resistor and the 50 Ω are in parallel, so they have equal potential. Even though the entire current transits the 10 Ω resistance, it has the lowest electrical potential difference since it is the smallest resistance in the circuit.

(L.O. 5.B.9.3)

45. **C** Kirchhoff's loop rule applies. Electrical energy developed in the battery is delivered to the external circuit. (Connections from the battery and the resistors are ideal conductors.) The electrical potential difference across the ends of the resistor multiplied by the current in the resistor is the power dissipated in the resistor, or the work per unit time. The conventional current flows into the 6.0 Ω resistor and passes through an ideal conductor into the 3.0 Ω resistor, where the rest of the energy developed by the battery is dissipated.

(L.O. 5.B.9.1)

46. **A and C** Newton's second law $F = ma$ applies; to increase the velocity of the mass $m$, a force with a component in the direction of the initial velocity vector must act on the mass, $m$. This force acting through a displacement will do work on the mass $m$, increasing its kinetic energy and its momentum. The work-kinetic energy theorem indicates that $W = \Delta K$.

(L.O. 3.B.1.4, 4.C.2.1)

47. **A and D** The gravitational force of attraction between the sun and the satellite is given by $F = G \dfrac{M_s m}{r^2}$. The force is larger at point A since the distance is smaller. The speed of the satellite is largest at point A since no net torque acts on the satellite; thus, its angular momentum must be conserved. Angular momentum is given as

$L = I\omega$. The satellite is a point mass relative to the sun ($I = mr^2$) and angular velocity can be expressed as $\omega = \dfrac{v}{r}$. Substitution into $I = mr^2$ gives $I = (mr^2)\dfrac{v}{r} = mvr$. The velocity is dependent on the distance $r$ from the sun: $mv_A r_A = mv_B r_B$.

(L.O. 3.C.1.2, 5.E.1.1)

48. **A and C** The work done on an object is the area under the curve. Since the net area is positive, the work done is positive and the kinetic energy will increase. Since the kinetic energy increases, the momentum will also increase. The correct momentum equation is $F \cdot \Delta t = m\Delta v$.

(L.O. 3.F.1.4, 4.B.2.1)

49. **B and C** The first harmonic is the fundamental wave set up in a string or a wire. The fixed ends are consecutive nodes; thus, the wavelength is $\lambda = 2L$. The distance between the nodes is half the wavelength. For the wire that is 1.20 m long, the wavelength is 2.40 m and the wave speed in the wire is 180 m/s. The frequency of vibration can be calculated from $v = \lambda f$, so 180 m/s = (1.20 m)$f$. The frequency is 75 Hz.

(L.O. 6.C.1.1, 6.D.4.2)

50. **B and C** In a transverse wave, the particles move perpendicular to the direction of the wave disturbance in the medium as shown in Figure 2. In a longitudinal wave, the particles move back and forth parallel to the direction of the wave front advance through the medium as shown in Figure 1.

(L.O. 6.A.1.2)

## ANSWERS TO FREE-RESPONSE PROBLEMS

## QUESTION 1

(a) The acceleration due to gravity can be determined from the acceleration of a multi-mass system accelerated by a net force by determining the time $t$ taken by the masses on the hanger to move a distance $h$.

    Set up masses on the mass hangers so that the mass hangers move up and down at constant speed when given a slight push. Record the total mass.

    Transfer a mass from the ascending side $m_2$ to the descending side $m_1$. The net force of the system will be the difference in weight between the two hangers.

Start the experiment with the ascending mass on the floor. Measure the distance between the mass hangers and the floor. Release the descending mass.

Time the masses hangers as they move, stopping the timer when the descending mass reaches the floor. Record the net force and the time.

Repeat for at least two more trials.

Determine the average of these times and calculate the acceleration of the system using $= v_0 t + \frac{1}{2} a t^2$.

Transfer additional mass from the ascending side to the descending side. Repeat the steps above, again making at least three separate determinations of the time.

Continue transferring mass until you have made several different runs. **(5 points)**

(b) Measurement of the height the mass hangers move and the time for each trial will permit calculation of the acceleration of the system using $x = v_0 t + \frac{1}{2} a t^2$. Comparison of the acceleration for each trial can be obtained by $\Sigma F_y = (m_1 + m_2) a_y$. **(2 points)**

(c) Plot a graph of the acceleration for each experiment run versus $\frac{(m_1 - m_2)}{(m_1 + m_2)}$. The slope of the line is the experimental value for the acceleration due to gravity, $g$. Compare the experimental value to the accepted value. **(3 points)**

(d) Assuming the pulley/wheel to be frictionless will change the experimental value determined for $g$ because friction will supply a net torque to the pulley/wheel, causing the pulley/wheel to have an angular acceleration. The pulley/wheel is not massless, but has a value $M$, which gives it a moment of inertia $I$. **(2 points)**

(L.O. 3.A.1.1, 2.B.1.1, 3.A.1.2, 3.A.1.3, 3.B.1.1)

## Question 2

(a) The torque acting on the meterstick is $\tau = Fr\sin\theta$. The $F$ is perpendicular to the moment arm $r$; thus, $\theta$ is 90°. The torque acting on the meterstick is $= \frac{mgL}{2}$. **(1 point)**

Since the moment of inertia is given in the problem, $= \frac{1}{3} mL^2$.

Substitution into the torque equation $\tau = I\alpha$ is $\frac{mgL}{2} = \frac{1}{3} mL^2 \alpha$. **(1 point)**

The initial angular acceleration is $\alpha = \frac{3 g}{2 L}$. **(1 point)**

(b) The correct line checked is ___✓___ increase.

The torque acting on the meterstick is one-half the initial torque since the moment arm is reduced by 2. The new torque is

$$\tau = \frac{mgL}{4}.$$

Changing the location of the point of rotation changes the moment of inertia. Both the torque and the moment of inertia have reduced in value, but the moment of inertia is much smaller. Thus, the initial angular acceleration is larger when the stick is released.

**(2 points)**

Mathematically, $\frac{mgL}{4} = \frac{7}{48}mL^2\alpha$. The angular acceleration is now $\frac{12\,g}{7\,L}$.

(c) To set the meterstick into rotational equilibrium, the net torque on the system must be zero.

The moment arm for the weight of the meterstick is the distance from the pivot point to the center of mass of the meterstick. Since the meterstick is uniform, the center of mass is at the 0.500 mark. Thus, its moment arm is (0.500 m – 0.250 m) = 0.250 m.

Substitution into the $\Sigma\tau = 0$ equation is (0.980 N)(0.250 m) = (1.50 N)$r$· $r$ = 0.1633 m = **0.163** m mark on the meterstick and is directed downward. **(2 points)**

(L.O. 3.F.1.1, 2.F.1.5, 3.F.2.1, 3.F.3.2, 4.A.1.1)

## QUESTION 3

(a) The current is greater when the switch is closed.

When the switch is open, the resistance is 60 ohms.
When the switch is closed, the resistance is

$$60\Omega + \left(\frac{1}{50} + \frac{1}{70}\right)^{-1} = 35\Omega.$$  **(2 points)**

With the switch closed, there is an additional path through which current can flow. There is less resistance, so there will be more current. **(3 points)**

(b) More power is consumed in the 60 Ω resistor than in the 50 Ω resistor. When the switch is closed, more current flows through the 60 Ω resistor than through the 50 Ω resistor. With greater resistance and greater current, $R_{60}$ dissipates more power because $P = I^2R$. **(3 points)**

(c) The ranking for the brightness of the lamps is **D > A > (B = C)**.
**(1 point)**

From brightest to dimmest, the order is the 60 Ω bulb, the 70 Ω bulb, and the 50 Ω bulb. The 60 Ω resistor carries double the current of the other two. Because they are in series, the same current runs through the 70 Ω and the 50 Ω resistors. The power dissipated is $P = I^2R$ so the one with the greater resistance is brighter. **(3 points)**

(L.O. 5.B.9.3, 5.C.3.1, 5.C.3.3)

QUESTION 4

(a) The period of vibration is found from $= 2\pi\sqrt{\dfrac{m}{k}}$. Substituting into

the equation and solving, the period is $T = 2\pi\sqrt{\dfrac{1.53 \text{ kg}}{15.0 \text{ N/m}}} = 2.00$ s.

**(1 point)**

(b) The maximum velocity occurs when the body passes through the equilibrium position. Since the surface is flat and frictionless, the mechanical energy is conserved. $\dfrac{1}{2}mv^2 = \dfrac{1}{2}kA^2$

$$v^2 = \frac{(15.0 \text{ N/m})(0.120 \text{ m})^2}{1.53 \text{ kg}}$$

The maximum velocity is 0.376 m/s.    **(1 point)**

(c) The maximum acceleration occurs at the amplitude $A$ and points toward equilibrium. Its magnitude is determined from $kx = ma$.

The maximum acceleration is $\dfrac{(15.0 \text{ N/m})(0.120 \text{ m})}{1.53 \text{ kg}} = 1.18$ m/s.

**(1 point)**

(d) The graphs are as follows:
  i.  displacement versus period of the oscillation    **(2 points)**

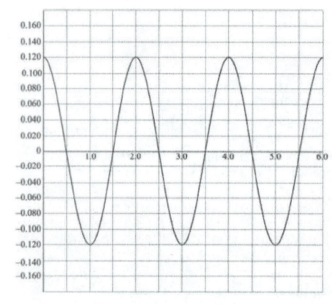

  ii.  sketch of velocity as a function of time    **(1 point)**

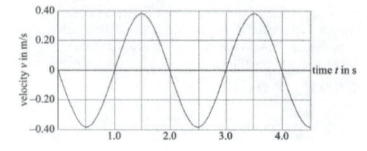

iii. acceleration as a function of time  (1 point)

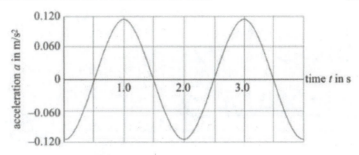

(L.O. 3.B.3.1, 3.B.3.4, 5.B.3.1, 5.B.3.2, 5.B.4.2)

## QUESTION 5

(a) Conservation of energy applies. Since the surface of the ramp is frictionless, the sum of the initial kinetic energy and gravitational potential energy is equal to the sum of the final kinetic energy and potential energy. At the highest point on the first ramp, the 2 kg block is stationary and has no kinetic energy. At the bottom of the first ramp, the 2 kg block has zero gravitational potential energy.

So $mgh = \frac{1}{2}mv^2$

$v = \sqrt{2\,gh} = \sqrt{2(9.8 \text{ m/s}^2)(2.0 \text{ m})} = 39 \text{ m/s}$.  (1 point)

b. Conservation of momentum applies for the completely inelastic collision. Setting up and solving for the velocity of the two-block system will give the needed value to solve for the height the composite mass rises on the second ramp.

$\Sigma p_{before} = \Sigma p_{after}$

$m_{2.0 \, kg} v_{2.0 \, kg} = (m_{2.0 \, kg} + m_{3.0 \, kg})v$

$(2.0 \text{ kg})(39 \text{ m/s}) = (5.0 \text{ kg})v$

$v = 15.6 \text{ m/s} = 16 \text{ m/s}$  (1 point)

c. All collisions conserve momentum. The only type of collision in which kinetic energy is conserved is a perfectly elastic collision. The velocity of each block before and after the collision would need to be measured. Using the velocities, the kinetic energy before and after the collision can be determined. If the kinetic energy before and the kinetic energy after the collision are equal, the collision is said to be perfectly elastic. If the kinetic energies are not equal, the collision is inelastic.  (5 points)

(L.O. 4.C.1.1, 5.A.2.1, 5.B.3.2, 5.B.4.2, 5.D.1.3, 5.D.2.3, 5.D.2.5)

# AP® PHYSICS 1
# PRACTICE EXAM B

AP® PHYSICS 1
Section I
50 Multiple-Choice Questions
Time—90 Minutes

**Note:** To simplify calculations, you may use $g = 10$ m/s$^2$ in all problems.

**Directions:** Each of the questions or incomplete statements below is followed by four suggested answers or completions. Select the one that is best in each case.

1. A car moves in one dimension along the x-axis. A velocity vs. time graph shows its motion over a period of 12.0 s.

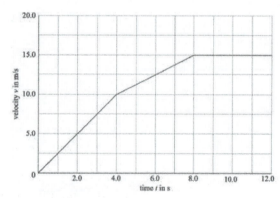

What is the car's displacement during the time interval 0–12 seconds?
(A) 90 m
(B) 100 m
(C) 110 m
(D) 130 m

2. A 2.00 kg object is supported by two cables attached to the ceiling as shown in the diagram. The tension in each cable is nearly

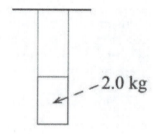

(A) 20 N.
(B) 10 N.
(C) 5 N.
(D) 1 N.

3. A 20.0 N force acts on two blocks, moving them horizontally across a frictionless floor. What force does the 6.00 kg block exert on the 4.00 kg block?

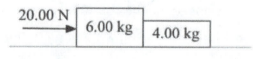

(A) 20.0 N
(B) 16.0 N
(C) 12.0 N
(D) 8.00 N

**574**

4. A basketball of mass *m* attached to a massless cord is used as a pendulum. The basketball is large enough that the force of air resistance $\vec{R}$ cannot be ignored. The basketball swings from point A to point B as shown in the diagram below.

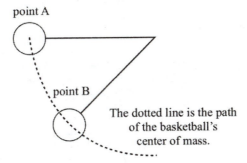

point A

point B

The dotted line is the path of the basketball's center of mass.

Which of the following is the correct free-body diagram of the forces acting on the center of mass of the basketball at point B?

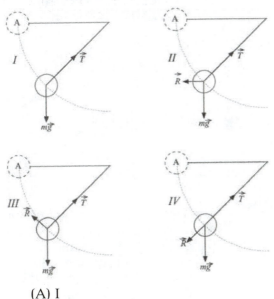

(A) I
(B) II
(C) III
(D) IV

5. A 0.20 kg sphere and a 0.40 kg sphere are dropped simultaneously from the top of a tower. Ignoring air resistance, when they reach the ground,
(A) they have the same acceleration and the same kinetic energy.
(B) the larger sphere has a larger acceleration and a larger kinetic energy than the smaller sphere.
(C) the smaller sphere has the same acceleration and a smaller kinetic energy than the larger sphere.
(D) they have the same kinetic energy but the larger sphere has the larger acceleration.

6. In the diagram shown, a 5.00 kg mass is supported by two cables $\vec{T_1}$ and $\vec{T_2}$. The tension $\vec{T_1}$ is most nearly

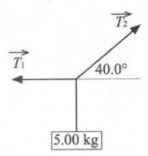

5.00 kg

(A) 30.0 N
(B) 48.4 N
(C) 50.0 N
(D) 60.0 N

7. Two waves travel through the same medium with a phase difference of 90° as shown below. As the waves travel over time, the resulting waveform will

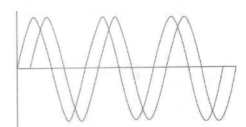

(A) vary in both amplitude and frequency.
(B) be constant in amplitude and frequency.
(C) vary in amplitude while maintaining constant frequency.
(D) be constant in amplitude only.

**GO ON TO NEXT PAGE**

8. A small charge is located in the hollow center of an uncharged thin conducting ring.

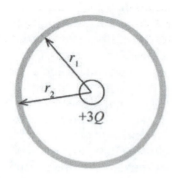

The charge on the outer surface $r_2$ is
(A) +3Q.
(B) –3Q.
(C) +1.5Q.
(D) –1.5Q.

9. Four identical blocks with the same initial velocity $\vec{v}_i$ experience a force as they move a distance on a frictionless surface in the right (+x-direction). At the end of the displacement, the blocks each have a different final velocity $\vec{v}_f$. The magnitude of the force $|\vec{F}|$ acting on the blocks is the same.

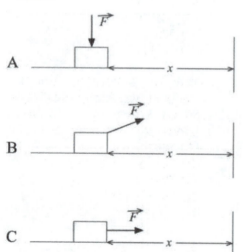

Rank the work done on the blocks by the applied force from the greatest to the least amount of work. Positive work is to be ranked higher than negative work.
(A) C > B > D > A
(B) C > B > A > D
(C) B > D > A > C
(D) C > A > (B = D)

10. A block moving at a constant speed of 1.20 m/s to the right enters a rough region where a frictional force acts on it.

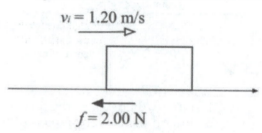

A graph of the block's velocity as a function of time is shown.

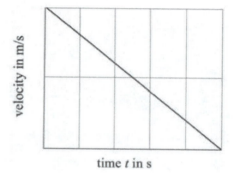

The velocity as a function of time graph suggests that
(A) the acceleration of the block is positive and constant and the graph of the displacement versus time is parabolic, opening upward.
(B) the acceleration of the block is positive and constant and the graph of the displacement versus time is parabolic, opening downward.
(C) the acceleration of the block is negative and constant and the graph of the displacement versus time is parabolic, opening upward.
(D) the acceleration of the block is negative and constant and the graph of the displacement versus time is parabolic, opening downward.

11. A block is at rest on a level tabletop. Which of the following statements is the correct statement describing the reaction to the pull of gravity on the book?
    The force of
    (A) Earth is pulling on the book.
    (B) the table is pushing on the book.
    (C) the book is pulling on Earth.
    (D) the book is pushing on the table.

12. A 10.0 kg block is uniformly accelerated to the right on a horizontal surface as shown below, where the coefficient of friction $\mu_k$ between the block and the surface is 0.010.

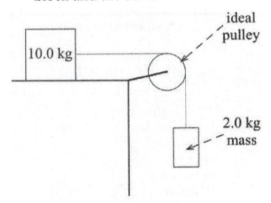

Which of the following is the correct free-body diagram for the 10.0 kg block?

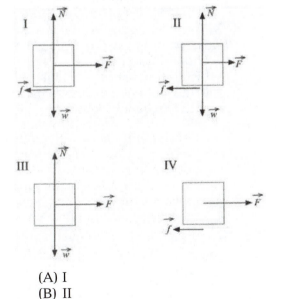

(A) I
(B) II
(C) III
(D) IV

13. Two blocks connected by a light metal rod, as shown in the diagram, are moved a distance to the right along the frictionless floor by a force F that acts at some angle $\theta$ below the horizontal. The kinetic energy of the blocks

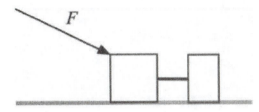

(A) remains the same since the force is applied to only the larger block.
(B) increases because the force applied to the two blocks will increase the velocity of the center of mass, increasing the kinetic energy.
(C) decreases since the force is applied to only one block.
(D) remains constant because the force acts downward below the horizontal and cannot change the velocity of the center of mass.

14. Three cars, A, B, and C, are pulled to the right by a 30.0 N force.

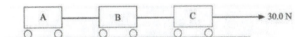

The mass of each car is 0.500 kg. Neglecting friction, determine the tension in the rope between cars A and B.
(A) 5.0 N
(B) 10.0 N
(C) 20.0 N
(D) 30.0 N

**GO ON TO NEXT PAGE**

15. A 100 N block is placed on a scale and is connected to a 20 N weight by a massless rope that passes over an ideal pulley as shown.

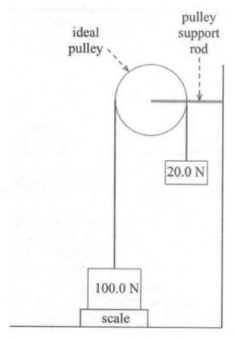

The normal force exerted on the block by the scale is
(A) 60 N.
(B) 80 N.
(C) 100N.
(D) 120 N.

16. The diagram below shows a 3.00 kg block moving on the +x-axis with a speed of 4.00 m/s colliding with and sticking to a 1.00 kg block moving in the opposite direction with a speed of 2.00 m/s. The surface is frictionless.

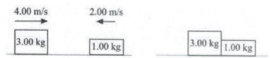

What is the speed of the composite system after the collision?
(A) 3.50 m/s
(B) 2.50 m/s
(C) 3.33 m/s
(D) 4.66 m/s

17. Which of the following statements correctly describes an elevator car rising at a constant velocity?
(A) The total mechanical energy of the elevator car is constant.
(B) The car has upward acceleration because it is rising.
(C) Both the force on the car and the acceleration of the car are constant.
(D) The net force on the elevator car is directed upward.

18. A 0.05 kg bullet with an initial velocity of 400 m/s strikes a 500 kg block at rest on a horizontal frictionless surface and emerges with a final velocity of 200 m/s.

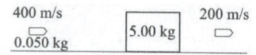

What is the velocity of the block after the interaction with the block?
(A) 2.0 m/s
(B) 3.0 m/s
(C) 4.0 m/s
(D) 6.0 m/s

19. A physics student swings on a rope from point A to point D as shown in the diagram below. Points A and D are the highest positions of the swing, and point C is the lowest position of the swing. Somewhere in front of the student a stationary whistle is blowing. At which position will the student hear the highest sound frequency from the whistle?

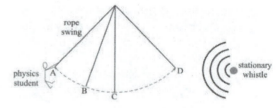

(A) At B when moving toward A
(B) At B when moving toward C
(C) At C when moving toward B
(D) At C when moving toward D

20. A student uses a very precise spring scale and a balance to determine the "weight" of an object at ground floor of the world's tallest building and later at the 163rd floor of the same building. Which of the following is the correct answer concerning the readings for the "weight"?
    (A) The readings on both the spring scale and the balance are the same.
    (B) The readings on both the spring scale and the balance are different.
    (C) The reading on the spring scale is the same, but the reading on the balance is different.
    (D) The reading on the spring scale is different, but the reading on the balance is the same.

21. A 1.00 kg mass attached to a horizontal massless string is released from rest at point A as shown in the diagram below.

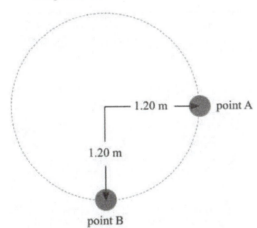

The mass then swings through a sector of a vertical circle of 1.20 m radius to point B. The moment of inertia of the 1.00 kg mass is $I = mr^2$. At point B, the angular momentum of the mass is
    (A) equal to the angular momentum at point A because momentum is conserved.
    (B) zero because there was no collision to transfer momentum.
    (C) at a minimum value because the moment of inertia decreases as the ball loses potential energy.
    (D) at a maximum value because the ball is at its maximum speed.

22. A particle of mass $m$ is moving to the right along the +$x$-axis with a velocity of 10.0 m/s when it suddenly breaks into two unequal pieces. One piece of mass $\frac{m}{3}$ continues in the forward direction with a velocity of 20.0. What is the velocity of the second piece?
    (A) –10.0 m/s
    (B) 5.00 m/s
    (C) 10.0 m/s
    (D) 15.0 m/s

**Questions 23 and 24**
Use the diagram below to answer questions 23 and 24.

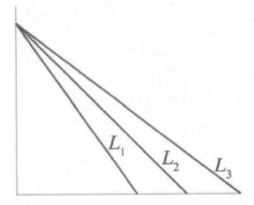

23. Three identical blocks are placed at the top of three frictionless inclines of lengths $L_1$, $L_2$, and $L_3$. The heights of all three inclines are the same. Rank the speed of the blocks at the bottom of the inclines from largest to smallest. The block sliding down $L_1$ has a speed of $v_1$. Blocks sliding down $L_2$ and $L_3$ have speeds of $v_2$ and $v_3$, respectively.
    (A) $v_1 > v_2 > v_3$
    (B) $v_3 > v_2 > v_1$
    (C) $v_1 = v_2 = v_3$
    (D) $v_1 > v_3 > v_2$

**GO ON TO NEXT PAGE**

24. Equal amounts of frictional forces are applied to each identical object on the three ramps. The blocks start from rest and slide down the ramp. Rank the speed at the bottom of the incline from greatest to least.
    (A) $v_1 > v_2 > v_3$
    (B) $v_3 > v_2 > v_1$
    (C) $v_1 = v_2 = v_3$
    (D) $v_1 > v_3 > v_2$

25. A variable resultant force $F$ acts on a 0.100 kg block initially at rest. The graph below shows the force as a function of displacement $x$. The block is accelerated to speed $v$ after being displaced 0.200 m by the force. Calculate the speed $v$ of the block.

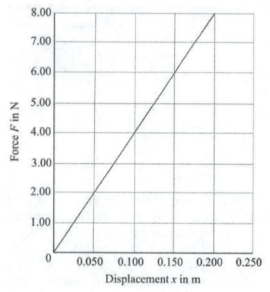

(A) 1.00 m/s
(B) 2.00 m/s
(C) 3.00 m/s
(D) 4.00 m/s

26. A 100 kg object sitting on the surface of Earth is lifted to a new position that is 1.50 Earth Radii, $R_E$, above the surface of Earth. The weight of the object at that location is
    (A) 157 N.
    (B) 437 N.
    (C) 543 N.
    (D) 980 N.

27. Four blocks experience the forces shown in the diagram below for 0.10 s. Rank the blocks in order of their change in momentum from largest to smallest.

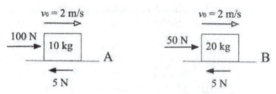

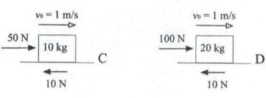

(A) A > B > C > D
(B) D > A > C > B
(C) A > D > B > C
(D) A > C > D > B

28. A force vs. time graph is shown for the net force acting on a block that is moving to the right along the +$x$-axis.

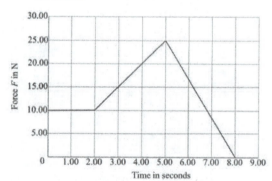

The momentum of the block over the 8.00 s interval will
(A) increase because the variable force acting on the block is always positive.
(B) decrease because the slope of the line is negative between 5.00 s and 8.00 s.
(C) decrease because the force decreases from 25.00 N to 0.00 N at 8.00 s.
(D) increase from 0 to 5.00 s and then decrease in the interval between $t = 5.00$ s to 8.00 s.

29. A 4.00 kg object is lifted vertically 1.20 m and then carried horizontally a distance of 2.50 m. The work done by the gravitational force is closest to
(A) −47.0 J.
(B) −51.0 J.
(C) −98.0 J.
(D) −145.0 J.

30. A 2.00 kg mass initially at rest on a horizontal frictionless surface is struck by an unknown mass $m$ initially traveling to the right with a speed of 0.300 m/s. The collision is a totally elastic collision with the 2.00 kg mass moving to the right with a speed of 0.200 m/s while the unknown mass $m$ moves to the left with a speed of 0.100 m/s. What is the mass $m$ of the unknown object?
(A) 0.500 kg
(B) 1.00 kg
(C) 1.20 kg
(D) 1.80 kg

**Questions 31 and 32**
Questions 31 and 32 refer to the waveform shown in the graph below.

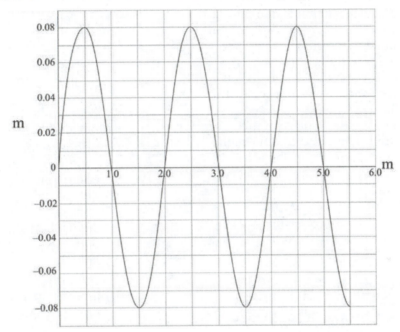

31. What is the amplitude of vibration of the wave in the graph?
(A) 0.08 m
(B) 0.16 m
(C) 2.0 m
(D) 4.0 m

32. What is the wavelength of the wave shown in the graph?
(A) 0.08 m
(B) 0.16 m
(C) 2.0 m
(D) 4.0 m

33. Two wave pulses approach each other on a string shown below. What is the amplitude of the superimposed waveform at point P?

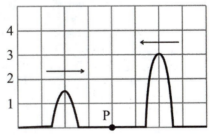

(A) 1.5
(B) 3.0
(C) 4.5
(D) 6.0

**GO ON TO NEXT PAGE**

34. A drive system consists of two pulleys connected by a drive belt. The radius of the output pulley $\vec{r}_o$ is twice the radius of the input pulley $\vec{r}_i$. The moment of inertia of the output pulley also is twice as large as the input pulley.

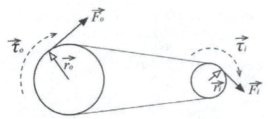

The belt wraps around both pulleys and rotates in a clockwise direction without slipping.
(A) The input pulley has twice the angular velocity of the output pulley since the input pulley has half the radius.
(B) The angular velocity of the output pulley is the same as the angular velocity of the input pulley since the belt connects them.
(C) The input pulley has half the angular velocity of the output pulley because rotational energy, $\frac{1}{2}I\omega^2$, is conserved.
(D) The output pulley has twice the angular velocity of the input pulley because it has twice the moment of inertia.

35. A cylinder is initially rotating about a frictionless axle in a counter-clockwise direction at $4.00 \; \text{rad}/\text{s}$ when $\vec{\tau}_1$ and $\vec{\tau}_2$ are applied to the cylinder as shown.

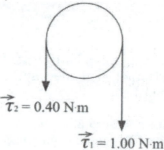

$\vec{\tau}_2 = 0.40 \; \text{N·m}$

$\vec{\tau}_1 = 1.00 \; \text{N·m}$

The cylinder will rotate with
(A) increasing angular speed since both of the torques are applied tangent to the cylinder pointing downward.
(B) decreasing angular speed because $\vec{\tau}_1 > \vec{\tau}_2$ and $\vec{\tau}_1$ is applied opposite the initial direction of rotation, causing the cylinder to come to rest.
(C) decreasing speed until it stops and then increasing speed as it changes its direction of rotation because $\vec{\tau}_1 > \vec{\tau}_2$ and $\vec{\tau}_1$ is applied opposite the initial direction of rotation.
(D) increasing speed because $\vec{\tau}_1 > \vec{\tau}_2$ and $\vec{\tau}_1$ is applied in the initial direction of rotation.

36. A 50.0 kg student stands 2.00 m from the central hub of a merry-go-round, whose moment of inertia is $I = 800$ kg · m². The student and the merry-go-round system are initially rotating at 1.50 $rad/s$. What is the angular velocity of the system when the student walks to a position that is 1.00 m from the hub of the merry-go-round? (Consider the student to be a point mass with a moment of inertia of $I = mr^2$.)
   (A) 1.50 $rad/s$
   (B) 1.76 $rad/s$
   (C) 2.36 $rad/s$
   (D) 2.50 $rad/s$

37. A variable torque is applied to a rotating platform for 4.00 s as shown in the graph.

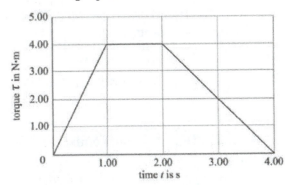

The increase in the angular momentum of the platform at the end of 4.00 s is
   (A) 2.00 N · m · s.
   (B) 4.00 N · m · s.
   (C) 10.0 N · m · s.
   (D) 16.0 N · m · s.

38. A tennis player hits a tennis ball with his racket, returning it to the player on the other side of the net. (Assume an elastic collision between the ball and the racket.) The racket increases the
   (A) time the force is applied to the tennis ball, decreasing the change in momentum.
   (B) force applied to the tennis ball by reducing the time of contact, reducing the change in momentum.
   (C) force by decreasing the time, thereby increasing the change in momentum.
   (D) time of the impulse, which reduces the force needed to change the direction of the ball, maintaining the magnitude of the momentum.

39. A standing wave is set up in a wire that is 1.80 m long. The oscillator is electrically driven with a frequency of 120 Hz.

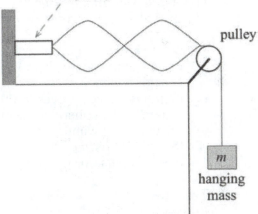

The speed of the wave in the wire is
   (A) 66.7 m/s.
   (B) 133 m/s.
   (C) 108 m/s.
   (D) 216 m/s.

GO ON TO NEXT PAGE

40. A circuit is shown below. The 22.0 Ω resistor $R_1$ is in series with the parallel combination of $R_2$ equal to 12.0 Ω and $R_3$ equal to 24.0 Ω.

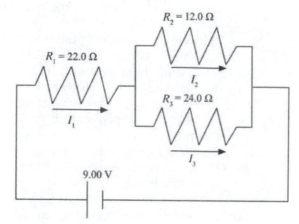

Rank the current in the resistors from largest to smallest.
(A) $I_1 > I_2 > I_3$
(B) $I_1 > (I_2 = I_3)$
(C) $I_3 > I_1 > I_2$
(D) $I_2 > I_1 > I_3$

41. A charging wand carrying a positive charge is brought near but does not touch an isolated neutral conducting sphere. During this action,
    (A) the sphere gains a net positive charge.
    (B) positive charge moves to the region of the sphere opposite the charging wand.
    (C) negative charge moves to the region of the sphere closest to the charging wand.
    (D) the sphere loses positive charge due to repulsion.

42. Two charges $q_1 = +6.50$ μC and $q_2 = -8.1$ μC are 0.100 m apart in air. The electrical force between them is
    (A) 47.4 N attractive.
    (B) 47.4 N repulsive.
    (C) 59.0 N attractive.
    (D) 59.0 N repulsive.

43. A circuit consists of three resistors as shown.

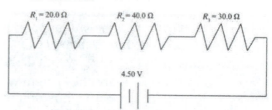

The current in the 15.0 Ω resistor is closest to
(A) 6.67 A.
(B) 1.57 A.
(C) 0.600 A.
(D) 0.150 A.

44. Two isolated spheres 0.100 m apart and carrying charges $q_1 = 5.00$ μC and $q_2 = -15.00$ μC exert a force $\left|\vec{F}_1\right|$ on each other. The spheres are brought into contact and then moved back to their original distance of separation. The new force they exert on each other is

(A) $\left|\vec{F}_2\right| = \dfrac{\left|\vec{F}_1\right|}{2}$ and attractive.

(B) $\left|\vec{F}_2\right| = \dfrac{\left|\vec{F}_1\right|}{2}$ and repulsive.

(C) $\left|\vec{F}_2\right| = \dfrac{\left|\vec{F}_1\right|}{3}$ and attractive.

(D) $\left|\vec{F}_2\right| = \dfrac{\left|\vec{F}_1\right|}{3}$ and repulsive.

45. A circuit consists of three resistors as shown in Figure 1 below.

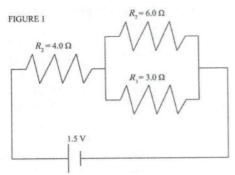

FIGURE 1

$R_2 = 6.0\ \Omega$

$R_1 = 4.0\ \Omega$

$R_3 = 3.0\ \Omega$

1.5 V

How will changing the circuit configuration, shown in Figure 2, change the current in the 4.0 Ω resistor?

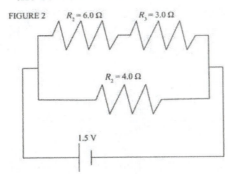

FIGURE 2  $R_2 = 6.0\ \Omega$  $R_3 = 3.0\ \Omega$

$R_1 = 4.0\ \Omega$

1.5 V

(A) The net resistance in the second arrangement is smaller, increasing the current in the circuit. Because the electrical potential difference across the parallel arrangement is the same, more current will go through the 4.0 Ω resistance in the lower branch.
(B) The entire current existed in the 4.0 Ω in the first arrangement. Current will divide in the parallel branch; thus, the current through the 4.0 Ω resistor will reduce.
(C) The 1.5 V battery was not changed; therefore, the current in the 4.0 Ω resistor will not change.
(D) The current in the 4.0 Ω resistor will be smaller since most of the current will go through the smaller 3.0 Ω resistor.

## Questions 46 to 50
**Directions:** For each of the questions or incomplete statements below, two of the suggested answers will be correct. For each of these questions, you must select both correct choices to earn credit. No partial credit will be earned if only one correct choice is selected. Select the two that are best in each case.

46. A 1.0 kg crate accelerates down an inclined plane at 0.60 m/s². Which of the following statements are true? Select two answers.
(A) Increasing the mass of the crate will decrease the acceleration down the plane since Newton's Second Law applies.
(B) Increasing the angle will increase the acceleration since the effective weight acting down the plane will increase.
(C) Increasing the angle will have no change on the acceleration since it will not change the net force.
(D) Increasing the mass will not change the acceleration down the plane since Newton's Second Law applies.

47. Students used various pieces of standard laboratory equipment to design an experiment to measure the motion of an object. Using the listed equipment, which of the following approaches will permit them to determine either the velocity or the acceleration of the object? Select two answers.
(A) A constant motion cart, a meterstick, and a stopwatch
(B) An inclined plane, a block of unknown mass, and a platform balance
(C) A glider on a tilted air track, a protractor, and a photogate
(D) A mass attached to a spring of known spring constant and a meterstick

**GO ON TO NEXT PAGE**

48. A uniform rod 1.00 m long is placed on a pivot at 0.60 m from the left end of the rod. For the rod to be in static equilibrium, which of the following statements must be true? Select two answers.
    (A) The sum of the torques applied to the rod must be zero.
    (B) The sum of the forces applied to the rod must be zero.
    (C) The length of the moment arms on each side of the pivot must be equal.
    (D) The amount of mass on each side of the pivot must be equal.

49. Which of the graphs below may apply to an object in translational equilibrium? Select two answers.

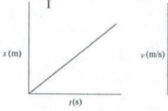
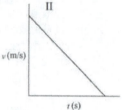
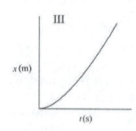
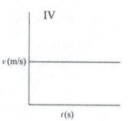

    (A) I
    (B) II
    (C) III
    (D) IV

50. An elevator car moves upward at a constant speed. Which of the following statements is correct? Select two answers.
    (A) The kinetic energy and the gravitational potential energy of the elevator car is constant.
    (B) The kinetic energy of the elevator car is constant but not the gravitational potential energy.
    (C) The upward force on the elevator car is constant.
    (D) The acceleration of the elevator car is equal to the gravitational acceleration.

**STOP**
**END OF SECTION I**

_**IF YOU FINISH BEFORE TIME IS CALLED, YOU MAY CHECK YOUR WORK ON THIS SECTION. DO NOT GO ON TO SECTION II UNTIL YOU ARE TOLD TO DO SO.**_

## AP® PHYSICS 1
### Section II
### 5 Free-Response Problems
### Time—90 Minutes

**Directions:** Solve each of the following problems. Unless the directions indicate otherwise, respond to all parts of each question.

1.  **(7 points)** A steel sphere of mass $m_1$ and radius $r$ rolls 0.3 meter down a ramp without slipping and collides with another steel ball of $m_2$. $m_2$ is double the mass of $m_1$. The moment of inertia for a solid sphere is $I = \frac{2}{5}mr^2$. A side view of the ramp-spheres system is shown below.

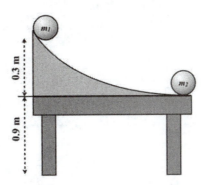

(a) Calculate the speed of $m_1$ right before it collides with $m_2$.

(b) The spheres have a glancing collision, and both spheres leave the table at the same time with different horizontal velocities. Which sphere will hit the floor first?

____ $m_1$      ____ $m_2$      ____ Both hit the floor at the same time.

Justify your answer.

(c) How, if at all, would your answer to part (a) have changed if sphere $m_1$ had been replaced with a sliding frictionless block of the same mass $m_1$?

2.  **(12 points)** You have two unlabeled bulbs. You know that one is "75 Watts" and the other is "40 Watts" although you're not sure which is which.

(a) Sketch a circuit that includes both bulbs, a 120 V source, and conducting wires that will help you identify which bulb is the 75 W and which is the 40 W. Explain how you could distinguish one bulb from the other without using any additional tools.

(b) In another experiment, you set up the 75 W bulb and the 40 W bulb in parallel with the 120 V DC voltage source. Your friend says, "More electrons go into the 75 W than into the 40 W bulb every minute. When electrons are in the bulb, they get used up and turned into light."

  i. What, if any, aspects of your friend's claim are correct?
  ii. What, if any, aspects of your friend's claim are incorrect?

(c) Sketch a circuit that includes both bulbs, a 120 V source, conducting wires, and any other tools or devices typically found in a physics laboratory that would allow you to demonstrate anything you mentioned in part (i) and refute anything you mentioned in part (ii).

(d) What measurements would you make, and how would you make these measurements?

(e) Clearly explain how the measurements you make would support or refute the aspects of your friend's argument that you referred to in part (b).

**GO ON TO NEXT PAGE**

3. **(12 points)** A planetary lander with an initial mass $m = 2.00 \times 10^4$ kg lands on a newly discovered planet whose radius is 1.10 $R_E$ and whose mass is 0.900 $M_E$. During the descent to the planet, thrusters provide a constant upward thrust of $1.30 \times 10^5$ N over a period of 40.0 s to slow the lander.

   (a) Using the information below, plot the data provided and draw a best-fit line through the data points on the grid below.

| V(m/s) | t(s) | Remarks |
|--------|------|---------|
| −20.0 | 0 | Begin Timing |
| −16.0 | 10.0 | Altitude $H$ |
| −12.5 | 15.0 | |
| −11.0 | 20.0 | |
| −5.0 | 30.0 | |
| 0 | 40.0 | $h = 0, v = 0$ |

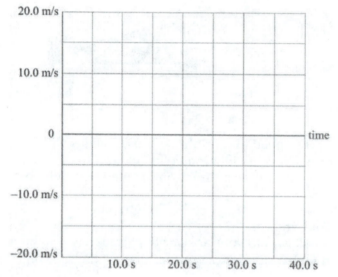

   (b) Using the best-fit line drawn on the graph, determine the acceleration of the lander during the descent.

   (c) i. Explain how you would calculate the height of the lander when the timing begins at time $t = 0$ seconds.

   ii. Explain how you would calculate the height of the lander when it is at altitude $H$ when $t = 10.0$ s.

   (d) Explain how you would determine the weight of the lander when it reached the surface of the planet.

|  | Earth | New Planet |
|--|-------|------------|
| Autumn | 90 | 198 |
| Winter | 89 | 158 |
| Spring | 93 | 147 |
| Summer | 93 | 183 |
| Total Days | 365 | 686 |

   (e) A day on the new planet lasts approximately the same number of hours as a day on Earth. Students reading a news report of the landing see the data tables above and make the following claims.

Student 1: The new planet is closer to the sun during the summer and farther away during the winter. Since the summer is so hot there, the new planet must be closer to the sun than Earth is. Earth's seasons all last about the same amount of time because Earth always goes the same speed, so the new planet must speed up and slow down.

Student 2: That can't be right. For a circular orbit, a planet is always the same distance from the sun. Gravity pulls the planet into orbit, but gravity doesn't speed the planet up or slow it down. Since a year on the new planet is about twice as long as a year on Earth, it must be about twice as far away from the sun as Earth is.

In a coherent paragraph length response, explain which parts of the students' statements are correct and in terms of Kepler's Laws and conservation of energy.

4. **(7 points)** Students added weights to an elastic strap and measured the elongation after each addition of weight. When they unloaded the weights, they obtained additional values for elongation. Their data are shown below for both loading and unloading the elastic strap.

| Force (N) | Elongation (m) Loading | Force (N) | Elongation (m) Unloading |
|---|---|---|---|
| 1.00 | 0.015 | 22.0 | 0.180 |
| 3.00 | 0.325 | 19.0 | 0.155 |
| 6.00 | 0.475 | 15.0 | 0.133 |
| 9.75 | 0.680 | 9.75 | 0.105 |
| 16.5 | 0.105 | 4.50 | 0.078 |
| 20.2 | 0.128 | 1.50 | 0.063 |
| 21.5 | 0.148 | 0.500 | 0.033 |
| 22.0 | 0.180 | | |

(a) Plot the data on the grid below using • for loading and ▲ for unloading.

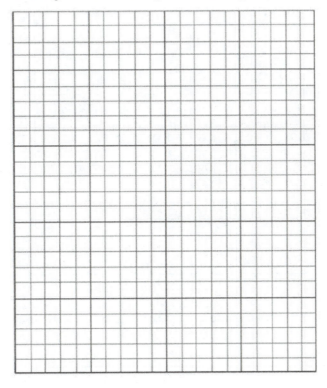

**GO ON TO NEXT PAGE**

(b) Does the elastic strap obey Hooke's law? Explain your reasoning.
(c) i.   What is the approximate area bounded by the loading and unloading readings?
    ii.   What does this tell you about the work done?

5.  **(7 points)** Two waves pulses approach each other as shown at time $t = 0$.

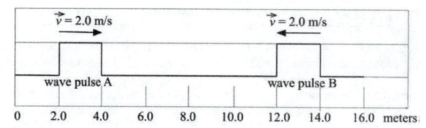

(a) Sketch the pulses on the grid for the following times.
    i.   $t = 1.00$ s

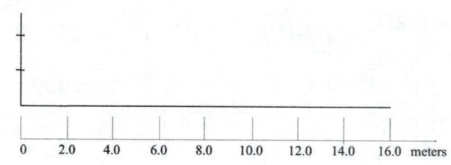

    ii.   $t = 2.00$ s

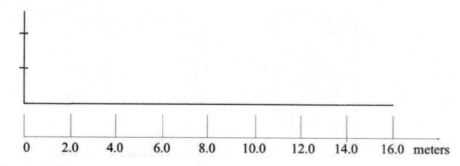

    iii.   $t = 2.50$ s

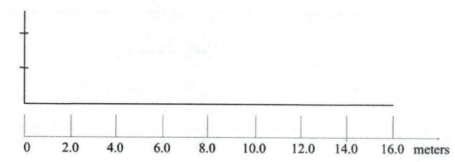

iv.  $t = 3.00$ s

v.  $t = 4.00$ s

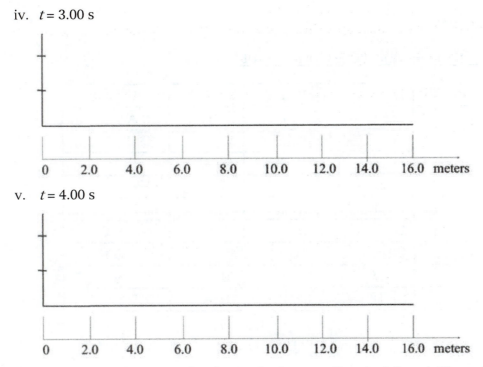

(b)  A pulse is set up in a rope that is attached to a wall on its left end. The pulse approaches the barrier as shown.

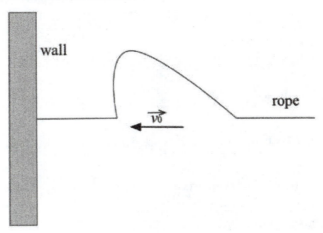

Draw the reflected pulse and describe any changes to the wave upon reflection.

*END OF EXAMINATION*

# Answers

## ANSWER KEY FOR MULTIPLE-CHOICE QUESTIONS

| | | | | |
|---|---|---|---|---|
| 1. D | 11. C | 21. D | 31. A | 41. C |
| 2. B | 12. A | 22. B | 32. C | 42. A |
| 3. D | 13. B | 23. C | 33. C | 43. D |
| 4. C | 14. B | 24. A | 34. A | 44. D |
| 5. C | 15. B | 25. D | 35. C | 45. A |
| 6. D | 16. B | 26. A | 36. B | 46. B & D |
| 7. C | 17. C | 27. C | 37. C | 47. A & C |
| 8. A | 18. A | 28. A | 38. D | 48. A & B |
| 9. B | 19. D | 29. A | 39. D | 49. A & D |
| 10. D | 20. D | 30. B | 40. A | 50. B & C |

## EXPLANATIONS FOR THE MULTIPLE-CHOICE ANSWERS

1.  **D** The displacement of the car during the time shown on the graph is the area under the curve. There are four areas on the graph:

$$A_1 = \frac{1}{2}(4.0\text{ s})\left(10.0\,\frac{\text{m}}{\text{s}}\right) = 20\text{ m}.$$

$$A_2\,(8.0\text{ s} - 4.0\text{ s})\left(10.0\,\frac{\text{m}}{\text{s}}\right) = 40\text{ m}.$$

$$A_3 = \frac{1}{2}(8.0\text{ s} - 4.0\text{ s})\left(15.0\,\frac{\text{m}}{\text{s}} - 10.0\,\frac{\text{m}}{\text{s}}\right) = 10\text{ m}.$$

$$A_4 = (12.0\text{ s} - 8.0\text{ s})\left(15\,\frac{\text{m}}{\text{s}}\right) = 60\text{ m}.$$

The total area is 130 m.

(L.O. 3.A.1.1)

2.  **B** The 2.00 kg object has a weight given by $\vec{F}_g = m\vec{g}$. The weight of the 2.00 kg object is 19.6 N. Because the stationary object $\Sigma\vec{F} = 0$ is supported by two identical vertical cables, the tension in the cables is the same—9.80 N.

(L.O. 2.B.1.1, 3.A.3.1)

3. **D** The force acts on the two blocks, giving them the same acceleration.

$\Sigma \vec{F} = (m_1 + m_2)\vec{a}$. The acceleration is $\vec{a} = \dfrac{20.0 \text{ N}}{10.0 \text{ kg}} = 2.00 \dfrac{\text{m}}{\text{s}^2}$.

The force accelerating the 6.00 kg block is

$\vec{F} = (6.00 \text{ kg})\left( 2.00 \dfrac{\text{m}}{\text{s}^2} \right) = 12.0 \text{ N}$.

The reaction force the 4.00 kg block exerts on the 6.00 kg object is 20.0 N – 12.0 N = 8.00 N. The reaction force the 6.00 kg object exerts on the 4.00 kg object is 8.00 N.

(L.O. 3.A.1.1, 3.A.4.1,)

4. **C** Three forces are acting on the center of mass of the basketball: the tension, the weight of the basketball, and the resistive force of air resistance. The resistive force is tangent to the motion of the center of mass of the basketball.

(L.O. 3.B.2.1)

5. **C** They are in the same gravitational field; therefore, the gravitational force acting on them is the same. In the absence of air resistance, the gravitational potential energy *mgh* is converted into kinetic energy $K = \frac{1}{2}mv^2$. Then since $mgh = \frac{1}{2}mv^2$, they hit the ground with the same velocity. The smaller object has the smaller kinetic energy since $K = \frac{1}{2}mv^2$.

(L.O. 3.A.1.1, 4.C.1.1)

6. **D** Because the sum of the forces in the *y*-direction is zero, the 50 N weight is equal the *y*-component of $T_3$ Because the sum of the forces in the *x*-direction is zero, $T_1$ equals the horizontal component of $T_3$. If the angle were 45°, both components would be equal. Since the angle is only 40°, we know that the *y*-component, 50 N, is slightly more than the *x*-component. The next nearest value is 60 N.

(L.O. 3.A.4.2, 3.A.4.3)

7. **C** As shown in the diagram, the waves are 90° out of phase, with the same frequency and wavelength. Since they are traveling in the same medium, the speed of the wave will not change. The superimposed waveform will vary in amplitude only.

(L.O. 1.A.5.1, 6.D.1.1)

8. **A** The only charge that can move on the surface of the conductor, inner and outer radii, is the negative charge, which will move toward the inner radius, attracted to the +3Q charge at the center. Thus, the charge on the outer surface is +3Q. Electrostatic charge comes in discrete amounts of charge ±e, ±2e, ±3e, etc., which eliminates answers C and D.

   (L.O. 1.B.1.1, 1.B.1.2, 1.B.3.1)

9. **B** The work done on the blocks is equal to the change in the kinetic energy $\left(\vec{F}\cos\theta\right)\vec{x} = \frac{1}{2}m\left(\vec{v}_f - \vec{v}_0\right)$. The blocks move in the positive x-direction, and at the end of the displacement, the blocks have some final velocity $\vec{v}$. In figure A, the force does no work since it is perpendicular to the displacement, cos90° = 0. In D, the force is opposite the displacement and there is a reduction in the kinetic energy since the work is negative and the velocity will be reduced. Since the force in B is applied at an angle $\theta$, the force acting in the direction of the displacement will be smaller than the force applied in C.

   (L.O. 3.E.1.1)

10. **D** Since the slope of the linear velocity vs. time graph is negative, the acceleration is constant and negative. The velocity vs. time graph has some initial positive value. Thus, the displacement is parabolic, opening downward.

    (L.O. 4.A.2.1)

11. **C** By Newton's third law, for each action, there is an equal but opposite reaction. Because we are told that Earth pulls on the book, the reaction is the book pulling on Earth.

    (L.O. 3.A.4.2)

12. **A** The correct free-body diagram for the forces acting on the 10.0 kg block indicates that the normal force, which is the reaction force exerted on the block from the surface, is in response to the downward force of the block and the surface. Since no other forces are exerted in the y-direction, the arrows representing the normal and the weight are equal in length. The block is accelerated to the right by the weight hanging from a cord passing over the pulley, and because it is accelerating, the tension in the rope is larger than the frictional force.

    (L.O. 3.A.4.3)

13. **B** The force acts on the center of mass of the two bodies in contact, increasing the velocity, since a component of the force and the displacement are in the same direction. Increasing the velocity increases the kinetic energy.

(L.O. 4.C.2.2)

14. **B** Newton's second law applies. A net force applied to the total mass will produce an acceleration.
$\Sigma\vec{F} = (m_1 + m_2 + m_3)\vec{a}$. The acceleration is $\dfrac{30.0\text{ N}}{15.0\text{ kg}} = 2.00\ \dfrac{\text{m}}{\text{s}^2}$.
Because the acceleration of all of the cars is the same, the tension in the rope connecting cars A and B is
$\Sigma\vec{F} = (5.00\text{ kg})\left(2.00\ \dfrac{\text{m}}{\text{s}^2}\right) = 10.0\text{ N}$.

(L.O. 3.A.3.1, 3.B.1.4)

15. **B** The normal force is determined by writing $\Sigma\vec{F}_y = 0$.
$\vec{N} - 100\text{ N} + 20\text{ N} = 0$.
The normal force is 80.0 N.

(L.O. 3.A.4.1)

16. **B** In a perfectly inelastic (sticking) collision, the bodies move away with a common velocity since they form a composite mass. The equation for conservation of momentum for this question is $m_1 v_1 + m_2 v_2 = (m_1 + m_2)V$. Substitution into the equation is
$(3.00\text{ kg})\left(4.00\ \text{m}/\text{s}\right) - (1.00\text{ kg})\left(2.00\ \text{m}/\text{s}\right) = (3.00\text{ kg} + 1.00\text{ kg})V$.
The final velocity is 2.50 $\text{m}/\text{s}$.

(L.O. 5.D.2.5)

17. **C** The two forces acting on the elevator car, tension acting upward and weight acting downward, add to zero. Neither force changes over time, so the net force remains zero. Because there is no net force, the acceleration remains constant at 0 m/s².

(L.O. 3.A.1.1, 3.E.1.1, 4.A.2.1, 4.C.1.2)

18. **A** Conservation of momentum for this question gives
$m_{\text{bullet}} v_{0_{\text{bullet}}} = m_{\text{bullet}} v_{f_{\text{bullet}}} + M_{\text{block}} v_{\text{block}}$. Substitution into the
equation is $(0.050\text{ kg})\left(400\ \dfrac{\text{m}}{\text{s}}\right) = (0.050\text{ kg})\left(200\ \dfrac{\text{m}}{\text{s}}\right) + (5.00\text{ kg})v$.

The velocity of the block is $\dfrac{10.0 \; \frac{kg \cdot m}{s}}{5.00 \; kg} = 2.0 \; \frac{m}{s}$.

(L.O. 5.A.2.1)

19. **D** The student's velocity is highest at point C. When the student is at point C an approaching the sound source, the perceived frequency will be Doppler shifted up the most.

(L.O. 6.B.5.1)

20. **D** The balance will measure the inertial mass of the object and thus will read the same on the ground floor and the 163rd floor. A spring scale measures weight. As elevation increases, the value for gravitational acceleration will decrease and the spring scale will give a different reading. At the top, the object will weigh only 99.97% of what it weighed at the bottom.

(L.O. 1.C.3.1)

21. **D** Angular momentum depends on moment of inertia and angular speed. The moment of inertia for the mass is the same throughout its motion, so the point with the highest speed has the greatest angular momentum.

(L.O. 5.B.4.2, 5.E.1.1, 5.E.1.2)

22. **B** The mass of the second piece is $m - \dfrac{m}{3} = \dfrac{2}{3}m$. Conservation of momentum applies after the mass breaks into two separate pieces.

$$m\left(10.0 \; \frac{m}{s}\right) = \frac{m}{3}\left(20.0 \; \frac{m}{s}\right) + \frac{2}{3}mv$$

$$\left(10.0 \; \frac{m}{s}\right) = \frac{1}{3}\left(20.0 \; \frac{m}{s}\right) + \frac{2}{3}v \text{ then } 3.33 \; \frac{m}{s} = \frac{2}{3}v$$

The velocity of the second piece is $5.00 \; \frac{m}{s}$.

(L.O. 5.A.2.1)

23. **C** The conservation of energy applies to this problem since there is no friction between the blocks and the inclines. $mgh_i + \frac{1}{2}mv_i^2 = mgh_f + \frac{1}{2}mv_f^2$. The initial kinetic energy is zero as is the final gravitational potential energy (using the bottom of the inclines as the reference point for the gravitational potential energy). The three blocks reach the bottom of the inclines with the same speeds since the inclines have the same height.

(L.O. 4.C.1.1)

24. **A** As the blocks move down the incline, an identical frictional force acts. The work due to friction is negative because the force and the displacement are anti-parallel. Work of friction is dependent on path. The distance along incline $L_3$ is the greatest; along $L_1$, the least. Therefore, the ranking for the speed at the bottom of the ramp is $v_1 > v_2 > v_3$. (The frictional force will reduce the kinetic energy in the system, converting it into some other form of internal energy, such as thermal energy or sound.)

(L.O. 4.C.1.2)

25. **D** The solution is found by using the work-kinetic energy theorem.

$W$ = area under the curve = $\Delta K$, and $\dfrac{1}{2}\text{base} \times \text{altitude} = \dfrac{1}{2}mv^2$.

Solving for $v$ gives $v = \sqrt{\dfrac{(0.200 \text{ m})(8.00 \text{ N})}{(0.100 \text{ kg})}} = 4.00 \text{ m/s}$.

(L.O. 5.B.5.3)

26. **A** The 100 kg object is lifted a distance of $R_E + 1.5R_E$ from the center of Earth.
The acceleration due to gravity at that location is found from

$\vec{g} = G\dfrac{M_E}{(R_E + 1.5R_E)^2}$.

A ratio for the acceleration can be written as follows:

$\dfrac{\vec{g}_{surface}}{\vec{g}_{altitude}} = \dfrac{(R_E + 1.5R_E)^2}{R_E^2} = 6.25 \cdot \vec{g}_{saltitude} = \dfrac{\vec{g}_{surface}}{6.25} = 1.57 \dfrac{m}{s^2}\ 45\text{ N}$

At that altitude, the 100 kg object has a weight $\vec{F}_g = m\vec{g}$ equal to 157 N.

(L.O. 2.B.1.1, 2. B.2.1)

27. **C** The net forces are applied to the blocks for the same amount of time. The larger the impulse applied to the blocks, $\vec{F}\Delta t = \Delta\vec{p}$, the larger the change in momentum. The net force applied to block A is 100 N – 5 N = 95 N. The net force applied to block D is 100 N – 10 N = 90 N. The net force of B is 50 N – 5 N = 45 N, and finally, the net force on C is 50 N – 10 N = 40 N.

(L.O. 4.B.2.1)

28. **A** The change in momentum is the area under the force vs. time graph. Since the variable force is always positive, there is an increase in momentum.

(L.O. 4.B.2.2)

29. **A** $W = Fx\cos\theta$. The work done by the gravitational force on the object is negative since the force is opposite the vertical displacement of the object. The gravitational force does no work as the object is moved horizontally since the angle between the force and the displacement is 90°.

(L.O. 4.C.1.1)

30. **B** Conservation of momentum applies. Since the 2.00 kg object is initially at rest, the equation is $m\left(0.300 \text{ m/s}\right) = (2.00 \text{ kg})$ $\left(0.200 \text{ m/s}\right) - m\left(0.100 \text{ m/s}\right)$ because the unknown mass moves to the left after the collision. The unknown mass is then found as $m\left(0.400 \text{ m/s}\right) = 0.400 \text{ kg} \cdot \text{m/s}$. The unknown mass is 1.00 kg.

(L.O. 5.D.1.3)

31. **A** The amplitude of the wave is the maximum displacement of the vibrating particles on the wave form from the equilibrium position. The amplitude of the wave from the graph is 0.08 m.

(L.O. 6.A.3.1)

32. **C** The wavelength is the distance from a point on a wave to the next point that has the same amplitude and the same phase (moving in the same direction).

(L.O. 6.B.2.1)

33. **C** Superposition applies. The pulses approach each other, and as they pass through each other at point P, their amplitudes add constructively.

(L.O. 6.D.1.1)

34. **A** The linear speed of the belt is the same on either pulley, so both pulleys have the same tangential velocity.

$r_o \omega_o = r_i \omega_i$

The output pulley has double the radius, so the input pulley has double the angular speed.

(L.O. 4.D.3.1)

35. **C** The net torque applied to the cylinder causes clockwise angular acceleration. The cylinder will decrease its speed in the counter-clockwise direction; stop momentarily; and reverse its direction of rotation, moving clockwise.

(L.O. 4.D.1.1)

36. **B** Conservation of angular momentum applies.

$$(I_{\text{merry-go-round}} + I_{0 \text{ student}})\omega_0 = (I_{\text{merry-go-round}} + I_{f \text{ student}})\omega_f$$

$$\left(800 \text{ kg} \cdot \text{m}^2 + (50 \text{ kg})(2.00 \text{ m})^2\right)\left(1.50 \text{ rad}/\text{s}\right)$$

$$= \left(800 \text{ kg} \cdot \text{m}^2 + (50 \text{ kg})(1.00 \text{ m})^2\right)\omega_f$$

$$\left(1000 \text{ kg} \cdot \text{m}^2\right)\left(1.50 \text{ rad}/\text{s}\right) = \left(850 \text{ kg} \cdot \text{m}^2\right)\omega_f.$$

The final angular velocity is $1.76 \text{ rad}/\text{s}$.

(L.O. 5.E.1.2)

37. **C** The change in the angular momentum is the area under the torque vs. time graph.

$$\Sigma\vec{\tau} = \frac{\Delta\vec{L}}{\Delta t}$$

The area is

$$\frac{1}{2}(4.00 \text{ N} \cdot \text{m})(1.00 \text{ s}) + (4.00 \text{ N} \cdot \text{m})(1.00 \text{ s}) + \frac{1}{2}(4.00 \text{ N} \cdot \text{m})(2.00 \text{ s})$$

$$= 10.0 \text{ N} \cdot \text{m} \cdot \text{s}$$

(L.O. 4.D.3.1)

38. **D** $\vec{F}\Delta t = m\Delta\vec{v}$. The tennis racket changes the direction of the momentum of the ball. The time of the impulse is increased, which reduces the force needed to maintain the magnitude of the momentum and change its direction.

(L.O. 3.D.2.2)

39. **D** The standing wave in the wire has three nodes and two antinodes. Since the end points of a standing wave in a wire must be nodes, the wavelength is 1.80 m. The velocity of the wave in the wire is given by $v = \lambda f$. $v = (120 \text{ Hz})(1.80 \text{ m}) = 216 \text{ m}/\text{s}$

(L.O. 6.D.4.2)

40. **A** Kirchhoff's junction rule applies. 0.300 A enters the junction and divides with $R_2$, being half as large as $R_3$ and getting twice the current as $R_3$, 0.30 A.

(L.O. 5.C.3.1)

41. **C** The charging wand is not in contact with the isolated sphere, and charge cannot be transferred. The sphere cannot gain charge from the ground since it is isolated. The side of the sphere closest to the positive charging wand will be negative because the negative charge will move to the side of the sphere closest to the charging wand. (The only charge that can move on the sphere is the negative charge.)

(L.O. 1.B.2.1)

42. **A** The electrical force acting between the two charges is found from $\left|\vec{F}_E\right| = \dfrac{kq_1q_2}{r^2}$.

Substitution gives

$$\left|\vec{F}_E\right| = \frac{\left(9 \times 10^9 \ \text{N} \cdot \text{m}^2 \middle/ \text{C}^2\right)\left(6.50 \times 10^{-6} \ \text{C}\right)\left(8.10 \times 10^{-6} \ \text{C}\right)}{\left(0.100 \ \text{m}\right)^2}$$

The magnitude of the force is 47.4 N. Since the charges are $q_1 = +6.50 \ \mu\text{C}$ and $q_2 = -8.1 \ \mu\text{C}$, the force is attractive.

(L.O. 3.C.2.1)

43. **D** The circuit is a simple series circuit. The total resistance in the circuit is 15.0 Ω + 13.0 Ω + 32.0 Ω = 60.0 Ω. The current in the circuit from Ohm's law is $I = \dfrac{9.00 \ \text{V}}{60.0 \ \Omega} = 0.150 \ \text{A}$. Kirchhoff's junction rule will show that the current in all three resistors is the same.

(L.O. 5.C.3.1, 5.C.3.3)

44. **D** The charges $q_1 = 5.00 \ \mu\text{C}$ and $q_2 = -15.00 \ \mu\text{C}$ initially attract each other. When they touch, charge moves from $q_2$ to $q_1$. The net charge will be $-10.00 \ \mu\text{C}$ and will redistribute until both spheres have the same charge of $-5.00 \ \mu\text{C}$; therefore, the force will be repulsive. Since the product of the charges $q_1 \times q_2$ has been reduced by $\dfrac{1}{3}$, the force between them will be reduced by $\dfrac{1}{3}$.

(L.O. 1.B.1.1, 1.B.1.2, 3.C.2.1)

45. **A** The net resistance in the first arrangement is larger than in the second arrangement. Therefore, the total current is larger in the second configuration. The 4.0 Ω resistor in the lower branch will receive the most current since it is in the branch with the least resistance.

(L.O. 5.C.3.1)

46. **B** and **D** Newton's second law applies. A general equation for motion down the plane where no other force is acting on the object is $m\vec{g}\sin\theta - \mu_k mg\cos\theta = ma$. Whether there is a frictional force, changing the mass of the object will not change the acceleration. Increasing the angle will increase the effective force down the plane $m\vec{g}\sin\theta$ and decrease the normal to the surface, decreasing the frictional force if a frictional force exists between the object and the surface of the plane on the object.

    (L.O. 2.B.1.1, 3.A.1.1)

47. **A** and **C** The speed of the constant motion cart can be determined from $V_{avg} = \dfrac{\Delta x}{\Delta t}$.

    The acceleration of the glider can be determined by $v_f^2 = v_0^2 + 2ax$ from the tilted air track by solving. Starting from rest, the glider moves on the marked air track a given distance. The photogate will permit the students to determine the final velocity.

    (L.O. 3.A.1.2)

48. **A** and **B** For an object to be in static equilibrium, it must be in translational and rotational equilibrium. Thus, <u>both</u> $\Sigma\vec{\tau} = 0$ and $\Sigma\vec{F} = 0$ apply.

    (L.O. 3.A.3.1, 3.F.1.1, 3.F.1.2)

49. **A** and **D** When the object is in equilibrium, either stationary or moving at a constant speed, the net force acting on it is $\Sigma\vec{F}_y = 0$. There is no acceleration. The slope of the distance vs. time graph gives velocity. The slope of graph I, $x = f(t)$ is constant; thus, the velocity is constant. Graph IV, $v = f(t)$ is constant and has a slope of zero, no acceleration.

    (L.O. 3.A.1.1, 4.A.2.1)

50. **B** and **C** The elevator is rising with constant speed; thus, its kinetic energy is constant. It is increasing its height in a gravitational field, thereby increasing its gravitational potential energy. There is no acceleration on the elevator car since it is rising at a constant speed, and that means the force acting on the elevator car is constant.

    (L.O. 3.A.1.1, 2.A.3.1, 4.C.1.2, 5.B.4.1)

## ANSWERS TO FREE-RESPONSE PROBLEMS

### QUESTION 1

(a) Initial potential energy becomes translational and rotational kinetic energy.

$$mgh = \frac{1}{2}mv^2 + \frac{1}{2}I\omega^2 \text{ and } I = \frac{2}{5}mr^2 \text{ and } \omega = \frac{v}{r}$$

$$mgh = \frac{1}{2}mv^2 + \frac{1}{2}(\frac{2}{5}mr^2)\frac{v^2}{r^2}$$

$$gh = \frac{1}{2}v^2 + \frac{1}{10}v^2$$

$$v = \sqrt{\frac{10gh}{7}} = \sqrt{\frac{10(9.8^m/_{s^2})(0.3\ m)}{7}} = 2.0\frac{m}{s} \qquad \textbf{(3 points)}$$

(b) Both spheres will strike the floor at the same time.

The sphere with twice the mass is pulled by gravity with twice the force.

By Newton's second law, the downward acceleration of $m_1$ is

$$a = \frac{m_1\ g}{m} = g.$$

The downward acceleration of $m_2$ is $a = \frac{2m_1\ g}{2m} = g.$

So both falling spheres have the same downward acceleration and travel the same vertical distance to land on the floor at the same moment. **(2 points)**

(c) If mass $m_1$ were only sliding, all of its initial potential energy would be converted entirely into kinetic energy. The mass would have a greater velocity. **(2 points)**

(L.O. 3.A.1.1, 4.C.1.1, 5.B.4.2, 5.D.1.5)

### QUESTION 2

(a) There are two possible solutions. Either solution 1 or solution 2 may earn points.
Solution 1: Connect the bulbs to the voltage source in parallel with each other. The brighter bulb is the 75 W bulb.
Solution 2: Connect the bulbs to the voltage source in series. The dimmer bulb is the 75 W bulb. **(2 points)**

A sketch for part (a) that is consistent with a method in part (b) for determining which bulb is brighter earns both points.

(b) i. It is correct to say that more electrons flow through the 75 W bulb per unit time. The 75 W bulb has less resistance than the 40 W bulb; thus, more current flows through the 75 W resistor when it is connected in parallel to the 40 W bulb.

ii. It is incorrect to say that electrons are used up or converted into light. Electrons are not used; some of their energy is transferred.

(c)

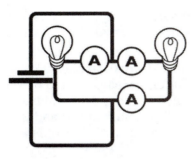

**(2 points)**

The circuit can be used to determine how much current each bulb receives because an ammeter is in series with each bulb.

The circuit can be used to determine whether the electrons going into one of the bulbs also come out because there are two ammeters in series with the bulb. **(2 points)**

The diagram should contain an ammeter in each of the two branches in order to determine which branch conducts the most current. **(1 point)**

(d) I would measure the current with each ammeter. I would look at the bulbs to determine which one was brighter. **(2 points)**

(e) If the brighter bulb has more current flowing through it, then my friend is correct. Also, if two ammeters in series with a bulb give the same reading, then the number of electrons entering and leaving that bulb in every unit of time is the same. **(3 points)**

## QUESTION 3

(a)     The correct graph is shown below.

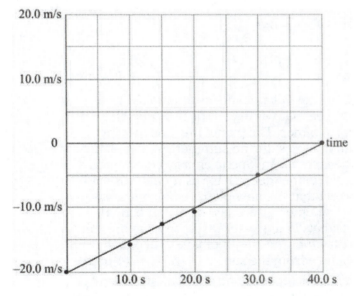

**(2 points)**

(b) The acceleration of the lander is the slope of the best-fit line. Taking two slope points, not data points, from the line, the slope is

$$\vec{a} = \frac{-2.50\ \frac{m}{s} - \left(-12.5\ \frac{m}{s}\right)}{(35.0\ s - 15.0\ s)} = 0.50\ \frac{m}{s^2}$$

**(1 point)**

(c) i. The displacement is the area under the velocity vs. time graph. This will give the height of the lander when the thrusters were fired at $t = 0$ s. **(1 point)**

ii. The displacement during the first 10.0 s is the sum of areas under the curve—the rectangular area bounded by the product of 10.0 s and −15.0 m/s as shown in the graph plus ½ of the product of 10.0 s × −5.0 m/s. The height of the lander would then be the displacement of the lander determined from part (c)i., the displacement found in this part. **(2 points)**

**Alternative explanation from kinematics equations**

Since the slope of the line gives the deceleration of the lander, the displacement for the first 10 s can be calculated by taking the final velocity squared minus the initial velocity squared, then dividing that answer by twice the acceleration from the slope of the line. The height can also be determined by multiplying one-half the slope of the line by the time squared and then subtracting that number from the product of the initial velocity times the time. In both cases, the height $H$ is the difference between the altitude when $t = 0$ and $t = 10$ s.

(d) The acceleration due gravity on the surface of the planet is found from the ratio of the gravitational values between Earth and the planet.

$$\frac{\vec{g}_{planet}}{\vec{g}_{Earth}} = \frac{(0.90 M_E)/(1.10 R_E)^2}{M_E/R_E^2} \quad \vec{g}_{planet} = 0.75 \vec{g}_{Earth} = 7.29 \text{ m}/\text{s}^2$$

The weight of the lander is $\vec{F}_{\vec{g}Planet} = m \vec{g}_{Planet}$

$$(2.00 \times 10^4 \text{ kg})\left(7.29 \frac{\text{m}}{\text{s}^2}\right) = 1.46 \times 10^5 \text{ N}$$

**(1 point)**

(e) Student 1 is correct that the new planet speeds up and slows down. From the table, we see that winter on the new planet is significantly shorter than summer. The planet must be moving faster during winter than it moves during summer. As the planet gets closer to the sun, the increase in kinetic energy comes with a loss of potential energy; the planet goes faster as it gets closer to the sun. Student 2 is correct that the new planet is farther away from the sun than Earth is. Kepler's third law says that orbital periods are greater for planets at a greater distance from the sun. Student 2 is incorrect to say that the length of a year depends linearly on a planet's distance from the sun. **(4 points)**

If the response has no incorrect or contradictory statements, it earns an additional point. **(1 point)**

(L.O. 2.B.1.1, 3.A.1.1, 3.A.1.3, 3.C.1.1, 3.D.2.2, 4.A.2.1)

## QUESTION 4

(a) The graph of the data is shown below.　　　　　**(2 points)**

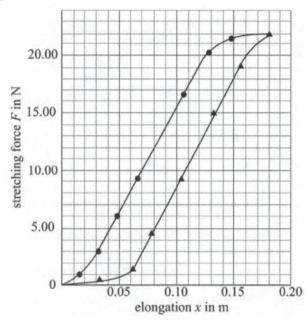

(b) The elastic strap does not appear to obey Hooke's law. Hooke's law is a linear relationship between the force and the displacement of the material.　　　　　**(1 point)**

(c) i.　Counting blocks bounded by the lines will give an internal energy production of approximately 77 blocks. The area of each block is (1.00 N)(0.100 m), or 0.100 J. Approximately 8 J of energy was produced as internal energy in the stretching and relaxing of the strap.　　　　　**(2 points)**

　　ii.　The area bounded by the loading and unloading of the strap represents the energy that was transferred to internal friction as the chemical bonds were stretched in the strap. This energy will leak to the environment as thermal energy.　　**(2 points)**

　　(L.O. 4.C.1.2, 5.A.2.1, 5.B.3.1, 5.B.3.3, 5.B.4.1, 5.B.5.3)

## QUESTION 5

(a) The pulses shown for the times are

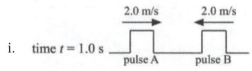

　i.　time $t = 1.0$ s

　ii.　time $t = 2.0$ s　　　　　The leading edges of the pulses start to move through each other with the same velocity and amplitude as at $t = 1.0$ s.

iii. **time** $t = 2.5$ s ⎍ The pulses have combined constructively giving an amplitude that is twice as great as the amplitude at $t = 1.0$ s and 2.0 s.

iv. **time** $t = 3.0$ s ⎏ The trailing edges of the pulse start to separate as the pulses move with their initial speeds in their original directions. The amplitude of each pulse is equal to its initial amplitude.

v. **time** $t = 4.0$ s ⊓⊓ The pulses have separated

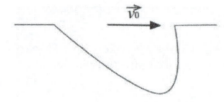

2.0 m/s     2.0 m/s

pulse B     pulse A

and continue in their original directions with the same speed and the same amplitude. **(5 points)**

(b) The correct reflected pulse is **(1 point)**

$\vec{v_0}$

The wave pulse is reflected from a fixed barrier, the wall, and is inverted. It travels with the same speed since the medium (the rope) determines the speed of the pulse or the wave. **(1 point)**

(L.O. 6.D.1.1, 6.D.2.1)

# AP® PHYSICS 2
# PRACTICE EXAM C

AP® PHYSICS 2
Section I
50 Multiple-Choice Questions
Time——90 Minutes

**Note:** To simplify calculations, you may use $g = 10$ m/s² in all problems.

**Directions:** Each of the questions or incomplete statements below is followed by four suggested answers or completions. Select the one that is best in each case.

1. Blood from a main vein ($r_{vein} = 1.00$ mm) enters into ten narrow capillaries. The radius of each capillary is $r_{capillary} = 0.100$ mm. The speed of the blood in the vein is 20.0 cm/s.

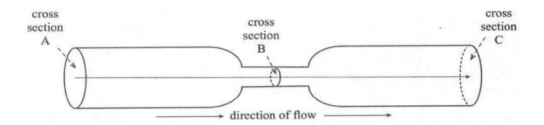

cross section A    cross section B    cross section C

→ direction of flow →

What is the speed of the blood in each of the capillaries?
(A) 2.00 cm/s
(B) 5.00 cm/s
(C) 20.0 cm/s
(D) 100 cm/s

2. An ideal fluid moves smoothly through the tube shown above, moving from point A to point B and then to point C. The cross-sectional area of A and C is the same.
   Which of the following is a correct statement?
   (A) The pressure drops as the fluid moves through the tube. The pressure at A is larger than the pressure at B, which is larger than the pressure at C.
   (B) The pressure at points A and C is the same. The pressure at B is smaller than the pressure at A and C.
   (C) The pressure at A is somewhat larger than the pressure at C. The pressure at A and C is larger than the pressure at B.
   (D) The pressure at points A, B, and C is the same since the fluid moves smoothly through the tube.

**GO ON TO NEXT PAGE**

3. An atom of $^{239}_{94}\text{Pu}$ undergoes a nuclear fission by a neutron as indicated in the reaction below.

$$^{239}_{94}\text{Pu} + {}^{1}_{0}\text{n} \rightarrow {}^{148}_{56}\text{Ba} + {}^{90}_{38}\text{Sr} + ?$$

What is the missing particle?

(A) $^{2}_{0}\text{n}$

(B) $2\left({}^{1}_{0}\text{n}\right)$

(C) $^{2}_{1}\text{H}$

(D) $2\left({}^{1}_{1}\text{H}\right)$

4. A sample of an ideal gas undergoes an isothermal compression in a container. As a result, there is an increase in the rate at which the molecules collide with the walls of the container going through a change in momentum during the collision. This produces

(A) an increase in the force exerted per unit area on the walls of the container and a corresponding increase in pressure.

(B) an increase in the force exerted per unit area on the walls of the container and a corresponding decrease in pressure.

(C) a decrease in the force since the time of interaction increases. This causes a drop in pressure.

(D) a decrease in the time of interaction, causing a corresponding decrease in pressure since the force is smaller.

5. The four spheres shown below are immersed in water. The volume $V$ of each sphere is given.

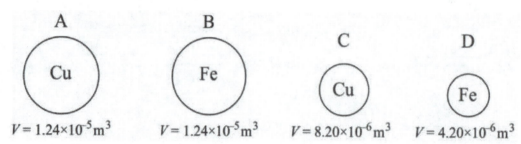

Rank the buoyant force acting on each sphere from the greatest magnitude to the least magnitude. The density of copper is $\rho_{Cu} = 8.92 \times 10^3 \ \frac{\text{kg}}{\text{m}^3}$, and the density of iron is $\rho_{Fe} = 7.86 \times 10^3 \ \frac{\text{kg}}{\text{m}^3}$.

(A) A > B > C > D
(B) A > C > B > D
(C) (A = B) > C > D
(D) B > A > D > C

6. A concave mirror with a radius of curvature of 30.0 cm produces an upright image located 40.0 cm behind the mirror. What is the position of the object?
   (A) 10.9 cm
   (B) 12.8 cm
   (C) 17.1 cm
   (D) 21.8 cm

7. A gas expands along the path ABC in the diagram below. The work done by the gas in this expansion is

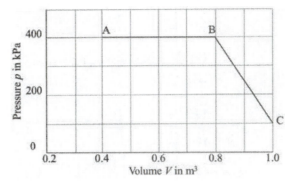

   (A) $1.6 \times 10^5$ J.
   (B) $1.9 \times 10^5$ J.
   (C) $2.1 \times 10^5$ J.
   (D) $2.4 \times 10^5$ J.

8. Water $\rho = 1000 \ \frac{kg}{m^3}$ enters the lower end of the tube shown below with a velocity of $2.00 \ \frac{m}{s}$. With an input pressure at the lower end of the tube of $2.10 \times 10^5$ Pa, what is the pressure at the top of the tube?

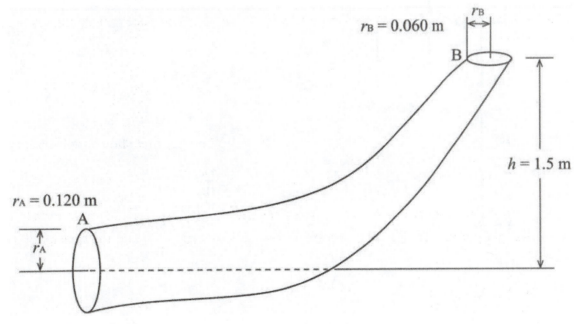

   (A) $2.10 \times 10^5$ Pa
   (B) $1.92 \times 10^5$ Pa
   (C) $1.65 \times 10^5$ Pa
   (D) $1.47 \times 10^5$ Pa

**GO ON TO NEXT PAGE**

9.  The diagram below shows the frequency ranges of different types of light in the electromagnetic spectrum.

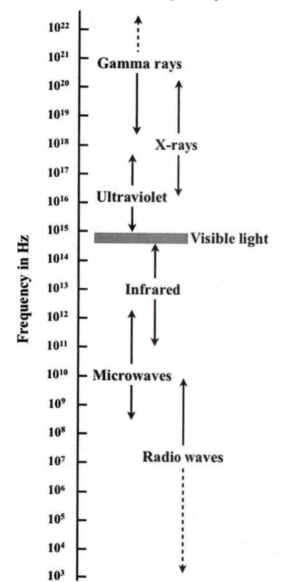

Identify the correct ranking that lists electromagnetic waves from longest wavelength to shortest wavelength.
(A) Infrared radiation > Ultraviolet radiation > X-rays > Gamma rays
(B) Ultraviolet radiation > Gamma rays > X-rays > Infrared radiation
(C) Gamma rays > X-rays > Ultraviolet radiation > Infrared radiation
(D) Ultraviolet radiation > Infrared radiation > Gamma rays > X-rays

10. The diagram below shows rays of light that originate in the lower medium of index of refraction $n_1$. The rays pass into a medium of index of refraction of $n_2$ where $n_2 > n_1$, then into air where the index of refraction of air is less than the index of refraction of the other two mediums $n_{air} < n_1 < n_2$.

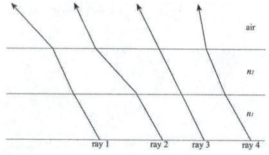

The ray that shows the correct path is
(A) ray 1.
(B) ray 2.
(C) ray 3.
(D) ray 4.

11. When a converging lens is placed in front of an object, its image is formed on the screen on the other side of the lens at a distance that is shown in the figure below.

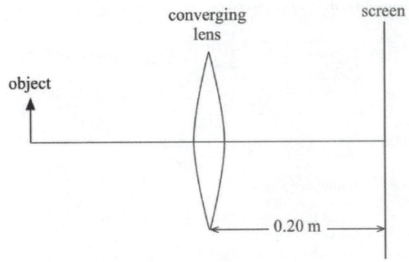

A diverging lens is then placed 0.05 m behind the converging lens. A new image is formed 0.30 m from the diverging lens on the screen as shown.

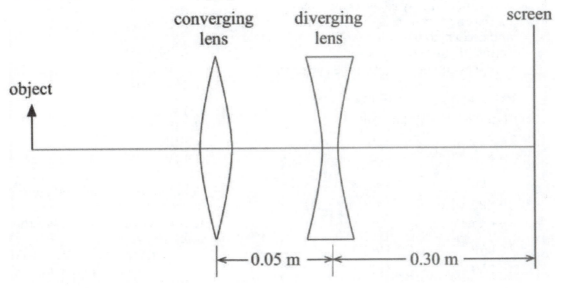

What is the focal length of the diverging lens?
(A) −0.15 m
(B) −0.30 m
(C) −0.45 m
(D) −0.60 m

**GO ON TO NEXT PAGE**

12. The diagram below shows the flow of an ideal non-compressible fluid flowing through a tube of three different cross-sectional areas from the left to right. Which of the following statements concerning the velocity and pressure is correct for the areas shown?

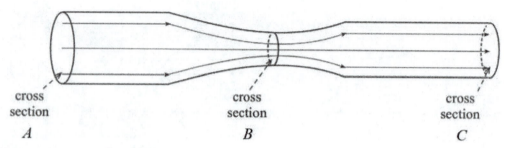

cross section *A*      cross section *B*      cross section *C*

(A) $v_A = v_B = v_C$ and $p_A = p_B = p_C$
(B) $v_A = v_B = v_C$ and $p_A \neq p_B \neq p_C$
(C) $v_A \neq v_B \neq v_C$ and $p_A = p_B = p_C$
(D) $v_A \neq v_B \neq v_C$ and $p_A \neq p_B \neq p_C$

13. A group of students obtained the following set of data from an experiment that was performed by increasing the pressure on a cylinder and recording the volume of the confined gas in the isothermal compression of the gas in the cylinder.

| Pressure | Volume |
|---|---|
| $(10^5$ Pa$)$ | $(10^{-3}$ m$^3)$ |
| 1.10 | 5.00 |
| 1.15 | 4.78 |
| 1.20 | 4.58 |
| 1.25 | 4.40 |
| 1.30 | 4.23 |
| 1.40 | 3.92 |
| 1.50 | 3.67 |

Which of the following graphs best illustrates the correct plot of the data above?

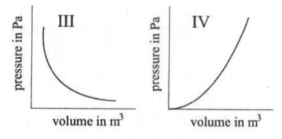

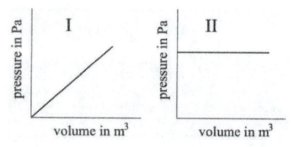

(A) I
(B) II
(C) III
(D) IV

14. A liquid in an insulated container is stirred and undergoes a rise in temperature. Considering the liquid to be a system,
(A) the temperature increased because work was done by the system.
(B) there was an increase in the internal energy of the system.
(C) there was no change in the internal energy of the system.
(D) the work done on the system is equal to the work done by the system.

15. An ideal gas goes through a cyclic process ABCDA as shown in the diagram.

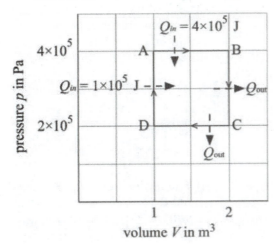

If the thermal energy entering the system along path AB is $5 \times 10^5$ J, what is the thermal energy that leaves the system during the cyclic process?
(A) $2 \times 10^5$ J
(B) $3 \times 10^5$ J
(C) $5 \times 10^5$ J
(D) $7 \times 10^5$ J

16. Two rays of light in air—red and violet—fall on the same side of a glass prism as shown.

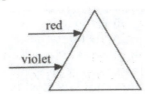

As the rays pass through and emerge from the prism, the ray of
(A) red light bends more than the ray of violet light because the violet light has a longer wavelength and higher frequency.
(B) violet light bends more than the ray of red light because the violet light has the shorter wavelength and higher frequency.
(C) red light bends more than the ray of violet light because the red light has a longer wavelength and higher frequency.
(D) red light bends more than the ray of violet light because the red light has the shorter wavelength and higher frequency.

17. An external magnetic field is applied to four ferromagnetic materials. The domains associated with these four ferromagnetic materials are shown below. Rank the materials from the one with the strongest magnetic field to the one with the least magnetic field.

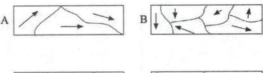

(A) A > D > C > B
(B) A > B > C > D
(C) D > (C = B) > A
(D) (C = B) > D > A

18. A wire that is 1.20 m long carrying a current of 5.00 A toward the top of the page is placed in a magnetic field of 0.100 T directed out of the page as shown in the diagram below.

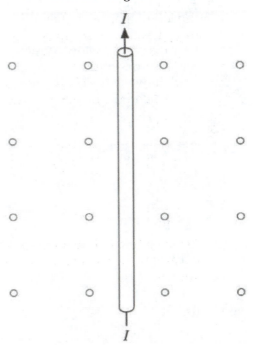

The force on the wire is
(A) 0.60 N to the right of the page.
(B) 0.60 N to the left of the page.
(C) 6.0 N to the right of the page.
(D) 6.0 N to the left of the page.

**GO ON TO NEXT PAGE**

19. Two wires are carrying currents of $I_1 = 4.0$ A and $I_2 = 8.0$ A, as shown in the figure below. The wire carrying $I_1$ exerts a force $\left|\vec{F}_{12}\right|$ on the wire carrying current $I_2$. The wire carrying current $I_2$ exerts a force $\left|\vec{F}_{21}\right|$ on the wire carrying current $I_1$.

The force between the wires is

(A) $\left|\vec{F}_{12}\right| = \left|\vec{F}_{21}\right|$ attractive.

(B) $2\left|\vec{F}_{12}\right| = \left|\vec{F}_{21}\right|$ attractive.

(C) $\left|\vec{F}_{12}\right| = 2\left|\vec{F}_{21}\right|$ repulsive.

(D) $\left|\vec{F}_{12}\right| = \left|\vec{F}_{21}\right|$ repulsive.

20. A positive point charge is located to the left of a plate containing a negative charge as illustrated in the four diagrams below. Which diagram best indicates the electric field between the positive point charge and the negative plate?

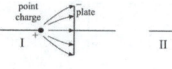

(A) I
(B) II
(C) III
(D) IV

21. The gases in two containers A and B shown below have the same volume and the same pressure. Container A has half the number of molecules as container B. Which of the following choices is a correct statement?

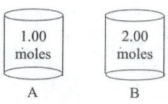

(A) The molecules in container A have a lower rms (root-mean-square) speed than the molecules in container B because the temperature in container A is related to the temperature in B by $T_A = \dfrac{T_B}{2}$.

(B) The molecules in container A have a higher rms speed than the molecules in container B because the temperature in A is twice as high as the temperature in B.

(C) The molecules in containers A and B have the same rms speed since $\dfrac{p_1 V_1}{T_1} = \dfrac{p_2 V_2}{T_2}$ and the only difference is the number of moles in the containers.

(D) Since we do not know what gases are in the containers, we do not know which container has molecules with the higher rms speed.

**Questions 22 and 23**
Use the circuit shown below to answer questions 22 and 23.

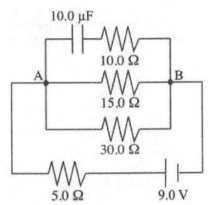

22. The current through the 15.0 Ω resistor when the circuit above has reached steady state is
(A) 0.150 A.
(B) 0.200 A.
(C) 0.300 A.
(D) 0.400 A.

23. What is the charge on the 10.0 $\mu$F capacitor when the circuit above has reached steady state?
(A) 30.0 $\mu$C
(B) 45.0 $\mu$C
(C) 60.0 $\mu$C
(D) 90.0 $\mu$C

24. Four charges are arranged in a square as shown below. The electric field $\vec{E}$ is zero

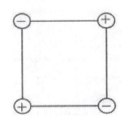

(A) at points on each side of the square midway between each pair of charges.
(B) at the midpoint of the square.
(C) midway between the top two charges and midway between the bottom two charges.
(D) midway between the two charges on the left and midway between the two charges on the right.

25. When monochromatic rays of red and blue light fall on a narrow slit producing a diffraction pattern,
(A) red light is diffracted more than blue light since it has the higher frequency.
(B) blue light is diffracted more than red light since it has the higher frequency.
(C) red light is diffracted more that blue light since it has the longer wavelength.
(D) blue light is diffracted more than red light since it has the longer wavelength.

26. An object placed 30.0 cm in front of a lens produces a real image that is three times the size of the object. What is the focal length of the lens?
(A) 22.5 cm
(B) 30.0 cm
(C) 45.0 cm
(D) 60.0 cm

27. A positive charge $+Q$ moves along the $+x$-axis with some velocity $\vec{v}$ into a region where a magnetic field of strength $\vec{B}$ is directed along the $+z$-direction as shown in the diagram below. The direction of the force on the positive charge is in the

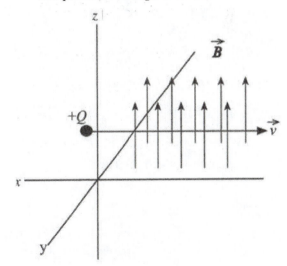

(A) positive $z$-direction.
(B) positive $y$-direction.
(C) negative $z$-direction.
(D) negative $y$-direction.

**GO ON TO NEXT PAGE**

28. A dipole molecule is aligned along the x-axis with the positive end $q_1 = 8.00 \times 10^{-9}$ C at 0.200 m and the negative end $q_2 = -8.00 \times 10^{-9}$ C at −0.200 m.

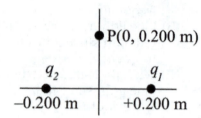

The magnitude and direction of the electric field at point P (0, 0.200 m) is closest to

(A) $1.27 \times 10^3 \dfrac{N}{C}$ along the −x-axis.

(B) $1.27 \times 10^3 \dfrac{N}{C}$ along the +y-axis.

(C) $1.80 \times 10^3 \dfrac{N}{C}$ along the −x-axis.

(D) $1.80 \times 10^3 \dfrac{N}{C}$ along the +y-axis.

29. Three charges are located on a line. What electrostatic force on $q_2$ is due to the other two charges?

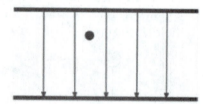

(A) $\dfrac{kq^2}{r^2}$ to the left

(B) $\dfrac{kq^2}{r^2}$ to the right

(C) $\dfrac{3kq^2}{r^2}$ to the left

(D) $\dfrac{3kq^2}{r^2}$ to the right

30. Consider the electrical circuit shown below. The battery has negligible internal resistance, and its emf is $\mathcal{E} = 10$ V. Given $R_1 = R_2 = R_3 = R_4 = R_5 = 10\,\Omega$, what is the resistance of $R_6$ so that the power delivered by the battery is 10 W?

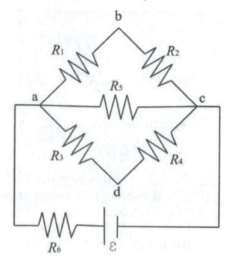

(A) 5 Ω
(B) 10 Ω
(C) 20 Ω
(D) 30 Ω

31. An electron, initially at rest, is placed in the region between two electrically charged parallel plates as shown below. The arrows in the diagram represent the direction of the electric field between the plates.

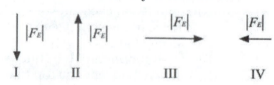

The direction of the force $|F_E|$ exerted on the electron by the field is best illustrated by which arrow?

(A) I
(B) II
(C) III
(D) IV

32. Two isolated identical spheres A and B containing charges as shown attract each other initially.

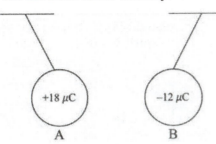

A piece of copper wire is connected to the spheres and then removed. Which of the following choices correctly indicate the resultant charge on A and B and the electrostatic force between them?

|  | Charge on A | Charge on B | Result |
|---|---|---|---|
| (A) | $+6 \ \mu C$ | $-6 \ \mu C$ | attraction |
| (B) | $+6 \ \mu C$ | $+6 \ \mu C$ | repulsion |
| (C) | $+3 \ \mu C$ | $-3 \ \mu C$ | attraction |
| (D) | $+3 \ \mu C$ | $+3 \ \mu C$ | repulsion |

33. A current-carrying wire of infinite length carries a current $I$ in the positive $z$-direction as shown in the diagram below. A negative charge moves with some velocity $\vec{v}$ parallel to the wire in a region very close to the wire. What is the direction of the magnetic force on the charge due to the magnetic field created by the current-carrying wire?

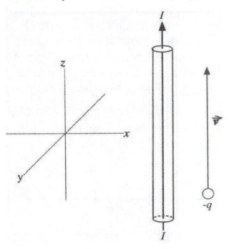

(A) Force is in the $-x$-direction.
(B) Force is in the $+x$-direction.
(C) Force is in the $-y$-direction.
(D) Force is in the $+y$-direction.

34. A proton with an initial horizontal velocity is injected into the region between two parallel plates as shown below. Four possible paths are shown for the motion of the proton when it enters the region between the plates. Which of the four possible paths will the photon follow?

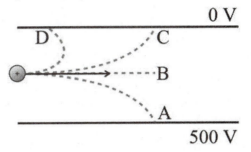

(A) Path A
(B) Path B
(C) Path C
(D) Path D

35. Rank the currents through the resistors in the circuit shown below from the largest current to the smallest current.

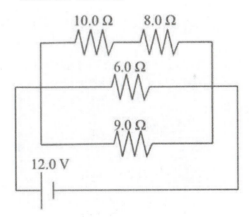

(A) $I_{6\Omega} > I_{8\Omega} > I_{9\Omega} > I_{10\Omega}$
(B) $I_{6\Omega} > I_{8\Omega} > I_{10\Omega} > I_{9\Omega}$
(C) $I_{10\Omega} = I_{8\Omega} > I_{9\Omega} > I_{6\Omega}$
(D) $I_{6\Omega} > I_{9\Omega} > (I_{8\Omega} = I_{10\Omega})$

**GO ON TO NEXT PAGE**

36. The diagram below shows a standing wave pattern corresponding to the de Broglie wavelength for the electron in a particular orbit of the Bohr atom of hydrogen. The energy level occupied by this electron is

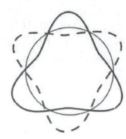

(A) $n = 1$.
(B) $n = 2$.
(C) $n = 3$.
(D) $n = 4$.

37. Identify the missing particle in the nuclear reaction shown below.
$$^{23}_{11}\text{Na} + \; ? \; \rightarrow \; ^{20}_{10}\text{Ne} + \; ^{4}_{2}\text{He}$$
(A) $^{0}_{-1}e$

(B) $^{1}_{0}n$

(C) $^{1}_{1}\text{H}$

(D) $^{2}_{1}\text{H}$

38. Two conducting loops have a common vertical axis as shown below.

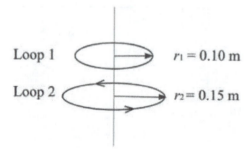

Loop 1 has a radius $r_1 = 0.10$ m, and loop 2 has a radius $r_2 = 0.15$ m. Current in loop 2 (lower loop) is in the counterclockwise direction as viewed from above and increasing at a constant rate. The induced current in loop 1 as viewed from above is
(A) clockwise and steady.
(B) counterclockwise and increasing.
(C) clockwise and decreasing.
(D) counterclockwise and decreasing.

39. A conducting rod of length 0.10 m is moving with a constant velocity of 10.0 m/s toward the right in a uniform magnetic field of 0.10 T directed out of the page as shown in the figure below.

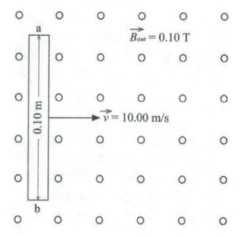

What is the electrical potential difference induced across the ends of the rod and which end of the rod is positive?
(A) 0.10 V and the top of the rod is positive.
(B) 0.10 V and the bottom of the rod is positive.
(C) 1.0 V and the top of the rod is positive.
(D) 1.0 V and the bottom of the rod is positive.

40. An electric field $\vec{E}$ is directed to the right as shown in the diagram below. Four points, A, B, C, and D, are in the region produced by the field. Rank the points in the diagram from the greatest to the least potential.

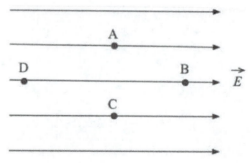

(A) D > C > B > A
(B) B > C > A > D
(C) D > (A = C) > B
(D) A = B = C = D

41. An energy level diagram is illustrated for a certain atom. Rank the transitions for the emitted electromagnetic radiation from longest wavelength to shortest wavelength.

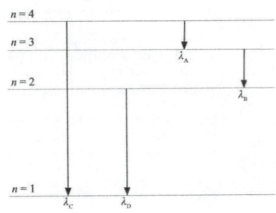

(A) $\lambda_C > \lambda_D > \lambda_B > \lambda_A$
(B) $\lambda_A > \lambda_B > \lambda_D > \lambda_C$
(C) $\lambda_A > \lambda_B > \lambda_C > \lambda_D$
(D) $\lambda_D > \lambda_C > \lambda_B > \lambda_A$

42. Dysprosium-152* has an unstable nucleus that undergoes gamma decay to become the stable Dysprosium-152.

$$^{152}Dy^* \rightarrow {}^{152}Dy + \gamma$$

The change undergone by $^{152}Dy^*$ to become $^{152}Dy$ is best explained by
(A) conservation of mass.
(B) conservation of charge.
(C) conservation of energy.
(D) conservation of momentum.

43. The isotope C–14 is radioactive and decays as

$$^{14}_{6}C \rightarrow {}^{14}_{7}N + {}^{0-}_{0}\nu + ?$$

What unknown particle (?) is needed to complete the decay equation?
(A) $^{0}_{+1}e$
(B) $^{0}_{-1}e$
(C) $^{1}_{0}n$
(D) $^{1}_{1}H$

44. The graph below shows the number of undecayed nuclei as a function of time. The half-life of the radioactive isotope is approximately
(A) 1 s.
(B) 2 s.
(C) 4 s.
(D) 10 s.

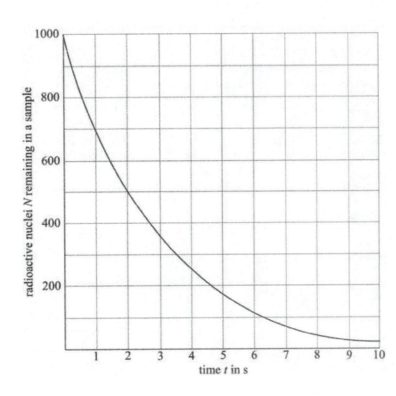

**GO ON TO NEXT PAGE**

45. The surface of a photoemissive metal plate has a threshold frequency of $1.50 \times 10^{15}$ Hz. When light of frequency $2.00 \times 10^{15}$ Hz strikes this metal plate,
    (A) an electron with zero velocity and a $0.50 \times 10^{15}$ Hz photon are emitted from the surface.
    (B) no electron is emitted from the surface.
    (C) an electron is emitted with a speed of 0.750 c since the ratio of the frequencies is $\frac{1.50 \times 10^{15}}{2.00 \times 10^{15}} = 0.750$.
    (D) an electron is emitted with a slightly lower speed since the new frequency is slightly larger than the threshold frequency.

**Questions 46 to 50**
**Directions:** For each of the questions or incomplete statements below, two of the suggested answers will be correct. For each of these questions, you must select both correct choices to earn credit. No partial credit will be earned if only one correct choice is selected. Select the two that are best in each case.

46. A positively charged rod is brought very close to the left side of an uncharged conducting sphere and held in place for several seconds. The rod is then removed and discharged. Which image below correctly depicts the charge distribution on the conducting sphere after the rod had been removed and discharged? Select two answers that are correct.

    (A)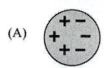

    (B)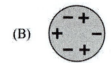

    (C)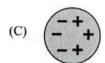

    (D)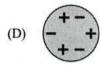

47. An electron drops from $n = 3$ to $n = 2$ in the Bohr atom of hydrogen. As a result, the (select two answers)
    (A) energy of the electron increases.
    (B) force on the electron increases.
    (C) energy of the electron decreases.
    (D) force on the electron decreases.

48. A ray of light traveling in a medium whose index of refraction is 1.33 strikes the interface between it and a second medium with an index of refraction of 1.52 at an angle of 30°. In the second medium, the ray of light (select two answers)
    (A) increases in speed.
    (B) decreases in speed.
    (C) has an increased wavelength.
    (D) has a decreased wavelength.

49. A proton with a velocity of $6.0 \times 10^5$ m/s enters a region in which there is a magnetic field of 0.10 T directed into the page. The proton follows the path shown.

    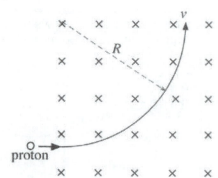

    Which of the following two statements are true?
    (A) The initial force on the proton is $9.6 \times 10^{-15}$ N directed to the top of the page.
    (B) The initial force on the proton is $9.6 \times 10^{-15}$ N directed to right, causing the velocity of the proton to increase.
    (C) The momentum of the proton is constant because the magnetic force is perpendicular to the velocity of the proton. The speed is the same when it leaves the field.
    (D) The momentum of the proton has changed since the direction of the velocity has changed.

50. Two rods of copper and aluminum are placed between two walls maintained at temperatures of 100°C and 0°C as shown in Figure 1. In Figure 1 the two pieces of metal are connected between the walls in parallel, and in Figure 2, the two pieces of metal are in series between the walls.

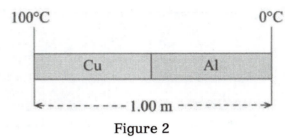

**Figure 2**

100°C                    0°C

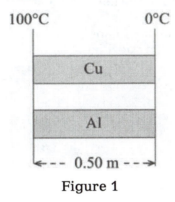

**Figure 1**

The rate at which thermal energy is conducted from one wall to the other is greater (select two answers)

(A) through Figure 2 than through Figure 1 since the rod connecting the two walls is longer.

(B) through Figure 1 since the effective cross-sectional area is larger.

(C) through Figure 1 since the change in the temperature $\Delta T$ is the same for both rods.

(D) through Figure 2 since there is a smaller temperature difference $\Delta T$ between the ends of the copper rod and the aluminum rod.

## STOP
## END OF SECTION I

*IF YOU FINISH BEFORE TIME IS CALLED, YOU MAY CHECK YOUR WORK ON THIS SECTION. DO NOT GO ON TO SECTION II UNTIL YOU ARE TOLD TO DO SO.*

## AP® PHYSICS 2
### Section II
### 4 Free-Response Problems
### Time—90 Minutes

**Directions:** Solve each of the following problems. Unless the directions indicate otherwise, respond to all parts of each question.

1. **(10 points)** A 2.0 cm tall object O is placed 60 cm to the left of a thin converging lens A of focal length $f_A = 20$ cm as shown in the figure below.
   (a) Use two rays from the tip of object O to locate the image.
       i.   Describe the type of image, its height, and its position.

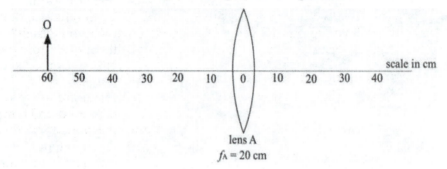

       ii.  Explain how you could determine the size and magnification of the image.
   (b) A second converging lens B with a focal length $f_B = 20$ cm is placed 20 cm to the right of lens A as shown below.

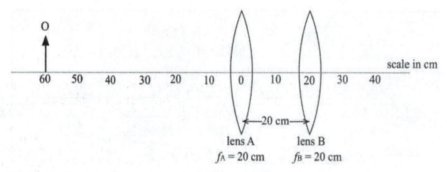

       i.   What is the location of the final image relative to lens B?
       ii.  Describe the final image formed in terms of the type of image, its magnification, and its location.
   (c) A concave mirror of $f_{Mirror} = 20$ cm is placed 40 cm to the right of lens A as shown.

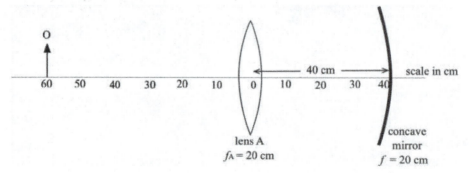

       i.   Explain how you could determine the type of image formed.
       ii.  Explain where the final image will be formed.

2. **(12 points)** When you walk into the classroom on a laboratory day, your teacher tells you that she attached one end of a coil of wire to simulate an internal resistance in an opaque box connected to a power supply. Using the equipment placed on the table, she tells you to design an experiment to determine the "internal resistance" of the power supply. On your laboratory table, besides the power supply with internal resistance, you have a voltmeter, a switch, a variable resistor in the range of 0.1 Ω to 110.0 Ω, and connecting wires. (The power supply was preset to give you a maximum value of 1.50 V.)

   (a) Design an experiment in sufficient detail that another student could perform your experiment and obtain the same results. Include the schematic of your setup in your procedure.

   (b) What measurements are you going to take in the experiment, and how are you going to use the data from the measurements to determine the "internal resistance" attached to the power supply?

   (c) When the switch is open, what is the value of the current in your circuit? What does this tell you about the value of the resistance $R$ in the circuit?

   (d) If you make a graph, what are you going to graph and how will your graph help you answer the experimental question?
      i.   What does the intercept of the graph suggest?
      ii.  Why does the voltmeter not indicate the exact emf of a cell?

   (e) Another group of students was given a power supply with an identical coil of wire in an opaque box but a different set of equipment with which they built the schematic shown below.

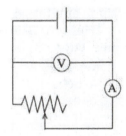

   The other group plotted a graph of their data and came up with an internal resistance that had a higher value than your "internal resistance." What are two reasons that could account for their higher value?

3. **(12 points)** A loop of wire 20.0 cm by 20.0 cm enters a region where a magnetic field of 1.60 T is directed out of the page and confined in a square of dimensions 70.0 cm by 70.0 cm.

   The loop of wire has a velocity of 10.0 cm/s perpendicular to the $\vec{B}$ field as shown in the diagram.

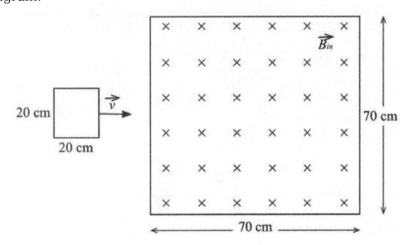

(a) In a short paragraph, use correct scientific terms to explain what happens to the emf induced in the loop as the loop enters the field, travels through the field, and then exits the field.

(b) On the grid below, graph the emf as a function of time as the loop travels through the field.

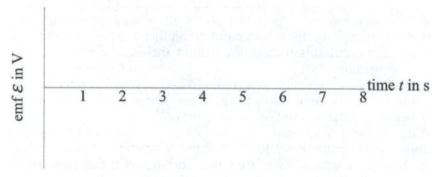

(c) If the loop of wire has a resistance of 40.0 Ω, what is the magnitude of the maximum induced current in the loop?

(d) Calculate the magnetic flux through the loop when the loop is completely immersed in the field.

4. **(10 points)**

(a) The isotope Pu-236 is an alpha emitter with a half-life of 2.86 years. It decays into U-232 as $^{236}_{94}\text{Pu} \rightarrow {}^{232}_{92}\text{U} + {}^{4}_{2}\text{He} + Q$. The alpha particles from a Pu-236 source are allowed to fall on a thin section of gold foil. An alpha particle, $^{4}_{2}\text{He}$, comes within $80.0 \times 10^{-15}$ m of a gold nucleus, $^{197}_{79}\text{Au}$. Assuming that both the gold nucleus and the alpha particle behave as point charges, determine the maximum electrostatic repulsive force between them at this distance.

(b) An alpha particle penetrates the electron cloud of a gold atom, colliding with and ejecting the inner electrons of the first energy level and creating vacancies. Explain what happens to these inner electron vacancies.

(c) The alpha particle is observed to have a momentum of $1.10 \times 10^{-19}$ kg·m/s at 0°. What is the momentum of the U-232 nucleus?

(d) A sample of Pu-236 containing $4.00 \times 10^{10}$ nuclei is prepared in a nuclear physics laboratory. How would you determine the number of Pu-236 nuclei that remain after four half-life periods?

*END OF EXAMINATION*

# Answers

## ANSWER KEY FOR MULTIPLE-CHOICE QUESTIONS

| 1. D | 11. B | 21. D | 31. B | 41. B |
|------|-------|-------|-------|-------|
| 2. C | 12. D | 22. D | 32. D | 42. D |
| 3. B | 13. C | 23. C | 33. B | 43. B |
| 4. A | 14. B | 24. B | 34. C | 44. B |
| 5. C | 15. B | 25. C | 35. D | 45. D |
| 6. A | 16. B | 26. A | 36. C | 46. B & D |
| 7. C | 17. A | 27. D | 37. C | 47. B & C |
| 8. C | 18. A | 28. A | 38. A | 48. B & D |
| 9. A | 19. D | 29. D | 39. B | 49. A & D |
| 10. A | 20. A | 30. A | 40. C | 50. B & C |

## EXPLANATIONS FOR THE MULTIPLE-CHOICE ANSWERS

1. **D** The equation of continuity is $A_1v_1 = A_2v_2$. Since area is $A = \pi r^2$, we write the equation as $r_1^2 v_1 = r_2^2 v_2$. Substitution is

$$(0.100 \text{ cm})^2 \left( 20.0 \ \frac{\text{cm}}{\text{s}} \right) = 10(0.010 \text{ cm})^2 v_2.$$

Solving yields 200 $\text{cm}/_\text{s}$.

(L.O. 5.B.10.3)

2. **C** The ideal fluid moves through the tube from a larger cross section into a smaller one. The equation of continuity relates the velocity in the region $A_1v_1 = A_2v_2$. The fluid will flow faster in the constriction. Bernoulli's equation, a statement of conservation of energy, for the horizontal tube is $p + \frac{1}{2}\rho v^2 = \text{constant}$. The pressure at B is smaller than the pressure at A and C. But the pressure at A must be slightly higher than the pressure at C because the fluid is moving in the tube from point A to B and then to C.

(L.O. 5.B.10.4)

3. **B** In the fission of $^{239}_{94}\text{Pu}$ by a neutron, charge number and mass number must be conserved. For charge, $94 + 0 \rightarrow 56 + 38$. The missing particle must have a charge of 0. The mass numbers are $239 + 1 \rightarrow 148 + 90 + ?$ The mass needed to balance is 2. A neutron is correctly written as $^1_0 n$; thus, two neutrons are required to complete the reaction $2(^1_0 n)$.

(L.O. 5.G.1.1)

4. **A** This is an isothermal process. The pressure will increase as the volume decreases at constant temperature. Molecules will collide elastically with the walls of the container more frequently. Since this takes place over a very short period of time, the force must increase. The increased force per unit area on the walls of the container results in an increase in pressure.

(L.O. 7.A.1.1)

5. **C** The fluid exerts the buoyant force on the object. The magnitude of the buoyant force $|\vec{F}_B|$ is equal to the weight of the water displaced.

Since spheres A and B have the same volume, they displace the same volume of water and the buoyant force on each is the same. The buoyant force is directed upward on all of the spheres.

(L.O. 3.C.4.1)

6. **A** The focal length of the mirror is $f = \dfrac{R}{2} = \dfrac{30.0 \text{ cm}}{2} = 15.0$ cm. The image formed is negative since it is formed behind the mirror. Substitution in $\dfrac{1}{f} = \dfrac{1}{s_o} + \dfrac{1}{s_i} = \dfrac{1}{15.0 \text{ } cm} = \dfrac{1}{s_o} + (-\dfrac{1}{40.0 \text{ } cm})$. The object is located at 10.9 cm.

(L.O. 6.E.4.2)

7. **C** The work done in the expansion of the gas along path ABC is the area under the pressure-volume curve. This is the sum of three areas. $A_1$ is the rectangular area of

$(4.0 \times 10^5 \text{ Pa} - 0)(0.8 \text{ m}^3 - 0.4 \text{ m}^3) = 1.6 \times 10^5 \text{ J}$.

$A_2$ is the triangular area equal to

$\dfrac{1}{2}(4.0 \times 10^5 \text{ Pa} - 1.0 \times 10^5 \text{ Pa})(1.0 \text{ m}^3 - 0.8 \text{ m}^3) = 3.0 \times 10^4 \text{ J}$, and

$A_3 = (1.0 \times 10^5 \text{ Pa} - 0)(1.0 \text{ m}^3 - 0.8 \text{ m}^3) = 2.0 \times 10^4 \text{ J}$.

The net work is $2.1 \times 10^5 \text{ J}$.

(L.O. 5.B.7.3)

8. **C** The equation of continuity $A_1 v_1 = A_2 v_2$ applies. Substitution into the equation is $\pi(0.120 \text{ m})^2 \left(2.00 \text{ } \dfrac{\text{m}}{\text{s}}\right) = \pi(0.060 \text{ m})^2 v_2$. The exit velocity at the top is $8.00 \text{ } \dfrac{\text{m}}{\text{s}}$.

Bernoulli's equation for this problem is $p_1 + \dfrac{1}{2}\rho v_1^2 = p_2 + \dfrac{1}{2}\rho v_2^2 + \rho g h_2$. If we take the initial height $h_1$ as the reference point, $h_1 = 0$. Substitution into the equation is

$$2.10 \times 10^5 \text{ Pa} + \frac{1}{2}\left(1000 \text{ } \frac{kg}{m^3}\right)\left(2.00 \text{ } \frac{m}{s}\right)^2$$

$$= p_2 + \frac{1}{2}\left(1000 \text{ } \frac{kg}{m^3}\right)\left(8.00 \text{ } \frac{m}{s}\right)^2 + \left(1000 \text{ } \frac{kg}{m^3}\right)\left(9.80 \text{ } \frac{m}{s^2}\right)(1.50 \text{ m})$$

$$2.10 \times 10^5 \text{ Pa} + \frac{1}{2}\left(1000 \text{ } \frac{kg}{m^3}\right)\left(4.00 \text{ } \frac{m^2}{s^2}\right)$$

$$= p_2 + \frac{1}{2}\left(1000 \text{ } \frac{kg}{m^3}\right)\left(64.0 \text{ } \frac{m^2}{s^2}\right) + \left(1000 \text{ } \frac{kg}{m^3}\right)\left(9.80 \text{ } \frac{m}{s^2}\right)(1.50 \text{ m})$$

$$p_2 = 1.65 \times 10^5 \text{ Pa}$$

(L.O. 5.B.10.1, 5.B.10.3)

9. **A** Gamma rays have a wavelength in the order of $10^{-12}$ m. X-ray radiation is on the order of 10 nm to about $10^{-4}$ nm. Ultraviolet radiation covers a range of about 400 nm to about 0.6 nm, and infrared "heat waves" are about 1 mm to around 700 nm.

(L.O. 6.F.1.1)

10. **A** Light travels in a medium of uniform optical density in straight lines. The light traveling from $n_1$ to $n_2$ slows as it enters the optically slower substance, bending toward the normal (perpendicular). The ray of light traveling from $n_2$ into air, an optically slower medium, will bend away from the normal.

(L.O. 6.E.3.1)

11. **B** For the diverging lens, the object distance is –0.15 m and the image distance is 0.30 m. Using $\frac{1}{f} = \frac{1}{s_o} + \frac{1}{s_i}$, substitution is $\frac{1}{f} = \frac{1}{0.30} + \frac{1}{-0.15}$ then $\frac{1}{f} = \frac{-2+1}{0.30 \text{ m}} = \frac{-1}{0.30 \text{ m}}$. The focal length of the diverging lens is –0.30 m.

(L.O. 6. E.5.1)

12. **D** The fluid flows from region A into B. The area decreases in B, and the speed increases from the equation of continuity $A_1 v_1 = A_2 v_2$. The fluid then flows into region C where the cross-sectional area is larger than B but smaller than A. Although the speed will decrease, it will be faster than the speed in area A because the cross-sectional area is smaller than in A.

When the speed changes in the horizontal pipe, Bernoulli's equation, a statement of conservation of energy that relates the pressure plus the kinetic energy to a constant, will show that the pressures must be different in all three sections since the speed in each section is different.

(L.O. 5.B.10.4)

13. **C** Since the experiment was done at constant temperature, this is an isothermal process. Boyle's law relates the relationship between the volume of the confined gas and its pressure. $p_1 V_1 = p_2 V_2$. A pressure-volume graph is hyperbolic.

(L.O. 5.B.7.2)

14. **B** The system is insulated, so heat does not flow in or out of the system. Work is done on the system when the liquid is stirred, so there is an increase in the internal energy of the system.

(L.O. 5.B.3.1, 5.B.4.1)

15. **B** In a cyclic process, the system returns to initial state conditions of pressure, volume, and temperature. Since the temperature returns to its initial value, $\Delta U = 0$ and $Q = -W$. The net work done per cycle is the area bounded by the cycle.

$W = -(4 \times 10^5 \text{ J} - 2 \times 10^5 \text{ J})(2 \text{ m}^3 - 1 \text{ m}^3) = -2 \times 10^5 \text{ J}$.

The thermal energy added along AB is $Q_{in} = 5 \times 10^5$ J. The thermal energy leaving the system is $Q_{out} = 5 \times 10^5 \text{ J} - 2 \times 10^5 \text{ J} = 3 \times 10^5 \text{ J}$.

(L.O. 5.A.2.1, 5.B.4.1, 5.B.7.3)

16. **B** The angle of refraction of the light falling on the prism depends on the wavelength of the light. The index of refraction usually decreases with increasing wavelength. Violet light has a shorter wavelength and will have a higher angle of deviation (it will spread further from the original path). It will refract more than the ray of red light.

(L.O. 6.F.1.1)

17. **A** The domains in the ferromagnetic material listed as A are aligned in somewhat the same direction by an external magnetic field, followed by the domains in D. While C and B are close, B is a little more random than the domains in C.

(L.O. 4.E.1.1)

18. **A** The direction of the force on the current in the wire is given using the right-hand rule. The conventional current is toward the top of the page. The magnetic field is out of the page; therefore, the force on the wire is toward the right.

The magnitude of the force is $\left|\vec{F}\right| = \left|\vec{B}\left(I\vec{L}\right)\right| \sin\theta$. Substitution into the equation is $\left|\vec{F}\right| = \left|0.10 \text{ T}(5.00 \text{ A} \cdot 1.2 \text{ m})\right| \sin 90° = 0.60 \text{ N}$.

(L.O. 2.D.1.1)

19. **D** Both wires carrying current produce magnetic fields that point into the page in the region between the two wires. This acts as though two N poles of magnet are near each other, causing a repulsive force. The forces the wires exert on each other are action-reaction pairs. They exert the same magnitude of force on each other.

(L.O. 2.D.2.1, 3.A.4.2)

20. **A** The electric field points away from the positive point charge and toward the negative charge. Diagram I is correct. Five lines leave the positive charge and terminate on the negative plate, indicating that the magnitudes of the charges are the same.

(L.O. 2.C.2.1)

21. **D** The rms speed of a gas molecule is given by $v_{rms} = \sqrt{\dfrac{3N_A kT}{M}}$.

The containers have the same volume and pressure, but because they have a different number of moles of the gas, they have different temperatures. Since $pV = NkT$, $N_A kT_A = N_B kT_B$. Since container B has twice the number of moles as A, it must have half the temperature if the pressure in A is to equal the pressure in B.

(L.O. 7.A.2.1)

22. **D** At steady state, no current flows through the 10.0 Ω resistor in the top branch of the parallel network because the capacitor is fully charged. The resistance in the parallel branch without including the 10 Ω resistor is $\dfrac{1}{R} = \dfrac{1}{15.0\ \Omega} + \dfrac{1}{30.0\ \Omega}$. This gives a resistance of 10.0 Ω, which is in series with a 5.00 Ω resistance for a total of 15.0 Ω.

The current from the battery that enters junction A is

$I = \dfrac{9.00\ V}{15.0\ \Omega} = 0.600\ A$.

By Kirchhoff's junction rule, the 15.0 Ω resistor has a current of 0.400 A and the 30.0 Ω resistor is 0.200 A.

(L.O. 5.C.3.4, 5.C.3.7)

23. **C** When the circuit has reached steady state, no current flows through the 10.0 Ω resistor. The entire electrical potential difference for the top branch is across the plates of the capacitor. The electrical potential difference across the parallel section of the circuit is

9.00 V = (0.600 A)(5.00 Ω) + $V_{parallel}$ = 6.00 V. Then $C = \dfrac{Q}{V}$. The charge

on the plates of the capacitor is (10.0 μF)(6.00 V) = 60.0 μF.

(L.O. 5.B.9.5, 5.C.3.7)

24. **B** $\vec{E}$ is zero wherever the net force acting on a test charge is zero. At the center of the square, the two positive charges alone would produce a resultant electric field of zero, as would the two negative charges alone. Thus, the resultant force acting on a test charge at the midpoint of the square will be zero.

(L.O. 2.C.2.1)

25. **C** The extent to which light is diffracted depends on wavelength. Blue light has a shorter wavelength, so it is diffracted least. Red light, with its longer wavelength, is diffracted most.

(L.O. 6.C.2.1)

26. **A** Magnification of the image is given by $|M| = \dfrac{s_i}{s_o}$. The image distance, which is positive, is 3(30.0 cm) = 90.0 cm. The equation $\dfrac{1}{f} = \dfrac{1}{s_o} + \dfrac{1}{s_i}$ will give the focal length of the convex lens (image is real). $\dfrac{1}{f} = \dfrac{1}{30.0\text{ cm}} + \dfrac{1}{90.0\text{ cm}} = \dfrac{1+3}{90.0\text{ cm}}$. $f = 22.5$ cm

(L.O. 6.E.5.1)

27. **D** By the right-hand rule, the force on the positive charge is in the negative $y$-direction.

(L.O. 3.A.2.1)

28. **A** The electric field points away from the positive charge and toward the negative charge. By symmetry, the vertical components of the electric field are equal in magnitude and opposite in direction.

The net electric field at (0, 0.200 m) is directed along the $-x$-axis, and its magnitude is given by $\left|\vec{E}_{x_1}\right| + \left|\vec{E}_{x_2}\right| = \left|\Sigma\vec{E}_x\right|$.

Since the dipole has the same magnitude of charge on its ends, $\left|\vec{E}\right| = \dfrac{2k_e q}{r^2}\cos 45°$. Since $r^2 = (0.200\text{ m})^2 + (0.200\text{ m})^2 = 0.080\text{ m}^2$, the field strength is $2\dfrac{\left(9\times10^9\ \frac{\text{N}\cdot\text{m}^2}{\text{C}^2}\right)(8.00\times10^{-9}\text{ C})}{(0.080\text{ m}^2)}\cos 45°$. The magnitude of the field is $1.27\times10^3\ \dfrac{\text{N}}{\text{C}}$.

(L.O. 2.C.1.2, 2.C.2.1, 2.C.4.2)

29. **D** The force on $q_2$ due to $q_1$ and $q_3$ is toward the right. The force between $\vec{F}_{12}$ is repulsive since both charges $q_1$ and $q_2$ are positive and will be toward the right. The force $\vec{F}_{23}$ between $q_2$ and $q_3$ is attractive and thus to the right. The magnitude of the force is $\dfrac{kq_1q_2}{r^2}+\dfrac{kq_2q_3}{(2r)^2}$. Substitution for the values of the charges $\dfrac{k(q)(2q)}{r^2}+\dfrac{k(2q)(2q)}{(2r)^2}=\dfrac{3kq^2}{r^2}$.

(L.O. 2.C.1.2)

30. **A** The 10 V battery delivers 10 W. We can use $P=\dfrac{V^2}{R}$ to find that the total resistance is $R=\dfrac{V^2}{P}=\dfrac{(10\,V)^2}{10\,W}=10\ \Omega$. We resolve the parallel network with $R_{eq}=\Big(\dfrac{1}{R_1+R_2}+\dfrac{1}{R_3+4}+\dfrac{1}{R_5}\Big)^{-1}=5\ \Omega$. If the total resistance is 10 Ω and the network is 5 Ω, then $R_6$ is 5 Ω.

(L.O. 5.B.9.6, 5.C.3.5)

31. **B** The direction of the electric field in the region between the plates is downward from the upper plate toward the lower plate as shown by the field arrows. A positive charge will move in the direction of the field. The force on the electron is opposite the field and will move the electron toward the upper plate.

(L.O. 2.C.1.1)

32. **D** When the copper wire connects A and B, negative charge will move and the net charge remaining between the two spheres will be +6 $\mu C$. Negative charge will redistribute so that each sphere will have the same charge, +3 $\mu C$, when the copper wire is removed. The force between the two spheres will then be repulsive.

(L.O. 1.B.1.2)

33. **B** The wire creates a magnetic field to the right of the wire that is in the page wire. A charge moving parallel to the wire experiences a force by the right-hand rule that is in the $-x$-direction. However, the charge is negative; therefore, the force acting on it is toward the right.

(L.O. 3.C.3.1)

34. **C** The electric field between the plates accelerates the proton toward the upper plate. The force is perpendicular to the initial velocity of the proton; therefore, the proton will follow a parabolic path toward the upper plate.

(L.O. 2.C.1.1, 2.C.5.3)

35. **D** Kirchhoff's junction rule applies. The total current entering the junction divides as it branches into three separate paths. The branch with the smallest resistance $R = 6.0\ \Omega$ will have the largest current, followed the $9.0\ \Omega$ resistor. The two resistors in the top branch are in series, and each has the same current.

(L.O. 1.B.1.2)

36. **C** There are whole number multiples of the de Broglie wavelengths for the electron in a stable orbit of hydrogen. In the diagram, there are three complete wavelengths. This gives $n = 3$ for the stable orbit.

(L.O. 7.C.2.1)

37. **C** In any nuclear reaction, charge and mass must be conserved. Conservation of charge is $11 + ? = 10 + 2$. The missing particle has a charge of $+1$. Conservation of mass (A number) is $23 + ? = 20 + 4$. The missing particle has a mass of $+1$. This is the proton $_1^1 H$.

(L.O. 5.C.1.1, 5.G.1.1)

38. **A** The increasing current in the lower loop produces a magnetic field pointing upward from loop 2, which behaves like the north pole of a magnet approaching the upper loop. According to Lenz's law, this induces a current in the upper loop (loop 1) that flows in the direction that creates a magnetic field to oppose what induced it. The current as viewed from above will be clockwise. Since the current in the lower loop is increasing at a constant rate, the current in the upper loop will be constant. The induced current depends on the rate of change of the current in loop 2.

(L.O. 4.E.2.1)

39. **B** The potential difference induced in the rod is determined by
$\mathcal{E} = Blv$.

The electrical potential difference is
$(0.10\ \text{T})(0.10\ \text{m})\left(10.0\ \tfrac{\text{m}}{\text{s}}\right) = 0.10\ \text{V}$. The direction of the field is out of the page, and the velocity of the rod is toward the right. Therefore, the force on the conventional charge carrier (positive) is by the right-hand rule toward the bottom of the rod. The rod will act like a battery of $\mathcal{E} = 0.10\ \text{V}$ with the positive terminal at the bottom of the rod.

(L.O. 3.C.3.1)

40. **C** The direction of the electric field is toward the right, out from the positive end of the field. Since $\left|\vec{E}\right| = \dfrac{\Delta V}{\Delta r}$, the electrical potential will decrease from left to right in the field. Points A and C are on the same isoline and thus have the same electrical potential. Point D is closest to the positive end of the field and therefore has the highest electrical potential. B has the least electrical potential since it is to the right of the other points.

   (L.O. 2.E.2.3)

41. **B** The transition between $n = 4$ to $n = 1$ results in the radiation with the most energy and therefore the highest frequency and the shortest wavelength $\lambda_C$ since $\Delta E = hf = \dfrac{hc}{\lambda}$. The lowest energy is the drop from $n = 4$ to $n = 3$ or $\lambda_A$. This is the lowest frequency and the longest wavelength. The correct ranking is $\lambda_A > \lambda_B > \lambda_D > \lambda_C$.

   (L.O. 6.F.1.1)

42. **A** In the annihilation process, matter is converted into pure energy. The electron and positron each produce one gamma ray photon.

   The energy release is $0.000\ 55\ \text{u}\left(\dfrac{931\ \text{MeV}}{1\ \text{u}}\right) = 0.51\ \text{Mev}$.

   (L.O. 4.C.4.1, 5.B.11.1)

43. **B** To conserve charge, the unknown particle must have a $-1$ charge. This is response B, the electron. $^{14}_{6}\text{C} \rightarrow\ ^{14}_{7}\text{N} +\ ^{0-}_{0}\nu +\ ^{0}_{-1}\text{e}$. The total charge on the left-hand side of the reaction is $+6$, and on the right side, it is $+7 - 1 = +6$.

   (L.O. 5.C.1.1)

44. **B** Half-life is the time for one-half the number of initial radioactive nuclei to decay to another isotope. The time on the curve where the number of remaining nuclei is one-half the initial sample is 500. This occurs at $t = 2$ s.

   (L.O. 7.C.3.1)

45. **D** The photoelectron has a low speed since the difference between the threshold frequency needed to overcome the work function is $1.50 \times 10^{15}$ Hz and the new frequency of $2.00 \times 10^{15}$ Hz is $0.50 \times 10^{15}$ Hz. This will give the photoelectron a lower energy, $3.31 \times 10^{-19}$ J, and a lower speed.

   (L.O. 6.7.3.1)

46. **B** and **D** Once the charged rod is removed, the charges on the conducting sphere will be evenly distributed. The sphere does not retain the polarization it had when the rod was nearby.

    (L.O. 1.B.1.1, 1.B.1.2, 1.B.2.3)

47. **B** and **C** The electron was raised to $n = 3$ by the absorption of a photon, and when it drops back to $n = 2$, it releases energy as a photon in the visible range of the electromagnetic spectrum $E = hf$. As it drops to a lower energy state, it is closer to the proton in the nucleus. Since $r$ has decreased, the electrical force between the proton and the electron has increased.

    (L.O. 3.A.3.4, 3.C.2.1, 6.F.1.1)

48. **B** and **D** As a wave travels from a medium of lower index of refraction to one of higher index of refraction, its speed decreases. Since the frequency of the wave does not change, the wavelength must decrease. $\lambda_1 n_1 = \lambda_2 n_2$.

    (L.O. 6.E.3.2, 6.E.3.3)

49. **A** and **D** The force on the proton is given by

    $$\left|\vec{F}_B\right| = 0.10\ \text{T}\left(1.6 \times 10^{-19}\ \text{C}\right)\left(6.0 \times 10^5\ \frac{\text{m}}{\text{s}}\right).$$

    The force is $9.6 \times 10^{-15}$ N. By the right-hand rule, the given direction of the magnetic field is toward the page and the velocity is to the right of the page. Therefore, the force is initially to the top of the page. The magnetic force is a centripetal force acting on the charge, continually changing its direction. Since the direction of the velocity vector has changed, its momentum has changed.

    (L.O. 2.D.1.1, 3.C.3.1)

50. **B** and **C** Thermal energy is conducted through the rods at a rate given by $\dfrac{Q}{\Delta t} = \dfrac{kA}{L}$ (then conversion of temperatures from °C to K by adding 273 to each temperature).

    Thermal energy will transit from the 373 K wall to the 273 K wall because molecules colliding with the rods have higher momentum and energy at the wall whose temperature is 373 K. Increasing the effective cross-sectional area in Figure 1 means that there are more molecules/unit areas transferring both energy and momentum to those particles, which have lower energies. Since the rate at which thermal energy is conducted depends on the temperature difference, the arrangement in Figure 1 has the same $\Delta T$ between the ends of each rod. In Figure 2, $\Delta T$ is smaller for each rod, the effective cross-sectional area is less, and there is a smaller rate of transfer of energy with increasing length.

    (L.O. 4.C.3.1, 5.B.6.1)

ANSWERS TO FREE-RESPONSE PROBLEMS

QUESTION 1

(a) As shown in the correct drawing, the image is

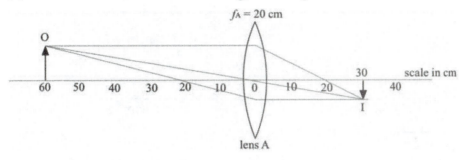

i.   real, inverted, and smaller than the object.        **(1 point)**
ii.  The height of the image can be determined from the magnification of the image. The absolute value of the magnification $|M|$ is determined from the ratio of the image distance $s_i$ to the object distance $s_o$. Since the drawing gives both $s_i$ and $s_o$, the magnification is 1/2 and the image size is 1/2 the object size and will be 1 cm tall.        **(1 point)**

(b) The image is determined from $\frac{1}{f} = \frac{1}{s_o} + \frac{1}{s_i}$. The object distance for the second lens B is 10 cm from the lens and is inverted.

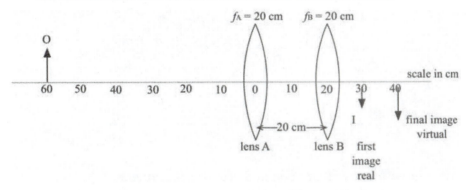

Solving for the position of the image of the first object,

$$\frac{1}{20.0 \text{ cm}} = \frac{1}{10.0 \text{ cm}} + \frac{1}{s_i}; \text{ then } \frac{1}{s_i} = \frac{1-2}{20.0 \text{ cm}}.$$

The image is located 20 cm to the right of lens B at the 40 cm mark on the scale. It is inverted, virtual, and the same size as the first object O.        **(4 points)**

(c) The first image from lens A in part (a) is 30 cm from the lens that places it 10 cm from the mirror. Since it is between the focal length and the mirror, the image formed by the mirror will be virtual. Because the initial image was inverted, the image formed by the mirror also will be inverted.

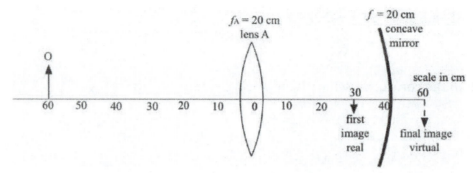

Using the equation $\frac{1}{f} = \frac{1}{s_o} + \frac{1}{s_i}$ the final image is $\frac{1}{20.0 \text{ cm}} = \frac{1}{10.0 \text{ cm}} + \frac{1}{s_i}$;

solving $\frac{1}{s_i} = \frac{1-2}{20.0 \text{ cm}}$ and $s_i = -20$ cm.

The initial image was half the size of the object. This image has a magnification of +2. The final image is virtual, inverted, and the same size as object O for the lens. **(4 points)**

(L.O. 6.E.4.2, 6.E.5.1)

## QUESTION 2

(a) One possible schematic diagram of the equipment is shown below.

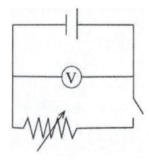

**(1 point)**

Read the voltmeter with the switch open and record this value in the data table as the emf of the power supply, $\mathcal{E}$.

Since the teacher set the power supply to read 1.50 V, adjust the resistance on the box to a value of 20.0 Ω. Close the switch and record the reading on the voltmeter and the variable resistance box. Open the switch. Repeat this procedure at least three times, making sure your readings are consistent.

Change the resistance on the variable resistance box and close the switch. Record the voltmeter reading and the resistance value for this set of readings. Repeat this procedure at least three times, making sure your readings are consistent.

Repeat this step until you have five or six different resistance/voltmeter readings.

Adjust the resistance box to zero and record the voltmeter reading. **(4 points)**

(b) The data taken and recorded will be the voltmeter reading in V and the resistance from the variable resistance box in Ω. The current in the circuit for each reading will be calculated using $I = \dfrac{V}{R}$. **(1 point)**

(c) When the switch is opened, the reading on the voltmeter is the emf, $\mathcal{E}$, of the power supply and the current is zero. This indicates that the resistance in the circuit is infinite. **(1 point)**

(d) The graph plotted will be the terminal potential difference VTPD in volts versus the current $I$ for each different resistance value from the variable resistance box. The graph should look similar to the one below.

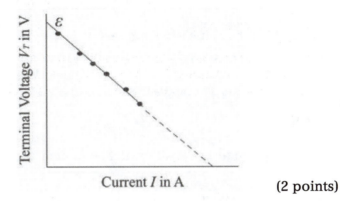

**(2 points)**

The slope of the line is the negative of the "internal resistance."

$$r = \frac{\mathcal{E} - V_{TPD}}{\Delta I}$$

**(1 point)**

(e) There are two meters in the second arrangement. Some current is needed to operate the voltmeter. The second arrangement has an ammeter that should be a pure conductor but will always suffer an IR drop across its ends.

There is no switch in the second arrangement, so the circuit may have heated slightly, changing the resistance of the coil of wire if the second group did not disconnect a lead wire between readings. However, the switch in the first arrangement has resistance that is difficult to measure. **(2 points)**

(L.O. 4.E.5.2, 5.B.9.6, 5.B.9.7, 5.B.9.8, 5.C.3.4)

## QUESTION 3

(a) Before the loop enters the field, there is no magnetic flux through the loop. As the leading edge enters the field, the flux increases until the trailing edge of the loop is in the field. The flux decreases as the loop leaves the field.

Before the loop entered the field, there was no motional emf in the loop because no field was present. As the right side of the loop enters the magnetic field, flux increases directed out of the page. According to Lenz's law, the induced current must be clockwise through the loop because it must produce a magnetic field that is directed into the page to counter the increase in flux out of the page. Since the current is clockwise, the induced emf is negative.

Once the loop is completely in the field, the change in the magnetic flux is zero. This occurs because when the left edge of the loop enters the field, the induced emf in it cancels the motional emf present in the right side of the loop.

As the right side of the loop leaves the field, the flux outward begins to decrease. This sets up a counterclockwise current in the loop as it tries to maintain the magnetic field. **(5 points)**

(b) Setting up the graph requires the correct value for the emf on the vertical axis.

Solving for the magnitude of the emf and marking the y-axis **(2 points)**

For a horizontal line from $t = 0$ until $t = 2$ s at some negative value **(1 point)**

For a horizontal line from $t = 2$s until $t = 5$ s at zero V **(1 point)**

For a horizontal line from $t = 5$ until $t = 7$ s at some positive value **(1 point)**

The time for the loop to enter and leave the field completely is

$$t = \frac{0.20 \text{ m}}{0.10 \text{ m}/\text{s}} = 2.0 \text{ s}$$

The time for the leading edge to transit the field completely is

$$t = \frac{0.70 \text{ m}}{0.10 \text{ m}/\text{s}} = 7.0 \text{ s}$$

The correct graph is shown below.

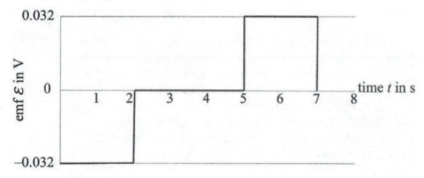

(c) $I = \dfrac{\varepsilon}{R} = \dfrac{3.20 \times 10^{-2} \text{ V}}{40.0 \, \Omega} = 8.00 \times 10^{-4} \text{ A}$ **(1 point)**

(d) The magnetic flux when the loop is in the field completely is

$\phi_M = \vec{B}A$.

$\phi_M = (1.60 \text{ T})(4.00 \times 10^2 \text{ m}^2) = 6.40 \times 10^{-2} \text{ Wb}$ **(1 point)**

(L.O. 3.C.3.1, 4.E.2.1)

## QUESTION 4

(a) The electrostatic force is found using Coulomb's law. Knowing that the charge of a proton is $e = 1.6 \times 10^{-19}$ C,

$$F = k\frac{q_{He}q_{Au}}{r^2}$$

$$= \left(9 \times 10^9 \ \text{N} \cdot \text{m}^2\middle/\text{C}^2\right)\frac{(2e)(79e)}{\left(80.0 \times 10^{-15} \ \text{m}\right)^2}$$

$$= 5.70 \ \text{N}$$    **(3 points)**

(b) Electrons from outer electron shells fall into the vacancies in the inner shells. In large atoms such a gold, these "quantum jumps" are followed by the emission of photons that carry considerable energy. The photons produced when electrons jump from the fifth or sixth electron shells into the lowest electron shells are in the X-ray range.    **(3 points)**

(c) By the law of conservation of linear momentum, the recoiling U-232 nucleus must be $1.10 \times 10^{-19} \ \text{kg} \cdot \text{m}\middle/\text{s}$ at 180°. The total momentum before and after decay must be zero.    **(2 points)**

(d) At the end of each half-life, one-half of the remaining sample will remain. Half of a half of a half of a half means that one-sixteenth of the original sample will remain.    **(2 points)**

(L.O. 3.C.2.1, 4.C.4.1, 5.B.8.1, 5.B.11.1, 5.D.1.1, 5.G.1.1, 7.C.3.1)

# AP® PHYSICS 2
# PRACTICE EXAM D

## AP® PHYSICS 2
## Section I
## 50 Multiple-Choice Questions
## Time—90 Minutes

**Note:** To simplify calculations, you may use $g = 10$ m/s² in all problems.

**Directions:** Each of the questions or incomplete statements below is followed by four suggested answers or completions. Select the one that is best in each case.

1. A convex lens produces a real image of an object that has a magnification of −4.0 when the object is placed 50.0 cm from the lens. What is the focal length of the lens?

   (A) 40 cm
   (B) 50 cm
   (C) 80 cm
   (D) 200 cm

## Questions 2 and 3
Use the diagram of the pipe below to answer questions 2 and 3. The radius of the pipe at cross section A is $r_A = 0.040$ m and at cross section B is $r_B = 0.025$ m. Water enters cross section A with a speed of $5.00$ m/s.

2. What is the volume flow rate of the water as it enters cross section A of the pipe?

   (A) $1.06 \times 10^{-3}$ m³/s

   (B) $2.33 \times 10^{-2}$ m³/s

   (C) $2.51 \times 10^{-2}$ m³/s

   (D) $6.28 \times 10^{-1}$ m³/s

3. What is the speed of the water as it enters cross section B of the pipe?

   (A) $1.60$ m/s

   (B) $3.13$ m/s

   (C) $8.00$ m/s

   (D) $12.8$ m/s

**640**

4. Three resistors $R_1 = 20.0\ \Omega$, $R_2 = 40.0\ \Omega$, and $R_3 = 30.0\ \Omega$ are connected to a 4.50 V power supply as shown in the diagram below.

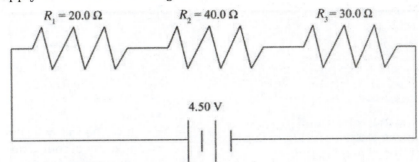

Which of the following graphs best represents the electrical potential as a function of position for a positive charge traveling around the circuit in the clockwise direction and starting at the right side of the battery?

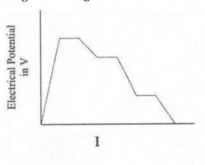

I

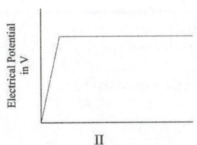

II

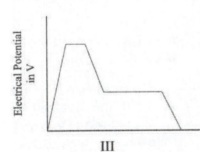

III

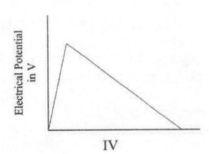

IV

(A) I
(B) II
(C) III
(D) IV

5. Thorium decays into radium through alpha decay as shown in the following equation:

$$^{232}_{90}\text{Th} \rightarrow {}^{228}_{88}\text{Ra} + {}^{4}_{2}\text{He}$$

The mass of an atom of each element is given below.

Thorium    $3.851832 \times 10^{-25}$ kg
Radium      $3.7853160 \times 10^{-25}$ kg
Helium       $6.6443421 \times 10^{-27}$ kg

How much energy is released in this decay?
(A) $2.184 \times 10^{-21}$ J
(B) $6.551 \times 10^{-13}$ J
(C) $6.933 \times 10^{-8}$ J
(D) $3.464 \times 10^{-8}$ J

**GO ON TO NEXT PAGE**

6. A sphere of mass 0.10 mg carries an electric charge of +1.00 $\mu$C. It is placed in the E-field existing between two parallel plates that are 0.05 m apart.

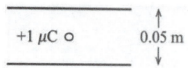

What are the magnitude and direction of the electric field between the plates if the charge remains stationary?

(A) 980 $\frac{N}{C}$ directed upward

(B) 980 $\frac{N}{C}$ directed downward

(C) 100 $\frac{N}{C}$ directed upward

(D) 100 $\frac{N}{C}$ directed downward

7. Water flows smoothly through the horizontal pipe illustrated below.

Which of the following statements is true?
(A) The pressure at opening A is greater than at opening B because the velocity of the water is greater at opening B.
(B) The pressure at opening A is greater than at opening B because the velocity of the water is greater at opening A.
(C) The pressure at openings A and B is the same because the pipe is horizontal.
(D) The pressure is less at opening A than at opening B because the cross-sectional area at A is larger than at B.

8. In the simple circuit shown below, an ammeter measures the current through a resistor while a voltmeter measures the voltage across the resistor.

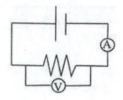

Which of the following graphs of voltage and current would indicate that the resistor follows Ohm's law?

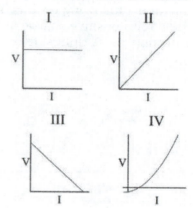

(A) I
(B) II
(C) III
(D) IV

9. An electron moves with velocity of $6.00 \times 10^7$ m/s in a magnetic field of strength 0.200 T. Which of the following choices gives the correct magnitude and the direction of the force acting on the electron?

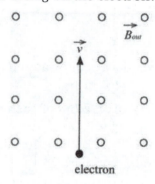

(A) $1.92 \times 10^{-12}$ N directed to the left
(B) $1.92 \times 10^{-12}$ N directed to the right
(C) $9.60 \times 10^{-12}$ N directed to the left
(D) $9.60 \times 10^{-12}$ N directed to the right

10. The probability of finding an electron in a hydrogen atom of Bohr radius $a_o$ is greatest in a region
    (A) a distance of $a_0$ from the nucleus when the electron has a de Broglie wavelength of $\pi a_0$.
    (B) a distance of $a_0$ from the nucleus when the electron has a de Broglie wavelength of $3\pi a_0$.
    (C) a distance of $2a_0$ from the nucleus when the electron has a de Broglie wavelength of $5\pi a_0$.
    (D) a distance of $2a_0$ from the nucleus when the electron has a de Broglie wavelength of $3\pi a_0$.

11. A charge of $+2q$ travels in the $+x$-direction with a velocity $\vec{v}$ above an infinitely long wire carrying a current $I$ also in the $+x$-direction, as shown in the diagram below.

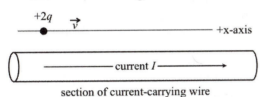

section of current-carrying wire

Which of the force vectors below correctly shows the force acting on the charge $+2q$ due to the current in the wire?

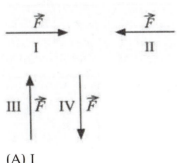

(A) I
(B) II
(C) III
(D) IV

12. Water of density 1000 $\frac{kg}{m^3}$ flows smoothly through the horizontal pipe below.

The water enters cross section A with a speed of 4.00 m/s and a pressure $p_A = 2.05 \times 10^5$ Pa. If the radius $R_A = 2R_B$, what is the pressure $p_B$ as the water enters cross section B?
    (A) $1.30 \times 10^4$ Pa
    (B) $6.90 \times 10^4$ Pa
    (C) $7.70 \times 10^4$ Pa
    (D) $8.50 \times 10^4$ Pa

13. A 19.6 N sphere is suspended from a spring scale and immersed in water in a beaker as shown below.

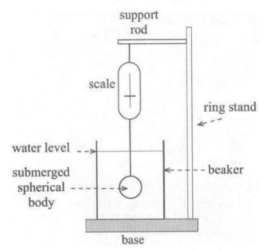

The spring scale reads 14.6 N when the body is placed in the water. The volume of the submerged body is closest to
    (A) $5.0 \times 10^{-4}$ m³.
    (B) $5.0 \times 10^{-3}$ m³.
    (C) $1.5 \times 10^{-3}$ m³.
    (D) $2.0 \times 10^{-3}$ m³.

**GO ON TO NEXT PAGE**

14. Which of the following observations support the theory of special relativity?
    (A) Light from a receding star appears to be red-shifted, while light from an approaching star appears to be blue-shifted.
    (B) During a solar eclipse, the path of light from distant stars is observed to bend when it passes near the sun.
    (C) A $\mu$ meson created in Earth's upper atmosphere can travel for 4 $\mu$s to Earth's surface even though a $\mu$ meson has a half-life of only 2 $\mu$s.
    (D) Electrons striking the surface of a crystal reflect with a wavelike interference pattern and not as the particle model would predict.

15. A charge of $+2Q$ is brought to a point outside but not touching an isolated metal sphere.

    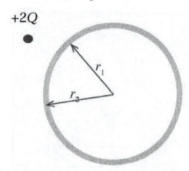

    +2Q

    Which of the following statements best describes what will occur?
    (A) A positive charge of $+2Q$ is induced on the outer surface of the sphere.
    (B) A positive charge of $+Q$ is induced on the outer surface, and a charge of $-Q$ is induced on the inner surface.
    (C) A negative charge of $-2Q$ is induced on the outer surface, and a charge of $+Q$ is induced on the inner surface.
    (D) A negative charge of $-2Q$ is induced on the outer surface, and a charge of $+2Q$ is induced on the inner surface.

16. A ray of polychromatic light composed of violet light and red light is incident on a crown glass plate 1.00 mm thick.

    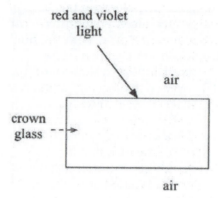

    red and violet light

    air

    crown glass

    air

    When the ray emerges from the other side of the plate, both wavelengths of light leave the plate at
    (A) the same point with the same speed.
    (B) the same point, but the violet light travels faster since it has a smaller wavelength.
    (C) different points with the same speed.
    (D) different points, but the violet light travels faster since it has a smaller wavelength.

17. Students are given an optical bench, a centimeter ruler, a 2.00 cm tall object, and a curved mirror. They perform an experiment to determine the radius of curvature of the mirror. Part of their data is shown below.

| $s_i$ (cm) | $s_o$ (cm) | $|M|$ |
|---|---|---|
| 40.0 | 13.3 | 0.33 |
| 30.0 | 15.0 | 0.50 |
| 20.0 | 20.0 | 1.00 |
| 5.0 | −10.0 | 2.00 |

Their results lead them to conclude that the mirror was
(A) concave with a radius of curvature of +20.0 cm.
(B) concave with a radius of curvature of +40.0 cm.
(C) convex with a radius of curvature of −20.0 cm.
(D) convex with a radius of curvature of −40.0 cm.

18. A beam of monochromatic light from a point source shining on a small coin will cast a shadow with alternating light and dark bands bordering a central dark spot. Which of the following statements best describes what happens?
    (A) Light reflected from the surface of the coin produces the bright bands when they meet at a single point.
    (B) Interference patterns are produced by the light reflected from the coin, which is 180° out of phase with the incoming light.
    (C) A small amount of light bends around the edge of the geometric shadow of the coin diffracting into the shadow, producing alternating light and dark bands.
    (D) The coin linearly polarizes the light in regions that are dark bordering the shadow.

19. A diffraction grating is used to determine the wavelength of light. In using the same grating illuminated by red light compared to blue light, the angle for the first-order maximum is greater for
    (A) red light than for blue light because red light has a smaller wavelength.
    (B) red light than for blue light because red light has a longer wavelength.
    (C) blue light than for red light since blue light has a smaller frequency.
    (D) blue light than for red light since blue light has a higher frequency.

20. An object is placed in front of three mirrors, and an upright image is produced in all three. The first mirror produces an image that is smaller than the object. The second mirror produces an image that is the same size as the object. The third mirror produces an image that is larger than the object. Respectively, the first, second, and third mirrors are
    (A) convex, plane, and concave.
    (B) convex, concave, and plane.
    (C) plane, concave, and convex.
    (D) plane, convex, and concave.

21. When the fire alarm goes off during a fire drill, it emits light and sound. Which of these statements is true of both light waves and sound waves?
    (A) Sound and light need a medium in order to travel.
    (B) Light waves can be diffracted, but sound waves cannot.
    (C) Sound and light can be polarized.
    (D) Sound and light can be Doppler shifted.

22. One mole of an ideal gas is confined in a fixed container. The pressure-temperature graph shown below indicates the response in temperature due to changes in pressure.

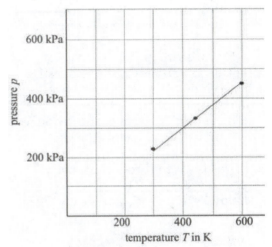

Extrapolation of the line on the graph will indicate
    (A) the amount of work done by the gas as the pressure is reduced.
    (B) that as the pressure approaches zero, the temperature of the gas becomes constant.
    (C) that as the temperature of the gas approaches absolute zero, the pressure approaches zero.
    (D) that the pressure of the gas approaches zero because some of the molecules have solidified.

**GO ON TO NEXT PAGE**

23. Four identical laser beams are passed through several diffraction gratings to produce the diffraction patterns below. In each case, the screen is the same distance from the grating.

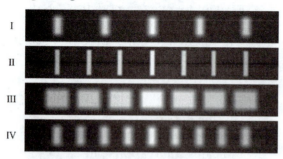

The grating with the closest slit spacing is
(A) grating I because the bright bands are spaced farthest apart.
(B) grating II because the lines are sharpest.
(C) grating III because the dark bands between maxima are smallest.
(D) grating IV because the distance between maxima is the least.

24. A gas is taken through the cyclic process shown in the $pV$ graph below. What work is done on the gas?

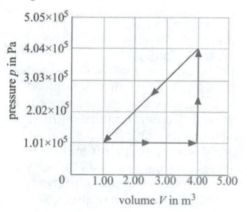

(A) 450 kPa
(B) 600 kPa
(C) 800 kPa
(D) 900 kPa

25. Glass prisms are used in many optical instruments in place of mirrors. Prisms can be used to change the direction of light beams by 90° because total internal reflection better preserves the brightness and sharpness of light beams. In the diagram below, the line N–N is the normal.

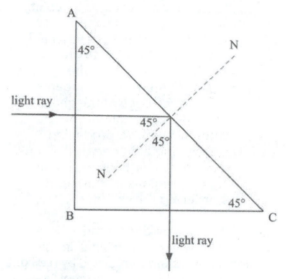

Which of the following best represents the minimum index of refraction for such a prism?
(A) 1.00
(B) 1.33
(C) 1.41
(D) 1.50

26. Three charges are located on the vertices of an equilateral triangle as indicated in the diagram below. The charges are $q_1 = 9.00$ nC, $q_2 = 9.00$ nC, and $q_3 = -9.00$ nC.

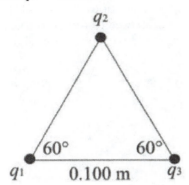

The net force acting on charge $q_2$ due to $q_1$ and $q_3$ is
(A) $7.28 \times 10^{-5}$ N directed toward the right.
(B) $14.6 \times 10^{-5}$ N directed toward the right.
(C) $18.1 \times 10^{-5}$ N directed toward the right.
(D) $24.3 \times 10^{-5}$ N directed toward the right.

27. A plastic rod is rubbed on a sweater, and the rod becomes positively charged. The rod is brought close to a piece of paper. What is true about the electric force between the rod and the paper?
(A) The paper is attracted to the rod because positive charges in the paper are repelled by the rod and leave to ground, leaving one side of the paper more negative than the other side.
(B) The paper is attracted to the rod because electrons in the paper move slightly toward the rod, causing one side of the paper to become more negative than the other side.
(C) The paper is not attracted to the rod because there is no net charge on the paper.
(D) The paper is not attracted to the rod because charges cannot move in an insulator.

28. The system below consists of a negative charge, a positive charge, and a small charge $q$. Both the negative charge and the positive charge are located on the $y$-axis, and the test charge is located along the $x$-axis.

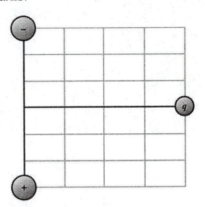

At the location of the test charge $q$, the electric field is
(A) directed left.
(B) directed up.
(C) directed left or right depending on the charge on $q$.
(D) directed up or down depending on the charge on $q$.

29. Two metal spheres connected to the ceiling of a room by silk threads are in contact as shown below. A positively charged wand is brought near but does not touch either sphere.

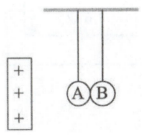

What are the charges on A and B?
(A) The left sides of A and B are negative, and the right sides of A and B are positive.
(B) The left sides of A and B are positive, and the right sides of A and B are negative.
(C) A is negative, and B is positive.
(D) A is positive, and B is negative.

**GO ON TO NEXT PAGE**

30. The electric field lines between two charged spheres are shown below. The sphere on the left has a charge of −2 μC.

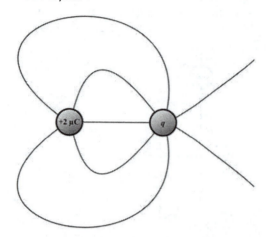

What is the charge, *q*, on the sphere on the right?
(A) −2.8μC
(B) −2.0μC
(C) −1.4μC
(D) +2.0μC

31. What is the current in the ammeter in the circuit shown?

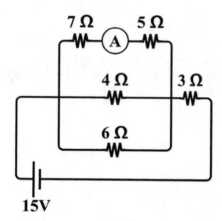

(A) 0.5 A
(B) 1.0 A
(C) 1.5 A
(D) 3.0 A

32. Two parallel plates are 4 cm apart and there is a 1000 V potential difference between them. A proton is placed next to the positive plate and released from rest. How much kinetic energy does the proton gain as it accelerates to the negative plate?

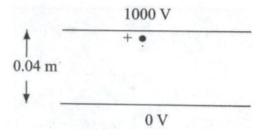

(A) 1.6 × 10⁻¹⁶ J
(B) 2.4 × 10⁻¹⁶ J
(C) 3.2 × 10⁻¹⁶ J
(D) 6.4 × 10⁻¹⁶ J

33. The ammeter in the circuit below reads 4 A. If another resistor of resistance *R* were added in parallel with the other two, the reading on the ammeter would

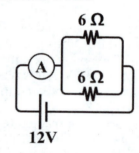

(A) decrease.
(B) stay the same.
(C) increase.
(D) either increase or decrease depending on the value of *R*.

34. Four identical point charges of $+q$ are arranged at the corners of a square as shown in the diagram below.

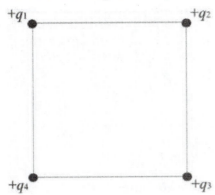

Which of the following vectors shows the direction of the net electrostatic force on $+q_3$ due to the other three charges?

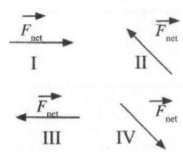

(A) I
(B) II
(C) III
(D) IV

35. A proton initially moves to the right with a velocity of $\vec{v}$ in a magnetic field that is directed into the page.

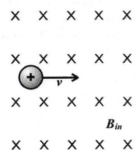

The direction of the magnetic force acting on the proton is
(A) toward the top of the page.
(B) toward the bottom of the page.
(C) into the page.
(D) out of the page.

36. A bar magnet falling toward a loop of wire induces a clockwise current in the loop as shown below.

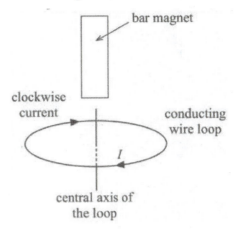

As the bar magnet approaches the loop, which pole is facing the loop?
(A) The north pole of the magnet is moving toward the loop.
(B) The north pole of the magnet is moving away from the loop.
(C) The south pole of the magnet is moving away from the loop.
(D) The south pole of the magnet is moving toward the loop.

37. Two wires carry currents $I_1 = 4.0$ A and $I_2 = 8.0$ A in the same direction as shown.

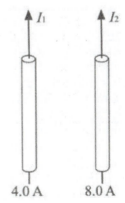

4.0 A     8.0 A

The force acting between the two wires is
(A) $\left|\vec{F}_{12}\right| = \left|\vec{F}_{12}\right|$ and is repulsive.
(B) $2\left|\vec{F}_{12}\right| = \left|\vec{F}_{12}\right|$ and is repulsive.
(C) $\left|\vec{F}_{12}\right| = \left|\vec{F}_{12}\right|$ and is attractive.
(D) $2\left|\vec{F}_{12}\right| = \left|\vec{F}_{12}\right|$ and is attractive.

**GO ON TO NEXT PAGE**

38. Rank from highest intensity to lowest intensity the strength of the electric field $|\vec{E}|$ surrounding the negative charge in the diagram below.

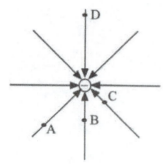

(A) C > B > D > A
(B) C > B > A > D
(C) A > (B = C) > D
(D) D > A > B > C

39. A metal surface is illuminated by light of a given frequency. If the light intensity falling on the metal is increased, the
(A) work function for the metal will decrease.
(B) work function for the metal will increase.
(C) photoelectron current will decrease.
(D) photoelectron current will increase.

40. The figure, a copper bar of mass $m$ and length $L$, is free to slide up or down an incline on frictionless metal strips that are a distance $L$ apart. The metal strips are connected to a voltage source and carry a current. A magnetic field $B$ penetrates this conducting loop. The inclined plane is tilted at an angle $\theta$ above the horizontal. What current, $I$, in the copper bar will hold the bar at rest?

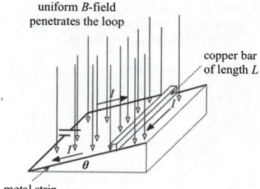

(A) $I = \dfrac{mg \sin \theta}{BL \sin \theta}$

(B) $I = \dfrac{mg \cos \theta}{BL \sin \theta}$

(C) $I = \dfrac{mg \cos \theta}{BL \cos \theta}$

(D) $I = \dfrac{mg \sin \theta}{BL \cos \theta}$

41. A wavelike characteristic of an electron is demonstrated in the
(A) ejection of a photoelectron from a metal surface when illuminated with light of a wavelength less than the threshold wavelength.
(B) diffraction of the electron from a crystal.
(C) Compton scattering of an electron from a crystal and the incident X-ray striking the crystal surface.
(D) transition of the electron from an upper orbital state to a lower orbital state in the Bohr atom as the energy is emitted as a photon of light.

42. What particle(s) will complete this nuclear reaction?

$$^{235}_{92}\text{U} + ^{1}_{0}\text{n} \rightarrow ^{144}_{56}\text{Ba} + ^{89}_{36}\text{Kr} + ?$$

(A) $^{4}_{2}\text{He}$
(B) $^{3}_{0}\text{n}$
(C) $3\left(^{1}_{0}\text{n}\right)$
(D) $^{3}_{1}\text{H}$

43. This reaction occurs when a neutron strikes the nucleus of uranium-235.

$$^1_2\text{n} + ^{235}_{92}\text{U} \rightarrow ^{141}_{56}\text{Ba} + ^{92}_{36}\text{Kr} + 3^1_0\text{n}$$

What type of nuclear reaction is represented above?
(A) fission
(B) fusion
(C) alpha decay
(D) beta decay

44. In the decay of $^{222}_{86}\text{Rn} \rightarrow ^{218}_{84}\text{Po} + ?$, what is the missing particle?
(A) $^0_{-1}\beta$
(B) $^0_{+1}\text{e}$
(C) $2(^2_1\text{H})$
(D) $^4_2\text{He}$

45. A $1.2 \times 10^{-6}$ kg sample of Co-60 is prepared in a nuclear physics laboratory for use in a medical center. Co-60 has a half-life of 5.24 years. What quantity will remain after 15.72 years?
(A) $6.0 \times 10^{-7}$ kg
(B) $3.0 \times 10^{-7}$ kg
(C) $1.5 \times 10^{-7}$ kg
(D) $7.5 \times 10^{-8}$ kg

**Questions 46 to 50**
**Directions:** For each of the questions or incomplete statements below, two of the suggested answers will be correct. For each of these questions, you must select both correct choices to earn credit. No partial credit will be earned if only one correct choice is selected. Select the two that are best in each case.

46. The kinetic energy of a photoelectron (select two answers)
(A) increases with increasing wavelength of light beyond the threshold wavelength.
(B) depends on the intensity of light falling on the metal surface.
(C) increases with increasing frequency of light beyond the threshold frequency.
(D) depends on the nature of the surface being illuminated.

47. A mole of an ideal gas is taken through a process ADCDA as shown.

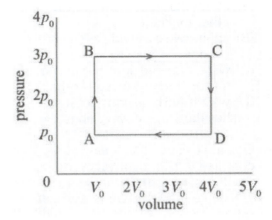

In the process, the net (select two answers)
(A) change in the internal energy of the system is $6P_0V_0$.
(B) change in the entropy of the system is 0.
(C) thermal energy that crossed the boundaries is $6P_0V_0$.
(D) work done during the process was 0.

48. A container holds 0.10 mole of $H_2$ and 0.10 mole of $O_2$ in thermal equilibrium. Which of the following two statements are true?
(A) The $O_2$ molecules and the $H_2$ molecules have the same average kinetic energy since they are in thermal equilibrium.
(B) The $O_2$ molecules and the $H_2$ molecules have different average kinetic energies since the $O_2$ molecule is more massive.
(C) The $O_2$ molecules and the $H_2$ molecules have the same rms speed since they are in thermal equilibrium.
(D) The $O_2$ molecules and the $H_2$ molecules have different rms speeds since the $O_2$ molecule is more massive.

**GO ON TO NEXT PAGE**

49. In a single slit diffraction pattern, the central maximum is (select two answers)
    (A) about the same width as the other maxima.
    (B) about twice as wide as the other maxima.
    (C) widest for red light because it has the longest wavelength.
    (D) widest for blue light because it has the longest wavelength.

50. One mole each of an ideal gas is confined in two identical cylinders. The temperature of cylinder 1 is $T_1 = 500$ K, and the temperature of cylinder 2 is $T_2 = 300$ K. The same amount of thermal energy $Q_{in}$ is added to both cylinders as they undergo an isothermal expansion.

Which of the following two statements are true?
(A) The work done by the gas in cylinder 1 is greater than the work done by the gas in cylinder 2 since the temperature of the gas in cylinder 1 is higher.
(B) The work done by the gas in cylinder 1 is the same as the work done in cylinder 2 since the same amount of thermal energy was added to the cylinders during the isothermal expansion.
(C) The change in the entropy is the same in both cylinders since the same amount of thermal energy was added.
(D) The change in the entropy is smaller in cylinder 1 because it had a higher initial temperature.

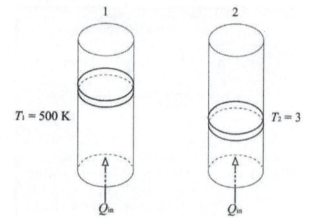

STOP
END OF SECTION I
*IF YOU FINISH BEFORE TIME IS CALLED, YOU MAY CHECK YOUR WORK ON THIS SECTION. DO NOT GO ON TO SECTION II UNTIL YOU ARE TOLD TO DO SO.*

## AP® PHYSICS 2
### Section II
### 4 Free-Response Problems
### Time—90 Minutes

**Directions:** Solve each of the following problems. Unless the directions indicate otherwise, respond to all parts of each question.

1. **(12 points)** An ideal gas has initial state conditions of $P_A$ = 2.00 MPa and $V_A$ = 0.001 m³ and an internal energy of $U_A$ = 20.0 kJ. The gas is taken through the following steps:

   A → B the pressure is increased at constant volume to $P_B$ = 5.00 MPa.

   B → C the volume is increased isobarically until it reaches $V_C$ = 0.005 m³.

   C → D the gas is compressed at constant volume to its initial pressure.

   D → A the gas is isobarically returned to its initial volume.

   (a) Graph the above steps on the pressure-volume graph shown below.

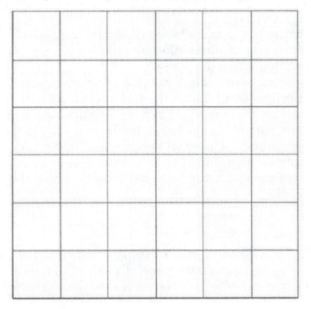

   (b) Explain how you will determine the amount of work done on each step and in the cycle without a mathematical solution.

   (c) i.  10 kJ of thermal energy enters the system along path A → B. What is the internal energy at point B?

   ii.  If the internal energy of the gas at point C is 70 KJ, how much thermal energy entered or left the system along the path B → C? Justify your answer. (Your answer may be mathematical.)

   (d) What is the net thermal energy entering or leaving the system as it returns to point A? Justify your answer.

**GO ON TO NEXT PAGE**

2.  **(12 points)** On your lab table, you have the apparatus shown below. In the setup, a length of wire passes through one large horseshoe magnet. A variable voltage source and an ammeter are connected to the wires as shown. The scale below the setup reads a force in Newtons and reads zero when there is no current in the wires. The support stands and clamps are not shown in this drawing.

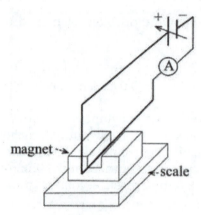

You and your lab partners are tasked with designing an experiment to measure the strength of the magnetic field in the middle of the horseshoe magnet. You have access to other standard equipment in the room.

(a) Explain your procedure in enough detail that another laboratory group could repeat your experiment and obtain the same results.

(b) What measurements are you going to take, and how are you going to use them to determine the strength of the magnetic field?

(c) When you graph your data, how will your graph be arranged so that the magnetic field can be determined by the slope of your graph?

(d) i.   A wire carries current to the right of the page. Show the magnetic field around the wire.

$I$

ii.  The drawing below shows a wire carrying a current out of the page. Four compasses have been placed the same distance from the core of the wire. Indicate the correct direction of the field that will be produced surrounding the wire.

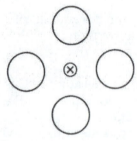

The wire produces a magnetic field surrounding it. What electric field is produced?

___   $\vec{E} = \vec{B}$

___   $\vec{E} > \vec{B}$

___   $\vec{E} < \vec{B}$

___   $\vec{E} = 0$

Justify your answer.

3. **(10 points)**

(a) In the circuit below, initially both $S_1$ and $S_2$ are closed. Which resistor will dissipate the least power? Justify your answer without calculations.

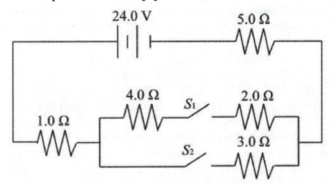

(b) Sometime later, switch $S_1$ is removed and a 5.0 $\mu$F capacitor is placed in the circuit between the 4.0 $\Omega$ resistor and the 2.0 $\Omega$ resistor as shown below. Then switch $S_2$ is closed.

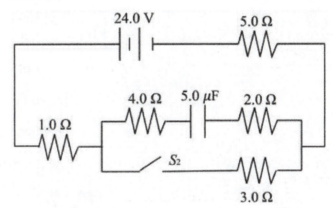

   i. Explain how you can determine the charge stored on the plates of the capacitor when the circuit has reached steady state.

  ii. What is the charge on the capacitor?

(c) In some cases, the resistance of some resistors does not obey Ohm's law. In a cohesive paragraph, explain specifically what it would mean if a resistor's resistance followed Ohm's law. Then for a resistor that does not obey Ohm's law, explain what factor changes the resistance and what causes that factor to change.

**GO ON TO NEXT PAGE**

4. **(10 points)**
   (a) A narrow beam of electrons is traveling in a vacuum. The stream is directed parallel to the page upward toward the top of the page. How could this beam of electrons be deflected 90° to the right without touching the electrons or changing their speed? Justify your answer. If necessary, use diagrams to help with your justification.
   (b) Each electron in the beam has a kinetic energy of 100.0 eV.
       i.   What is the de Broglie wavelength of one of the electrons?
       ii.  The beam strikes a crystal lattice with spacing between the crystal planes of 0.263 nm. Do the electrons produce a diffraction pattern? Justify your answer.
   (c) A separate beam of electrons is directed at the surface of a sample of metal. The apparatus allows the researchers to adjust the kinetic energy of the beam between 50 V and 500 V. Throughout this range of energies, a diffraction pattern can be seen on a phosphorescent screen.

       As the researchers increase the kinetic energy of the electrons in the beam, what happens to the width of the diffraction pattern shown on the screen? Explain your answer.

## *END OF EXAMINATION*

# Answers

## ANSWER KEY FOR MULTIPLE-CHOICE QUESTIONS

| | | | | |
|---|---|---|---|---|
| 1. A | 11. D | 21. D | 31. A | 41. B |
| 2. C | 12. D | 22. C | 32. A | 42. C |
| 3. D | 13. A | 23. A | 33. C | 43. A |
| 4. A | 14. C | 24. A | 34. D | 44. D |
| 5. B | 15. D | 25. C | 35. A | 45. C |
| 6. A | 16. C | 26. A | 36. D | 46. C & D |
| 7. A | 17. B | 27. B | 37. C | 47. B & C |
| 8. B | 18. C | 28. B | 38. B | 48. A & D |
| 9. A | 19. B | 29. C | 39. D | 49. B & C |
| 10. A | 20. A | 30. A | 40. D | 50. B & D |

## EXPLANATIONS FOR THE MULTIPLE-CHOICE ANSWERS

1. **A** The magnification of an image is given by $|M| = \dfrac{s_i}{s_o}$. $M = -4$ and $s_o$ is 50 cm, so $s_i$ is 200 cm.

   $\dfrac{1}{f} = \dfrac{1}{s_i} + \dfrac{1}{s_o} = \dfrac{1}{200 \text{ cm}} + \dfrac{1}{50 \text{ cm}}$, so the focal length is 40 cm.

   Alternatively, we know from experience that converging lenses produce magnified real images when the object distance, $s_i$, is slightly greater than $f$.

   (L.O. 6.E.5.1)

2. **C** The volume flow is $Av$. The volume flow rate at end A is

   $\pi(0.040 \text{ m})^2 \left(5.00 \ ^m\!/_s\right) = 2.51 \times 10^{-2} \ ^{m^3}\!/_s$.

   (L.O. 5.F.1.1)

3. **D** We use the equation of continuity to find the speed.

   $A_1 v_1 = A_2 v_2$

   $\pi(0.040 \text{ m})^2(5.00 \ ^m\!/_s) = \pi(.0250 \text{ m})^2 v_2$

   The velocity at end B is $12.8 \ ^m\!/_s$.

   (L.O. 5.B.10.3)

4. **A** Because the resistors are in series, we know that the same current runs through all three. Since $V = IR$ and current is the same, $V$ is proportional to $R$. The order of greatest to least potential difference is the same as the order of greatest to least resistance.

   (L.O. 5.B.9.8)

5. **B** Find the mass difference between the reactants and the products.

$\Delta m = (3.85186 \times 10^{-25}$ kg$) - (3.7853160 \times 10^{-25}$ kg $+ 6.644321 \times 10^{-27}$ kg$)$

$\Delta m = 1.008 \times 10^{-29}$ kg

Using $E = mc^2$, $E = (1.008 \times 10^{-29}$ kg$)(3.00 \times 10^8$ m/s$)^2 = 6.551 \times 10^{-13}$ J

(L.O. 4.C.4.1, 5.C.1.1, 5.G.1.1)

6. **A** The net force acting on the charge must be zero.

$\Sigma \vec{F} = 0$, $m\vec{g} = \vec{E}q$.

Weight pulls the charge downward, so the electric field must be directed upward. The lower plate is positive.

$$\left|\vec{E}\right| = \frac{(0.10 \times 10^{-3} \text{ kg})\left(9.80 \ \frac{\text{m}}{\text{s}^2}\right)}{1.00 \times 10^{-6} \text{ C}} = 980 \ \frac{\text{N}}{\text{C}}$$

(L.O. 2.B.1.1, 2.C.1.1, 2.C.1.2)

7. **A** The pressure is lower in constriction B because the velocity is greater. This is a direct application of Bernoulli's principle. At constant height, there is a tradeoff between kinetic energy and pressure.

(L.O. 5.B.10.4, 5.F.1.1)

8. **B** If a resistor obeys Ohm's law, the graph of potential difference across the ends of the resistor versus the current is linear, where current and potential difference are directly proportional. Graph II indicate the proper linear relationship for an ohmic material.

(L.O. 5.B.9.7)

9. **A** Since the electron moves perpendicular to the field, by the right-hand rule, the force on the electron is maximum and is directed to the left of the page (opposite the direction of the force on a proton). The magnitude of the force is found with $\left|\vec{F_B}\right| = q\vec{v}\vec{B}\sin\theta$.

$\left|\vec{F_B}\right| = (1.60 \times 10^{-19}$ C$)(6.00 \times 10^7 \ \frac{\text{m}}{\text{s}})(0.200$ T$) = 1.92 \times 10^{-12}$ N

(L.O. 2.D.1.1, 3.A.2.1)

10. **A** The highest probability of finding an electron in the hydrogen atom is in a region where an integer number of de Broglie wavelengths fit into the circumference of the electron orbit. In option (A), the circumference is $2\pi a_0$, which is exactly one de Broglie wavelength.

(L.O. 7.C.2.1)

11. **D** The wire creates a magnetic field that is into the page below the wire and out of the page above the wire in the region where the +2$q$ charge is moving to the right.

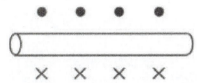

The force on the charge is given by the right-hand rule. The magnetic field is out of the page, and the velocity is toward the right of the page; thus, the force is toward the bottom of the page.

(L.O. 2.D.2.1, 3.C.3.1)

12. **D** Use the equation of continuity to find the speed of the water in the constriction.

$A_A v_A = A_B v_A$

Since $R_B$ = half of $R_A$, $A_B$ is a quarter of $A_A$. That means $v_B$ is four times $v_A$, or 16 m/s.

Bernoulli's equation $P_1 + \dfrac{1}{2}\rho v_1^2 + \rho g h_1 = P_2 + \dfrac{1}{2}\rho v_2^2 + \rho g h_2$ gives the pressure in the constriction. Since the pipe is horizontal, both potential energy terms are equal and subtract out of both sides.

Substitution into the equation is

$\left(2.05 \times 10^5 \text{ Pa}\right) + \dfrac{1}{2}\left(1000 \text{ }^{kg}\!/_{m^3}\right)\left(4.00 \text{ }^{m}\!/_{s}\right)^2 =$

$P_2 + \dfrac{1}{2}\left(1000 \text{ }^{kg}\!/_{m^3}\right)\left(16.00 \text{ }^{m}\!/_{s}\right)^2$, and then

$(2.05 \times 10^5 \text{ Pa}) + 8.00 \times 10^3 \text{ Pa} = P_2 + 1.28 \times 10^5 \text{ Pa}.$

This gives $P_2 = 8.50 \times 10^4 \text{ Pa}.$

(L.O. 5.B.10.1)

13. **A** The difference in apparent mass, 5.0 N, is caused by the buoyant force. That means that 5.0 N of fluid was displaced. 5 N of water has a mass of 500 grams. 500 grams of water has a volume of 500 cm³. 500 cm³ is equivalent to $5.0 \times 10^{-4}$ m³.

(L.O. 3.C.4.2)

14. **C** From our frame of reference, a $\mu$ meson takes 4 $\mu$s to get to Earth's surface, but for the meson traveling close to the speed of light, the trip takes only 2 $\mu$s.

(L.O. 1.D.3.1)

15. **D** Since the charge outside the sphere is +2$Q$, a charge of −2$Q$ will be induced on the outer surface. Since the only charge that can move in a conductor is the negative charge, this leaves a deficit of

negative charge on the inner surface. The charge on the inner surface will be $+2Q$.

(L.O. 1.B.1.1, 1.B.1.2, 1.B.2.3, 1.B.2.2)

16. **C** The rays diffract at different angles and travel with different speeds in the glass plate since the index of refraction depends on the wavelength of the light. When the rays emerge on the other side of the glass plate, they are at different points. Since the red and violet rays emerge into the same medium, they will have the same speed after leaving the glass.

(L.O. 6.E.3.1)

17. **B** The mirror cannot be convex because convex mirrors produce a reduced image. The third data point shows that when the object is 20 cm from the mirror, the image is also 20 cm from the mirror. This only happens at the focal point, so the focal point is 20 cm. The radius is double the focal length, or 40 cm.

(L.O. 6.E.4.1, 6.E.4.2)

18. **C** The diffraction of light is the ability of light to bend around obstacles in its path, causing the light to spread into regions that would be in shadow. In the same manner you hear sound that bends around corners, the light wave diffracts around the edges of the coin.

(L.O. 6.C.4.1)

19. **B** The spacing of the grating and the order of the bright spot are the same, so $m\lambda = d\sin\theta$ tells us that $\sin\theta$ is proportional to $\lambda$. Red light has a longer wavelength than blue light, so the diffraction angle for red light is greater than the diffraction angle for blue light.

(L.O. 6.F.1.1)

20. **A** In all cases, the image is virtual and upright. A convex mirror always produces a virtual upright image that is smaller than the object. The image in a plane mirror is always virtual, upright, and the same size as the object. A concave mirror will produce a virtual upright enlarged image when the object is between the focal point and the mirror.

(L.O. 6.E.4.2)

21. **D** The apparent wavelength of sound waves and lights waves depends on the motion of the observer and the source. In terms of sounds, this is the familiar effect of fast-moving objects seeming to

change pitch as they pass. In terms of light, the Doppler effect is apparent in the shifting of apparent wavelength in the light from distant stars.

(L.O. 6.A.1.3)

22. **C** As the temperature approaches zero, the average speed of the gas particles approaches zero. As $v_{rms}$ approaches zero, fewer collisions with the walls of the container occur. Those collisions are the cause of pressure, so pressure also approaches zero.

(L.O. 7.A.3.1)

23. **A** The spread of the maxima is related to slit spacing by $d\sin\theta = m\lambda$. Because the wavelength is the same in all cases, the slit spacing, $d$, varies inversely with $\sin\theta$. The diffraction grating with the smallest spacing will produce the most separated maxima.

(L.O. 6.C.3.1)

24. **A** The work done in the cyclic process shown on the $pV$ graph is the area bounded by the cycle. The work done is

$$\frac{1}{2}bh = \frac{1}{2}\left(4.04\times10^5 \text{ Pa} - 1.01\times10^5 \text{ Pa}\right)\left(4.00 \text{ m}^3 - 1.00 \text{ m}^3\right) = 450 \text{ kPa}.$$

(L.O. 7.A.3.3)

25. **C** The angle of incidence and the angle of reflection are both 45°, so the critical angle of the prism must be smaller than 45° if the ray is to be totally reflected at side AC. From $n_1\sin\theta_C = n_2\sin90°$ and with $n_2 = 1.00$, we can write the minimum $n_1$ as $n_1 = \dfrac{1}{\sin45°} = 1.41$.

(L.O. 6.F.1.1)

26. **A** The force between $q_1$ and $q_2$ is repulsive because both of them are positive charges. The force between $q_2$ and $q_3$ is attractive since they are unlike charges. $\left|\vec{F}_{12}\right| = \dfrac{kq_1q_2}{r_{12}^2}$. Substitution gives

$$\left|\vec{F}_{12}\right| = \frac{\left(9\times10^9 \dfrac{\text{N}\cdot\text{m}^2}{\text{kg}^2}\right)\left(9.00\times10^{-9} \text{ C}\right)^2}{\left(0.100 \text{ m}\right)^2} = 7.28\times10^{-5} \text{ N}.$$ Since $q_2$ and $q_3$ have the same magnitude of charge, $\left|\vec{F}_{23}\right|$ is also $7.28 \times 10^{-5}$ N.

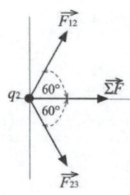

The vertical components of the forces are equal in magnitude and opposite in direction, giving a vector sum of zero. The resultant force is then

$$2\left|\vec{F}\right|\cos 60° = 2\left(8.10\times 10^{-6}\ \text{N}\right)\cos 60° = 7.28\times 10^{-5}\ \text{N}.$$

(L.O. 3.A.4.1, 3.C.2.3)

27. **B** The paper is neutral, but electrons can be slightly displaced from the lattice structure, making one side of the paper more negative than the other side. The paper will be attracted to the positively charged rod. This is referred to as polarization of charge.

(L.O. 1.B.2.2)

28. **B** The resultant electric field is caused by the two charges on the $y$-axis, not by charge $q$, so the sign of the charge on $q$ does not affect the field. The direction of the electric field is found below.

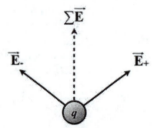

(L.O. 2.C.2.1, 2.C.4.1)

29. **C** The spheres are uncharged initially, containing equal amounts of negative and positive charges. The wand causes some of the negative charge from spheres A and B to move to sphere A, which is positioned closer to the positive wand. This gives sphere A a negative charge and sphere B a positive charge only as long as the positive charging wand is near.

(L.O. 1.B.1.1, 1.B.2.2)

30. **A** Five electric field lines originate on the $+2.0\ \mu C$ charge, and seven lines terminate on $q$. Field lines never go from positive to positive, so q must be negative. The magnitude of $q$ is $\frac{7}{5}(+2.0\ \mu C) = 2.8\ \mu C$.

(L.O. 2.C.2.1, 5.C.2.1)

31. **A** In the upper branch of the parallel arrangement, the series resistors have an equivalent resistance of $7\ \Omega + 5\ \Omega = 12\ \Omega$. The net resistance for the parallel arrangement is $\frac{1}{R_{eq}} = \frac{1}{12\ \Omega} + \frac{1}{4\ \Omega} + \frac{1}{6\ \Omega}$, which yields a resistance of $2\ \Omega$. The total resistance for the circuit is $2\ \Omega + 3\ \Omega = 5\ \Omega$.

The current entering the junction is $I = \frac{V}{R} = \frac{15V}{5\ \Omega} = 3$ A.

The voltage drop across the parallel arrangement is $V = IR = (3\text{ A})(2\ \Omega) = 6V$.

The current in the ammeter is then $I = \frac{6V}{12\ \Omega} = 0.5$ A.

(L.O. 5.C.3.4)

32. **A** The gain in kinetic energy comes from the work done by the field. $W = q\Delta V = (1.6 \times 10^{-19}\text{C})(1000\text{V}) = 1.6 \times 10^{-16}\text{J}$

(L.O. 5.B.5.5)

33. **C** Adding even a small resistor in parallel lowers the equivalent resistance by providing another path for current to follow. Decreasing a circuit's resistance allows more current to flow.

(L.O. 5.C.3.3)

34. **D** All of the charges repel one another. Charge $q_3$ is pushed right by $q_4$, down by $q_2$, and diagonally down and right by $q_1$. The sum of these forces is a vector down and to the right.

(L.O. 2.C.4.2, 3.A.2.1, 3.A.4.3, 3.C.2.1)

35. **A** By the right-hand rule, the force on the proton is directed to the top of the page.

(L.O. 2.D.1.1)

36. **D** The clockwise current induced in the loop causes flux down through the loop. By Lenz's law, this flux opposes the change that caused it, meaning that the magnet's motion is increasing the flux up through the loop. Magnetic field lines point toward a south pole, so the face closest to the loop must be a south pole.

(L.O. 3.C.3.1)

37. **C** By the right-hand rule, the magnetic field around the wires is the same as if the north and south poles of two magnets were brought near each other. The force is attractive. By Newton's third law, the force must be equal.

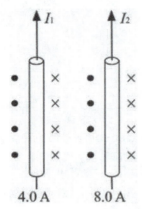

(L.O. 2.D.2.1, 3.A.4.2)

38. **B** The electric field intensity $|\vec{E}|$ is an inverse square law. $|\vec{E}| = k_e \frac{Q}{r^2}$. Ranking the magnitude of the field gives C > B > A > D.

(L.O. 2.C.3.1)

39. **D** The work function is a property of the metal and is not affected by the light falling on the surface. Increasing the intensity of the light means increasing the number of photons that collide with the metal every second, which increases the number of electrons liberated from the metal every second.

(L.O. 6.F.3.1, 6.G.1.1)

40. **D** The bar is in equilibrium, so the sum of all forces acting on it must be zero. In the plane of the incline, the only two forces acting are the component of the weight vector that points down the ramp, $mg\sin\theta$, and the component of the magnetic force that points up the ramp, $BIL\cos\theta$.

Setting these equal gives us $BIL\cos\theta = mg\sin\theta$.

Solving for $I$ gives us $I = \dfrac{mg\sin\theta}{BL\cos\theta}$.

(L.O. 3.A.1.1, 3.C.3.1)

41. **B** The diffraction of the electron from a crystal experimentally confirmed that an electron-matter had wave properties. The diffraction occurs because of the wave nature of the electron.

(L.O. 6.G.1.1, 6.G.2.1)

42. **C** The missing particles are $3\left(_0^1 n\right)$. In any nuclear reaction, equation charge and mass are conserved. Conservation of charge Z is $92 + 0 = 56 + 36$. The missing particle has zero charge. Conservation of mass A is $235 + 1 = 144 + 89 + 3$. Since the missing particles have no charge and a mass of 3, it is three neutrons that are correctly written as $3\left(_0^1 n\right)$.

(L.O. 5.C.1.1, 5.G.1.1)

43. **A** When a slow neutron enters a nucleus of a fissionable material like U-235, the neutron is captured and the nucleus splits. The nucleus fissions with a large release of energy. The reaction represents a fission reaction.

(L.O. 5.B.11.1)

44. **D** In the alpha decay reaction $_{86}^{222}\text{Rn} \rightarrow\ _{84}^{218}\text{Po} + ?$, charge and mass must balance. There are 88 protons in the parent nuclei and 86 protons in the daughter nucleus. The missing particle must have a charge of $+2q$. Mass must balance as well. The mass number for the parent is 222, and the mass number from the daughter is 218. That gives the missing particle a mass of 4. This is an alpha particle, $_2^4\text{He}$.

(L.O. 5.C.1.1)

45. **C** 15.72 years/5.25 years = 3 half-lives

$$\frac{1}{2} \times \frac{1}{2} \times \frac{1}{2} \times 1.2 \times 10^6 \text{kg} = 1.5 \times 10^{-7} \text{kg}$$

(L.O. 7.C.3.1)

46. **C and D** The kinetic energy depends on the frequency of the light falling on the surface beyond the threshold frequency required to produce a photoelectron $E = hf$. The kinetic energy is dependent on the surface being illuminated since a certain amount of energy is needed to free the photoelectron from the surface, $E = hf - \phi$.

(L.O. 1.D.1.1, 6.G.1.1)

47. **B and C** The process is a cyclic process, taking the working substance through a series of state changes and returning to the initial conditions of $p_0 V_0 T_0$. Since it returned to its initial temperature, the change in the entropy during the process ABCDA is zero. In the cyclic process, since $\Delta U = 0$, the work done must equal the thermal energy added to the system.

(L.O. 5.B.4.1, 5.B.7.1, 7.B.2.1)

48. **A and D** Since they are in thermal equilibrium, they have the same average kinetic energy. $K_{avg} = \frac{3}{2} k_B T$. They have different average speeds since they have different masses. $v_{rms} = \sqrt{\dfrac{3RT}{M}}$. The $H_2$ molecule will have the higher speed since its molar mass is 1/16 as large as $O_2$.

(L.O. 7.A.2.1)

49. **B and C** The width of a single slit diffraction pattern depends on the wavelength of light. $d \sin\theta = m\lambda$, so red light, with its longer wavelength, will diffract more.

(L.O. 6.C.2.1, 6.F.1.1)

50. **B and D** The work done in the isothermal expansion of the gas is equal to the amount of thermal energy added to each cylinder. The change in the internal energy was zero since the temperature did not change. $Q = W$. The entropy of the system will change since thermal energy was added and the gases occupied a greater volume, but the change in the entropy of cylinder 1 is smaller since it has the higher temperature.

(L.O. 7.B.2.1)

## ANSWERS TO FREE-RESPONSE PROBLEMS

## QUESTION 1

(a) The correct graph of the process A → B → C → D → A is shown below.

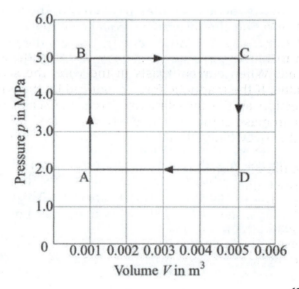

(3 points)

(b) The work done is the area under the curve. A → B is isochoric, and there is no change in volume; therefore, the work done is zero along this path. B → C is isobaric, so the work done is the product of the pressure and the volume change. Since work is defined as $W = -p\Delta V$, the work done on the gas along this path is negative. Along path C → D, the volume again is constant; therefore, the work is zero. The work on the path D → A is given as $W = -p\Delta V$. Since the volume decreases along this path, the work on path D → A is positive. The net work is the sum of the work done on each path. **(4 points)**

(c) i. The change in the internal energy on path A → B can be determined from the first law of thermodynamics $\Delta Q = \Delta U + \Delta W$. Along path A → B, no work is done since the volume is constant. Thus, the change in the internal energy is equal to the thermal energy that crosses the boundary.
$U_B - U_A = Q_{A\to B} U_B = 20.0 \text{ kJ} + 10.0 \text{ kJ} = $ **30.0 kJ.** **(2 points)**

ii. The internal energy of the gas at point C is 70 kJ with an isobaric expansion in which the gas did work against the piston as the gas expanded. If the internal energy increased, then thermal energy entered the system. $\Delta Q = \Delta U + \Delta W$

iii. $\Delta Q = (70.0 \text{ kJ} - 30.0 \text{ kJ}) - 5.00 \text{ MPa}(0.005 \text{ m}^3 - 0.001 \text{ m}^3)$. The thermal energy crossing into the system is
$\Delta Q = 40.0 \text{ kJ} - 20.0 \text{ kJ} = $ **20.0 kJ.** **(2 points)**

(d) The net change in the thermal energy is equal to the net work that is done in the cycle since the gas returns to its initial temperature and therefore to an internal energy at $U_A = 20.0$ kJ. Thermal energy will enter the system. **(1 point)**

(L.O. 5.A.2.1, 5.B.4.1, 5.B.5.5, 5.B.7.1, 5.B.7.2, 5.B.7.3, 7.A.3.3)

## QUESTION 2

(a) Determine the mass of the system before current exists in the wire. Record this value.

Connect a lead to the variable power supply and introduce a current in the circuit. When current exists in the wire, the scale reads a different value. If the magnetic force produced by the wire is in the same direction as the gravitational force, the scale will indicate an increase in mass. If the magnetic force produced by the wire is opposite the gravitational force, the scale will read a decrease in mass.

The net force is the difference between the two scale readings multiplied by the gravitational acceleration.
$$\Sigma \left| \vec{F}_{net} \right| = (\Delta m_{scale}) \vec{g} = \left| \vec{F}_B \right|.$$
Disconnect the lead.

Start the experiment by connecting a lead to the variable power supply and adjust the potential difference across the power supply to give a small current in the wire. Record the ammeter reading. Record the mass reading on the scale. Disconnect the lead.

Repeat this reading by connecting the lead wire to make sure your values are consistent.

Connect the lead to the power supply and increase the current in the circuit. Record the new values for the scale reading and the current.

Repeat the procedure for at least six more readings, recording both the mass indicated on the scale and the ammeter reading.

Measure the length of the wire in the U-shaped region of the magnets. **(3 points)**

(b) The measurements taken will be the initial mass, $m_{\text{zero current}}$; the mass, when current exists in the loop, $m_{\text{with current}}$; and $\Delta m$, the difference between these two values. Convert this reading to force units

$$\Sigma \left| \vec{F}_{net} \right| = (\Delta m_{scale}) \vec{g} = \left| \vec{F}_B \right|.$$

The current in amperes will be recorded as well as the length of the wire in the U-shaped region. **(1 point)**

(c) The graph plotted should be force on the wire versus the current $I$ in the loop.

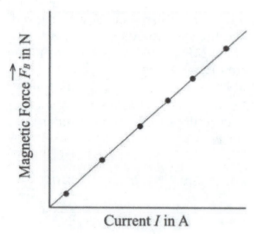

Current $I$ in A

This graph should show a linear relationship between the force and the current.

The slope of the line is $\left| \dfrac{\vec{F}_B}{I} \right| \propto \vec{B}$. Since the length of the wire in the

magnetic field has been measured, the strength of the field is

$$\left| \dfrac{\vec{F}_B}{I\vec{L}} \right| = \vec{B}.$$ **(2 points)**

(d) Since $\left| \vec{F}_B \right| = I\vec{L}\vec{B}$, keeping the current constant in the experiment and measuring different lengths of the wire in the same magnetic field will give the data needed to plot a graph of force versus the

length of the wire. The slope of this line will be $\left| \dfrac{\vec{F}_B}{\vec{L}} \right| \propto \vec{B}$. Dividing

the slope by the current I will give the strength of the field. $\left| \dfrac{\vec{F}_B}{I\vec{L}} \right| = \vec{B}.$

Comparison of the two should give values for the magnetic field that are consistent. **(2 points)**

(e) i.

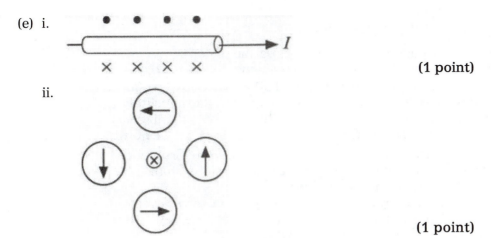

(1 point)

ii.

(1 point)

iii. The correct line checked is $\vec{E} = 0$ ___✓___. The wire contains an equal number of positive and negative charges. Since there is no net charge, there is no electric field outside the wire. The charges are moving in the wire, producing a current. The motion of the charges produces the magnetic field surrounding the wire. **(2 points)**

(L.O. 2.B.1.1, 2.D.2.1, 2.D.3.1, 3.C.3.1)

## QUESTION 3

(a) When both switches are closed, the parallel section contains two branches. The top branch consists of two resistors 4.0 Ω and 2.0 Ω in series. This resistance is larger than the 3.0 Ω resistance in the lower branch. The net resistance for the parallel arrangement is less than the smallest resistance in the lower branch. The total resistance for the circuit is the sum of the two external resistors 5.0 Ω and 1.0 Ω and the equivalent resistance of the parallel arrangement.

   The total current will pass the two resistors 5.0 Ω and 1.0 Ω that are in series with the parallel arrangement, giving them larger power dissipations than the resistances in the parallel part of the circuit. Kirchhoff's junction rule will apply to the current entering and dividing in the parallel section. The upper branch will get the smaller current. The two resistors in the upper branch 4.0 Ω and 2.0 Ω will receive the same current, and the smaller resistor, 2.0 Ω, will dissipate the least power. (Power is the product of the current squared times the resistance.) Although not needed for the problem, the ranking of the power dissipated is P5 Ω > P3 Ω > P1 Ω > P4 Ω > P2 Ω. **(3 points)**

(b) i. When the capacitor in the upper branch reaches steady state, the potential difference across its plates equals the electrical potential difference across the 3.0 Ω resistor in the lower branch since they are in parallel. The charge on the plates of the capacitor is equal to the product of the capacitance times the electrical potential difference. **(1 point)**

ii. At steady state, the current in the circuit is $I = \dfrac{V}{R} = \dfrac{24.0\text{ V}}{9.0\ \Omega} = 2.67$ A.

The electrical potential difference across the 3.0 Ω in the lower branch equals the electrical potential difference across the plates of the capacitor since they are in parallel. $V_{3\,\Omega} = (2.67\text{ A})(3.0\ \Omega) = 8.0$ V. The charge on the capacitor is given by $Q = CV$, and substitution into this equation gives the charge on the plates. $Q = (5.0\ \mu F)(8.0\text{ V}) = 40\ \mu C$ **(1 point)**

(c) Since the resistance of a resistor depends on four factors—length, cross-sectional area, material, and temperature—temperature is the factor that affects the resistance the most.

Use each resistance by itself to see if it obeys Ohm's law by varying the potential difference across the resistor to determine the current through the resistor. A voltage-current graph will be linear if the resistance is ohmic.

If a switch connecting the resistive element is closed for some time, thermal energy dissipated in the resistor will cause the molecules that make up the resistor to gain a larger kinetic energy, resulting in dimensional changes as the molecules with higher kinetic energy move through a greater range.

The resistivity of a resistor generally is considered a constant over a wide temperature range, but this is not always true. **(5 points)**

(L.O. 4.E.4.1, 4.E.5.1, 4.E.5.2, 5.C.3.4)

## QUESTION 4

(a) The electron beam must be directed into a uniform magnetic field that will direct the electrons to the right. Since the electrons are negative, the magnetic field must be directed into and perpendicular to the page. The magnetic force will bend the electron beam to the right without changing its speed.

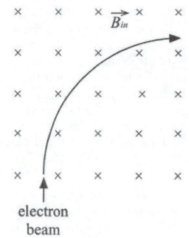

electron
beam **(3 points)**

(b) i. To determine the de Broglie wavelength of one of the electrons, its speed must be found. The kinetic energy is 100.0 eV. Kinetic energy is defined as $K = eV = \dfrac{1}{2}mv^2$. Speed then is

$$v = \sqrt{\frac{2Ve}{m_e}} = \sqrt{\frac{2(100.0 \text{ eV})(1.60 \times 10^{-19} \text{ J}/\text{eV})}{9.11 \times 10^{-31} \text{ kg}}} = 5.93 \times 10^6 \text{ m}/\text{s}.$$

The de Broglie wavelength is

$$\lambda = \frac{h}{mv} = \frac{6.63 \times 10^{-34} \text{ J} \cdot \text{s}}{(9.11 \times 10^{-31} \text{ kg})(4.19 \times 10^6 \text{ m}/\text{s})} = 1.23 \times 10^{-10} \text{ m}.$$

**(3 points)**

ii. The electron beam will undergo diffraction when it interacts with the crystal since the de Broglie wavelength is approximately the size of the spacing between the planes of the crystal. **(2 points)**

(c) As researchers increase the energy, the speed of the electrons increases. This increases the electrons' momentum. The electrons' de Broglie wavelength decreases with increasing momentum. Decreasing the wavelength causes the diffraction pattern to be less wide. **(2 points)**

(L.O. 3.A.2.1, 3.C.3.1, 5.D.1.6, 6.F.4.1, 6.G.1.1, 6.G.2.1, 6.G.2.2)